D0874262

RUBBER TECHNOLOGY

RUBBER TECHNOLOGY

SECOND EDITION

Edited by

MAURICE MORTON

Director
Institute of Polymer Science
The University of Akron
Akron, Ohio

Sponsored by the Rubber Division of the

AMERICAN CHEMICAL SOCIETY

ROBERT E. KRIEGER PUBLISHING COMPANY
MALABAR, FLORIDA

Original Edition 1973
Reprint 1981

Printed and Published by
**ROBERT E. KRIEGER PUBLISHING COMPANY, INC.
KRIEGER DRIVE
MALABAR, FLORIDA 32950**

Printed in the United States of America

Library of Congress Cataloging in Publication Data
Main entry under title:

Rubber technology.

Reprint. Originally published: 2nd ed. New York: Van Nostrand Reinhold, 1973.
Includes index.
1. Rubber industry and trade. I. Morton, Maurice.
II. American Chemical Society. Rubber Division.
TS1890.R86 1981 678'.2 81-8317
ISBN 0-89874-372-9 AACR2

PREFACE

This book was originally intended as the second edition of *Introduction to Rubber Technology*. However, in the course of planning and editing a revised edition, several changes occurred which really made it desirable to adopt a more fitting name for this book.

First of all, the character of the book has changed. The original volume was based on a compilation of lectures prepared by experts in the field for presentation before various Rubber Groups in the United States. Hence the material was more or less ready-made for the Editor. For this edition, however, each chapter was solicited by the Editor from the most suitable source, and was prepared specifically for this book.

Secondly, the time of publication of the original book was somewhat unfortunate in that it occurred on the eve of the commercial development of the new "stereo" rubbers resulting from the great breakthrough in organometallic polymerization. Hence the present volume contains entirely new material, as shown in Chapters 8, 9, 11, 16, 17, and 20.

Finally, this volume benefits from new organization of the material, based on the experience obtained in using it as the text for the Rubber Division Correspondence Course for the past nine years. This is mainly reflected in the first five chapters, in which the general area of rubber compounding, vulcanization, and testing is covered prior to any discussion of the individual characteristics of the various elastomers.

All of the above changes eventually resulted not only in an increase in size but also in degree of sophistication of this volume. This is probably not surprising in the light of the marked changes which have taken place in this field since 1959. Care was taken, however, not to render this material out of reach for the reader with little or no scientific background. Thus it is hoped that readers at all levels of scientific understanding will find herein the information they seek in the various areas of rubber technology.

The Editor wishes to express his sincere appreciation to all the contributors for their cooperation in preparing the material for this volume and in making the necessary revisions. He would especially like to acknowledge the assistance of Dr. H. L. Stephens in the planning of this book, and of Mrs. Barbara Zimmerman for her dedicated and painstaking secretarial work.

MAURICE MORTON

CONTENTS

RUBBER TECHNOLOGY

1

INTRODUCTION TO POLYMER SCIENCE

MAURICE MORTON
Regents Professor of Polymer Chemistry
The University of Akron

Polymer science is concerned with the composition and properties of a large number of substances classed as "polymers," which include rubbers, plastics, and fibers. The meaning of the term "polymer" will become clear during the course of this chapter, but it does involve some understanding of basic chemistry. The science of chemistry concerns itself with the composition of matter and with the changes that it undergoes. To the layman, the methods and processes of chemistry appear quite baffling and difficult to comprehend. Yet these are essentially based on a simple, logical development of knowledge about the substances which comprise our physical world. Since this book is not intended exclusively for readers trained in chemistry, the subject of polymer chemistry will be developed from the type of first principles comprehensible to anyone.

ATOMS AND MOLECULES

In this "atomic" age, it is certainly common knowledge that the atom is the basic unit of all matter. There are as many different kinds of atoms as there are elementary substances, i.e., elements, and these number about 100. It is the myriads of combinations of these atoms which make possible the hundreds of thousands of different substances comprising our world. Just as in the case of the 100-odd elements it is the atom which is the smallest unit, so too in the case of all other substances (*compounds*), it is the particular *group of atoms*, or *molecule*, which is the smallest possible unit. If the molecule of a compound substance is broken up, the substance ceases to exist and may be reduced to the elementary substances which comprise it. It is the various possible combinations of element atoms to form different molecules which is the basis of chemistry.

Now, in this "nuclear" age, we hear much about "atomic fission" processes, in which atoms are split apart, with the accompaniment of considerable energy release. How then can an atom be split if it is supposedly a basic unit of matter?

Well, we know today that atoms are not the indivisible particles they were originally believed to be. Instead, they, too, are made up of even smaller particles such as protons, electrons, neutrons, and others. The differences between atoms can be explained by the different numbers and arrangements of these subatomic particles.

The thing to remember, however, is that the different substances of our world make their appearance once atoms have been formed. This becomes obvious when we consider the fact that all protons are the same, all neutrons are the same, and so on. There is no such thing as a "proton of iron" being different from a "proton of oxygen." However, an atom of iron is considerably different from an atom of oxygen. That is why the chemist is concerned mainly with the arrangement of atoms into molecules and considers the elemental atoms as his basic building blocks.

VALENCE

So far nothing has been said about the manner in which the subatomic particles are held together or what holds atoms together in molecules. Although this is not our immediate concern here, it should be understood that there are electromagnetic forces which hold all these particles together. Thus most of the subatomic particles carry either positive or negative electrical charges, which act as binding forces within the atoms as well as between different atoms. Because each atom contains a different number and arrangement of these charged particles, the forces acting between different atoms also are different. These forces, therefore, result in certain well-defined rules which govern the number and type of atoms which can combine with each other to form molecules. These rules are called the rules of *valence*, and refer to the "combining power" of the various atoms.

From the beginning of the nineteenth century, chemists have been carefully weighing, measuring, and speculating about the proportions in which the element substances combine to form compounds. Since they found that these proportions were always the same for the same substances, this was most easily explained by assuming combinations of fixed numbers of atoms of each element. This was the basis for Dalton's Atomic Theory, first proposed in 1803. In accordance with this hypothesis, an atom from the element hydrogen is never found to combine with more than one atom of any other kind. Hence, hydrogen has the lowest combining power, and was assigned a valence of 1. On the other hand, an atom of the element oxygen is found capable of combining with two atoms of hydrogen, hence it can be said to have a valence of 2.

In this way, the kinds and numbers of atoms which make up a molecule of a given compound can be deduced and expressed as a *formula* for that compound. Thus the compound water, whose molecules are each found to contain 2 atoms of hydrogen and 1 atom of oxygen, is denoted by the formula H_2O, where the letters represent the kinds of atoms, and the numbers tell how many of each atom are present within one molecule. This formula could also be written as H–O–H to show how 1 oxygen atom is bonded to 2 hydrogen atoms. In the case of the well-known gas, carbon dioxide, the molecule would be represented as CO_2, where the carbon atom would show a valence of 4, since it is capable of combining with 2 oxygen atoms (each of which has a valence of 2). This formula could also be written as O=C=O, which is known as a *structural formula*, since it shows exactly how the carbon and oxygen atoms are bonded. Such structural formulas must be written in a way which will satisfy the valence of all atoms involved.

MOLECULES AND MACROMOLECULES

We have seen how the chemist is concerned with the structure of the molecules which comprise the substances in our world. At this point it might be of interest to consider the actual size of these molecules. It is now known that a molecule of water, for instance, measures about ten-billionths of an inch, that is, one million such molecules could be laid side-by-side to make up the thickness of the type used on this page. Putting it another way, one drop of water contains 1,500,000,000,000,000,000,000 molecules! Most of the ordinary substances have molecules of this approximate size. Thus the molecule of ordinary sugar, which has the rather impressive chemical formula, $C_{12}H_{22}O_{11}$, is still only a few times larger than the water molecule.

Such substances as water, sugar, ammonia, gasoline, baking soda, etc., have relatively simple molecules, whose size is about the same as indicated above. However, when chemists turned their attention to the chemical structure of some of the "materials" like wood, leather, rubber, etc., they encountered some rather baffling questions. For instance, *cellulose* is the basic fibrous material which comprises wood, paper, cotton, linen, and other fibers. Chemists knew long ago that cellulose has the formula $C_6H_{10}O_5$, as shown by chemical analysis, yet their measurements did not indicate that the molecule of cellulose existed in the form shown by the formula. Instead, they found evidence that the cellulose molecule was unbelievably large. Hence they did not really believe that it was a molecule, but considered the true molecule to be $C_6H_{10}O_5$, and that hundreds of such molecules were bunched together in aggregates.

Rubber, like cellulose, was also classed as a *colloid*, i.e., a substance which contains large aggregates of molecules. Over 100 years ago, the formula for rubber was established as C_5H_8, but here again the evidence indicated that the molecule was much larger than shown by the formula. It was only as recently as the 1920's that Staudinger advanced the revolutionary idea that the rubber molecule was indeed a giant molecule, or *macromolecule*. Today, it is generally accepted that the class of materials which form the useful fibers, plastics, and rubbers are all composed of such macromolecules. Hence the formula for cellulose, if it is to represent one molecule, should more correctly be written as $(C_6H_{10}O_5)_{2000}$, while rubber should be represented as $(C_5H_8)_{20,000}$.

Such molecules are, of course, many thousands times larger than the molecules of ordinary chemical substances, hence the name *giant* molecules. However, it should be remembered that even these giant molecules are far from visible in the best microscopes available. Thus the cellulose or rubber molecule may have a diameter several hundred times larger than that of a water or sugar molecule, but its size is still only several millionths of an inch! Even so, however, their relatively enormous dimensions, in comparison with ordinary molecules, give them the unusual properties observed in plastics, rubbers, and fibers.

POLYMERS AND MONOMERS

The molecular formula for rubber, $(C_5H_8)_{20,000}$, at first glance appears formidable indeed. There can obviously be a vast number of possible arrangements of 100,000 carbon atoms and 160,000 hydrogen atoms! To deduce the exact way in which they are arranged seems an impossible task. Yet it is not as difficult as it would seem. Fortunately – for the chemists – these huge molecules are generally composed of a large number of simple repeating units, attached to each other in long chains. Thus the structural formula for the molecule of natural rubber may be represented by the simple unit, C_5H_8, multiplied many thousand times. It is actually shown thus:

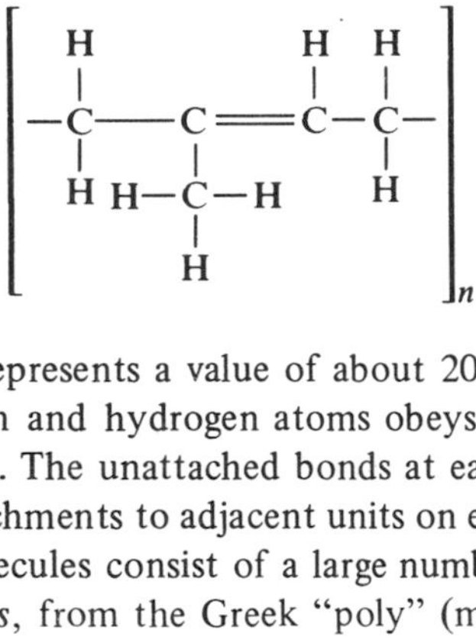

$$\left[\begin{array}{ccccccccc} & \mathrm{H} & & & & \mathrm{H} & & \mathrm{H} & \\ & | & & & & | & & | & \\ - & \mathrm{C} & - & \mathrm{C} & = & \mathrm{C} & - & \mathrm{C} & - \\ & | & & | & & & & | & \\ & \mathrm{H} & & \mathrm{H-C-H} & & & & \mathrm{H} & \\ & & & | & & & & & \\ & & & \mathrm{H} & & & & & \end{array}\right]_n$$

In the above structure, n represents a value of about 20,000, as stated previously. The arrangement of carbon and hydrogen atoms obeys the valence rules of 4 for carbon and 1 for hydrogen. The unattached bonds at each end of the unit are, of course, meant to show attachments to adjacent units on each side.

Because these giant molecules consist of a large number of repeating units, they have been named *polymers*, from the Greek "poly" (many) and "meros" (parts). The repeating unit shown above would, therefore, be called the *monomer*. It is actually very similar to a simple compound, known as isoprene, a low-boiling liquid. In 1860, Greville Williams first isolated this compound as a decomposition product obtained when rubber was heated at elevated temperatures. The relationship between the monomer and polymer structures, in this case, is illustrated below.

monomer

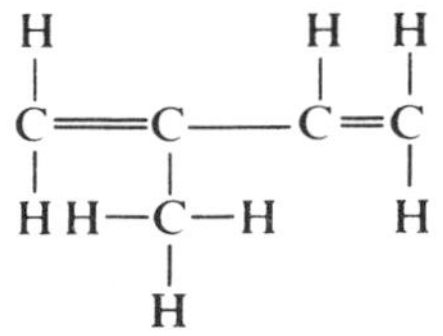

$$\begin{array}{ccccccc} \mathrm{H} & & & & \mathrm{H} & & \mathrm{H} \\ | & & & & | & & | \\ \mathrm{C} & = & \mathrm{C} & - & \mathrm{C} & = & \mathrm{C} \\ | & & | & & & & | \\ \mathrm{H} & & \mathrm{H-C-H} & & & & \mathrm{H} \\ & & | & & & & \\ & & \mathrm{H} & & & & \end{array}$$

isoprene

polymer

$$\left[\begin{array}{ccccccccc} & \mathrm{H} & & & & \mathrm{H} & & \mathrm{H} & \\ & | & & & & | & & | & \\ - & \mathrm{C} & - & \mathrm{C} & = & \mathrm{C} & - & \mathrm{C} & - \\ & | & & | & & & & | & \\ & \mathrm{H} & & \mathrm{H-C-H} & & & & \mathrm{H} & \\ & & & | & & & & & \\ & & & \mathrm{H} & & & & & \end{array}\right]_n$$

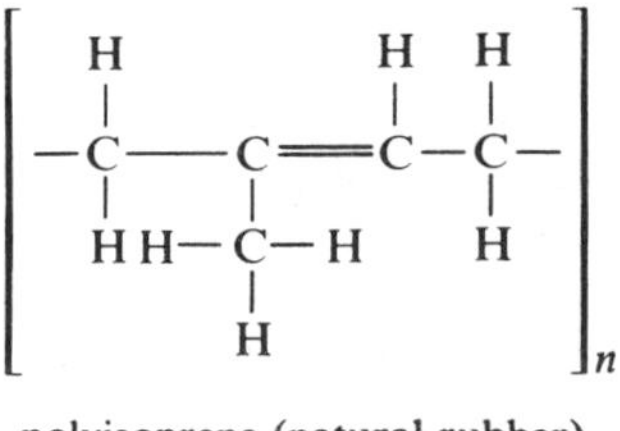

polyisoprene (natural rubber)

It can be seen, here, that the unit in the chain molecule corresponds exactly with the uncombined monomer molecule, except for the necessary rearrangements of the bonds between the carbon atoms.

THE SYNTHESIS OF MACROMOLECULES

So far we have been discussing the macromolecules, or polymers, which occur ready-made in nature. Although the chemist may surmise, he cannot be entirely certain just how these large molecules are built up in nature. However, over the past half-century, chemists have become very proficient in producing synthetic polymers from a wide variety of possible monomers. It is from these processes that there have arisen all the new and varied synthetic rubbers that we know today.

In order to form a polymer molecule, it is, of course, necessary to find a way in which individual small molecules may join together in large numbers. This process is, for obvious reasons, called *polymerization*. There are a number of ways in which this process can be effected, but these can all be broadly classed under two main types of chemical reactions. One of these is known as addition polymerization, since it involves a simple addition of monomer molecules to each other, *without the loss of any atoms* from the original molecules, The other type is called condensation polymerization, since it involves a reaction between monomer molecules during the course of which a bond is established between the monomers, but some of the atoms present are lost in the form of a by-product compound. In this case, then, the polymer molecule is a *condensed* version of the original monomer molecules which reacted.

Addition Polymerization

The simplest example of addition polymerization is the formation of polyethylene from ethylene. This can be shown by means of the following molecular structures:

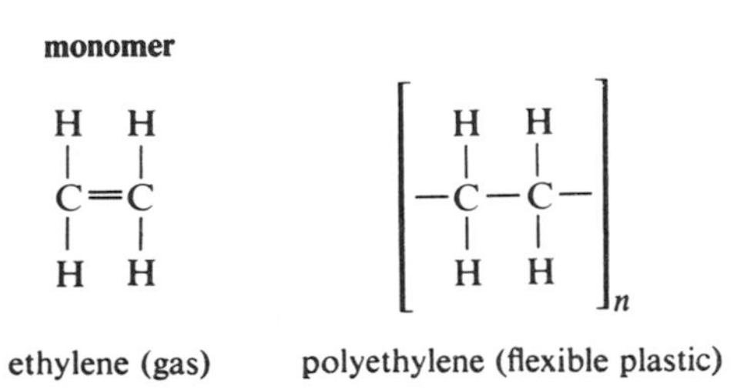

ethylene (gas) polyethylene (flexible plastic)

It can be noted that, in the ethylene molecule, after due regard is given to the requirements of valence, there is a "double bond" between the two carbon atoms. Furthermore, to cause polymerization, *one* of these two bonds must *open* and become available as *two bonds*, ready to unite with similarly available bonds from adjacent ethylene molecules. For this reason, ethylene is one of the compounds known as "unsaturated," due to the presence of this extra bond between the carbon atoms, a bond which can be "saturated" by bonding to *two* other atoms, carbon or otherwise. This makes it possible for an "unsaturated" molecule to form a long-chain polymer molecule, as shown.

The well-known line of "vinyl" and "acrylic" polymers in the plastics and fibers field are all based on this activity of the carbon–carbon double bond. As a matter of fact, the term "vinyl" is a generic chemical name for this bond. The following monomers are the starting materials for the better-known polymers in this field, as the names will indicate.

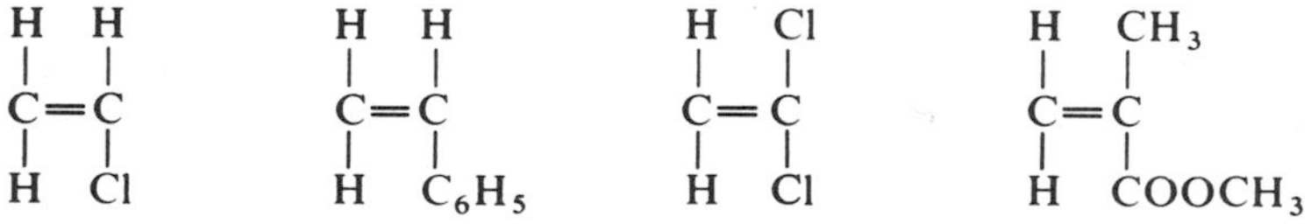

```
H  H        H  H         H  Cl       H  CH3
|  |        |  |         |  |        |  |
C=C         C=C          C=C         C=C
|  |        |  |         |  |        |  |
H  Cl       H  C6H5      H  Cl       H  COOCH3
```

vinyl chloride — styrene — vinylidene chloride — methyl methacrylate

A somewhat more complex form of addition polymerization occurs with the unsaturated monomers of the "diene" type. These differ from the vinyl monomers in possessing *two* double bonds. An important and typical diene is *butadiene*, which forms the basis of the most important synthetic rubber today. The structures of the monomer and polymer are shown below.

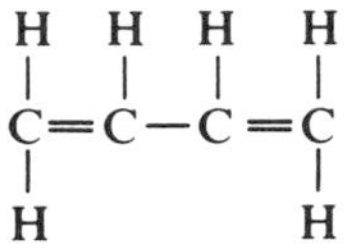

```
H  H  H  H            [    H  H  H  H    ]
|  |  |  |            [    |  |  |  |    ]
C=C–C=C               [  –C–C=C–C–       ]
|        |            [    |        |    ]
H        H            [    H        H    ]n
```

Butadiene (gas) 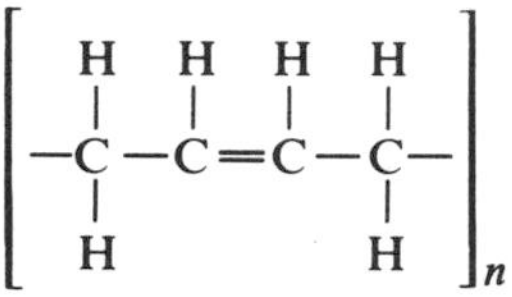 — Polybutadiene (a rubbery solid)

The similarity between butadiene and isoprene is immediately apparent. Isoprene is actually the same as butadiene, except for the presence of a "methyl" group (CH_3) on one of the carbon atoms, instead of a hydrogen atom. As a matter of fact, isoprene has the chemical name of 2-methyl butadiene.

It can be seen at once that when a diene is polymerized, the polymer obtained differs in one important respect from the vinyl polymers, viz., it still contains a *double bond* in each unit of the chain. This is a natural outcome of the fact that the monomer has *two double bonds* at the outset. The residual unsaturation in these polymers plays a very important role in the process of vulcanization, which will be discussed later. It is quite obvious, however, from the foregoing discussion, that the presence of these double bonds in the polymer makes it possible for the latter to react further with agents, like sulfur, which can add to these available double bonds.

Condensation Polymerization

Polymers made by a condensation reaction are not as important in the rubber field as the addition polymers discussed above. However, there are one or two types which are of some importance. One of these is the class of polymers known as polysulfide polymers, and bearing the trademark of "Thiokol." This was one of the first synthetic rubbers developed in the United States, as an oil- and solvent-resistant elastomer. Various grades of this material are prepared using variations in starting materials, but the following will illustrate the process:

$$Cl-CH_2-CH_2-Cl + Na-S_x-Na \longrightarrow [-CH_2-CH_2-S_x-]_n + NaCl$$

ethylene dichloride	sodium polysulfide	poly(ethylene polysulfide)	salt (sodium chloride)

where $x = 2–4$.

The above equation shows that this polymerization proceeds by means of the reaction between the sodium (Na) atoms and the chlorine (Cl) atoms, which combine to form salt (NaCl). This results in a bond being formed between the sulfur and carbon atoms, and a long-chain molecule is thus formed. Here the salt is the by-product of the condensation reaction and is not included in the polymer chain. In all such condensation reactions, provision must be made to remove such by-products in order to permit the reaction to proceed smoothly and to avoid contamination of the polymer. In the above case, the salt is conveniently removed by performing the reaction in the presence of water, which dissolves the salt but not the polymer.

THE PHYSICAL BEHAVIOR OF POLYMERS

Up to this point, we have discussed the basic principles involved in the synthesis and structure of polymers, and it would, therefore, be appropriate to consider, in more detail, the individual rubber polymers with which the technologist is concerned. However, before doing so, it would seem highly desirable to consider the special features which make it possible for polymers to have their unique physical properties. We have seen how these macromolecules are composed of the same kinds of atoms as ordinary "chemicals," and are distinguished only by their enormous size. How does this difference in size convert a simple *chemical* into a strong, tough, durable *material*? That is indeed a question which scientists have been studying with a good deal of interest.

The most reasonable answer to the above question may be obtained by considering the two following aspects, viz., the type of forces which act on atoms and molecules in general and the structure and behavior of an individual polymer molecule.

Interatomic and Intermolecular Forces

We have already seen that there are electromagnetic forces operating between atoms, i.e., "valence" forces which bind atoms together. These forces, operating as they do at relatively short ranges, are very powerful indeed, so that a considerable amount of energy is required to break up a molecule of a compound. This is usually evident by the fact that high temperatures are usually necessary to decompose substances. The physical strength of interatomic bonds can best be exemplified by considering the basis for the strength of metals and minerals. It is generally agreed that such materials do not consist of molecules at all but contain a three-dimensional crystalline structure of atoms bonded to each other by the powerful valence forces mentioned above. Hence, to rupture the metal or mineral, it is necessary to actually break these valence bonds, and this usually requires forces of many thousands of pounds per square inch.

On the other hand, a consideration of the forces operating *between molecules* soon shows that these are very much smaller than the interatomic forces. Hence it can be stated that those substances where atoms exist in small groupings, i.e., molecules, are either liquids or weak solids, since the intermolecular bonds are very weak. It is well-known that the "states" of matter depend on the balance between the intermolecular forces and the heat energy of the molecules, as indicated by temperature. Thus heat tends to make atoms or molecules move faster, i.e., gives them a greater "kinetic" energy, while the intermolecular forces pull them closer together and tend to restrain their motions. This is the distinction between gases, where the intermolecular forces are far too weak to overcome the kinetic energy of the molecules, and liquids or solids where the opposite is the case. There is a further distinction between liquids, which flow because these forces are too weak to hold the molecules firmly in place, and true solids, where such molecules (or atoms) are held in a rigid pattern, i.e., crystals.

Structure and Behavior of Macromolecular Chains

In this regard, the behavior of macromolecules is further influenced by their long-chain character. Although the chemical formulas shown thus far for these molecules indicate that they are very long chains, they do not show that the chains are *flexible*, by virtue of the ability of the chain carbon atoms, held together by single bonds, to *rotate* around their axis. Hence each individual macromolecular chain can be expected to be capable of twisting into various convolutions. Depending on the temperature, therefore, it could be expected that these chains would be in a constant twisting motion and that they would thus be badly entangled with each other. In such a disordered state, if the intermolecular (interchain) forces are not very strong, the polymer could be considered to be a *liquid*, but a very *viscous* liquid because of the long-chain, entangled state of the molecules. In fact, such a material would be so viscous as to have the appearance of a solid, i.e., an *elastic* solid or a rubber.

The reason for the "rubbery state" is, of course, that any deformation will tend to "straighten out," or uncoil, the entangled mass of contorted chains, and these will tend to coil up again when the restraining force is released. Thus the elastic retractive force is really due to the violent contortions of the long, flexible chains.

It should be remembered that the long-chain character of polymer molecules does not make it impossible for them to form crystals, just as small molecules do.

However, as in the case of simple compounds, this crystallization will only occur if the intermolecular (interchain) forces are strong enough to overcome the kinetic energy of the contorting chains. This can happen in two ways, i.e., either the chain atoms, or groups of atoms, can exert very powerful attractive forces, and/or the chain sections can "fit together" so well that they can come close enough for the interchain forces to take over. The only real distinction between the crystalline, solid state of simple compounds and polymers is that the crystallization is much less complete in the case of the latter. Thus, whereas sugar can be said to crystallize almost "perfectly," polyethylene is only a partly crystalline solid. This is, of course, due to the fact that it is obviously very difficult, if not impossible, to form "perfect" crystals from a mass of entangled chains.

There are a large number of polymers which exist as partly crystalline solids at normal temperatures, of which polyethylene is an excellent example. The reason why this well-known, flexible plastic material is opaque is, in fact, due to the presence of very fine crystallites which refract and scatter light. These crystallites also "tie together" and restrict the motions of the polymer chains so that the latter lose their elastic character. Hence the material has only limited elasticity but retains its flexibility, as in the case of polyethylene and other flexible plastics.

The ability that some macromolecular chains have of crystallizing under the right conditions plays a dominant role in the properties of two types of materials: *fibers and rubbers*. Thus, in the case of a fiber-forming polymer such as nylon, the polyamide chains can exert powerful attractive forces toward each other. Hence, when these chains are oriented, e.g., by cold drawing, the interchain forces are strong enough to cause the oriented chains to crystallize into elongated crystals containing bundles of rigid chains, i.e., *fibers*. These forces thus act to give the fibers very high strength, and the fiber "crystals" cannot be melted again even at temperatures high enough to cause chemical decomposition.

In the case of the rubbers, the macromolecular chains may or may not be capable of crystallizing on stretching, depending on the chemical structure and regularity of the chains. Thus natural rubber, which has a very regular chain structure, can undergo a high degree of crystallization on stretching and therefore becomes a "fiber" at high elongations. This results in a high tensile strength. However, the interchain forces which cause such strain-induced crystallization are not sufficiently powerful to maintain this fiber-like structure once the applied force is removed, so that the fiber-like crystals "melt" and the rubber chains retract to their normal configuration. This phenomenon of "temporary crystallization" therefore plays a very important role in controlling the mechanical properties of a rubber, especially the strength, and these properties vary greatly depending on whether the polymer does or does not undergo crystallization on stretching.

It is important to note, at this point, that the ability of an elastomer to crystallize on stretching, which is advantageous for strength, also means that the elastomer will crystallize, at least partially, at some low temperature, *without stretching*. This temperature is referred to as its *crystal melting point*, T_m, and is, of course, generally below room temperature, otherwise the rubber will crystallize on storage and harden considerably. As a matter of fact, this can sometime happen to natural rubber, whose T_m is just below normal ambient temperatures.

At lower temperatures, elastomers also exhibit another phenomenon, known as the "glass transition." This occurs regardless of whether the polymer is capable of crystallization, and results in a transformation of the elastomer into a rigid, brittle plastic. Thus the "glass transition temperature," T_g, also known as the "glass point," of natural rubber is −72°C, that of SBR is about −50°C, while the synthetic polybutadienes exhibit T_g values as low as −100°C. Although the glassy "state" makes these materials into rigid solids, they are still not considered as being in the true solid state, as defined by scientific criteria, since they have not really crystallized. Instead, such "glasses" are defined as "supercooled liquids," just as a rubber is defined as a "liquid." The best known example of such a "glass" is, of course, ordinary glass itself, which is a supercooled silicate. Other examples are the well-known glass-like plastics, such as polystyrene and poly(methyl methacrylate), which owe this property to the fact that their T_g is well *above room temperature*, i.e., about 100°C, above which they, too, become rubbery in character.

Finally, there is one more aspect of elastomers that needs careful consideration. Since these materials have been defined, in a scientific sense, as very viscous elastic *liquids*, it is not surprising that they can *flow*, especially as the temperature increases. This is due, of course, to the ability of the entangled long-chain molecules to *slip* past each other when under a distorting force. Hence it is necessary to "anchor" these elastic chains to each other in order to have the elastomer behave as a truly elastic material, i.e., to exhibit a high degree of elastic recovery and a minimum of "set." It is for this reason that elastomers have to be *vulcanized* for optimum properties, and this process consists simply of introducing *crosslinks* between the long-chain molecules to obtain a continuous *network* of flexible, elastic chains. This process is accomplished by means of various chemical reactions, depending on the chemical structure of the macromolecule, the most common process involving sulfur and its compounds, which work very well with the unsaturated elastomers such as natural rubber and the synthetic polydienes.

NATURAL RUBBER AND SYNTHETIC ELASTOMERS

On thc basis of the foregoing general discussion of polymers, it should now be possible for the reader to examine the chemical structure of the better-known elastomers, and to understand the various methods used in the vulcanization of these materials. Hence the remainder of this chapter will be concerned with a presentation of the structures of these important elastomers, together with their behavior on stretching and their method of vulcanization. Those elastomers which do not have the ability to crystallize on stretching exhibit inferior tensile strength, as might be expected. However, when mixed with "reinforcing" pigments, such as carbon black, these elastomers develop high strengths, equal to that of natural rubber. It is thought that the polymer chains actually form attachments to the surface of these reactive pigments, so that the pigment particles act just like crystallites in increasing the tensile strength. All of this information is summarized in Table 1.1 at the end of this chapter.

Natural Rubber

It has already been stated, in an earlier section, that natural rubber has the chemical name of polyisoprene. However, it is important to note that there is a special feature about its structure, which accounts for its special properties. This concerns the possible *isomers* that can occur in a polyisoprene chain as follows:

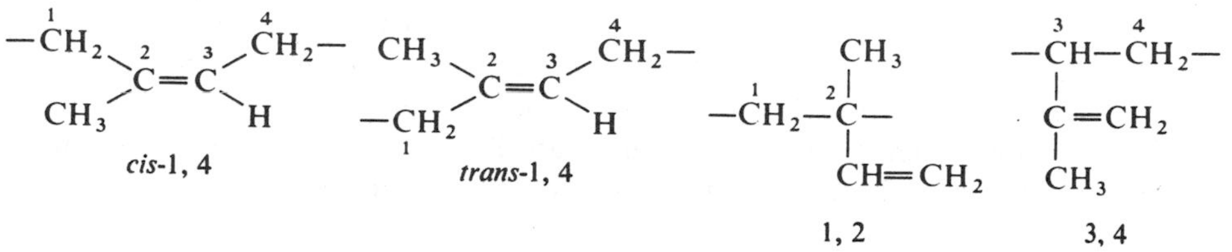

cis-1, 4: $$-\overset{1}{C}H_2(CH_3)\overset{2}{C}{=}\overset{3}{C}(H)\overset{4}{C}H_2-$$

trans-1, 4: $$-\overset{1}{C}H_2(CH_3)\overset{2}{C}{=}\overset{3}{C}(H)\overset{4}{C}H_2-$$

1, 2: $$-\overset{1}{C}H_2-\overset{2}{C}(CH_3)(CH{=}CH_2)-$$

3, 4: $$-\overset{3}{C}H(C(CH_3){=}CH_2)-\overset{4}{C}H_2-$$

All of the above structures could, in theory, occur in a polyisoprene chain. The numbers refer to the particular carbon atoms in each unit which are attached to adjacent units. Thus, a 1,4 structure means that carbon atoms 1 and 4 are joined in forming the chain. The terms *cis* and *trans* refer to the positions of the various carbon atoms with reference to the carbon–carbon double bond. Since a double bond is considered to prevent rotation of the attached atoms, it follows that other atoms or groups of atoms may occupy positions on *either side* of the double bond. Thus it can be seen that, in the *cis*-1,4 structure, carbon atoms 1 and 4 are *both* on the *same side* of the double bond, while, in the *trans*-1,4 structure, these two carbon atoms are on opposite sides of the double bond.

It turns out that natural rubber consists of polymer chains all having an almost perfect *cis*-1,4 structure, hence the true chemical name for this polymer is *cis*-1,4-polyisoprene. When the chain units in a macromolecule all consist of the *same isomer*, the polymer is said to be *stereoregular*. Because of this remarkable regularity, the natural rubber chains can attain a good regularity, especially when the rubber is stretched. Hence natural rubber crystallizes on stretching, resulting in high gum tensile strength.

Natural rubber is vulcanized with sulfur compounds which can crosslink the chains because of the presence of the reactive double bonds (unsaturation).

Synthetic "Natural Rubber" (Cis-1,4-polyisoprene)

Although the polymerization of isoprene dates back over one hundred years, all attempts to synthesize the *cis*-1,4-polyisoprene structure were unsuccessful until the advent of the "stereospecific" catalysts during the decade of the 1950's. Prior to that, any of the synthetic polyisoprenes had a "mixed" chain structure, containing a random arrangement of the four isometric units. When *cis*-1,4-polyisoprene was finally synthesized, it was found to virtually duplicate the behavior and properties of natural rubber, e.g., crystallization on stretching and high gum tensile.

Polybutadiene

Stereospecific catalysts can also be used to polymerize butadiene to a high *cis*-1,4 structure, as shown below.

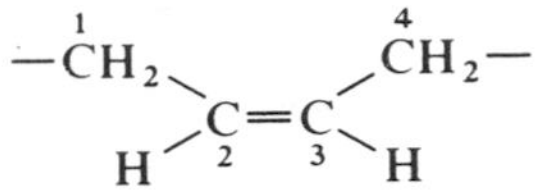

$$\begin{array}{ccccc} \overset{1}{-CH_2} & & & & \overset{4}{CH_2-} \\ & \diagdown & & \diagup & \\ & & \underset{2}{C}=\underset{3}{C} & & \\ & \diagup & & \diagdown & \\ H & & & & H \end{array}$$

This is an entirely new polymer which has been developed commercially during the past decade. Most of the polybutadienes produced today are of the *cis*-1,4 type, but some have a mixed chain structure. Although *cis*-1,4-polybutadiene, like its polyisoprene counterpart, is capable of crystallizing when stretched, it does not exhibit as high a gum tensile strength and is usually compounded with a reinforcing filler. Being an unsaturated elastomer it is easily vulcanized with sulfur.

SBR Polymers

Styrene-butadiene rubbers (SBR) are the general-purpose synthetic rubbers today, and were originally produced by government-owned plants as GR–S. They are "copolymers," i.e., polymer chains obtained by polymerizing a *mixture* of two monomers, butadiene and styrene, whose structures have been shown previously. The chains therefore contain random sequences of these two monomers, which gives them rubberlike behavior but renders them too irregular to crystallize on stretching. Hence these rubbers do not develop high tensile strengths without the aid of carbon black or other reinforcing pigments.

Nitrile Rubber

This, too, is a copolymer of two monomers, butadiene and acrylonitrile.

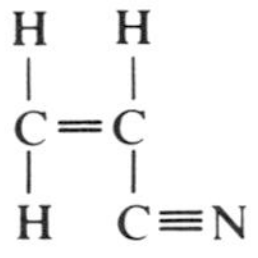

$$\begin{array}{ccc} H & & H \\ | & & | \\ C & = & C \\ | & & | \\ H & & C\equiv N \end{array}$$

acrylonitrile

It is prepared as a solvent-resistant rubber, the presence of the nitrile group (C≡N) on the polymer being responsible for this property. Like SBR, it also has an irregular chain structure and will not crystallize on stretching. Hence nitrile rubber requires a reinforcing pigment for high strength. Vulcanization is achieved by means of sulfur, as for SBR and natural rubber.

Butyl Rubber

This also is a copolymer, containing mostly isobutylene units, with just a few percent of isoprene units. Hence, unlike the butadiene rubbers or natural rubber, this polymer contains only a few percent double bonds (due to the small proportion of isoprene).

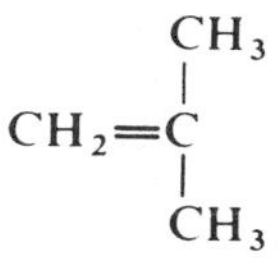

$$CH_2{=}\underset{\displaystyle CH_3}{\overset{\displaystyle CH_3}{\underset{|}{\overset{|}{C}}}}$$

isobutylene (gas)

This small extent of unsaturation is introduced to furnish the necessary sites for sulfur vulcanization, which is used for this rubber. The good regularity of the polymer chains makes it possible for this elastomer to crystallize on stretching, resulting in high gum tensile strength.

Ethylene-Propylene Rubbers

Like butyl rubber, the ethylene-propylene rubbers contain only a few percent double bonds, just enough for sulfur vulcanization. They are copolymers of ethylene and propylene, containing in addition a few percent of a diene for unsaturation. These elastomers are also among the newer ones made possible by the advent of the stereospecific catalysts. Because the various chain units are randomly arranged in the chain, these elastomers do not crystallize on stretching and require a reinforcing filler to develop high strength.

Neoprene

This elastomer is essentially a polychloroprene, as shown. Chloroprene monomer is actually 2-chlorobutadiene, i.e., butadiene with a chlorine atom replacing one of the hydrogens. Since the polymer consists almost entirely of *trans*-1,4 units, as shown,

$$CH_2{=}\underset{\displaystyle Cl}{\underset{|}{C}}{-}CH{=}CH_2 \longrightarrow \begin{matrix} -CH_2 & & & & H \\ & \searrow & & \nearrow & \\ & & C{=}C & & \\ & \nearrow & & \searrow & \\ Cl & & & & CH_2- \end{matrix}$$

chloroprene (liquid) | trans-1, 4-polychloroprene

the chains are sufficiently regular in structure to crystallize on stretching. Hence neoprene exhibits high gum tensiles and is used in the pure gum form in many applications.

The vulcanization of neoprene is quite different from the elastomers considered so far. Unlike the others, it is not vulcanized by means of sulfur. Instead, use is made of the fact that the chlorine atoms on the chain can react to some extent with active metals or metal oxides. Hence, zinc oxide or magnesium oxide are used to combine with some of the chlorine and interlink the polymer chains at those vacant sites.

Polysulfide Elastomers

These rubbers, best known under the trademark of "Thiokol," have already been described previously as condensation polymers. These polymer chains do not crystallize on stretching, hence this material is used with fillers for reinforcement.

Here again the vulcanization is not achieved by means of sulfur – which would not work, since this is not an unsaturated chain. Instead, use is made of the ability of the sulfur atoms to react with active metal oxides, like zinc oxide, which thus interlink the chains into a network.

Silicone Rubber

The silicone rubbers represent a completely different type of polymer structure from any of the others. This is because their chain structure does *not* involve a long chain of carbon atoms but a sequence of silicon and oxygen atoms, as shown.

polymer

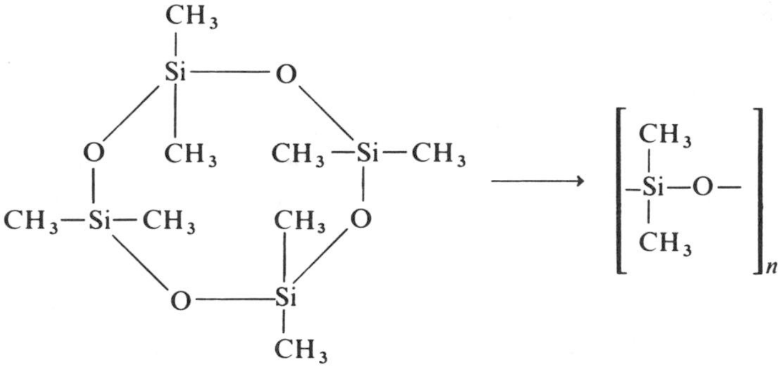

$$
\begin{array}{ccccc}
 & CH_3 & & & \\
 & | & & & \\
 & Si & — & O & \\
 / & | & & & \backslash \\
O & CH_3 & & & CH_3-Si-CH_3 \\
| & & & & | \\
CH_3-Si-CH_3 & & & CH_3 & O \\
\backslash & & & | & / \\
 & O & — & Si & \\
 & & & | & \\
 & & & CH_3 &
\end{array}
\longrightarrow
\left[
\begin{array}{c}
CH_3 \\
| \\
-Si-O- \\
| \\
CH_3
\end{array}
\right]_n
$$

cyclic siloxane polysiloxane

This siloxane structure results in a very flexible chain with extremely weak interchain forces. Hence the silicone rubbers are noted for showing little effect over a range of temperature. As might be expected, they show no tendency to crystallize on stretching and must be reinforced by a pigment, usually a fine silica powder.

Silicone rubber is customarily vulcanized by means of peroxides. These presumably are able to remove some of the hydrogen atoms from the methyl (CH_3) groups on the silicon atoms, thereby permitting the carbon atoms of two adjacent chains to couple and form crosslinks:

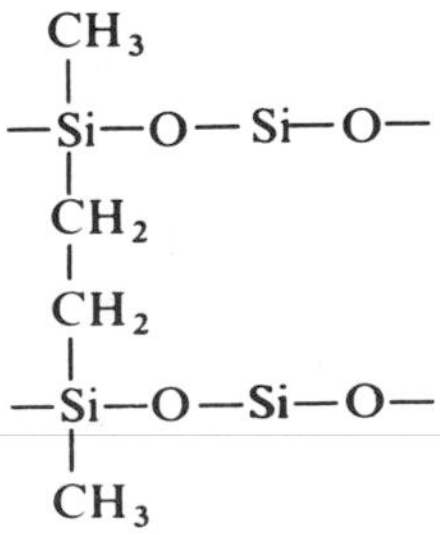

$$
\begin{array}{l}
\;\;CH_3 \\
\;\;\;| \\
-Si-O-Si-O- \\
\;\;\;| \\
\;\;CH_2 \\
\;\;\;| \\
\;\;CH_2 \\
\;\;\;| \\
-Si-O-Si-O- \\
\;\;\;| \\
\;\;CH_3
\end{array}
$$

Urethane Polymers

A novel modification in the polymerization field is represented by the urethane polymers. This is a novel method insofar as it involves a "chain extension" process rather than the usual polymerization reaction. In other words, these systems are able to make "big" macromolecules from "small" macromolecules, rather than from the monomer itself. Aside from being special processes for the preparation of polymers, these systems have special advantages, since the properties of the final polymer depend both on the *type* of original short-chain polymers used, as well as on their *chain length*. Hence a wide variety of polymers can thus be synthesized, ranging from rigid to elastic types.

The short-chain polymers used are generally two types, i.e., polyethers and polyesters. These can be shown as follows:

$$\mathrm{HO[-R-O-]_nH} \qquad\qquad \mathrm{HOR'O}\left[-\underset{\underset{\displaystyle \mathrm{O}}{\|}}{\mathrm{C}}-\mathrm{R}-\underset{\underset{\displaystyle \mathrm{O}}{\|}}{\mathrm{C}}-\mathrm{O}-\mathrm{R'}-\mathrm{O}-\right]_n\mathrm{H}$$

polyether — polyester

In the above formulas, the letter R or R′ represents a group of one or more carbon atoms. The value of n will vary from 10 to 50.

The chain extension reaction, whereby these short chains are linked together, is accomplished by the use of a reactive agent, i.e., a diisocyanate. For this reaction, the above short-chain polymers must have terminal hydrozy (OH) groups, as shown, so that the following reaction can occur between these groups and the diisocyanate:

$$\underset{\text{polyether or polyester}}{\mathrm{HO-P_n-OH}} + \underset{\text{diisocyanate}}{\mathrm{O{=}C{=}N-R-N{=}C{=}O}}$$

$$\downarrow$$

$$\mathrm{HO}\left[\mathrm{P}_n-\mathrm{O}-\underset{\underset{\displaystyle \mathrm{O}}{\|}}{\mathrm{C}}-\mathrm{NH}-\mathrm{R}-\mathrm{NH}-\underset{\underset{\displaystyle \mathrm{O}}{\|}}{\mathrm{C}}-\mathrm{O}\right]_x\mathrm{P-OH}$$

urethane polymer

Since the isocyanate group reacts vigorously with the hydroxy groups, a long-chain polymer results, with x having values up to 50 or 100.

In addition to this chain extension process, the diisocyanates are also capable of other reactions which can be utilized. Thus the diisocyanates can also react with some of the active hydrogen atoms attached to the polymer chain, leading to a cross-linking, i.e., vulcanization process. In this way, the urethane polymers lend themselves to a one-stage casting and curing process. Another reaction of the diisocyanates involves water, which reacts vigorously to form carbon dioxide gas, as

TABLE 1.1. ELASTOMERS AND THEIR CHARACTERISTICS

Name	*Chemical Name*	*Structure*	*Vulcanization Agent*	*Stretching Crystallization*	*Gum Strength*
Natural rubber	*cis*-1,4-polyisoprene	$[-CH_2-\underset{\displaystyle CH_3}{\underset{\vert}{C}}=CH-CH_2-]_n$	sulfur	good	good
Polyisoprene ("synthetic natural")	*cis*-1,4-polyisoprene	$[-CH_2-\underset{\displaystyle CH_3}{\underset{\vert}{C}}=CH-CH_2-]_n$	sulfur	good	good
Polybutadiene	polybutadiene	$[-CH_2-CH=CH-CH_2-]_n$	sulfur	depends on structure	poor to fair
SBR	poly(butadiene-co-styrene)	$[(-CH_2-CH=CH=CH_2-)_5(-CH_2-\underset{\displaystyle C_6H_5}{\underset{\vert}{CH}}-)]_n$	sulfur	poor	poor
Nitrile	poly(butadiene-co-acrylonitrile)	$[(-CH_2-CH=CH-CH_2-)_3(-CH_2-\underset{\displaystyle CN}{\underset{\vert}{CH}}-)]_n$	sulfur	poor	poor

Butyl	poly(isobutylene-co-isoprene)	$[(-CH_2-C(CH_3)_2-)_{50}(-CH_2-C(CH_3)=CH-CH_2-)]_n$	sulfur	good	good
EPR(EPDM)	poly(ethylene-co-propylene-co-diene)	$[-(-CH_2-CH_2-)_{37}(-CH_2-CH(CH_3)-)_{\overline{13}}\text{diene}-]$	peroxides (or sulfur)	poor	poor
Neoprene	polychloroprene	$[-CH_2-C(Cl)=CH-CH_2-]_n$	mag. oxide or zinc oxide	good	good
Silicone	polydimethylsiloxane	$[-Si(CH_3)_2-O-]_n$	peroxides	poor	poor
Thiokol(TM)	polyalkylenesulfide	$[-CH_2-CH_2-S_{\overline{2-4}}]_n$	zinc oxide	fair	poor
Urethane	polyester or polyether urethanes	$HO[-P-OCONHRNHCOO-]_nP-OH$	diisocyanates	depends on structure	good

Note: The fluorocarbon rubbers have not been included since they are simply the partially fluorinated analogs of various synthetic elastomers (see Chap. 16).

follows:

$$OCN-R-NCO + 2\,H_2O \longrightarrow H_2N-R-NH_2 + \underset{\text{carbon dioxide}}{2\,CO_2}$$

Hence, by mixing a little water together with the short-chain polymer and diisocyanate, a process of simultaneous foaming, polymerization, and vulcanization results, leading to the rapid formation of foamed elastomers or plastics.

As elastomers, the urethane polymers show some outstanding physical properties, including high gum tensile strength.

2

THE COMPOUNDING AND VULCANIZATION OF RUBBER

HOWARD L. STEPHENS
Associate Professor of Polymer Science
The University of Akron
Akron, Ohio

In the rubber industry, the problem of selecting the basic raw materials for the preparation of a specific commercial product usually is assigned to the compounder. Traditionally, the compounder has been a trained chemist or chemical engineer. This background is necessary since some of the processes involve complicated chemical reactions, of which vulcanization is the most important. In addition, chemical analysis of the raw materials and of the completed products may be required. Thus, this knowledge is necessary in order to select the proper test methods.

The compounder must be capable of describing these processes, and the problems involved to the engineers, development chemists, and the sales-service personnel, as they are concerned with the production of a serviceable product at a reasonable cost.

Compounding Recipes and Their Use

In order to aid in the development of a rubber compound, the various ingredients to be used are compiled into a "recipe." Every recipe contains a number of components, each having a specific function either in the processing, vulcanization, or end use of the product. Two typical tire tread recipes adapted from Vanderbilt's 1968 *Rubber Handbook* are given in Table 2.1.

In general, from the data given in Table 2.1, the following information can be obtained concerning compounding recipes.

(1) All the ingredients used are normally given in amounts based on a total of 100 parts of the rubber or combinations of rubbers (or masterbatches) used. This notation is generally listed as PHR (parts per hundred of rubber). Thus, when comparing different recipes, the effects of varying any ingredient used is easily recognized when the physical properties or processing characteristics are compared.

TABLE 2.1. TYPICAL TIRE TREAD RECIPES

	phr[a]		
Ingredient	*Natural Rubber*	*Synthetic*	*Function*
Smoked sheet	100	–	elastomer
Styrene-butadiene/oil masterbatch	–	103.1	elastomer-extender masterbatch
cis–polybutadiene	–	25	special purpose elastomer
Oil soluble sulfonic acid	2.0	5.0	processing aid
Stearic acid	2.5	2.0	accelerator-activator
Zinc oxide	3.5	3.0	accelerator-activator
Phenyl-beta-naphthylamine	2.0	2.0	antioxidant
Substituted N,N'–p–phenylene-diamine	4.0	4.0	antiozonant
Microcrystalline wax	1.0	1.0	processing aid and finish
Mixed process oil	5.0	7.0	softener
HAF carbon black	50	–	reinforcing filler
ISAF carbon black	–	65	reinforcing filler
Sulfur	2.5	1.8	vulcanizing agent
Substituted benzothiazole–2–sulfenamide	0.5	1.5	accelerator
N–nitrosodiphenylamine	0.5	–	retarder
Total weight	173.5	220.4	
Specific gravity	1.12	1.13	

[a]Parts per hundred parts of rubber, by weight.

(2) Although the function of each component, as shown in this example, is never indicated in industrial or laboratory recipes, it is apparent that many different materials with specific purposes are used in every recipe.

(3) In many recipes, the materials are listed in the general order that they are mixed into the rubber during processing. This method aids the compounder in setting up his mixing schedules for processing various compounds and for the preparation of special masterbatches which may be used in many different products.

(4) From the total amount of materials used (whether in grams, ounces, pounds, or other method of measurement), the cost of the total compound can be computed rather simply, as follows:

$$\text{Cost per pound} = \frac{\text{Total cost of all ingredients}}{\text{Total recipe weight}} \quad \text{or}$$

$$\text{Cost per volume-pound} = \underline{\text{Cost per pound}} \times \text{Specific gravity}$$

Generally, the cost is based on a volume-pound figure since this method produces a comparable value for all rubber vulcanizates, especially when the compounding ingredients are changed.

Components of the Recipe

Although the examples used in the above illustration are not typical of all recipes, in general, the materials utilized by the rubber compounder can be classified into nine major categories, which are defined as follows:

Elastomers: The basic component of all rubber compounds, it may be in the form of rubber alone, or "masterbatches" of rubber-oil, rubber-carbon black, or rubber-oil-carbon black, or reclaimed rubber. Combinations or blends as given in the synthetic tire tread recipe are quite common. The elastomers are selected in order to obtain specific physical properties in the final product.

Processing Aids: Materials used to modify rubber during the mixing or processing steps, or to aid in a specific manner during extrusion, calendering, or molding operations.

Vulcanization Agents: These materials are necessary for vulcanization, since without the chemical crosslinking reactions involving these agents, no improvement in the physical properties of the rubber mixes can occur.

Accelerators: In combination with vulcanizing agents, these materials reduce the vulcanization time (cure time) by increasing the rate of vulcanization. In most cases, the physical properties of the products are also improved.

Accelerator Activators: These ingredients form chemical complexes with accelerators, and thus aid in obtaining the maximum benefits from an acceleration system by increasing vulcanization rates and improving the final products properties.

Age-Resistors: Antioxidants, antiozonants, and other materials that are used to reduce aging processes in vulcanizates. They function by slowing down the deterioration of rubber products. The deterioration occurs through reactions with materials that catalyze rubber failure, i.e., oxygen, ozone, light, heat, radiation, etc.

Fillers: These materials are used to reinforce or modify physical properties, impart certain processing properties, or reduce cost.

Softeners: Any material that can be added to rubber to either aid mixing, promote greater elasticity, produce tack, or extend (or replace) a portion of the rubber hydrocarbon (without a loss in physical properties), can be classified as a softener.

Miscellaneous Ingredients: Materials that can be used for specific purposes but are not normally required in the majority of rubber compounds can be included in this group. It includes retarders, colors, blowing aids, abrasives, dusting agents, odorants, etc.

All of the classes of components listed above will be discussed in greater detail in this and subsequent chapters.

Processing Methods

Both in the laboratory and factory, the most common methods for incorporating the compounding ingredients into the raw rubber involve either the use of a mill or a Banbury (internal) mixer. There are many sizes of each type, and some typical examples are given in Table 2.2 and shown in Figs. 2.1 through 2.4.

For the purpose of illustrating how these two methods of mixing are used, two typical laboratory recipes and mixing schedules which have been developed by the Committee D-11 on Rubber and Rubber-Like Materials of the American Society for Testing and Materials (ASTM) have been chosen. These with other standards and test methods are published annually in part 28 of the ASTM Book of Standards.

TABLE 2.2. TYPICAL SIZE AND CAPACITY OF MILLS AND BANBURY MIXERS (SOURCE: FARRELL COMPANY BULLETINS 173E AND 215)

2 Roll Mills

Roll Size (in.)[a]	*Batch Size (lb)*[b]	*Motor (H.P.)*	*Weight (lb)*
6 x 13	1.25 to 2	7.5	6,200
8 x 16	2.5 to 4	10–15	8,000
10 x 20	5 to 8	15–20	10,000
12 x 24	10 to 18	30–40	13,800
14 x 30	20 to 30	40–50	18,000
16 x 42	30 to 50	70–75	24,000
18 x 48	45 to 70	75–100	30,000
22 x 60	75 to 125	125–150	39,500
24 x 72	125 to 200	150–200	50,000
26 x 84	150 to 250	150–200	65,000
28 x 84	175 to 300	200–250	70,000

Banbury Mixers

Size	*Capcity (lb)*[b]	*Motor (H.P.)*	*Weight (lb)*
Midget	0.67	7.5	3,300
BR	2.6	8.5–25	5,000
OOC	7	15–30	6,000
1	27	50–100	9,500
1A and 1D	27	50–100	12,900
3A and 3D	105	150–300	30,000
3A and 3D Unidrive	105	200–400	38,000
9	265	200–400	65,000
9D Unidrive	290	250–500	66,800
11 and 11D	370	300–600	90,000
11 and 11D Unidrive	370	400–800	106,000
27 and 27D Unidrive	930	1500	251,400

[a]Diameter x length.
[b]Based on 1.00 specific gravity stock.

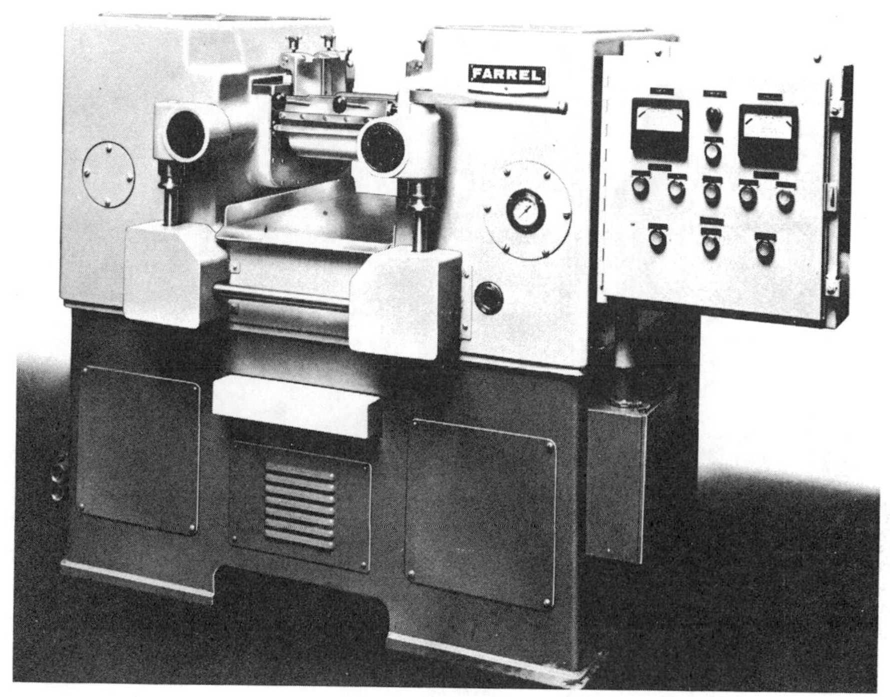

Fig. 2.1 6″ x 13″ laboratory mill. (*Courtesy Farrel Company*)

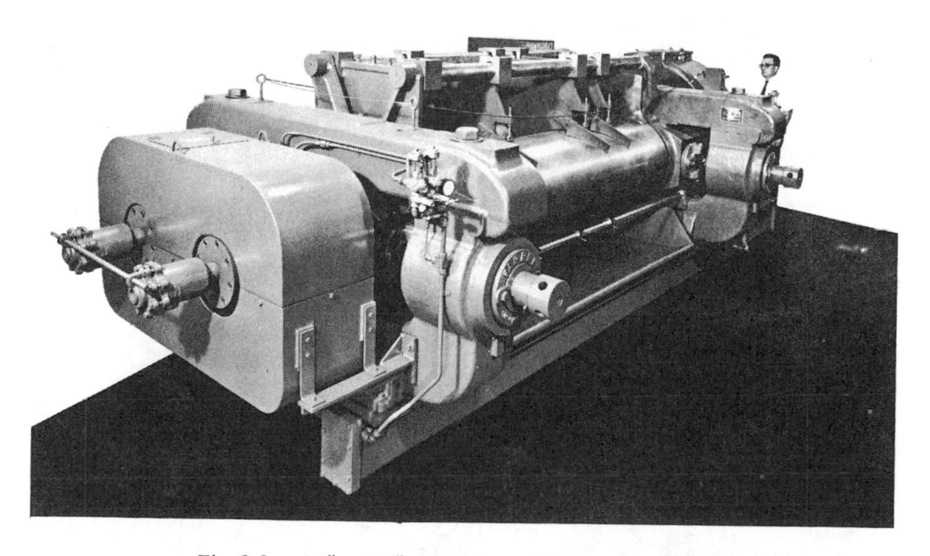

Fig. 2.2 26″ x 100″ mill. (*Courtesy Farrel Company*)

Fig. 2.3 BR laboratory Banbury mixer. (*Courtesy Farrel Co.*)

This particular part covers procedures adopted for rubber, carbon black, and gaskets; and all rubber technologists should become well acquainted with its contents.

The procedures are taken from Methods of Sample Preparation for Physical Testing of Rubber Products, ASTM Designation D 15–66T. The recipes in Table 2.3 using natural rubber were chosen for showing the use of mixing schedules during processing.

TABLE 2.3. NATURAL RUBBER RECIPES

	Mill Mixing (7A)	*Banbury Mixing (1G)*
Natural rubber	100	100
Zinc oxide	5	5
Sulfur	2.5	2.5
Stearic acid	1	3
Benzothiazyl disulfide	1	0.6
Phenyl-beta-naphthylamine	1	–
Gas furnace black	45	50
Total	155.5	161.1
Specific Gravity	1.12	1.13

Fig. 2.4 27D Banbury mixer with uni-drive. (*Courtesy Farrel Co.*)

The batch size for a 6″ laboratory mill is four times the recipe; and for the BR Banbury (internal) mixer with a capacity of approximately 1200 cm^3, the batch size is equal to the volume of the mixer times the specific gravity of the stock. In this case it is approximately eight times the recipe weight. The mixing cycles used are as shown in Table 2.4.

TABLE 2.4. MIXING CYCLES

	Mill	*Banbury*
Temperature, °C(F)	*70 ± 5 (158 ± 9)*	*110–125 (230–257)*[a]
Mixing Speed, slow roll	24 rpm	77 rpm
Roll ratio (slow to fast roll)	1 to 1.4	1 to 1.125

Mixing Steps	Time (minutes)	
1. Pass rubber through rolls twice without banding, at a mill roll opening of 0.008 in. at 70°C Regulate the temperature on the rotors and shell to obtain a dump temperature of 110 to 125°C	1	

TABLE 2.4. *(continued)*

Mixing Steps	*Mill*	*Banbury*
	Time (minutes	
2. Band with mill opening at 0.055 in. and break down, opening the mill to 0.075 in. as the band becomes smooth	4	
Add rubber	–	0.5
3. Add main pigment evenly across the rolls at a constant rate and open the mill at intervals to maintain an approximately constant bank. When about half the pigment is incorporated and all dry pigment in the bank has disappeared, make one 3/4 cut from each side. Then continue with the balance of the pigment. Be certain to add the pigment that drops through the mill pan	8	–
Add benzothiazyl disulfide		0.5
4. Add stearic acid	2	1
5. Add other ingredients	4	–
Add the zinc oxide and one-half of black	–	1.5
6. Make three 3/4 cuts from each side	2	–
Add remainder of the black	–	1.5
7. Cut the batch from the mill. Set the opening at 0.033 in. and pass the rolled stock endwise through the mill six times	2	–
Add sulfur	–	1
8. Sheet the batch to a minimum thickness of 0.25 in. and weigh	1	–
Dump	–	1
9. Sheet immediately from the mill set at a mill opening of 0.085 in. and a roll temperature of 70 ± 5°C. Band, cut, roll up, and weigh	–	2
10. Pass endwise six times at a 0.030 in. opening	–	2
11. Sheet off at 0.085 in. and cool	–	2
Total Time (min)	24	13

[a]Dump temperature.

From the above examples, it is apparent that the following information concerning mixing cycles and times can be extended to practical factory or laboratory processes:

(1) There are definite temperature ranges for mixing rubbers, each specific rubber having an optimum temperature at which the desired dispersion of compounding ingredients in the mix is obtained.

(2) Some rubbers require an initial breakdown period before the ingredients are added.

(3) A specific order of incorporation of the compounding ingredients is necessary.

(4) The time of mixing for each step in the process is important.

(5) A finishing step and control of the final temperature of the mix is necessary to prevent prevulcanization, and,

(6) Banbury (internal) mixes require less time and handling and mix larger batches than the corresponding mill mixers.

Although the Banbury mixer is capable of handling larger factory batches at a faster rate than mill mixing, there are other processing problems involved, which will be discussed in detail in Chap. 4.

It is only necessary here to point out that factory techniques require much stricter control tests than corresponding laboratory methods since large quantities of material are used. Hence the labor, processing, and equipment costs are a more important factor in production than in compounding research, especially if a large factory batch must be scrapped due to improper mixing or compounding.

Vulcanization

After rubber compounds have been properly mixed and shaped into blanks for molding, or calendered, extruded, or fabricated into a composite item (as a tire), they must be vulcanized by one of many processes. During vulcanization, the following changes occur:

(1) The long chains of the rubber molecules become crosslinked by reactions with the vulcanization agent to form three-dimensional structures. This reaction transforms the soft weak plastic-like material into a strong elastic product.

(2) The rubber loses its tackiness and becomes insoluble in solvents and is more resistant to deterioration normally caused by heat, light, and aging processes.

These changes generally occur with the use of the following vulcanization systems.

Vulcanization Systems

1. Sulfur Vulcanization It is quite apparent from the data presented in Table 2.5, that the most common rubbers used are the general purpose type, with the remaining types representing only about 16% of the total usage. Since these rubbers (general purpose) contain unsaturation, vulcanization with sulfur is possible, and it is in general the most common vulcanizing agent used. With sulfur, crosslinks and cyclic structures of the following type are formed:

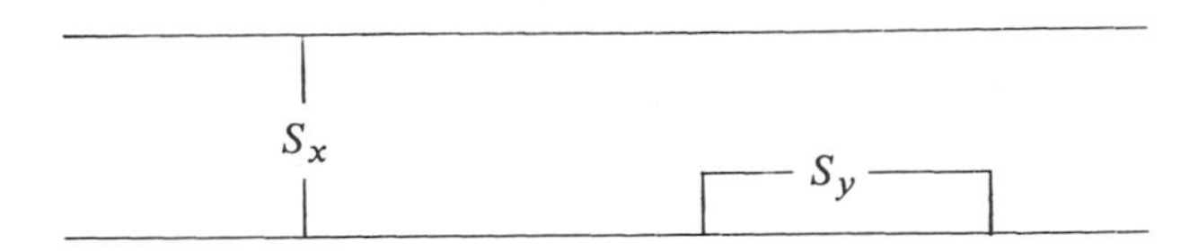

Generally, x in an efficient accelerated curing system is about 1 or 2, with little or no cyclic groups formed. In inefficient systems x equals up to 8 and many cyclic structures are formed. The total amount of sulfur combined in these networks is usually called the "coefficient of vulcanization" and is defined as the parts of sulfur combined per one hundred parts of rubber. For most rubbers, one crosslink for about each 200 monomer units in the chain is sufficient to produce a suitable vulcanized product (molecular weight between crosslinks equals ca. 8,000 to 10,000).

It is these amounts of cyclic sulfur (y) and the excessive sulfur in the crosslinks (x) which contribute to the poor aging properties of the vulcanizates.

2. Sulfurless Vulcanization Vulcanization effected without elemental sulfur, by the use of thiuram disulfide compounds (accelerators) or with selenium or tellurium, produces products which are more resistant to heat aging. With the thiuram disulfides, efficient crosslinks containing only 1 or 2 sulfur atoms are found, and in addition the accelerator fragments act as antioxidants.

3. Peroxide Vulcanization The saturated rubbers cannot be crosslinked by sulfur and accelerators. Organic peroxides are necessary for the vulcanization of these rubbers. When the peroxides decompose, free radicals are formed on the polymer chains, and these chains can then combine to form crosslinks, as shown below:

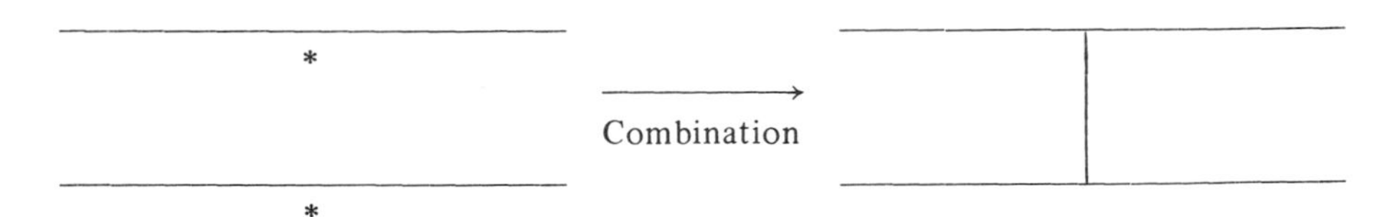

Crosslinks of this type only involve carbon-to-carbon bonds and are quite stable; they are also formed by gamma radiation and X-radiation.

4. Other Systems Some elastomers can be vulcanized by the use of certain nonsulfur bifunctional compounds which form bridge type crosslinks: for example, neoprene with metal oxides or butyl rubber with dinitrosobenzene. More detailed descriptions of the processes involved will be discussed in later chapters dealing specifically with each rubber.

Vulcanization Conditions

In vulcanization (curing) processes, consideration must be made for the difference in the thickness of the objects involved, the vulcanization temperature, and the thermal stability of the rubber compound.

1. Effect of Thickness Rubbers are poor heat conductors, thus it is necessary to consider the heat conduction, heat capacity, geometry of the mold, heat exchange system, and the curing characteristics of a particular compound when articles thicker than about one-quarter of an inch are being vulcanized. This effect is best shown by immersing thermocouples at various depths in a rubber compound and measuring the time required to reach the vulcanization temperature as indicated by the press temperature. This effect is illustrated in Fig. 2.5.

Since these effects are complicated, generally an estimate of the time required can be determined by adding an additional 5 minutes to the cure time for every one-quarter inch of thickness. In exceptionally thick or complicated articles, the item may be built up using sections with different curing characteristics, or by controlling the rate at which the mold is heated or cooled.

2. Effect of Temperature The vulcanization temperature must be chosen in order to produce a properly cured product having uniform physical properties in the shortest possible molding time. The "temperature coefficient of vulcanization" is a term used to identify the relationship that exists between different cure times at different temperatures. With information of this type, optimum cure times at higher or lower temperatures can be estimated for many rubber compounds with

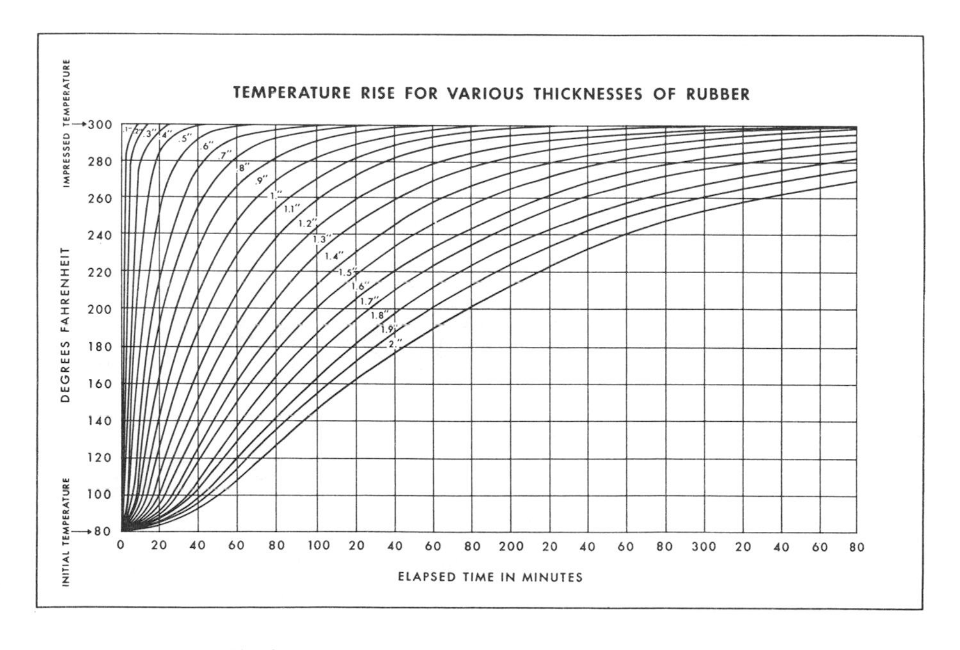

Fig. 2.5 Effect of thickness on temperature rise.

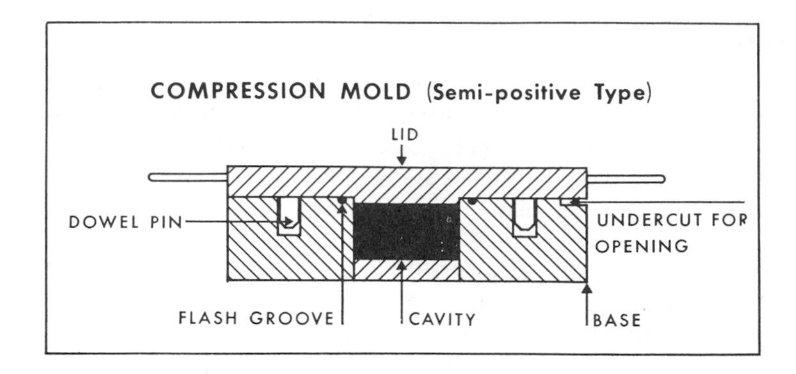

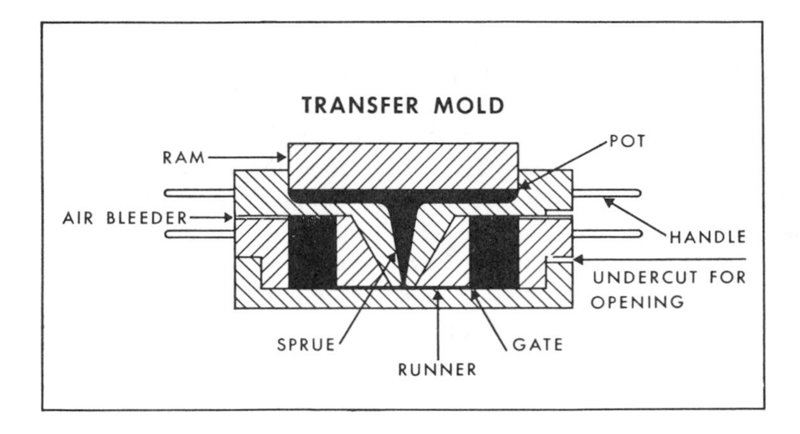

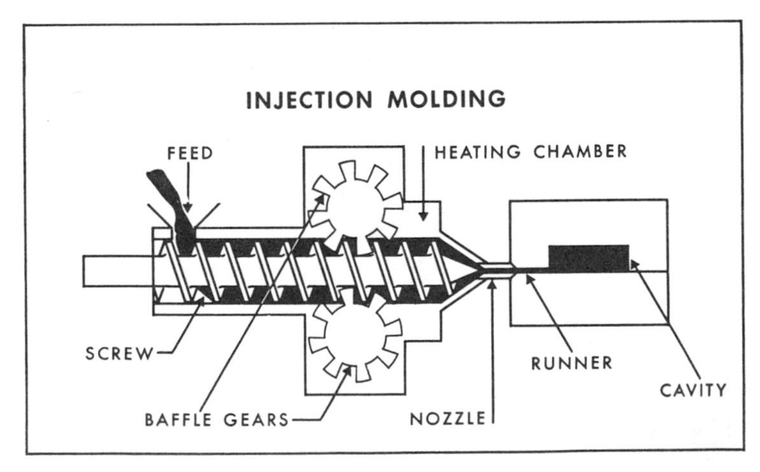

Fig. 2.6 Examples of compression, transfer, and injection molding techniques. (*Courtesy E. I. du Pont de Nemours and Co., Inc.*)

known coefficients of vulcanization. For example, most rubber compounds have a coefficient of approximately 2. This indicates that the cure time must be reduced by a factor of 2 for each 18°F (10°C) increase in cure temperature, or, if the temperature is reduced 18°F, the cure time must be doubled.

3. Effect of Thermal Stability Each type of rubber has a definite range of temperatures which may be used for vulcanization. These temperatures may vary somewhat but it is quite important not to exceed the maximum for each, as some form of deterioration will occur. This effect is either shown by the appearance of the finished product or by its physical properties. Many of these "thermal" effects will be discussed in the following chapters.

Vulcanization Techniques

Finished articles can be prepared by numerous vulcanization techniques; however, the specific methods used in most industries are usually based on producing suitable commercial goods utilizing standard techniques. These methods are briefly outlined as follows.

1. Compression Molding This method, or a modification of it, probably utilizes the most common type of mold used in the rubber industry. Essentially, it consists of placing a precut or shaped "slug" (or blank), or a composite item, into a two-piece mold which is closed. The pressure applied by the press forces the material to fit the shape of the mold, and the slight excess present flows out of the rim of the mold or through special vents. This excess is known as "mold flash." A simple mold of this type is shown in Fig. 2.6.

Tires are normally cured in a modification of the compression mold. A bladder, or air-bag, forces and holds the "green" tire against the mold surface during vulcanization. This force reproduces the design of the tread, and heat (in the form of steam) is normally introduced into the bladder to aid the curing process. An excellent history on the development of tires and their vulcanization is given in reference 3. Figure 2.7 shows a typical automatic tire curing press.

2. Transfer Molding As shown in Fig. 2.6, transfer molding involves the distribution of the uncured stock from one part of the mold (the pot) into the actual mold cavity. This process permits the molding of complicated shapes or the imbedding of inserts in many products; these procedures are difficult with the usual compression molds.

Although the molds are relatively more expensive than compression molds, the actual process permits shorter cure times through the use of higher temperatures and better heat transfer which is obtained due to the higher pressure applied to force the compound into the mold.

3. Injection Molding In recent years, injection molding processes, which are normally used for the production of plastics, have been developed so that small rubber compounds can be molded and vulcanized by this method. A diagram of the process is shown in Fig. 2.6 and a commercial unit is illustrated in Fig. 2.8.

By careful temperature control of the feed stock, items can be vulcanized in less than several minutes (cure times are generally reported in seconds). This method

Fig. 2.7a Automatic tire curing press showing curing bladder in place. (*Courtesy McNeil Corp.*)

can be completely controlled by programmed feed, injection, and demolding cycles resulting in low rejection rates and lower finishing costs. The initial cost of both the molds and equipment has hindered the adoption of this type of molding under compression.

Estimates have been made indicating the use of the injection method will expand from the present use of 5% to 25% (of the molding under compression type of vulcanization) by 1975.

4. Open Cures a. *Hot air ovens* can be used to vulcanize thin articles (balloons, etc.), items which have been preshaped (by extrusion) or by a combination of precuring in a mold, followed by post-vulcanization in an oven. The last process is used to remove peroxide decomposition products from items cured with peroxides. The system is not too efficient due to the poor heat transfer of hot air, and longer cure times at lower temperatures are necessary to prevent the formation of porosity or deformation of the unvulcanized products.

b. *Open steam* can be used in closed containers called "Autoclaves" (which resemble home pressure cookers); the process involves using saturated steam under pressure. The saturated steam acts as an inert gas, better heat transfer is obtained, thus higher temperatures can be used and shorter cure times are possible, making this process more desirable than the air oven. Hose, cables, built-up footwear, and tires in pot-heater molds are cured by this method.

c. *Water cures* can be used for articles that are not affected by immersion. The method is useful for large items (for example, large containers or rubber-lined

Fig. 2.7b Automatic tire curing press showing the removal of a cured tire. (*Courtesy McNeil Corp.*)

Fig. 2.8 75-ton injection molding machine. (*Courtesy McNeil Akron, Division of McNeil Corp.*)

containers) and is especially useful for hard rubber compositions. Direct contact with water produces better heat transfer than with air or open steam; consequently, with this system, less deformation and faster cures are obtained.

d. *Lead sheathing* can be used to cover soft, large extruded sections with a protective cover for vulcanization in steam. The process is used for garden and other hose; the lead sheath is usually applied by hydraulic pressure or extrusion immediately as the hose emerges from its extruder.

5. Continuous Vulcanization Processes (C.V.) Continuous vulcanization generally involves the use of some form of heating in a manner such that the vulcanization step usually occurs immediately after the rubber article is formed. The process is normally used for extruded goods, coated wiring, conveyor belts, and flooring.

a. *Liquid Curing Methods (L.C.M.)* involve the use of suitable hot liquid baths in which extrusions may be vulcanized in a continuous process. Items can be cured rapidly at temperatures from 200 to 300°C; however, the compounds must be modified to prevent porosity, as this is always a problem with any extrudate.

Suitable materials for the curing medium include bismuth-tin alloys; a eutectic mixture of potassium nitrate, sodium nitrite, and sodium nitrate; poly-glycols; and certain silicone fluids.

b. *Fluidized beds* consisting of small particles (glass beads) suspended in a stream of heated air are efficient vulcanization systems. They are normally used for continuous vulcanization of extrusions. The heat transfer is approximately 50 times greater than with hot air alone.

c. *Continuous hot air tunnels* can be used in the same manner that hot air ovens are used. Thin articles, for example, dipped goods, can be vulcanized after dipping by passing an endless conveyor carrying the items through a series of heated tunnels (ovens) at a rate that will complete vulcanization by the time articles are stripped from the forms.

d. *Steam tubes* are used for the continuous vulcanization of sheathing. After extrusion of the rubber cover onto the wire, the cable passes into a jacketed steam tube containing steam under pressure. Depending on the thickness of the cover, the period of time in the tube can be controlled to complete vulcanization, or conversely, the length of the steam tube can be extended.

e. The *Rotacure* process can be used to vulcanize large conveyor belts and continuous flooring strips. The process involves the use of an endless steel band which presses the article against a large heated drum. Slow rotation of the drum permits the vulcanization to occur after approximately ten minutes contact time. Belt curing presses are also used; however, this system is not completely continuous, although long lengths of belting are made by this method.

6. Cold Vulcanization Thin articles may be vulcanized by treatment with sulfur monochloride (S_2Cl_2) by dipping in a solution or exposure to its vapors. The process has been essentially replaced by using ultra accelerators which are capable of curing at room temperature.

7. High Energy Radiation Systems using either gamma radiation from cobalt

60 or electron-beams have been used for vulcanization. The electron beam method has been used to cure both polyethylene and silicone rubbers. This is generally accomplished by passing the materials through the beam on a conveyor.

8. Microwave Vulcanization Ultrahigh frequency fields (UHF) developed by alternating electromagnetic circuits can be used to warm up or vulcanize articles with large or uneven cross sections. The process requires polar rubber mixtures since nonpolar materials will not absorb the energy produced. It is possible to warm articles up to 200°C within 30 seconds with some UHF systems. Continuous processes utilizing extruders and microwave heaters are available.

COMPOUNDING INGREDIENTS

1. Elastomers (Rubber and Rubber-Like Materials)

Today's rubber compounder has many types of rubber and rubber-like materials available for use. In most instances, he must first decide what particular use the end product will be subjected to, before choosing the proper rubber. For this choice, rubbers are generally classified into three major classes, as follows:

	Former ASTM Classification	
General Purpose	R:	For services where specific resistance to the action of petroleum-base fluids is not required.
Solvent Resistant	S:	For services where specific resistance to the action of petroleum-base fluids is required.
Heat Resistant	T:	For services where specific resistance to the effects of prolonged exposure to abnormal temperatures or compounded petroleum oils, or both, is required.

A listing showing the types in general usage is given in Table 2.5; included is the general ASTM code, current usage, and price data.

For each rubber, there are many types, depending upon the manufacturing process, grade, and polymer composition. The problem of selecting a specific type is somewhat simplified by the use of publications available from major suppliers, manufacturers, and associations.

Natural Rubber The grades of natural rubber and the various types of each are listed in the "Green Book," which is the common name given to "The International Standards of Quality and Packing for Natural Rubber Grades," published by The Rubber Manufacturers Association, Inc.*

In addition, The Rubber Research Institute of Malaya has introduced innovations including "Standard Malaysian Rubber" (SMR). Information on these

*444 Madison Ave., New York, N.Y. 10022.

TABLE 2.5. TYPICAL COMMERCIAL RUBBERS

Common Name	*ASTM Designation*	*Consumption (U.S.) in long tons (1970)*[a]	*Price*[b] *($/lb)*
	General Purpose		
Natural	NR	559,300	0.26
Polyisoprene	IR	80,000	0.24
Styrene-butadiene	SBR	1,216,300	0.23
Butyl	IIR	92,300	0.25
Ethylene-propylene	EPDM	45,700	0.26
Polybutadiene	BR	277,400	0.25
	Solvent Resistant		
Polysulfides (Thiokol)	T	*	0.75
Nitrile	NBR	59,600	0.51
Polychlorprene (Neoprene).	CR	120,000	0.41
Polyurethanes			1.40
Polyester	AU	N.A.*	
Polyether	EU	N.A.*	
Epichlorohydrin	CO	*	0.70
Epichlorohydrin-ethylene oxide	ECO	*	0.69
	Heat Resistant		
Silicone	MQ	*	2.80
Chlorosulfonated-polyethylene (Hypalon)	CSM	*	0.50
Polyacrylates	ACM	*	1.15
Fluororubbers	CFM	*	10.00

*Total for all rubbers, excluding urethanes 146,000[a]

Total Consumption,[a]		
	synthetic rubbers	1,917,852
	natural rubber	559,315
	reclaimed rubber	199,571
	Total Consumption	2,676,738

[a] U.S. Deparment of Commerce, Bureau of Census Current Industrial Reports. Rubber: Supply and Distribution for the United States, June, 1971.
[b] Average prices for most common grade, May 1970.

newer types of natural rubber is available from the Natural Rubber Bureau* and is discussed in detail in a later chapter.

Synthetic Rubbers It is quite apparent from the listing given in Table 2.5 that there are many types and compositions available when a synthetic rubber is considered for use. During the development of the synthetic rubber industry during

*15 Atterbury Blvd., Hudson, Ohio 44236.

World War II, a numbering system was used by the government for the identification of SBR rubbers. However, when the plants were sold to private companies many of these numbering codes were changed. In 1960, the International Institute of Synthetic Rubber Producers Inc. (IISRP)** was formed in order to organize a systematic listing of all the available synthetic rubbers and to promote the interests of all concerned in the manufacture and use of synthetic rubbers. Over twenty-five producers have memberships in the organization and it currently publishes editions listing the procedures, the nomenclature developed for the classification of the polymers, and data on the various types of rubbers and latices actually available.

Information on styrene-butadiene, high-styrene resins, butadiene, isoprene, ethylene-propylene, butyl, chloroprene, and nitrile rubbers, and their latices is included in the tabulation. Since these rubbers will be discussed in detail in later chapters, no further classification will be necessary at this time.

Reclaimed Rubbers Many rubbers can be reclaimed and utilized as a partial or complete replacement for new rubber in many articles. This usage represents about 8.6% of the total consumption of all rubber in the U.S. during 1969. A complete description of the reclaim processes and product uses will be given in Chap. 19.

2. Chemical Plasticizers (or Peptizing Agents)

Some rubbers, especially natural and high viscosity synthetics, require an initial breakdown period during mixing in order to soften the material for processing or to increase building tack after compounding. This softening effect can be catalyzed by the addition of small amounts of chemical plasticizers (up to 2 phr) which help control the amount and speed of breakdown and aid in the dispersion of the other compounding ingredients. Since the process also can reduce the nerve and shrinkage of the compound, improved stock preparation (as in extrusion or calendering) and molding operations are achieved from the reduction of the molecular weight (chain length) of the rubber through oxidative chain scission during mastication.

For proper usage, the plasticizer should only function during the initial mixing period (nonpersistent) and its action should be stopped by the addition of either carbon black, sulfur, or accelerators. Chemical plasticizers are normally used with natural and styrene-butadiene rubbers; typical examples are given: xylyl mercaptan (thioxylenols); oil soluble sulfonic acids; zinc salt of pentachlorothiophenol; pentachlorothiophenol; 2-naphthlenethiol; phenylhydrazine salts.

It is apparent from the above listing that the majority of the peptizers have active –SH groups which function as chain terminating agents by reactions with the free radicals formed when the rubber chains rupture during mastication.

3. Vulcanization Agents

As stated earlier, these are the chemicals that are required to crosslink the rubber chains into the three-dimensional network which gives the desired physical

**45 Rockefeller Plaza, New York, N.Y. 10020.

properties in the final product. The type of crosslinking agent required will vary with the type of rubber used; however, they can usually be grouped in the following categories.

Sulfur and Related Elements The most common agent used is sulfur, as it enters into reactions with the majority of the unsaturated rubbers to produce vulcanizates. In addition, two other elements in the same periodic family, namely selenium and tellurium, are capable of producing vulcanization.

Two forms of sulfur, the rhombic and amorphous (or insoluble sulfur), are compared with selenium and tellurium in Table 2.6.

TABLE 2.6. COMPARISON OF ELEMENTAL VULCANIZATION AGENTS

	Sulfur			
	Rhombic	*Amorphous*	*Selenium*	*Tellurium*
Atomic Weight	32.06	32.06	78.96	127.61
Appearance	yellow powder	yellow powder	metallic powder	metallic powder
Specific gravity	2.07	1.92	4.80	6.24
M.P. °C	112.8–119	>110	217.4	449.8
Price $/lb	0.03	0.21	6.50	7.00

The rhombic form is normally used for vulcanization; it exists as a cyclic (ring) structure composed of eight atoms of sulfur, S_8. The amorphous form is actually polymeric in nature; it is a metastable high polymer with a molecular weight of 100,000 to 300,000. It is insoluble in most solvents and rubber, hence the name "insoluble sulfur." The amount of insolubility is usually determined by using carbon disulfide as the solvent. Because of this property, amorphous sulfur is used to prevent "blooming" on uncured rubber surfaces where it is necessary to maintain "building tack." Insoluble sulfur must not be processed above 210–220°F or it will revert to the rhombic form.

In general, about 1.0 to 3.0 phr of sulfur is used for most rubber products. Commercially both forms of sulfur are available in forms that have been treated with small amounts of a material (carbon black, magnesium carbonate, etc.) which produces free-flowing, noncaking powders. Oil-sulfur mixtures are used occasionally to improve dispersion. Masterbatches of sulfur with rubbers or rubber-like polymers are also used where processing safety and ease of dispersion are important.

Selenium and tellurium are used in place of sulfur where excellent heat resistance is required. They generally shorten cure time and improve some vulcanizate properties. Selenium is somewhat more active than tellurium.

Sulfur-Bearing Chemicals Accelerators and similar compounds can be used as a source of sulfur for the vulcanization of natural and styrene-butadiene rubbers in recipes using very small amounts of elemental sulfur. Generally in these "low-sulfur" cures, less than 1 phr of sulfur is used in combination with 3 to 4 phr of the sulfur donor and in some cases no elemental sulfur is added to the recipe.

The compounds used decompose at the vulcanization temperature and release radicals which combine with the chains to form crosslinks. With these systems, efficient crosslinking occurs as most of the sulfur is combined in crosslinks containing one or two sulfur atoms with little or no cyclic sulfur present. Consequently, this form of vulcanization produces products which resist aging processes at elevated temperatures much more effectively than those produced with normal curing systems. However, due to the large amounts of the sulfur donors used, these systems are more expensive than normal sulfur cures and are only used when necessary. Some typical compounds used in low-sulfur cures are shown in Table 2.7.

TABLE 2.7. TYPICAL COMPOUNDS USED FOR "LOW-SULFUR" VULCANIZATION

Compound	*Sulfur Content (%)*	*Price $/lb*
Tetramethylthiuram disulfide	13.3	0.45
Dipentamethylenethiuram hexasulfide	35.0	1.75
Dimorpholinyl disulfide	31.4	1.54
Dibutylxanthogen disulfide	21.4	1.95
Alkylphenol disulfide	23.0	0.47
	28.0	0.51

Nonsulfur Vulcanization Most nonsulfur vulcanization agents belong to one of three groups: a) metal oxides, b) difunctional compounds, or c) peroxides. Each will be discussed here separately; however, much more detail will be given in the chapters that follow.

Metal Oxides Carboxylated nitrile, butadiene, and styrene-butadiene rubbers may be crosslinked by the reaction of zinc oxide with the carboxylated groups on the polymer chains. This involves the formation of zinc salts by neutralization of the carboxylate groups. Other metal oxides are also capable of reacting in the same manner.

Polychloroprenes (Neoprenes) are also vulcanized by reactions with metal oxides, zinc oxide being normally used. The reaction involves active chlorine atoms and is described in a later chapter. Chlorosulfonated polyethylene (Hypalon) is also crosslinked in the same general way. Litharge (PbO), litharge/magnesia (MgO), and magnesia/pentaerythritol combinations are used.

In many of these systems, the metal oxides are used in combinations for the purpose of controlling the vulcanization rate and absorbing the chlorides formed.

Difunctional Compounds , Certain difunctional compounds form crosslinks with rubbers by reacting to bridge polymer chains into three-dimensional networks. Epoxy resins are used with nitrile; quinone dioximes with butyl; and diamines or dithio compounds with fluororubbers. Other examples are given later in the text.

Peroxides Organic peroxides are used to vulcanize rubbers that are saturated or do not contain any reactive groups capable of forming crosslinks. This type of vulcanization agent does not enter into the polymer chains but produces radicals which form carbon-to-carbon linkages with adjacent polymer chains.

Typical examples of nonsulfur compounds are given in Table 2.8.

TABLE 2.8. NONSULFUR VULCANIZATION COMPOUNDS

Compound	*phr Usage*	*Price $/lb*
	Metal Oxides	
Zinc oxide	5 (Neoprene)	0.12
Litharge	25 (Hypalon)	0.17
Magnesia/Pentaerythritol	4/3 (Hypalon)	0.21
	Difunctional Compounds	
Phenolic resins	12 (Butyl)	0.32
p-Quinonedioxime	2 (Butyl)	2.51
Hexamthylenediamine carbamate	< 1.5 (Fluororubber)	6.00
	Peroxides	
Dicumyl peroxide (40%)	2 (Silicone), 5 (Urethane)	0.73
2,5-bis(t-butylperoxy)-2,5-dimethylhexane	2 (Polyethylene or EPM)	2.05

4. Accelerators

As stated earlier, the main reason for using accelerators is to aid in controlling the time and/or temperature required for vulcanization and thus improve the properties of the vulcanizate.

The reduction in the amount of time required for vulcanization is generally accomplished by changing the amounts and/or types of accelerators used. This usage will be quite evident from the many examples given in the following chapters. However, there are some common practices in use by compounders in order to establish suitable recipe changes without extensive research. They are as follows:

(a) single accelerator systems (primary accelerators) which are of sufficient activity to produce satisfactory cures within specified times;

(b) combinations of two or more accelerators, consisting of the primary accelerator which is used in the largest amount, and the secondary accelerator which is used in smaller amounts (10 to 20% of the total) in order to activate and to improve the properties of the vulcanizate. Combinations of this type usually produce a synergistic effect as the final properties are somewhat better than those produced by either accelerator separately;

TABLE 2.9 RELATIVE ACTIVITY OF ACCELERATORS IN NATURAL RUBBER (Reference 4)

Type	*Relative Vulcanization Time at 284° F*	*Examples*
Slow	90 to 120 minutes	Aniline
Moderately Fast	ca. 60 min	Diphenylguanidine Hexamethylene-tetramine
Fast	ca. 30 min	Mercaptobenzothiazole Benzothiazyl disulfide
Ultra-accelerators	Several minutes	Thiurams Dithiocarbamates Xanthates

(c) delayed action accelerators – these are not affected by processing temperatures (thus providing some protection against scorching) but produce satisfactory cures at ordinary vulcanization temperatures.

Classification Although accelerators can be grouped by their acidic or basic nature, or by their activity in certain rubbers (as shown in Table 2.9), no single classification system is suitable since they behave differently in each rubber compound. Consequently, the chemical group classification given in Table 2.10 will

TABLE 2.10. CHEMICAL CLASSIFICATION OF ACCELERATORS

Type	*Example*	*Price $/lb*	*Typical Use*
Aldehyde-amine reaction products	Butyraldehyde-aniline condensation product	0.58	Self-curing adhesives
Amines	Hexamethylene tetramine		Delayed action for NR
Guanidines	Diphenyl guanidine	1.17	Secondary accelerator
Thioureas	Ethylenethiourea	1.42	Fast curing for CR
Thaizoles	2-Mercaptobenzothiazole	1.50	Fast curing general purpose w/broad curing range
	Benzothiazyl disulfide	1.51	Safe processing, general purpose, moderate cure rate

TABLE 2.10. (*continued*)

Type	*Example*	*Price $/lb*	*Typical Use*
Thiurams	Tetramethylthiuram disulfide	1.42	Safe, fast curing
Sulfenamides	N-cyclohexyl-2-benzothiazyl-sulfenamide	1.24	Safe processing, delayed action
Dithiocarbamates	Zinc dimethyldithio-carbamate	1.71	Fast, low temperature use
Xanthates	Dibutylxanthogen disulfide	1.15	General purpose, low temperature use

be used to illustrate the types in general use. Reference should be made to which type is in use for the specific rubbers given later in the text.

The classification listed in Table 2.9 has normally been used to describe the activity of the various chemical types in current use; however, grouping by chemical type as shown in Table 2.10 is more important since each specific type may produce different crosslinks in the resulting vulcanizates. Reference 3 should be consulted for more detailed information.

Production The production data for accelerators and related compounding ingredients are given in Table 2.11. From the information given, it is quite apparent that the thiazole derivatives are the most common type in use as they represent 67% of the total production.

5. Accelerator Activators

These components are used to increase the vulcanization rate by activating the accelerator so that it performs more effectively. It is believed that they react in some manner to form intermediate complexes with the accelerators. The complex thus formed is more effective in activating the sulfur present in the mixture, thus increasing the cure rate.

Accelerator activators are grouped as follows:

i. *Inorganic Compounds* (mainly metal oxides) used include zinc oxide, hydrated lime, litharge, red lead, white lead, magnesium oxide, alkali carbonates, and hydroxides. Zinc oxide is the most common and it is generally used in combination with a fatty acid to form a rubber-soluble soap in the rubber matrix.

Zinc oxide is manufactured by two processes, the French or indirect process, and the American or direct process. In the French process, zinc ore and coal are heated to produce zinc metal which is vaporized and oxidized to the oxide. This two-stage process is more effective than the American process where the zinc ore and coal are heated in air to form the oxide. Impurities from the impure zinc ore and coal are lead, cadmium, iron, and sulfur compounds. Typical ASTM specifications for zinc oxides are shown in Table 2.12.

TABLE 2.11. 1968 PRODUCTION OF RUBBER PROCESSING CHEMICALS (SOURCE: U.S. TARIFF COMMISSION REPORT, OCT. 1969)

	Pounds	*Unit Value ($/lb)*
Grand Total	312,647,000	0.64
Accelerators, activators, and vulcanization agents		
Aldehyde-amine reaction products	1,352,000	0.99
Dithiocarbamic acid derivatives (cyclic)	237,000	2.13
Dithiocarbamic acid derivatives (acyclic)	8,411,000	0.77
Thiazole derivatives	70,078,000	0.53
Thiurams	10,378,000[a]	0.46
All other cyclic accelerators	11,305,000	0.84
All other acyclic accelerators	5,375,000	1.10
Antioxidants, antiozonants, and stabilizers		
Amino compounds	124,598,000	0.67
Phenolic and phosphite compounds	41,137,000	0.74
Retarders, tackifiers, physical-property improvers, and blowing agents	8,007,000	0.48
Peptizers	6,840,000	0.58
Polymerization regulators, shortstops, conditioning, and lubricating agents	24,929,000	0.42

[a]Incomplete data (Sales 10,673,000).

The majority of metal oxides are used in coated or treated forms in order to disperse more readily in the rubber mixtures. From 2 to 5 phr are normal usages.

ii. *Organic acids* are normally used in combination with metal oxides; they are generally high molecular weight monobasic acids or mixtures of the following types: stearic, oleic, lauric, palmitic, and myristic acids, and hydrogenated oils from palm, castor, fish, and linseed oils.

TABLE 2.12. ASTM SPECIFICATIONS

	American	*French*
Zinc Oxide, min. (%)	98.0	99.0
Total Sulfur, max. (%)	0.2	0.1
Moisture and other volatile matter, max. (%)	0.5	0.5
Total Impurities, max. (%)	2.0	1.0
Coarse Particles, Residue on 44 micron sieve (No. 325), max. (%)	1.0	1.0

TABLE 2.13. TYPICAL ACCELERATOR ACTIVATORS

Type	*Sp. Gr.*	*Price $/lb*
Metal Oxides		
Zinc oxide, lead free (American Process)	5.6	0.1575
Zinc oxide, lead free (French Process)	5.6	0.1625
Red lead (98% Pb_3O_4)	9.0	0.18
Magnesium oxide	3.38	0.21
Litharge	9.5	0.17
Organic Acids		
Hydrogenated stearic acid	–	0.11
Lauric acid	0.87	0.30
Stearic acid	1.02	0.17
Oleic acid	0.89	0.14
Amines		
Diethanolamine	1.09	0.23
Triethanolamine	1.12	0.22

Note: The specific gravity and price vary with the purity of the substances, and in the case of the oxides, whether or not the particles have been surface treated to improve dispersion.

The usage of each particular type depends on the accelerator used and the amounts of other compounding ingredients present. Normally from 1 to 3 phr are used.

iii. *Alkaline substances* will increase the pH of a rubber compound and in most instances increase the cure rate. As a rule of thumb, in the majority of recipes, any material which makes the compound more basic will increase the cure rate since acidic materials tend to retard the effect of accelerators. Typical examples of these ingredients include ammonia, amines, salts of amines with weak acids, and reclaim rubbers made by the alkali process.

The phr usage would depend only on how the above materials are to be used in any specific recipe. Typical examples are given in Table 2.13.

6. Age Resistors

All rubbers are sufficiently affected by natural or accelerated aging processes so that it is necessary to add materials which are capable of retarding this type of deterioration.

The loss in physical properties, associated with aging processes, is normally caused by either chain scission, crosslinking, or some form of chemical alteration of the polymer chains. Consequently, the age-resistors used must be capable of reacting with the agents causing aging (ozone, oxygen, pro-oxidants, heat, light, weather, and radiation) to prevent or slow the polymer breakdown, to improve the aging qualities, and to extend the service life of the product involved.

(a) Chemical Protectants There are three general types of compounds used for their protective qualities:

Secondary amines	$R_2N{-}H$ (R\N–H / R/)
Phenolics	$R(OH)_x$
Phosphites	$(RO)_3P$

In general, the amines tend to discolor (staining) and are used only where color is not important. The phenolics are nonstaining and are used mainly in light colored goods where color retention is important. Phosphites are mainly used as stabilizers for SBR.

(b) Physical Protectants Products used in installations where little or no movement is involved can be protected with waxy materials which migrate (bloom) to the surface of the rubber part and form a protective coating which shields the part from the effects of oxygen, ozone, etc.

(c) Classification The phr usage of any age-resistor would depend on the type of service the rubber part is subjected to; however, normal usage is about 2–3 phr. Table 2.14 fives a classification of the types used, and production data are given in Table 2.11. It is important to note that the substituted amines represent over 75% of the total production.

Many examples of the use of age-resistors are given in the following chapters, and test methods for determining the effectiveness of any particular component in aging studies are given in chap. 5.

TABLE 2.14. EXAMPLES OF AGE-RESISTORS

Chemical Type	*Example*	*Use*	*S.G.*	*Price $/lb*
	Antioxidants			
Hindered Phenol	Styrenated phenol	Non-staining	1.08(liq.)	0.56
Hindered Bis-phenol	2,2′-Methylene-bis-(4 methyl-6-t.butylphenol)	”	1.08	1.00
Amino-phenol	2,6′-Di-t.butyl-α-dimethylamino-p-cresol	”	0.97	0.70
Hydroquinone	Hydroquinone mono-benzyl ether	”	1.26	3.00
Phosphite	Tri (mixed mono and di-nonylphenyl) phosphite	”	0.99(liq.)	0.59
Diphenylamine	Octylated diphenylamine	Semi-staining	0.99(liq.)	0.56

TABLE 2.14. (*continued*)

Chemical Type	*Example*	*Use*	*S.G.*	*Price $/lb*
Antioxidants				
Naphthylamines	Phenyl-β-naphthylamine	staining	1.24	0.56
Alkyldiamine	N,N'-Diphenylethylene diamine	,,	1.14	0.55
Aldehyde-amine condensation product	Aldol-alpha-naphthyl-amine	,,	1.16	0.87
Quinoline	Polymerized 2,2,4-trimethyl-1,2-dihydroquinoline	,,	1.08	0.56
Phenylenediamine	N,N'-Diphenyl-p-phenylene diamine	,, and flex crack resistant	1.28	1.11
Antiozonants				
Dialkyl-phenylene diamine	N,N'-Bis-(1-methylheptyl)-p-phenylenediamine		0.90(liq.)	0.90
Alkyl-aryl-phenylene diamine	N-Isopropyl-N'-phenyl-p-phenylenediamine		1.17	1.14
Carbamate	Nickel dibutyldithio-carbamate		1.26	1.78
Physical Type				
Waxes	Blended petroleum waxes		0.90	0.30
	Microcrystalline waxes		0.90	0.20

7. Softeners (Physical Plasticizers)

Physical plasticizers of this type do not react chemically with the rubbers involved but function by modifying the physical characteristics of either the compounded rubber or the finished vulcanizate. In all cases, whether the softener is used as a processing aid (usually 2 to 10 phr) or to alter the finished product (up to 100 phr), it must be completely compatible with the rubber and the other compounding ingredients used in the recipe. Incompatibility will result in producing poor processing characteristics and/or bleeding in the final product.

Drogin has rated the various classes of materials used as physical plasticizers according to the effects they produce. This rating is shown in Table 2.15. In addition, the following breakdown given in Table 2.16 gives examples of properties obtained with materials of this type. The most general use is probably for the preparation of oil-extended rubbers of the SBR type.

The reader should refer to the chapters dealing with each specific rubber in order to determine what material (including amounts) is in general use. It is quite typical for many of these materials to act as dual purpose ingredients, i.e., processing aids can also increase elongation, reduce hardness, improve tack, etc., depending on the amount and type used and the rubber involved.

TABLE 2.15. PROPERTIES OBTAINED FROM PHYSICAL PLASTICIZERS[a]

	Properties
Fatty Acids	
Cotton Seed	1
Rincinoleic	1
Lauric	1
Vegetable Oils	
Gelled Oils	1, 6, 12, 13
Solid Soya	4
Tall Oil	4,5,13
Soya Polyester	13
Petroleum Products	
Unsaturated	1
Mineral Oils	3, 4, 6, 7, 9, 11
Unsaturated Asphalt	3
Certain Asphalts	7,10,11
Coal Tar Products	
Coal Tar Pitch	1
Soft Cumars-Tars	3
Soft Coal Tar	5, 6
Cumar Resins	5, 11
Pine Products	
Crude Gum Turpentine	2, 4, 5, 12, 13
Rosin Oil	2, 5, 6
Rosin	2, 8, 12
Pine Tar	3, 4, 5, 6, 7
Dipentene	6, 13
Certain Rosins	13
Esters	
Dicapryl Phthalate	3
Butyl Cuminate	9
Dibutyl Phthalate	9
Butyl Lactate	10
Glycerol Chlorobenzoate	10
Chlorodibutyl Carbonate	13
Methyl Ricinoleate	2

Code

1. improved tubing
2. better tack
3. increased plasticity
4. low modulus
5. increased tensile
6. improved elongation
7. softer cured stocks
8. harder cured stocks
9. higher rebound
10. better tear
11. low hysteresis
12. high hysteresis
13. improved flex life

[a] Compiled from a lecture given by Dr. I. Drogin, N.Y. Rubber Group, Elastomer Technology Course, Oct. 17, 1955.

TABLE 2.15 (*continued*)

	Properties	*Code*
Resins		
Shellac	8	
Miscellaneous		
Amines	6	
Wool Grease	7	
Pitches	8 12	
Diphenyl oxide	9	
Benzoic acid	10	
Benzyl Polysulfide	10	
Waxes	11	
Fatty Acids	11	

8. Miscellaneous Ingredients

Materials of this type are used whenever some particular effect or property is desired in a vulcanizate. Some typical uses are as follows; examples of each are given in Table 2.17.

Abrasives Erasers, grinding and polishing wheels require some type of abrasive for proper usage. Mineral ingredients such as ground silica and pumice are suitable for this purpose.

Blowing Agents Some type of gas-generating chemical is necessary for preparing blown sponge and microporous rubber. Suitable agents must be capable of releasing gas during the vulcanization period. Azo compounds and carbonates are suitable gas releasing chemicals.

Colorants Materials used for coloring nonblack goods utilize either inorganic pigments or organic dyes. They must be stable, color fast, and reasonably priced.

Flame Retardants Chlorinated hydrocarbons, phosphate and antimony compounds may be added to reduce flamability.

Internal Lubricants Certain amines, amides, and waxy materials act as internal lubricants providing good mold release and fidelity.

Odorants Aromatic compounds are capable of screening out or masking odors from rubber compounds. These components are normally used for wearing apparel and drug sundries. Some are effective as germicides.

Promoters Nitroso and dioxime compounds promote improved reinforcement when added to certain rubber-carbon black mixtures during mastication under controlled conditions.

Retarders These ingredients should reduce the accelerator activity during processing and storage. Their purpose is to prevent scorch during processing and prevulcanization during storage. They should either decompose or not interfere with the accelerator during normal curing at elevated temperatures. In general, these materials are organic acids which function by lowering the pH of the mixture thus retarding vulcanization.

TABLE 2.16. EXAMPLES OF TYPICAL PHYSICAL PLASTICIZERS

Use	*Example*	*S.G.*	*Cost $/lb*
Extenders			
	Extender Oils		
	Naphthenic	0.92	0.24
	Paraffinic	0.90	0.24
	Aromatic	0.98	0.26
	Mineral Rubber	1.04	0.04
Processing Aids			
	Castor Oil	0.96	0.23
	High molecular weight oil-soluble sulfonic acid	0.90	0.54
	Tall Oil	0.94	0.08
Reduced Stock Hardness			
	Mineral Oil	0.93	0.58
	Pine Tar	1.03	0.04
	Vulcanized Vegetable Oil (Factice)	1.04	0.21
Tackifiers			
	Coumarone-indene Resins	1.04	0.09
	Ester Gum	1.10	0.19
	Oil-Soluble Phenolic Resin	1.01	0.29

The use of retarders should be avoided if possible by the proper selection of accelerator-sulfur combinations and careful control of processing conditions. Careful temperature control and use of proper storage (controlled temperature and humidity) are important especially during the summer months when factory temperatures are abnormally high.

TABLE 2.17. TYPICAL MISCELLANEOUS INGREDIENTS.

Type	*Example*	*S.G.*	*Price $/lb*
Abrasive	pumice (ground)	2.35	0.05
Blowing agent	azodicarbonamide	1.63	1.47
Colorants	titanium dioxide (white)	4.20	0.11
	cadmium oxide (red)	5.30	3.19
Flame retarders	antimony oxide	5.20	1.65
Internal lubricant	primary tallow amine	0.80	0.29
Odorant	methyl salicylate	1.18	0.60
Promoter	p-dinitrosobenzene	0.96	1.84
Retarder	salicylic acid	1.37	0.45

9. Fillers

Fillers may either reinforce, extend, dilute, or impart certain processing properties to rubbers. The types used and their specific effects are discussed in detail in the next chapter.

SUMMARY

Many ingredients are used to prepare rubber products. In general, they may be classified according to their specific use; however, many are capable of functioning in more than one manner in certain rubber compounds. Typical of this versatility is zinc oxide which may either function as an accelerator-activator, vulcanizing agent, filler, or colorant depending on which use it is selected for.

There are many references available giving listings of compounding ingredients. *Materials and Compounding Ingredients for Rubber and Plastics*, edited by Rubber World, New York, N.Y. was used in compiling the examples given in this chapter.

REFERENCES

(1) *The Vanderbilt Rubber Handbook*, edited by G. G. Windspear, R. T. Vanderbilt Co., Inc., 1968.
(2) *1968 Book of ASTM Standards*, Part 28: Rubber; Carbon Black; Gaskets, Amercan Society for Testing and Materials, 1916 Race St., Philadelphia, Pa. 19103.
(3) G. Alliger and I. J. Sjothun, *Vulcanization of Elastomers*, Reinhold Publishing Corp., N.Y., 1964.
(4) J. LeBras, *Rubber: Fundamentals of its Science and Technology*, Chemical Publishing Co., N.Y., 1957.

3

FILLERS: CARBON BLACK AND NONBLACK

B. B. BOONSTRA
Cabot Corporation
Billerica, Massachusetts

HISTORY

The use of fillers in rubber is almost as old as the use of rubber itself. The Amazon Indians in the Spanish times were known to use black powder in the rubber latex probably to improve light aging.

One aspect of filler addition has been improvement of properties. Since it was known that the tackiness which would develop in the early rubberized fabrics (such as made by MacIntosh) could be removed by sprinkling with talcum powder, it seemed logical to incorporate talcum powder in the rubber so that it would be present from the start. Another aspect was extension of the rubber with less expensive materials.

After Hancock's "pickle" machine and the development of rubber mixing machinery, incorporation of inert fillers in finely divided particulate form became standard practice. Fillers such as ground limestone, barytes, clay, kaolin, etc. were used in order to extend and cheapen the compounds since it was found that in natural rubber quite a bit of filler could be added without detracting too much from the final vulcanizate properties.

Zinc oxide was originally used for its whiteness. Carbon black which was known as a black pigment may have been used originally for that purpose. Lampblack was available in America in the time of Goodyear, and Hancock, who took out a patent in 1830 on its use. A 1868 British patent describes the use of 10% of lampblack for its "stiffening" action. The main reason for the predominant use of fillers, particularly carbon black, in elastomers, is the reinforcement they impart to the vulcanizates.[1, 2, 3, 4]

Systematic studies of the effect of fillers had been reported by Heinzerling and Pahl in Germany in 1891. This study showed some reinforcing effect of zinc oxide which became known as an "active" filler, particularly after the work of Ditmar in 1905. The "magic" action of zinc oxide obtained such a hold on the imagination of the more empirical compounders that it was not until the acceptance of the synthetic rubbers in the 1940's that the "magic" effects of zinc oxide as an active filler were dispelled. Part of this may be due to its activating effect on many vulcanization accelerators for which zinc oxide is still utilized.

In 1904, S. C. Mote,[5] working for the India Rubber, Gutta Percha, and Telegraph Works in Silvertown, England, discovered the reinforcing effect of carbon black. Mote reported tensile strength values of 293 kg/cm^2, a very high value for the then existing mixing, curing, and testing techniques. At the time there was no particular need for the reinforced rubber and the invention lay dormant for six years. Although automobiles had been around and running on rubber tires for more than a decade, the rubber always outlasted the other components of the tires, in particular the canvas, which was the weakest part of the construction.

With the introduction of the tire cord in 1910 based on patents by Palmer, the situation changed, the carcass was no longer the weakest part of the tire and there was room for improvement of the resistance to wear of the rubber on the tire. When the invention of carcass cord was bought later in 1912 by J. D. Tew of the B. F. Goodrich Company, the information on improvement of rubber by carbon black was given in the bargain. The importance of this discovery was rapidly recognized and developed in the United States but it took until about 1925 before the general public could be convinced that black tires wore better than the white ones which contained mainly zinc oxide as a filler. Carbon black is now the most important filler used in rubber.

The designation of types of carbon black has developed gradually. Before World War II the predominant type of reinforcing black was made from natural gas by burning in small flames impinging on iron channels. The deposited black scraped from these channels obtained the name channel black (1892). The finer particle size (about 24 nm* average diameter) used for rubber was called Hard Processing Channel (HPC) since it imparted stiff stocks in rubber because of its high surface area (that is, small particle size). The somewhat larger particle size (26 nm average diameter) was called Medium Processing Channel (MPC) and at 29 nm average diameter it was called EPC (or Easy Processing Channel). A much coarser black was made starting in 1922 by burning natural gas in large (4 x 10 x 14 ft) furnaces. The particle size of this black is about 60–80 nm and was indicated as SRF or Semi-Reinforcing Furnace black. Since 1942 other grades were also made by the gas furnace process: HMF (High Modulus Furnace) with a particle size of about 60 nm and higher structure than SRF, and a grade FF (Fine Furnace) with an average particle diameter of 40 nm. All gas furnace blacks are nowadays duplicated by oil furnace blacks for economic reasons. Of the channel blacks for rubber use only EPC is available but fast disappearing. Channel blacks are still being produced for the paint, lacquer, and ink trade where the high cost, due to low yield (0.3–3% of theoretical carbon), is not a major deterrent.

The thermal process, introduced in 1922, makes the largest particle size blacks.

In the thermal process natural gas impinges on hot brickwork and decomposes thermally into hydrogen and carbon. The types of black made by this process are FT (Fine Thermal), particle size about 180 nm, and MT (Medium Thermal) with a particle size of about 250–350 nm. Most rubber blacks are made by the oil furnace process where the main source of carbon is a heavy aromatic tar oil, and gas is used as an auxiliary fuel to obtain the necessary high temperature. The process became commercial in 1943 and has since been developed to become the major process today. The main types are:

FEF (Fast Extruding Furnace) – particle size about 40 nm, surface area about 40 m^2/g;
HAF (High Abrasion Furnace) – particle size 28 nm, surface area about 65–70 m^2/g;
SAF (Super Abrasion Furnace) – particle size 19 nm, surface area 110–120 m^2/g. This black is used on a limited scale.
ISAF (Intermediate Super Abrasion Furnace) – particle size about 23 nm, surface area 100 m^2/g. This carbon black is the one most used for treads of passenger car and truck tires.

According to ASTM D-2516-68, a numbering system is recommended in which only the first number has meaning and relates to particle size*; so 1 as the first digit means particle size 10–10 nm, a 2 means a size of 20–25 nm; 3 means 26–30 nm; 4 means 31–39 nm; 5 means 40–48 nm; 6 means 49–60 nm; 7 means 61–100 nm; 8 means 101–200 nm; 9 means 101–500 nm. The arbitrary addition of two more digits makes the system rather confusing in view of the many grades of oil furnace blacks that have emerged in the last few years.

The U.S. consumption of fillers in 1971 was[6] about 2.1 million long tons, of which carbon black accounted for 1.33 million long tons or 65%, that is more than all other fillers together.

The U.S. rubber consumption in 1971 was 2.65 million long tons or just about double the amount of carbon black. For the world (except for U.S.S.R. and Communist block countries) the figures are: carbon black, 2.97 million long tons; rubber, 8.10 million long tons. Calculated as a percentage of the reinforcing fillers, carbon black takes an even higher rank and therefore much of the following treatise relates to carbon black, although the white reinforcing fillers are also discussed when characteristic properties are involved.

In 1939 the first reinforcing siliceous filler was introduced – a calcium silicate prepared by wet precipitation from sodium silicate solution with calcium chloride. In further development of the process the calcium was leached out by hydrochloric acid to yield a reinforcing silica pigment of comparable particle size. About ten years later direct precipitation of silica from sodium silicate solution had developed to a commercial process and this is a major process today. In 1950 a different type of anhydrous silica appeared which was made by reacting silicon tetrachloride or "silico chloroform" (trichlorosilane) with water vapor in a hydrogen-oxygen flame. Because of the high temperature at formation (about 1400°C), this pyrogenic silica has a lower concentration of hydroxyl groups on the surface than the precipitated silicas. The latter contain about 85–90% SiO_2 and have ignition losses of 10–14%, whereas the pyrogenic silica contains 99.8% silica. Because of its much higher price, pyrogenic silica is mainly used as a filler for high cost compounds such as silicone rubber.[7] At present there are about sixty different types of silica and silicates. Bentonite clay has been used as a filler[8] in the form of an aqueous slurry added to the latex. Under these conditions, considerable stiffening of the final rubber vulcanizate is observed, whereas conventional mixing of powdered bentonite into dry rubber results in poor dispersion and poor vulcanizate properties.

Organic fillers have been announced at various times without obtaining a significant place on the market. In most cases they were incorporated in the latex phase rather than in dry rubber. Reinforcing resins were introduced around 1947. Piccini and LeBras[9] added a resorcinol-formaldehyde condensate, in its intermediate resol stage, as an aqueous solution to rubber latex which also contained dispersed curing ingredients. After drying and curing at low temperature a large reinforcement was observed in tensile strength, modulus, and elongation. Before that, van der Meer and Wildschut[10] in 1939 found that cresol-formaldehyde condensation products in early stages of condensation (paracresol dialcohol) could be added to dry natural rubber and that this combination would cure at normal vulcanizing temperatures without the addition of sulfur or accelerator.

Alkaline lignin, a byproduct from the paper manufacture, was introduced as a reinforcing filler[11] around 1947. Also this material had to be added to natural or synthetic latex in the form of an aqueous solution to observe a reinforcing effect in the vulcanized final product.

Starch derivatives have been reported as having certain reinforcing effects when added to latices[12] and after a coagulation and mill mixing step.

In 1954 the use of aminoplasts, aniline formaldehyde, and melamine- or urea-formaldehyde condensation products formed in stabilized acidified latex as reinforcing fillers was reported by van Alphen.[13] Higher tensile strength and elongation than for 50 parts of EPC black per 100 rubber were observed in some cases with natural rubber. None of these organic fillers, formed *in situ* in the latex or in the rubber, have become of commercial importance.

Fillers such as high styrene content resins, copolymers with butadiene added to general purpose rubbers, or phenol-formaldehyde resins in nitrile rubbers[14] form a different group. These resins are milled into the dry rubber at a temperature above their softening point and are mechanically dispersed and mixed as much as possible. They impart some of their stiffness at room temperature to the rubber and should not be considered as real reinforcing fillers but rather as polyblends similar to blends of polyvinyl chloride with nitrile rubber as a softener.

Lately resin fillers are reported in the form of particulate thermoset melamine- and urea-formaldehyde resins produced by Ciba[15] in Switzerland. These have fine particle size and high surface areas (up to 200 m^2/g) and are incorporated in rubber by dry mixing. They give higher modulus than inorganic white pigments at equal volume loading and have the advantage of lower specific gravity (about 1.5 vs. about 2.0 for silica).

REINFORCEMENT CONCEPTS

Reinforcement basically relates to composites built from two or more structural elements or components of different mechanical characteristics and whereby the strength of one of these elements is imparted to the composite combined with the set of favorable properties of the other component. This set must include easy shaping into the required form of the final article and stabilization of this shape within a reasonable time. One typical example of reinforcement is that of concrete with embedded steel rods or cable where the high tensile strength of the steel is imparted to the concrete to give it increased flexural and impact strength.

In a glass-fiber reinforced polyester, the enormous tensile strength of the glass fibers is combined with the easy processing of the polyesters, where solid glass would create problems of pouring, etc., and the polyester alone would be too weak. A strong bond must develop between the reinforcing members and the embedding matrix. In these cases strongly anisometric members, fibers, or rods with L/D ratios of many thousands are distributed in all directions overlapping each other over large sections and bonded together by the matrix so that their strength is transmitted from one region to another, as in the case with felt or leather. A loose network of strong, long members is bonded together by a high viscosity or solid matrix so that the members will support each other. However, the reinforcement imparted to elastomers by particulate solids which, although not actually spherical in shape, still are not so strongly anisometric that they can be said to overlap each other over large proportions of their length, must have a different explanation. The particles are not so strong that they would impart additional strength to the composite and besides, practically the same flexibility as the matrix is maintained.

A definition of reinforcement of rubbers by fillers is required first. A pure gum vulcanizate of general purpose styrene-butadiene copolymer has a tensile strength of no more than about 22 kg/cm^2 (350 psi); after compounding with 50% of its weight of carbon black the tensile level is up to 3500 psi, hence reinforcement is evident in tensile strength. However, in natural rubber, pure gum vulcanizates can be made from latex, with proper precautions, having a tensile strength of 450 kg/cm^2 (6500 psi) at 700% elongation. The best natural rubber SAF black compound will give only about 350 kg/cm^2 (5000 psi) at 550% elongation. It is obvious that there is no improvement in tensile in this case so that tensile strength is not a general criterion for reinforcement. Modulus at 300% elongation is not a good measure either since many relatively inert fillers will raise this modulus without improving any failure properties. The best definition may be: *A reinforcing filler improves the modulus and failure properties (tensile strength, tear resistance, and abrasion resistance) of the final vulcanizate.* The energy at rupture introduced by Wiegand[16] as "resilient energy" is the best single criterion for reinforcement. The energy at rupture can be obtained from the stress-strain curve as the area between the curve and the elongation axis. The effect of increasing loadings of carbon black and of an inert filler (Barytes) on natural rubber is demonstrated in Fig. 3.1, taken from van Rossem.[17]

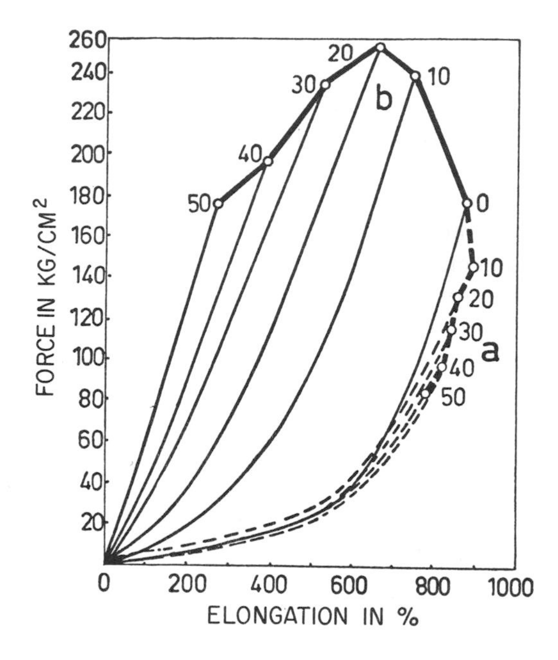

Fig. 3.1 Stress-strain curves of basic compound of natural rubber with increasing volume % of (a) barytes, and (b) EPC black (after van Rossem)

BASIC FACTORS INFLUENCING ELASTOMER REINFORCEMENT

Figure 1 shows the increase of strength and energy at rupture obtained with carbon black as a function of loading. The appearance of an optimum loading indicates that there are two opposing factors in action when a reinforcing filler such as carbon black is added:

(1) There is an improvement of modulus and tensile strength. This is very much dependent on the particle size of the filler; small particles have a much greater effect than coarse ones. Particle size is directly related to the reciprocal of surface area per gram of filler; thus the effect of smaller particles actually reflects their greater extent of interface between polymer and solid material. This will be discussed later.

(2) The reduction in properties at higher loading is a dilution effect, general to all fillers, merely due to a diminishing volume fraction of polymer in the composite. If the volume percentage of filler becomes so high that there is not enough rubber matrix to hold the filler particles together, strength approaches zero. Before this stage of loading is reached, the compound attains a level of stiffness where it becomes brittle and, at the normal rate of testing (e.g. 50 cm per minute) such a brittle compound would show poor strength. At much lower rates of stretching the decline in strength with higher loading would be less; the height and place of the maximum in the strength vs. loading curve are rate dependent. The maximum

occurs at higher loading when testing at slower rates. The place of the maximum is also dependent on the particle size of the filler.

TYPICAL FILLER CHARACTERISTICS

The action of particulate fillers on an elastomer is dependent on factors that can be classified as extensity, intensity, and geometrical factors. After a brief summary, they will be discussed in more detail below.

(a) The *extensity factor* is the total amount of surface area of filler per cm^3 of compound in contact with the elastomer.

(b) The *intensity factor* is the specific activity of this solid surface per cm^2 of interface, determined by the physical and chemical nature of the filler surface in relation to that of the elastomer.

(c) *Geometrical factors* are (1) the "structure" of the filler, determined by its void volume under standardized packing conditions, and (2) the porosity of the filler, usually a minor factor, which can be varied over a wide range with carbon blacks. Since the weight of individual spongy particles is lower than that of solid particles, the number of particles per cm^3 of compound at constant weight loading is greater.

Total Interface

The total outside surface area of a particulate solid is directly coupled to its particle size. If all particles were spheres of the same size it can be easily shown that the following relationship should exist for carbon black assuming its density to be 1.85 g/cm^3:

$$A_S = \frac{3200}{d}$$

where A_S is the surface area in m^2/g and d is the diameter in mm(10^{-9} m).

This rule of thumb gives a good idea of the order of magnitude of the relation between surface area and particle size.

In actual fillers there is always a distribution of sizes that can be averaged various ways[18] and particles are usually far from round.[19] Particle size or surface area is a factor of the greatest importance in reinforcement because it can vary over such a wide range. Coarse inorganic fillers may have about one square meter per gram surface area, whereas fine silicas are made up to 400 m^2/g and carbon blacks to 1000 m^2/g. The rubber grade carbon blacks vary from 6 m^2/g for medium thermal blacks (ASTM designation N-991) to 250 m^2/g for conductive blacks (ASTM designation N-472); so there is a factor of 40 between the highest and the lowest surface area. None of the other factors in reinforcement varies over such a large range.

What counts is the area of interface between solid and elastomer per cm^3 of compound and this is dependent on the surface area per gram of the filler, and on

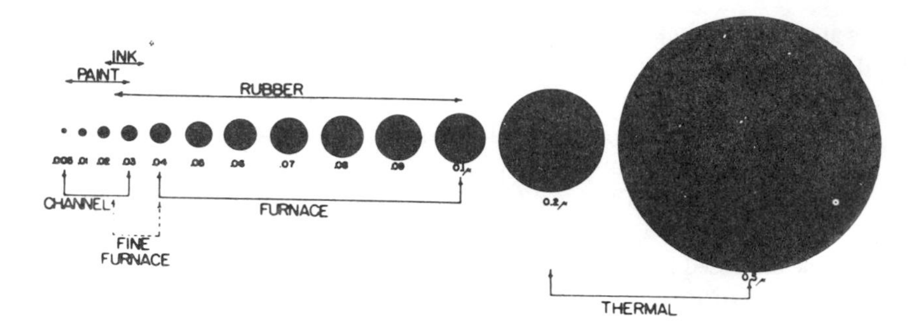

Fig. 3.2

the amount of filler in the compound, which brings in another large factor to the already wide range of variation. A tread compound containing 50 phr of an ISAF (N-220) black has about 35 m^2 of interface per cm^3 of compound and an HAF (N-330) about 25 m^2 per cm^3; and these values give a practical idea of the reinforcement obtained. Below about 6 m^2 per cm^3, not much reinforcement, as defined under "Reinforcement Concepts" on page 55, is obtained. A very schematic picture of the carbon black particles and the effect of particle size on mechanical properties is given in Fig. 3.2 and Table 3.1, and for white fillers in Table 3.2.[20]

It is interesting to consider that 25 m^2 per cm^3 of surface in a cube of 1 cm can be obtained by dividing the cube into 125×10^{12} small cubes of 200 nm side.

Measurement of Particle Size and Surface Area *Particle size* of powdery material can be determined by direct observation under a microscope or electron microsope at known magnification. Many particles must be viewed to obtain a representative average; their definition is often vague and the method is extremely time consuming.[21] With carbon black two other methods are available, both giving

TABLE 3.1. EFFECT OF PARTICLE SIZE OF CARBON BLACK AT 50 PHR ON MAIN PROPERTIES OF SBR

Type	*ASTM Designation*	*Average Particle Size Millimicron*	*Tensile (psi)*	*Tensile kg/cm*	*Relative Laboratory Abrasion*	*Relative Roadwear Resistance*
SAF	N110	20–25	3600	250	1.35	1.25
ISAF	N220	24–33	3300	230	1.25	1.15
HAF	N330	28–36	3200	225	1.00	1.00 standard
EPC	S300	30–35	3100	220	0.80	0.90
FEF	N550	39–55	2600	185	0.64	0.72
HMF	N683	49–73	2300	160	0.56	0.66
SRF	N770	70–96	2100	150	0.48	0.60
FT	N880	180–200	1800	125	0.22	
MT	N990	250–350	1400	100	0.18	

TABLE 3.2. PARTICLE SIZE OF FILLERS. EFFECT ON TENSILE AND TEAR OF NATURAL RUBBER VULCANIZATES[20]

		11.5 Volumes		23 Volumes		46 Volumes	
Filler	*Particle diam. mμ*	*Tens. psi*	*Crescent Tear lb/in.*	*Tens. psi*	*Crescent Tear lb/in.*	*Tens. psi*	*Crescent Tear lb/in.*
Whiting	550	3350	62	2870	67	1950	90
Barytes	500	3430	118	2840	106	2000	–
$MgCO_3$	410	3830	90	3130	112	2360	168
Lithopone	230	3920	123	3550	157	2800	258
Dixie Clay	200	3840	78	3410	112	2640	151
ZnO (ordinary)	130	3920	179	3630	342	2700	330
ZnO (fine part.)	90	3900	325	4010	420	3160	448
Frantex B[a]	65	4340	39	4410	62	3770	78
Silene EF[b]	43	4020	319	3630	398	–	–
Hi-Sil[c]	25	4170	409	3800	470	–	–
Cab-O-Sil[d]	14	4550	–	–	–	–	–

[a] Modified clay.
[b] Precipitated calcium silicate.
[c] Precipitated silica.
[d] Pyrogenic silica.

values relative to a standard black:

(1) *With light reflectance*, the amount of reflected light diminishes with smaller particle size and this has led to classification of carbon blacks according to the Nigrometer scale.[22] This scale runs from about 58 for the smallest particle size lacquer blacks to well over 100 for thermal blacks. The original standard was SRF black which rated 100.

(2) *Light absorption measurements* are made by determination of the tinting strength, that is, the relative reflectance of a mix of 0.100 g of black with a standard amount of zinc oxide (usually 3.75 g) mulled together in a vehicle such a linseed oil or soybean oil with a stabilizer, in comparison with a standard black. The traditional standard again has been SRF = 100. SAF black (N-110) has a tinting strength of about 240, while the others rate as follows: HAF, 190; SRF, 100; FT, 35. The tinting strength is dependent on the degree of dispersion of the black and on its level of aggregation, that is, its "structure."

Surface area is usually determined by gas adsorption measurement according to the method of Brunauer, Emmett, and Teller[23] (BET). The amount of gas adsorbed as a function of gas pressure (often nitrogen is used) is determined at temperatures near its boiling point. From the shape of this isotherm the amount of gas necessary to form a monolayer can be estimated. From the size of the nitrogen molecule in liquid form the extent of the surface can be calculated. For fast determinations the amount of I_2 adsorbed from a 0.1 N solution is measured by titration. An empirical equation shows a fairly constant relationship to the BET surface area.

Specific Surface Activity The nature of the solid surface may be varying in a chemical sense, having different chemical groups, e.g., hydroxyl, metaloxide, in nonblack inorganic fillers, and organic carboxyl, carboxyl, quinone, or lactone groups in carbon black, etc. In a physical sense surfaces can be different in adsorptive capacity and in energy of adsorption.

Elastomers of a polar nature such as neoprene, nitrile rubber, etc., will interact more strongly with filler surfaces having dipoles such as OH and COOH groups or chlorine atoms. With the general purpose hydrocarbon rubbers, however, no dramatic influences on reinforcement are noticeable when carbon black surfaces are chemically modified.

Chemical surface groups play an important role because of their effect on the rate of cure with many vulcanizing systems. Only in specific cases (heat treatment of butyl compounds and antioxidant action in polyethylene) are the composite properties strongly affected by the concentration of hydroxyl or other oxygen-containing groups on the surface of carbon black. On the whole, the physical adsorption activity of the filler surface is of much greater importance for the mechanical properties of the general purpose rubbers than its chemical nature. Fillers having about the same small particle size (equivalent surface area) will give about the same order of magnitude in reinforcement if there are no great differences in particle shape.

The adsorptive activity of reinforcing carbon blacks is not homogeneously distributed but concentrated at a number of sites of much greater activity than the majority of the surface. These active sites only represent a small percentage, less than 5%, of the total surface.[24] The first few percent of the surface to be covered is the most active part characterized by much higher heat of adsorption.

The importance of these active sites can be demonstrated by heat treatment ("graphitization") of the carbon black to a temperature of 1600°–3000°C. Through this treatment the higher activity of the sites is lost, recrystallization occurs, and the surface flaws in the lattice, which (probably) constitute these more active spots, disappear. The surface becomes homogeneous in adsorptive activity. The loss of active sites has a most profound effect on the mechanical properties of the vulcanizate made with this black. This is shown in Table 3.3,[24,25] which contains analytical data and corresponding rubber properties for a high structure black in its original state and after graphitization. The latter causes reorientation of the crystalline structure at the surface and, as the analytical data indicate, no great reduction in surface area; structure, if anything, seems to have increased. If one looks at the vulcanizate properties of these fillers in SBR, dramatic changes become apparent. Through graphitization of the black, the optimum vulcanizate modulus has dropped to about one quarter of the value of the untreated black compound while tensile strength has lost only some 10–15% of its value. The most important failure property, abrasion resistance, has declined parallel with the modulus to about one quarter of that obtained with untreated black. What has changed is the adsorptive activity of the black surface as illustrated by the drop in bound rubber, propane, and water absorption. These data show convincingly the

TABLE 3.3. EFFECT OF BLACK-GRAPHITIZATION ON MAJOR ANALYTICAL AND RUBBER PROPERTIES IN SBR

	High Structure ISAF		*ISAF*	
	Original	*Graphitized*	*Original*	*Graphitized*
Surface Area, N_2	116	86	108	88 m^2/g
Oil Absorption	1.72	1.78	1.33	1.54 cc/g
H_2O Adsorption, 55 R.H.	2.4	0	1.85	0%
Density (dry black)	0.283	0.295	0.353	0.392 g/cc
Propane Adsorption (cc/g) @ P/P_0 = 0.001	1.03	0.25	0.93	0.31
Bound SBR	25.1	0–2	18	0.4–2%
Extr. Shrinkage	30	37.5	39.6	43.5%
Mooney Viscosity, 212°F	83	87	73	76
Scorch, 275°F	10.5	17	18	20 minutes
Dispersion	99	99	99	98.2%
Tensile Strength	270	240	280	230 kg/cm^2
Modulus, 300%	150	36	105	30 kg/cm^2
Abrasion Loss	62	181	67	142 $cc/10^6$ rev.
Elongation	450	730	630	750%
Hardness	73	68	68	65 Shore A
Hysteresis	0.238	0.315	0.204	0.297

overpowering effect of the surface activity of a filler on final vulcanizate properties. One might then expect that still higher energy bonds between filler surface and elastomer chains would result in still higher reinforcement and better properties. The highest energy bond would be a chemical one where bonding energies of 100 kg/cal/mol are common and filler particles chemically bound to the elastomer, would seem the ideal situation for reinforcement. However, it appears that chemical bonds between filler and elastomer do not have that effect. There is an instance in which one can be reasonably sure that chemical bonds actually formed and that is when brass powder is used as a filler in a vulcanizate containing sulfur. Although no very small particle size, and therefore no large surface area can be obtained and rather large amounts of sulfur must be used to effect curing, the result is evident. One obtains a very high modulus but poor tensile strength and elongation.[26] Low swelling of this vulcanizate in good rubber solvents indicates high crosslink density either in the rubber matrix directly or via the brass particles.

Model fillers consisting of polystyrene particles and styrene butadiene copolymer particles have been studied as fillers in SBR vulcanizates by Morton et al.[27] These authors showed that in the case of the polystyrene particles no chemical bonds with the matrix were formed during vulcanization; this was to be expected since the polystyrene chain molecule has no olefinic double bonds and hence is not reactive. Particles made from a copolymer of styrene and 11% butadiene having the necessary olefinic bonds showed evidence of being chemically bound to the elastomer. Remarkably, higher tensiles were found with the vulcanizate having polystyrene particles as a filler than with the one containing the corresponding volume of styrene-butadiene copolymer particles as a filler. The latter, however, had a higher modulus. Strong chemical bonding between filler and elastomer does not lead to desirable vulcanizate strength properties but causes high moduli (load at 300% elongation).

In some cases (butyl rubber) there seems to be insufficient interaction between carbon black fillers and elastomer. Vulcanizate strength and rebound properties can be improved by oxidation of the carbon black surface or heat treatment of the compound with mild bonding agents.[28,29] The conclusion is that both total surface area and its specific activity are important factors in reinforcement and one could put them together in a product: total area x specific surface activity = reinforcement factor. If either one of the two factors is zero, the whole product is zero; in other words, neither one is of value without the other.

There are no precise and rapid methods to measure surface activity. The measurement of the differential heat of adsorption is too cumbersome to be of practical use. Water absorption[30] at various relative humidities has been used but is dependent on the polarity of surface groups. More appropriate is propane adsorption but no routine test method has been developed. Also indicative of surface activity is the amount of bound rubber formed during mixing of filler and elastomer. This will be discussed in the section on mixing.

Geometrical Characteristics

Primary Particles, Void Volume, and Anisometry In carbon black the particulate entities appear under the electron microscope as aggregates of primary particles fused together and roughly spherical in shape, which as such do not exist separately. It is the size and shape of this aggregate, often called "primary aggregate," that determines "structure" and its effect on rubber properties. The more these aggregates deviate from a solid spherical shape and the larger they are, the higher is the "structure." Nonspherical particles have a volume packing which is less dense than that of spheres (74.01%), leaving a greater volume of voids in between the particles. This void volume is used as a measure of structure through determination of oil absorption (mixing with oil or dibutylphthalate (DBP) according to a standard procedure) or by measuring compressibility of dry filler. The end point of the oil absorption test, when all voids are filled with oil (or dibutylphthalate), is rather difficult to observe sharply and therefore machines such as the Absorptometer[31,32] have been developed to reduce the error. The Absorptometer consists of a small mixing chamber in which a weighed amount of filler, for black usually 30 grams, is stirred around by a set of rotors while dibutylphthalate is being added at a constant rate from an automatic burette. The torque at the rotors is recorded and when all the voids between the filler particles or aggregates are filled the whole mass toughens up rather abruptly and the torque rises to a sharp peak. The machine is set to shut off at this point, the automatic burette is read and DBP absorption computed as cc of DBP per 100 g of black.

Structure has been analyzed in a different way by Medalia[33] through the analysis of electron micrographs.

In the inorganic or mineral fillers there is a great deal of difference in geometry of particles depending on the type of crystals the mineral forms. The minimum anisometry is found with materials that form crystals with approximately equal dimensions in the three directions, in short, round particles. More anisometric are particles in which one dimension is much smaller than the two others, in other words, platelets. Most anisometric are particles which have two dimensions much smaller than the third, so that they are rod-shaped. When components are compared with fillers of equivalent surface area and chemical nature but different anisometry, then the modulus increases with increasing anisometry. This is illustrated by comparison of three mineral fillers: (1) natural ground $CaCO_3$, round particles; (2) Kaolin clay, platelets; (3) acicular clay (Attacote), needles. The viscosity and modulus data at 20 vol. % are shown in Table 3.4.

TABLE 3.4. PROCESSING PROPERTIES AND MODULUS OF RUBBER COMPOUNDS CONTAINING INORGANIC FILLERS OF DIFFERENT PARTICLE ANISOMETRY

Shape	*Round*	*Platelets*	*Acicular*
Viscosity in SBR	40	43	66
Viscosity in NR	42	46	too high
Extrusion Shrinkage %	58	56	46
Modulus 300% (NR)	505	704	1040

In rubber technology it is customary to associate high structure of a filler with high modulus of the vulcanizate. That this is incidental was shown by Table 3.3, which illustrated the effect of "graphitization" of black on its surface adsorption activity and modulus of the vulcanizate. It has already been observed that neither total surface area nor structure (as indicated by oil absorption) were drastically changed by the heating of the black. Still the modulus dropped by a factor 4.

Again it can be stated that specific surface activity and structure are also interrelated and cooperate to cause reinforcement. It is the product that counts; high structure without surface activity does not result in (high) reinforcement as the data of Table 3.3 show. High surface activity without structure should not result in any higher modulus increase than dictated by the hydrodynamic factor of Einstein, Guth, and Gold.* Although the graphitized high-structure black still gives somewhat higher modulus than the corresponding regular structure material, the

*See Section on "Formation of Bound Rubber."

effect of surface activity on reinforcement completely dominates this small effect of structure as such.

The interpretation of these results may be that the main effect of anisometric, that is highly structured, particles on modulus is the resistance they offer against orientation of the matrix molecules in the direction of the flow lines, during extension or other deformation. This resistance is stronger, the greater the interaction between elastomer and filler surface. When the sliding of polymer molecules along this surface is unhampered because of the absence of strongly absorbing sites as is the case with graphitized black, the resistance against orientation of the particle is minimal – the matrix flows freely around it and a low modulus is the result. It is of interest to point out here that this high modulus is apparently a viscous phenomenon, an observation that will be used again in the theory of reinforcement.

In Table 3.5, the oil absorption values of a number of different blacks and white fillers are listed.[20]

It is remarkable that silicas and silicates or other white inorganic fillers have lower moduli at intermediate elongations (100–300%) than carbon blacks of corresponding particle size and structure. This is shown in Table 3.6. As a consequence the white filler vulcanizates have a lower energy at rupture than the corresponding black vulcanizates (integrated area under the stress-strain curve) and

TABLE 3.5. OIL ABSORPTION (VOID VOLUME) OF VARIOUS REINFORCING FILLERS

Carbon Black Types	*Oil Absorption Manual Test cc/g*	*DBP (Automatic Test) cc/100 g*	*BET S.A. Area m^2/g*
SAF	1.50	110	130
ISAF	1.35	117	115
HAF	1.30	105	75
FEF	1.35	119	29
GPF	1.1	74	20
HMF	0.85	76	30
SRF	0.7	70	20
FT	0.4	40	12
MT	0.3	30	6
XCF	2.50	225	190
White Fillers			
Pyrogenic Silica	1.50		190
Precipitated Silica	1.95	190	150
(Ultrasil VN_3)	1.95		±160
Santocel C	3.0		130
Calcium Silicate (Silene EF)	1.2		80
Calcium Carbonate (Microcal)	2.0		±75

TABLE 3.6. MECHANICAL PROPERTIES OF WHITE- AND BLACK-FILLED SBR VULCANIZATES AT EQUIVALENT VOLUME LOADINGS[a]

ASTM	*Carbon Black Types*	*S.A.*	*Structure (DBP)*	*Modulus 300% kg/cm²*	*Tensile kg/cm²*	*Elongation*	*Hardness*
N220	ISAF	115	110	105	225	520	59
N330	HAF	75	117	110	210	500	59
N770	SRF	20	70	70	140	540	55
S300	EPC	100	90	85	255	560	63
	White Fillers						
	Cab-O-Sil[b]	190	120	46	325	660	69
	Hi-Sil[c]	150	190	33	250	690	59
	Silene EF[d]	80	110	32	160	590	58
	Suprex Clay	5–10	25	45	180	590	62

[a]20 volumes per 100 parts of rubber by weight = 37 parts for black.
[b]Pyrogenic silica.
[c]Precipitated silica.
[d]Precipitated calcium silicate.

in most cases abrasion resistance is lower than that of corresponding black vulcanizates.

Persistent and Transient Structures Reinforcing fillers, carbon black in particular, undergo a number of mechanical treatments to make them easily processible in elastomers. Carbon black is in a very fluffy state immediately after formation and may have a pour density of 5–10 lb./cu. ft (0.08–0.16 g/cc) (the weight of material in grams that can be poured into a 1000 cm^3 beaker according to a standard procedure). In this condition it is difficult to incorporate in rubber or plastic and therefore it is "pelletized,"[34] that is, mildly agitated until it agglomerates, forming small balls that are flowing freely and can be readily redispersed in rubber. The density increases to about 23 lb cu. ft, that is, 0.4 g/cm^3, whereas the specific gravity of carbon black is about 1.85 g/cm^3, so 4/5 of the bulk still is air. During this relatively mild process the most fragile structures are broken and when the pellets are subjected to the operations necessary to determine the structure or void volume by oil absorption, some more of the "structure" is broken down. The hand operation is milder than the automated method with dibutylphthalate in the Absorptometer so that the latter always give somewhat lower values for the void volume. The most rigorous test, however, that the structure has to pass is the final mixing with high viscosity elastomer on the roll mill or in an internal mixer. The distinction between primary (nonbreakable or persistent) and secondary (breakable or transient) structure is rather arbitrary. It is possible to get an idea of this breakdown of structure during rubber mixing by recovering the black from the compound (rubber is distilled off in N_2 stream) and measuring oil absorption on

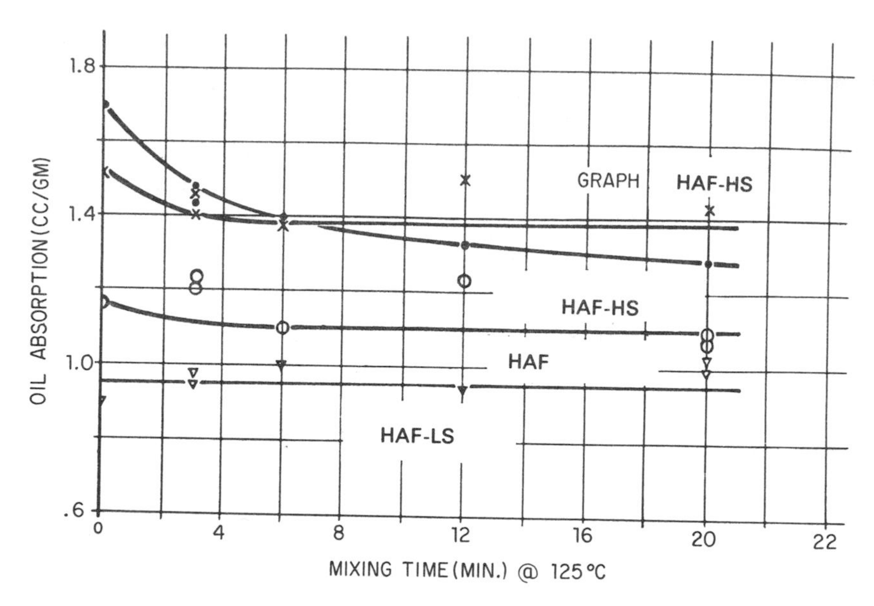

Fig. 3.3

the recovered black. As Fig. 3.3 shows, a considerable reduction in structure is found with HAF black of high structure[35] after mixing in rubber for increasing periods of time. Electron microscopic evidence for this breakdown has been supplied by Heckman and Medalia.[36]

Gessler[37] interpreted the breakdown of carbon black structure as a main source of active sites at the fracture surfaces. Voet[38] failed to observe structure breakdown in experiments in which he used a colloid mill to homogenize the black rubber solution before the rubber milling process. The more vigorous the mechanical treatment of a reinforcing filler such as carbon black, the less of its structure will be left. Important to the practical compounder is the structure as it exists in the rubber compound because it is this structure that influences the rheological behavior of the compound, in particular, its extrusion.

To judge the rheological behavior, an extrusion shrinkage test is run; a length of compound is extruded under standardized conditions (ASTM D-2230-63T). After a fixed time of recovery, usually one hour, one meter (100 cm) of extrudate is measured off, cut, and weighed (W). From the specific gravity of the stock and the cross-sectional area A of the extrusion die (A = usually 0.178 cm^2, diameter = 0.48 cm = 3/16″), we can calculate the theoretical length L_t that the extrudate should have had, if it had exactly the same diameter as the die ($L_t = W/\rho A$). This theoretical length L_t (in cm) then is used to calculate the extrusion shrinkage:

$$E = \frac{L_t - 100}{L_t} = 1 - \frac{100}{L_t}, \text{ in \%: } \left(1 - \frac{100}{L_t}\right) 100 = E$$

This entity can vary from 0 to 1 (100%). The same property is sometimes expressed as "die swell," given by the relation:

$$D = \frac{\text{Actual Cross-Sectional Area of Extrudate} - \text{Cross-Sectional Area of Die}}{\text{Cross-Sectional Area of Die}} = \frac{\frac{W}{100} - \frac{W}{L_t}}{\frac{W}{L_t}} = \frac{L_t}{100} - 1$$

Evidently this entity varies from 0 to ∞ and therefore is a more sensitive measure; the relation between the two entities is

$$D + 1 = \frac{1}{1 - E}, \quad \text{or} \quad D = \frac{E}{1 - E} \quad \text{or} \quad E = \frac{D}{D + 1}$$

Die swell or extrusion shrinkages are good practical measures for the filler structure in the final compound.

Filler structure is also broken down on deformation of the final vulcanizate, as was shown by Payne[39,40] in his experiments measuring dynamic modulus and electrical conductivity at increasing amplitudes of shear. The picture of a network of carbon particles penetrating the entire rubber sample is probably not a correct one, but there is undoubtedly some structure (aggregates or conglomerates) that is broken down as the amplitude of the cyclic shear is increased. Voet[41] found that at extremely small amplitudes the dynamic modulus G' diminishes – again probably because the amplitude is too small to influence existing agglomerates; the electrical conductivity follows a similar course.

Porosity Porosity is mainly a characteristic of carbon black or at least it can be relatively easily controlled with carbon blacks although a special type of porosity can be found with many particulate fillers. The nature of the carbon black particles[42] is such that crystallite regions at the surface alternate with short disordered sections and the interior is definitely much less ordered. By controlled oxidation it is possible to remove part of the less oriented material from the particle so that more or less porous particles remain. In the extreme case just an empty shell remains. Figure 3.4 illustrates this situation.

In most cases the pores are too small for elastomer chains to enter although some smaller molecules in the compound may do so; the specific gravity of a porous black compound is usually not significantly lower than that of the corresponding compound with solid particles. A direct consequence of the porosity is that the diameter-surface area relation mentioned previously is no longer valid. To determine the effective outside surface area one can use the so-called "t" method of de Boer et al.,[43] whereby the thickness of adsorbed gas layers is determined as a function of partial pressure. Only the outside surface area is effective in reinforcement.[44,45] In essence, all the rubber fillers are nonporous except the conductive furnace black XF-72, and the channel blacks. Color blacks like Monarch 81 and 71 are much more porous. Large internal surface areas may be

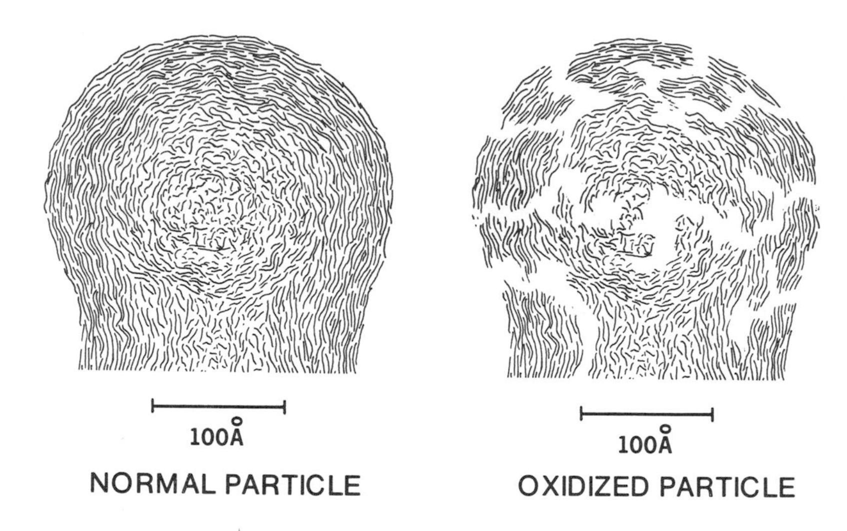

Fig. 3.4 Concentric layer orientation.

detrimental in so far that a certain proportion of accelerator may become immobilized and inactivated.

Another consequence already mentioned previously is the number of particles per gram weight. If the internal pore volume per particle is on the average 15%, then each particle is 15% lighter than a solid particle of that size; so the number of such porous particles in one gram will be 15% higher than for solid particles. Properties dependent on volume loading and interparticle distance are therefore influenced by porosity. A porous black will give higher viscosities and higher electrical conductivity than solid blacks. Polley and Boonstra[46] have shown that if the chemical nature of types of carbon black of vastly different particle size are the same, then the electrical conductivity is governed by the average particle distance, or rather the average distance of nearest neighbors.

Since porosity in carbon blacks is obtained by oxidation, these blacks often contain many oxygen groups on the surface. This would counteract the easy passage of conductance electrons and increase the resistance of the rubber compound made with such a black. Porous carbon blacks therefore do not necessarily give highly conductive compounds. This is only true if the outside surface chemistry is approximately the same as for the solid particle type carbon black.

Although porosity is a factor that cannot be overlooked in assessing a filler's influence on vulcanizate properties, its effect on reinforcement is a secondary one.

FILLER CHARACTERISTICS AND VULCANIZATE PROPERTIES

Although rubber properties are interconnected and relate to the combination of all filler properties, a brief summary of the main influence of each of the four filler characteristics is given below:

(1) *Smaller particle size* (larger external surface area) Results: higher tensile strength, higher hysteresis, higher abrasion resistance, higher electrical conductivity for carbon blacks, higher Mooney viscosity. Minor effects on: extrusion shrinkage and modulus.

(2) *Increase in surface activity* (physical adsorption)
Results: higher modulus at higher extensions (300% up); higher abrasion resistance, higher adsorptive properties, higher "bound rubber," lower hysteresis.

(3) *Increase in persistent structure and anisometry*
Results: lower extrusion shrinkage, higher modulus at low and medium extension (up to 300%), higher Mooney viscosity, higher hysteresis, longer incorporation time. Better dispersion, low electrical resistivity for carbon blacks.
This property particularly is interrelated with surface activity; structure changes on fillers without surface activity (graphitized black) show the effects indicated above only rather faintly. When a filler's surface activity is high (and constant), variations in its structure have the greatest effect on its rubber properties.

(4) *Porosity*
Results: higher viscosity, lower electrical resistivity for carbon blacks.

INFLUENCE OF FILLERS ON THE CROSSLINKING PROCESS

Fillers influence the crosslinking reaction; there is, for instance, the retardation in cure by channel blacks as compared to furnace blacks, of hard clay as compared to whiting, of some silicas as compared to silicates, all corresponding in particle size. In most cases the cause of this retardation can be traced to the greater or lesser acidity of the filler (indicated to some extent by the pH of its aqueous slurry) which influences the kinetics of the crosslinking reaction. Thus the slurry pH of a channel black is 4–4.5; that of a furnace black, 7–9; for clay and whiting the values are approximately 4.5–5.5 and 8–10; for silica it varies from 3.5 to 7, whereas the silicates approach 10.

Precipitated silicas reportedly give better properties in an accelerator-sulfur cured rubber-silica compound when the zinc oxide is omitted; this component which is actually an activator for the curing reaction has a large reducing influence on the viscosity of the unvulcanized compound. The effect has been explained by assuming that the ZnO neutralized the most active spots on the filler surface by forming zinc silicate.

Work in this laboratory[47] and elsewhere has shown that the chemistry of the carbon black surface plays an important part in the initial steps preceding the actual crosslinking reaction as well as in the crosslinking rate itself. A convenient technique to study effects on rate of cure and crosslinking is by means of curometers such as the Oscillating Disk Rheometer which is also used to evaluate fillers.[48,49,50]

The rheometer curve, taken at curing temperature, indicates the induction period, the course of the crosslinking reaction, the occurrence of a plateau or reversal, and at what time optimum cure is reached. Arbitrarily the time required to attain 90% of the maximum increase in torque over the minimum (ΔL_{max}) has been set as the time for optimum cure, although sometimes other percentages are used. ΔL_{max} will also indicate characteristic properties of the filler.[51]

THE MIXING PROCESS

This stage in the production of a vulcanizate introduces more variance in its mechanical properties than any other step on the road to the final vulcanizate; it probably introduced more variance than all of the other steps (curing, sample preparation, and testing) together. Although this has been known for some time, few studies have been made on the subject.

Microscopic evaluations of early stages of mixing[52,53] show that one must visualize the primary process in mixing of carbon black and rubber as a penetration of the voids between the aggregates by rubber, and the primary products are concentrated agglomerates held together by the rubber vehicle. When all voids are filled with rubber the black is considered "incorporated" but it is not yet dispersed. Immediately after and even during their formation these concentrated agglomerates are subjected to high shear forces that tend to break them down again into smaller and smaller units until the final dispersion is reached. This picture is representative of the general process of mixing of a particulate filler with a high viscosity vehicle, although the effect will not be so evident with coarse fillers.

It is of interest to look at the mechanical properties of compounds and vulcanizates obtained from very short mixing cycles and compare them with adequate mixing procedures. Table 3.7 shows the rubber properties of such mixes made in oil-extended SBR.

After 1.5 minutes of mixing, a tensile strength of about 2500 psi is reached (up from about 300 for the pure gum vulcanizate). Even though the microscopic rating gives only 23.6% black dispersed in aggregates smaller than 6 microns, the tensile strength has already attained 2/3 of its maximum value. The abrasion resistance is about 40% of that of the best dispersion (two-stage mix). The change in 100% modulus, which drops from 480 psi in the briefly mixed vulcanizate to less than half that value in the vulcanizate from the fully-mixed compound, is very meaningful for the theory of mixing. For details, reference is made to the original article but it should be mentioned here that the major part of this reduction in modulus is due to further breakdown of the carbon black aggregates in this early

TABLE 3.7. SINGLE-STAGE MIX OF ISAF BLACK IN SBR-1712
Recipe: SBR-1712; 137.5; Vulcan 6, 69; Stearic acid, 1.5; ZnO, 3; Hexamine, 1; Sulfur 2; Santocure, 1.1. Curing time: 60 min at 292° for sheets, 70 min at 292°F for thicker specimens.

	Mixing Time (min)							
Mechanical Properties	*1.5*	*2*	*2.5*	*3*	*4*	*8*	*16*	*Two-Stage*[a]
Tensile strength (kg/cm^2)	173	220	245	265	260	265	255	265
Modulus (psi)								
100%	34	31	27	20	17	15	12	15
200%	86	87	86	69	63	57	56	54
300%	130	146	142	127	128	122	119	123
Elongation (%)	380	460	490	540	530	540	530	540
Hardness (Shore A2)	65	65	64	62	61	59	59	57
Tear strength (kg/cm)	40	40	40	43	41	41	39	41
Abrasion (Akron, volume loss,								
$cc/10^6$ rev)	289	194	142	133	136	–	–	122
Cut growth (DeMattia, kc to 1 in.)	5	6.5	9	8.5	11	–	–	27
Torsional hysteresis "*K*," 212°F	0.273	0.278	0.266	0.238	0.226	0.228	0.213	–
Goodrich Flexometer:								
Static comp. (%)	24.4	24.8	26.6	28.6	28.5	28.5	28.6	–
Perm. set (%)	4.6	4.2	4.2	3.8	3.4	2.9	2.8	–
Heat build-up (°F)	92	89	87	83	80	74	69	–
Mooney viscosity ML 4 min 212°F	133	122	114	97	83	68	63	35
Extrusion shrinkage (%)	29.1	39.7	44.2	46.8	45.7	41.7	36.1	43.2
DC resistivity (ohm cm)	124	88	108	175	300	440	760	728
Specific gravity	1.148	1.152	1.152	1.152	1.153	1.153	1.152	–
Dispersion rating (%)[b]	23.6	71.4	86.4	96.9	99.3	100	100	100

[a]Stock prepared by two-stage "high viscosity" mix with 69 phr black in first stage
[b]Rating calculated by revised procedure assuming that A = s and v = 0.40 (see ref. 68).

stage of mixing. This idea is supported by the very low extrusion shrinkage of the shortest mixed compounds and the increase of this entity at continued mixing. Torsional hysteresis also shows a steady decrease as mixing progresses.

These observations show that reinforcing abilities are profoundly influenced by variations in mixing procedure. This consideration is still apart from the degradation of the elastomer. The heat degradation is considerable in natural rubber, relatively small in SBR, and far less than SBR and all other elastomers in polybutadiene, at least at temperatures below 148°C.

The composition of the primary formed agglomerate can be determined from the oil absorption (or DBP) test since the voids between particles and aggregates which in this test are filled with oil, are filled with rubber during mixing with rubber on the mill. For a filler with an oil absorption of 1.25 cm^3 per gram, the primary aggregate would have a composition of 125 cm^3 (118 grams) of rubber per 100 grams of filler, or per 100 grams of rubber, 85 grams of filler, if no breakdown of structure occurred. This, therefore, is the maximum loading of filler that is still

Fig. 3.5a Example of a good dispersion of ISAF black in SBR (magn. 0000X).

dispersible. It is possible to make higher loadings, but then much structure breakdown takes place, resulting in such intensive particle to particle contacts and interparticle interaction that the agglomerates formed are no longer dispersible within a practical time. Masterbatches intended for dilution to arrive at better dispersions than by direct mixing should not exceed this critical concentration, or else poor dispersion with large lumps will result.

A micrograph of such a poorly dispersed masterbatch is shown in Fig. 3.5 in comparison with a well-dispersed black. The lower the void volume (oil absorption) of a filler the higher is the critical loading that can be tolerated in a masterbatch before the masterbatch becomes indispersible. As a consequence of the mixing mechanism, one expects and finds:

(1) Low viscosity rubbers will penetrate faster (incorporation times are shorter) than highly viscous ones, but subsequent dilution to good dispersions is slow, due to small shearing forces.

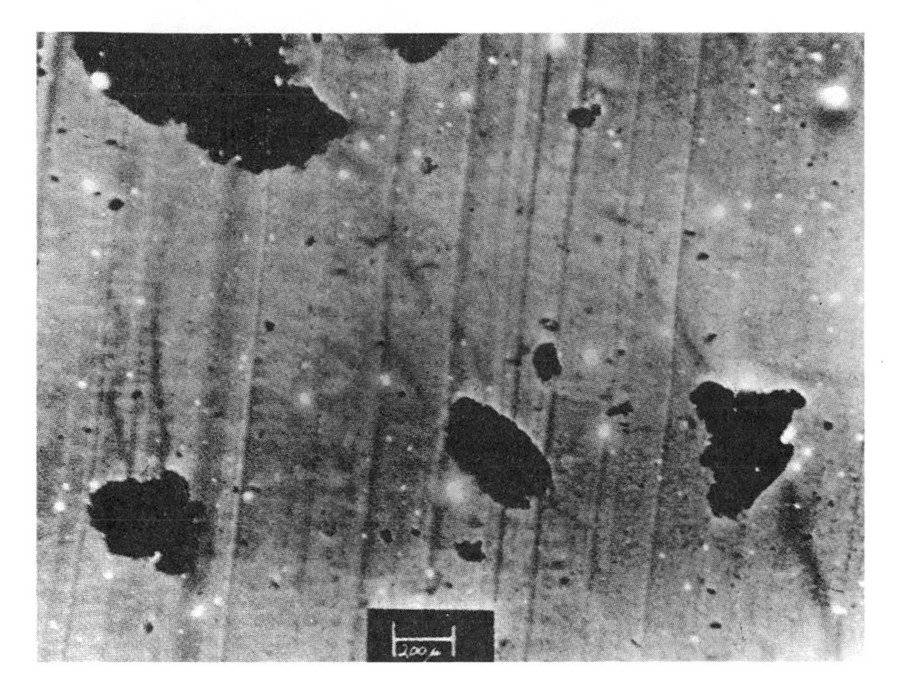

Fig. 3.5b Example of a very poor dispersion of ISAF black in SBR.

(2) High structure fillers incorporate slower than low structure fillers (more void volume to fill), but once incorporated the former disperse more easily and rapidly than their low-structure counterparts.

FORMATION OF BOUND RUBBER[54]

During the milling process, part of the rubber becomes attached to reinforcing fillers so that it cannot be extracted with regular rubber solvents. This insoluble rubber is the so-called "bound rubber" or rubber-black gel. The unfortunate custom has developed to express bound rubber quantitatively as a percentage of the rubber originally present in the compound with the black.

Highly reinforcing fillers bind high percentages of rubber, while coarse fillers bind practically none. An ISAF black, compounded at 50 phr in SBR or natural rubber, will bind about 35% of the rubber in the mix immediately after mixing; this percentage will rise somewhat when the stock is heat-treated near vulcanizing temperature (of course, without any vulcanizing ingredients). The gel remaining after rubber extraction (usually with benzene) is extremely rich in solvent and highly fragile; it usually contains 20 to 30 times more solvent than rubber. Bound rubber can be considered as a measure of the surface activity of the black or white filler.[55]

The high swelling ratio (low V_r = rubber fraction) would indicate a very low crosslink density in the bound rubber although work by Endter[56] has shown the bound rubber to consist of an open network that might well occlude a considerable amount of solvent so that the actual swelling may be much less than the apparent swelling and the crosslink density may be higher than calculated from the apparent swelling. Cotten,[57] however, showed that after destruction of the gel by mechanical milling, the bound rubber is still present in the form of small flocs that can be concentrated by ultracentrifuging. After this procedure the total bound rubber volume is about the same as before the mechanical disruption of the carbon gel. The conclusion is that bound rubber as such does not contribute appreciably to the crosslink density of the final vulcanizate.[58] However, as mentioned previously, it can be considered as a measure of surface activity, and because of this the percentage of bound rubber may run parallel to rubber properties related to surface activity such as modulus, abrasion resistance, and hysteresis.

The formation of bound rubber is usually explained assuming that mechanical breakdown of elastomer chain molecules results in the appearance of free radicals at the newly formed chain ends. Reactive sites on the filler surface then combine with these free radicals to form the bound rubber. Since there are many sites, a filler particle can act as a giant crosslink. The amount of bound rubber first increases with milling and then goes down as the polymer (natural rubber) breakdown becomes the dominating factor. Figure 3.6 illustrates this effect for carbon black. Many white fillers also form bound rubber.

Gessler[59] suggests a somewhat different concept, namely that the reaction between polymer and carbon black occurs when carbon black aggregates are mechanically broken during mixing. The newly exposed fracture surfaces are so active that they react with either normal chains or chain ends activated by mechanical breakdown and form the insoluble carbon black gel. There are a number of arguments that speak for this concept and it does not alter the conclusions on the influence of bound rubber on reinforcement. The amount of bound rubber on a filler with 100 m^2/g at 50 parts loading may be about 30%. This would amount to a layer of about 6 $m\mu$ surrounding each particle and contributing to the surface area.

It has been suggested[55] that this bound rubber should be considered as part of the volume occupied by the filler. This would influence the hydrodynamic properties, such as viscosity according to the Einstein, Guth, and Gold equation:

$$\eta_f = \eta_u \times (1 + 2.5C + 14.1C^2) \tag{1}$$

In this equation η_f and η_u are the voscosities of the filled and unfilled compound, C is the volume fraction of filler. Fifty parts of black in 100 parts of rubber would constitute a volume fraction of 0.2, which makes the term in parentheses equal to 2. If the 30% bound rubber is added to the filler fraction, C rises to 0.45 and the factor $(1 + 2.5C + 14.1C^2)$ rises from 2 to 4.9, so it becomes about 2½ times as high as without the bound rubber. Reinforcing blacks give a much greater contribution to the viscosity than expressed by the hydrodynamic equation (1)

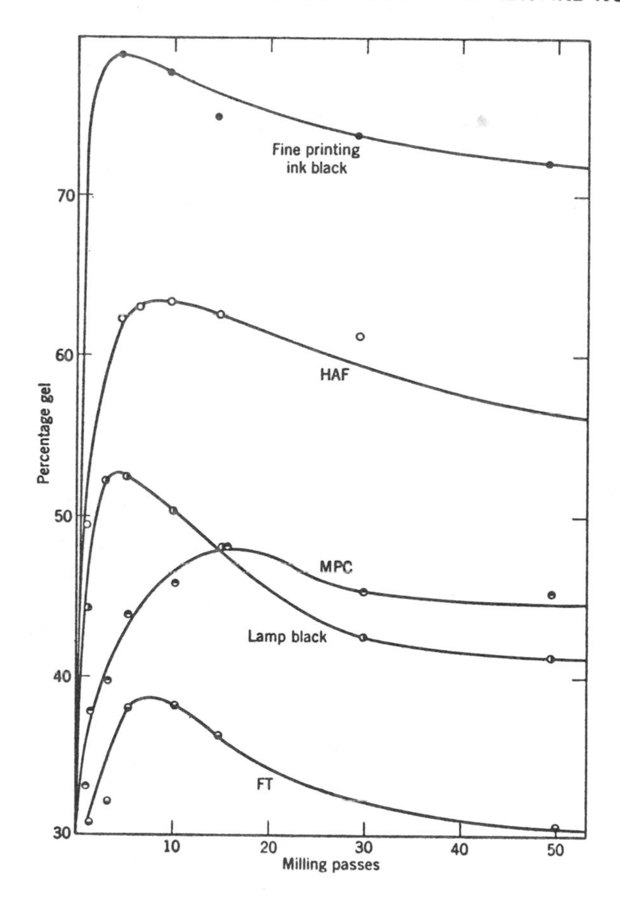

Fig. 3.6 Gel production on milling with different grades of carbon black.

using "C" the volume fraction of the filler only. In some cases this contribution is just about double that expected from the computation and of the same order of magnitude as would be due to bound rubber as part of the filler. In other cases this agreement does not exist: for instance, for graphitization of carbon black. This process reduced the bound rubber to almost zero; still the effect on viscosity is negligible – actually a higher viscosity may be found for the graphitized black compound due to lesser breakdown of the elastomer matrix during milling.

According to Guth[60] and to Cohen,[61] a relationship similar to the one for the viscosities is valid for Young's modulus of the vulcanizates crosslinked to the same degree with various amounts of filler. A shape factor "f" was also introduced to fit the experimental curve. This "f" is the ratio of length over diameter of rodshaped particles. For some reinforcing carbon blacks, $f = 6$ gives good agreement with the experimental values,[62] but according to electron microscopic observations, this value is not a realistic one. Aggregates of such a high degree of anisometry are hardly ever seen.

REINFORCEMENT AND CROSSLINK DENSITY

Effect of Reinforcing Fillers on Crosslink Density

Comparison of filled vulcanizates with pure gum vulcanizates of otherwise identical formulations and cured to their optimum shows two important characteristic differences: (1) strongly increased "modulus" at 300% elongation; (2) reduced swelling in solvents for the elastomer. Both modulus and swelling are used as means to determine crosslink density.

Modulus The modulus is related to crosslink density by the well-known formula of the kinetic theory of elasticity, in its simplest form:

$$\sigma = RT\nu\left(\lambda - \frac{1}{\lambda^2}\right) \qquad (2)$$

in which ν = the number of crosslinks per cm^3; $\nu = \frac{1}{2} N_a$ (N_a = number of active chains per cm^3 since there are, not counting loose ends, twice as many chains as there are crosslinks); λ is the extension ratio (at 100% elongation the extension ratio is 2, at 200% elongation $\lambda = 3$, etc.) According to equation (2), the modulus at a certain extension ratio and at a given temperature can only be increased by increasing "ν," the number of crosslinks per cm^3 (crosslink density). The addition of a reinforcing filler increases modulus, so its effect is the same as an increase in crosslink density.

Swelling An uncrosslinked elastomer dissolves in a suitable solvent, but if the rubber is held together by crosslinks between the molecular chains, it cannot dissolve. Rather it swells to an extent determined by the solvent power of the liquid which tends to extend the rubber gel, on the one hand, and the crosslinks which hold the molecular chains in the gel together, on the other hand. Evidently, for a given solvent, the higher the crosslink density of the rubber the lower is the swelling, and conversely for a given degree of crosslink density a more powerful solvent will give a higher degree of swelling. This relationship is quantitatively expressed by the Flory-Rehner equation:

$$\nu = \frac{1}{V_s} \cdot \frac{\ln(1 - V_r) + V_r + XV_r^2}{V_r^{1/3} - (1/2)V_r} \qquad (3)$$

which is used quite often to calculate ν, the crosslink density, from swelling measurements.

V_s is the molar volume of the solvent, V_r = the volume fraction of rubber in the swollen gel; χ = the interaction constant (for natural rubber usually about 0.4 in good solvents) dependent on the cohesive energy density of solvent, polymer, and swollen gel.

In the case of rubbers containing reinforcing fillers one finds that $V_{r(f)}$ of the rubber phase in the swollen gel (corrected for the volume of filler since the filler is assumed not to swell) is always much higher than for the corresponding pure gum

$V_{r(o)}$. Equation (3) then indicates a higher crosslink density. The value of $V_{r(f)}$ increases with loading so that the ratio

$$\frac{V_{r(o)}}{V_{r(f)}}$$

decreases with filler loading, as indicated in Fig. 3.7. This ratio represents the degree of restriction of the swelling of the rubber matrix due to the presence of filler. Figure 3.7 shows that, as the volume fraction C of filler in the vulcanizate increases, the restriction of swelling increases and more so at higher solvent power of the swelling medium (characterized by its decreasing $V_{r(o)}$). Nonadhering fillers show an increase in $V_{r(o)}/V_{r(f)}$ because of the pockets of solvent forming around the particles. Graphitized blacks constitute a unique class of fillers in that they neither restrict nor enhance but leave the swelling unaffected at any loading in any solvent. In the case of peroxide cures of natural rubber[63] it has been shown that the crosslink density in the rubber matrix is independent of the presence of fillers such as carbon black. Still, even in that case, considerable restriction of swelling due to the presence of reinforcing fillers occurs. Therefore the influence of reinforcing

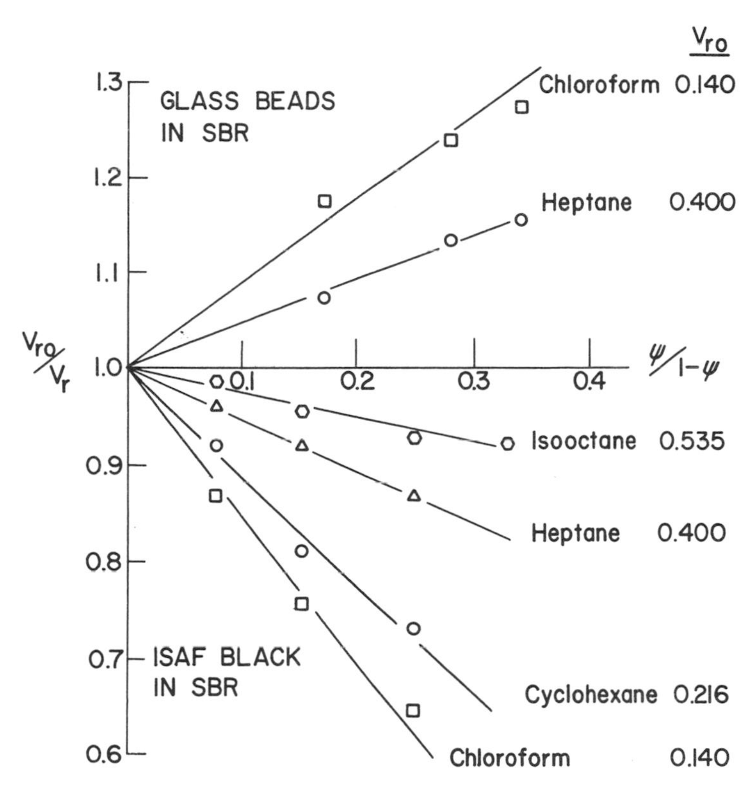

fillers on swelling points in the same direction as their influence on modulus; namely, they seem to increase crosslink density in a special way.

The apparent additional crosslinks in the matrix must be of a quite different type than the chemical crosslinks consisting of sulfur bridges or covalent carbon-to-carbon bonds. The contrast with chemical crosslinks can be summed up as follows:

	Chemical Crosslinks	*Reinforcing Fillers*
Creep and cold flow	Reduced	Increased
Hysteresis and heat build-up	Reduced	Increased
Change in modulus at high temperature	Positive	Negative
Effect on abrasion resistance	Negative	Positive

The "crosslinks" introduced by reinforcing fillers must be of a mobile nature allowing more creep and involving frictional energy dissipation.

Attachments of this type can be explained by the concept of mobile adsorption,[64] which is well established in adsorption physics. The principle is demonstrated by a plot of the energy of adsorption at various points along the surface of the carbon black, as in Fig. 3.8. The adsorptive heterogeneity of the surface of a reinforcing carbon black is depicted in the top part of Fig. 3.8. The active sites are represented by energy wells of various depths representing levels of energy of adsorption. In general, these wells are caused by defects in the crystal lattice or, in the case of carbon black, also by end effects at the edges of the crystallographic parallel layers.

By heat treatment lattice defects are eliminated and the plane layers grow together to a large extent; the result is illustrated in the bottom part of Fig. 3.8. Little energy is involved in moving a molecule or rubber segment from one place on the surface to another, but considerable energy is necessary to remove it completely from the surface. The swelling behavior reflects these situations: whereas the regular black restricts the swelling because the molecular chains are held at these sites of high energy, the graphitized black does not affect swelling at all. This is a very unusual behavior since many white fillers, so-called nonadhering fillers, will form pockets filled with fluid around the particles which make the swelling seem to increase with loading, as was shown by the glass beads of Fig. 3.7. The graphitized black neither lets go of the polymer chains adsorbed on its surface, nor puts any restriction on the swelling. This must mean that the adsorbed rubber segments move along the surface with hardly any restriction on their mobility, except that a definite average number of segments stays adsorbed at the surface. It is very

NORMAL CARBON BLACK SURFACE

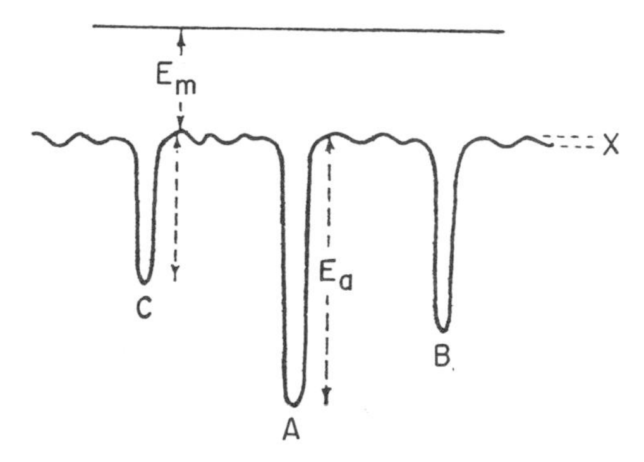

GRAPHITIZED CARBON BLACK SURFACE

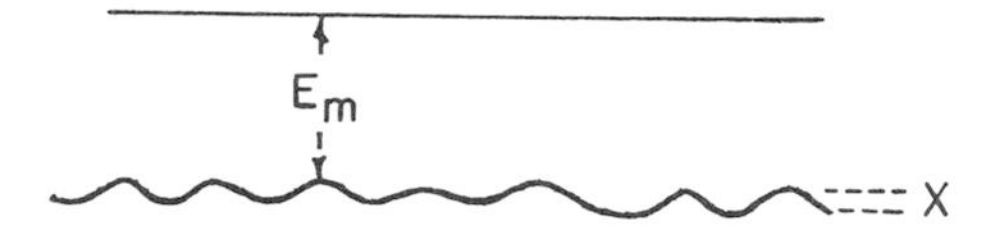

Fig. 3.8 Potential energy profiles of a normal carbon black surface (I) and a graphitized carbon black surface (II). Points A, B, and C represent sites of high adsorptive energy in decreasing order (After Ross and Olivier).

significant that untreated reinforcing carbon blacks and some reinforcing white fillers do not restrict swelling in solvents of low solvent power ($V_{r(o)} > 0.63$).[48] This indicates that with reinforcing fillers some limited lateral mobility is allowed but larger movements are restricted by attachments at the active sites.

THE REINFORCEMENT MECHANISM

The previous discussions have been leading up to a picture of reinforcement that can now be completed. The tensile strength of noncrystallizing rubbers such as SBR, nitrile rubber, ethylene-propylene terpolymer, etc., is poor since there is unequal distribution of stress in the test piece. The crosslinks are introduced at random at higher temperature; the moving molecular chains at the moment of formation of the crosslink may not be in their most probable positions. The result is a random distribution of the chain lengths between crosslinks and actually local microstresses in some of these chains.

As a consequence, when the sample is subjected to the tensile (or other) test, a number of chains, the shortest or most strained ones, will break early in the test, then the next most strained ones will go, until at the moment just before break only a few chains effectively carry the load. This is true for pure gum rubber vulcanizates as well as for all solid materials. All are lower in experimental tensile strength by a factor of 20–1000 than is computed by adding the strengths of all

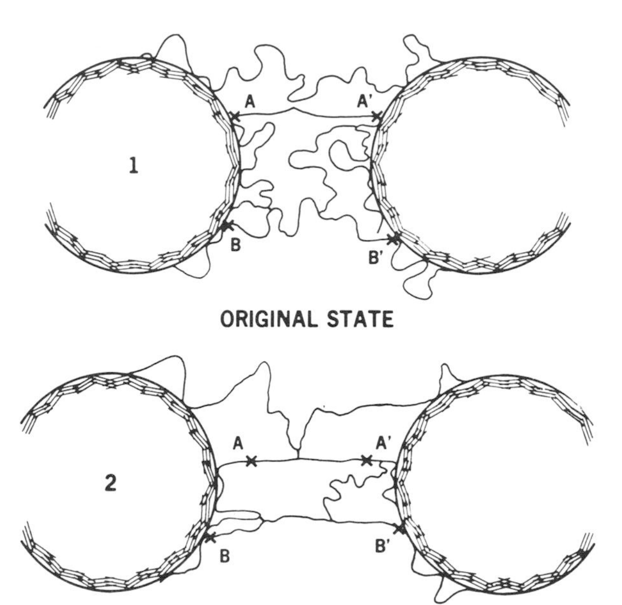

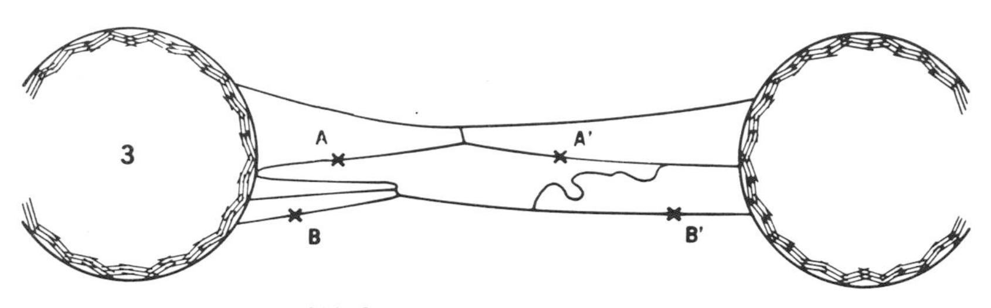

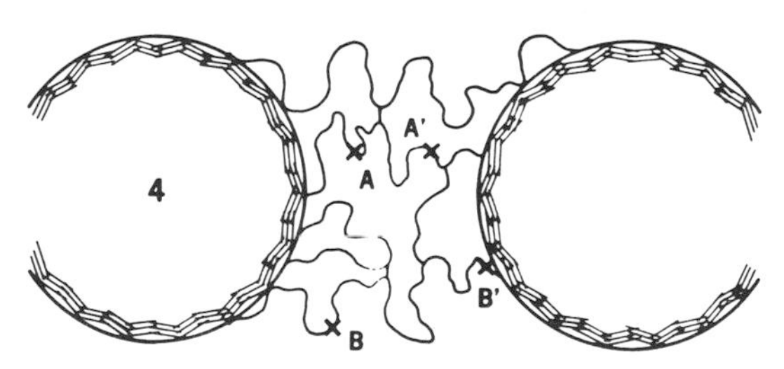

Fig. 3.9 Molecular slippage model of reinforcement mechanism.

chemical bonds running through a cross-section. If the most highly strained chains are given a chance of slippage to relieve the tensions caused by stretching, and those built in already, they would not break prematurely in the earlier part of the strain experiment and survive to the very moment before rupture. More chains effectively carry the load and a higher strength results.

A very schematic picture of such a slippage process is drawn in Fig. 3.9 a and b, which shows three chains of different lengths between two carbon black particles in the direction of stress. As the stretching process proceeds from stage 1, the first chain slips at the points of connection A and A′ until chain 2 is also taut between B and B′ (stage 2), and on continuing elongation, starts to slip there so that finally a stage 3 is reached in which all three chains are stretched to their maximum and share the imposed load. The homogeneous stress distribution causes a high improvement in strength. In stage 4 the tension is relieved and the test piece has retracted. There is now still a difference from the original situation (stage 1); due to the slippage the three chains now have about equal lengths and a repeated elongation will show a lower modulus than the first cycle, since the energy of slippage does not have to be furnished as was the case originally. This explains the so-called Mullins effect, or stress softening that will be discussed in the next section. It also identifies part of the modulus value at the initial cycle to be of a viscous nature and the slippage process as one of energy dissipation; this energy would otherwise be used to break bonds in or between molecular chains and particles. From this picture it also becomes evident that high modulus is due to the sites of high (adsorptive) energy, and high tensile strength is mainly due to the energy necessary for slippage. The last is also important for abrasion resistance, but Table 3.3 has shown that one loses most of the abrasion resistance if the sites of high adsorptive activity, the points of attachment, are eliminated. The graphitized (inactivated) black had only 1/4 of its abrasion resistance left. The influence of the active sites is probably dependent on the rate of abrasion, i.e., the severity of wear.

Support for this point of view is given by calculations made by Wake,[65] who computed the entropy changes hydrocarbon gas molecules undergo when they became adsorbed on surfaces of carbon black or other fillers. Wake found smaller changes in entropy for adsorption on carbon black than on ionic white fillers, indicating residual two-dimensional mobility of these molecules in the adsorbed state on carbon black and localized adsorption on the white fillers.

Stress Softening

Criticism of this mechanism to explain the stress-softening (Mullins effect) comes from Mullins et al., who showed that if elongated to the same stress, pure gum natural rubber shows the same softening phenomenon; so that the effect is not due to the filler but to the polymer.[66] However, to obtain the same stress as a carbon black containing vulcanizate, natural rubber gum has to be stretched to such a high elongation that crystallization occurs. The rubber crystallites act in the same way as reinforcing fillers, i.e., as stress homogenizers, so that one can expect a similar stress softening as with reinforcing fillers. Of course, the picture is schematic and breakage of chains on deformation is not completely avoided. This is shown by work of Brennan[67] and of Peremsky.[68] These authors found that the amount of stress softening expressed as the percent of strain energy retained at repeated cycles is generally speaking a function of the energy input and nearly the same for all rubber-filler combinations. The higher the energy input, the higher is the percent of energy lost in softening. Only in very precise experiments made under highly standardized conditions do differences between fillers, rubbers, and types of crosslinks become apparent.

Stress softening is a temporary effect; after a resting period in the unstrained state the stress-softened sample will recover some of its modulus to approach the original value, except for a percentage which is permanently lost. The proportion of the permanent and the recoverable modulus or energy loss depends again on the energy input, and evidently represents the chains or bonds that are actually broken. As the level of the energy at rupture is approached, the permanently lost proportion becomes larger. In one case even after 24 hours recovery only 20% of the input energy is recovered.[67]

The effect is also dependent on the amount and type of filler; higher energy losses are experienced at 50 phr loading of black than at 20 phr, and peroxide cures give lower losses than sulfur-accelerator cures. The last is understandable since peroxide cures give more permanent crosslink bonds than the polysulfide links obtained by sulfur-accelerator combinations and the stability of the crosslinks plays a part in the stress softening effect.

Graphitized black causes more stress softening than its nongraphitized counterpart. Again this is expected since graphitized black no longer has the sites of high adsorptive capacity on its surface, to restrain the adsorbent molecules from easy slipping. The importance of stress softening for reinforcement, particularly for resistance to wear of actual tires, is based on the following consideration.

Stress-Softening, Recovery, and Tire Wear

The wear of an automobile tire is a cyclic process. After each contact with the road the tread segment is at rest for the remainder of the travel, tracing a cycloid curve until it hits the road surface again. During the contact with the road, abrasion and stressing occur and part of the energy is dissipated in the form of the otherwise unwanted heat. After this process, stress-recovery must take place to such a degree that at the next contact, energy dissipation can occur again. The rate of recovery is dependent on the (number of) persistent bonds in the elastomer network and between network and filler surface; evidently the presence of many strong bonds runs against the large energy dissipation concept which is the basis of high abrasion resistance. It is impossible to have both energy dissipation and recovery at their maximum since one impairs the other. Therefore a compromise must be found between these two opposing factors, and it depends on the conditions of the road test as to which side is the more important. For use at higher speeds, which leave a shorter time cycle for recovery, more strong bonds between filler surface and polymer are necessary at the expense of energy dissipation. At higher speeds tire wear increases rapidly. Under these conditions one would want a filler with a more active surface and in many cases these more active surfaces are present with higher structure carbon blacks. The reason for the higher activity may be the breakdown of carbon black aggregates as proposed by Gessler, which would generate fresh and active surfaces *in situ*, or it may have incidental manufacturing causes. Under high severity conditions treads containing high structure black are superior to those with normal structure blacks, whereas under mild conditions differences are hardly noticeable.

Reinforcement of Elastomer Blends by Fillers

All previous general considerations for reinforcement of single elastomers hold as well for blends of rubbers. However, blends have some inherent special characteristics. Polymers are usually not miscible and blends consist of microdispersions of one into the other or of intermingled microregions of both often having dimensions around 0.1 to 1.5 microns. When fillers are mixed into such a blend, a situation may develop in which the filler is unevenly distributed over the two different phases and by this distribution affects the compound properties.

This behavior is found in blends of natural rubber and polybutadiene rubber. Hess, Scott, and Callan[69] showed evidence that the best mechanical properties of the blend were obtained when the polybutadiene contained the larger proportion of reinforcing carbon black as determined by electron microscopic techniques. If black was mixed into a previously prepared 50/50 blend of natural rubber and polybutadiene rubber, the black would preferably incorporate in the natural rubber.[70]

The explanation of this phenomenon is probably that during the mixing of the preblend with filler, the soft polymer will penetrate the voids between black particles first so that there will be a higher concentration of black in the softer rubber. Because of the uptake of filler, the viscosity of the softer rubber increases until it nearly matches that of the more viscous polymer, after which the second polymer participates in the black penetration at about the same rate so that the viscosity of the two phases is roughly the same. The other effect, which is that better properties are found when the polybutadiene phase contains the largest proportion of black, is due to the fact that polybutadiene as a gum vulcanizate is weak and needs carbon black for reinforcement whereas natural rubber by itself is already strong. It has the built-in reinforcement of its crystallites which fulfill a similar function as a reinforcing filler.

In other cases of blends the effects are not so pronounced as cited here[71] and even in this case one has to consider that the method of preparation may have an influence on the properties of the final composition. Besides there may be an influence of the degree of crosslinking, since the one component will not cure at the same rate as the other in identical formulations. Therefore one phase may be stiffer than the other.

In conclusion, it can be stated that the theoretical concept of reinforcement is well developed and in agreement with the main experimental results; there are still a number of unexplained phenomena and unanswered questions. To mention just a few: Do black and/or other fillers actually participate with their active surface groups in the vulcanization reaction, binding sulfur or forming primary bonds under peroxide curing? Blends of fillers such as silica and black can produce better vulcanizates than either filler separately for certain applications (off the road tires): Why? Does each filler particle create its own environmental domain, i.e., hard and soft regions of different crosslink density? Do blends of fillers result in better vulcanizate properties? Can one attain still higher levels of reinforcement by "super fillers"? These, and similar, questions may be answered in the years to come and these studies may lead to new and unexpected developments in the field of elastomer reinforcement.

REFERENCES

(1) G. Kraus, *Reinforcement of Elastomers*, Interscience: John Wiley & Sons, New York (1965).
(2) M. Studebaker, "The Chemistry of Carbon Black and Reinforcement," *Rubber Chem. & Tech.*, **30**, 1400 (1957).
(3) W. B. Wiegand, *India R. J.*, **60**, 423 (1920); *Trans. IRI*, **1**, 14 (1925).
(4) W. R. Smith, *Encyclopedia of Chemical Technology*, 3rd ed., Interscience, New York (1964), pp. 243–247, 280–281.
(5) H. J. Stern, *Rubber Natural and Synthetic*, MacClaren and Sons, Ltd., London–New York 277 (1967). Schidrowitz and Dawson, *History of the Rubber Industry*, Heffer and Sons, Cambridge, 71 (1952).
(6) U.S. Bureau of Mines, *Mineral Industry Survey*, Carbon Black Annual, June 19, 1969; also private communication.
(7) J. W. Sellers and F. E. Toonder, *Reinforcement of Elastomers*, edited by G. Kraus, Interscience: John Wiley & Sons (1965), Chap. 13, p. 405.
(8) A. van Rossem and G. van Nederveen, *Rev. Gen. Caout.*, **19**, 255 (1942).
(9) J. LeBras and I. Piccini, *Bul. Soc. Chim.* France, 215 (1950); *Ind. Eng. Chem.*, **43**, 381 (1951); *Rubber Chem. & Tech.*, **24**, 649 (1951).
(10) H. J. Wildschut, *Receuil. Trav. Chim.*, **61**, 898 (1942).
(11) J. J. Keilen and A. Pollak, *Ind. Eng. Chem.*, **39**, 480 (1947).
(12) R. A. Buchanan, O. E. Weisvogel, C. R. Russel, and C. E. Rist, *Ind. Eng. Chem., Res. & Dev.*, **7** (2), 155 (1968). Paper presented at ACS Div. of Rubber Chemistry Meeting, Cleveland, April 1968.
(13) J. van Alphen, *Proc. Third Rubber Technol. Conf.*, 670 (1954). R. Houwink and J. van Alphen, *J. Pol. Sci.*, **16**, 121 (1955).
(14) O. W. Burke, Jr., *Reinforcement of Elastomers*, edited by G. Kraus, Interscience, N.Y. (1965), Chap. 15, p. 494.
(15) A. Renner, B. B. Boonstra, and D. F. Walker, Lecture presented at the IRI Conference in Loughborough, September, 1969.
(16) W. B. Wiegand, *Can. Chem. J.*, **4**, 160 (1920); *Ind. Rubber J.*, **60**, 397, 423, 453 (1920).
(17) A. van Rossem, *Rubber* (in Dutch) *Servire*, The Hague (1958).
(18) E. M. Dannenberg and B. B. Boonstra, *Ind. Eng. Chem.*, **47**, 339 (1955).
(19) A. I. Medalia and F. A. Heckman, Paper to the Eighth Conference on Carbon, Buffalo, June 1967, *J. Carbon*, **7**, 567 (1969). A. I. Medalia and F. A. Heckman, *J. IRI*, **3**, 66 (1969).
(20) J. H. Bachmann, J. W. Sellers, M. P. Wagner, and R. F. Wolf, *Rubber Chem. & Tech.*, **32**, 1286 (1959).

(21) A. C. Hardy, U.S. Pat. 1.780.231, November 4, 1930.
(22) *Cabot Carbon Blacks Under the Electron Microscope*, 2nd ed., Cabot Corp., Boston, Mass. 1950.
(23) S. Brunauer, P. H. Emmett, and E. Teller, *J. Am. Chem. Soc.*, **60**, 309 (1938).
(24) G. L. Taylor and J. H. Atkins, *J. Phys. Chem.*, **70**, 1678 (1966).
(25) W. D. Schaeffer and W. R. Smith, *Ind. Eng. Ch.*, **47**, 1286 (1955); E. M. Dannenberg, *Rubber Age*, **98**, Sept., 82–Oct., 81 (1966).
(26) Z. Rigbi, *Rev. Gen. Caoutch.*, **33**, 243 (1956); *Bul. Res. Counc. Israel*, 6C (I), 67 (1957)
(27) M. Morton, J. C. Healy, and R. L. Denecour, Paper to the International Rubber Conference, Brighton, England, May 15–18, 1967. M. Morton and J. C. Healy, *Appl. Pol. Symp.*, **7**, 155 (1968).
(28) H. M. Leeper, C. L. Cable, J. J. D'Amico, and C. C. Tung, *Rubber World*, **135**, 413 (1956).
(29) A. M. Gessler, *Rubber Age*, **94**, 598, 750 (1964).
(30) E. M. Dannenberg and W. H. Opie, *Rubber World*, **138**, 85, 98 (1958).
(31) E. R. Eaton and J. S. Middleton, *Rubber World*, **157**, 94 (1965).
(32) W. H. Opie, V. A. Sljaka, D. L. Petterson, E. M. Dannenberg, and H. M. Cole, paper presented at the ACS, Div. of Rubber Chem., San Francisco, May 1966.
(33) A. I. Medalia, paper to Colloid Symposium, Buffalo, June 1967; *J. Colloid and Interface Sci.*, **24**, 393 (1967). *ibid.*, **32**, 115 (1970).
(34) This process was invented by Wiegand in 1927, improved by Cabot Corporation in 1934 and Huber in 1938, and was a big step forward in processing of carbon blacks.
(35) Unpublished data, Cabot Corporation, Rubber and Plastics Research Laboratory.
(36) F. A. Heckman and A. I. Medalia, *J. IRI*, **3**, 66 (1969).
(37) A. M. Gessler, paper presented at the International Conference, Brighton, 1967.
(38) A. Voet, P. Aboytes, and P. A. Marsh, paper presented at the ACS Div. of Rubber Chem. Meeting, Los Angeles, May 1969; *Rubber Age*, **101**, 78 (1969).
(39) A. R. Payne and W. F. Watson, *Trans IRI*, **39**, T125 (1963).
(40) A. R. Payne, *Rubber Chem. & Tech.*, **39**, 365 (1966); *39*, 915 (1966).
(41) A. Voet and F. R. Cook, *Rubber Chem. & Tech.*, **40**, 1367 (1967); **41**, 1215 (1968); **41**, 1207 (1968).
(42) D. F. Harling and F. A. Heckman, *Matiri Plastiche ed. Elastomeri*, **35**, 80 (1969), (in English).
(43) J. H. deBoer, B. G. Linsen, and T. J. Osinga, *J. Catalysis*, **4**, 643 (1965).
(44) B. B. Boonstra and E. M. Dannenberg, *Ind. Eng. Chem.*, **47**, 339 (1955).
(45) W. R. Smith and G. A. Kasten, presented at the Div. of Rubber Chem., ACS, Washington, May 1970; *Rubber Chem. & Tech.*, **43**, 960 (1970); see also Ref. 25.
(46) M. Polley and B. B. Boonstra, *Rubber Chem. & Tech.*, **30**, 170 (1957).
(47) G. R. Cotten, B. B. Boonstra, D. Rivin, and F. R. Williams, *Kautsch. u. Gummi K.*, **22**, 477 (1969).
(48) S. Wolff, lecture to the S. and S.W. German Rubber Group in Wurzburg, March 1969, ref. *Kautsch. u. Gummi K.*, **22**, 367 (1969).
(49) A. Y. Coran, *Rubber Chem. & Tech.*, **37**, 679 (1964).
(50) S. Wolff, paper to the International D.K.G. Rubber Conf., Berlin, May 1968; *Kautsch. u. Gummi K.*, **23**, 7 (1970).
(51) B. B. Boonstra and G. L. Taylor, *Rubber Chem. & Tech.*, **38**, 943 (1965).
(52) B. B. Boonstra and A. I. Medalia, *Rubber Age*, **92**, 892 (1963); *92*, 82 (1963); *Rubber Chem. & Tech.*, **36**, 115 (1963).

(53) Dispersion of Carbon Black in Rubber, Revised Calculation Procedure, A. I. Medalia, *Rubber Chem. & Tech.,* **34** (4) 1134 (1961).
(54) For a more complete coverage of filler-rubber attachments, see W. F. Watson, *Reinforcement of Elastomers*, edited by G. Kraus, Interscience, New York (1965).
(55) J. J. Brennan and T. E. Jermyn, *J. Appl. Poly. Sci.,* **9**, 2749 (1965).
(56) F. W. Endter, *Kautsch. u. Gummi,* **5**, WT17 (1952); *Rubber Chem. & Tech.,* **27**, 1 (1954).
(57) G. R. Cotten, *Rubber Chem. & Tech.,* **39**, 1553 (1966).
(58) B. B. Boonstra and E. M. Dannenberg, *Rubber Age,* **82**, 838 (1958).
(59) A. M. Gessler, paper presented at the International Conference, Brighton, 1967.
(60) E. Guth, International Conference on Rubber Technology, London (1948), paper 20.
(61) L. H. Cohen, *ibid.,* paper 35.
(62) T. D. Bolt, E. M. Dannenberg, R. E. Dobbin, and R. P. Rossman, *Rubber and Plastics Age,* **41**, 1520 (1960).
(63) C. R. Parks and O. Lorenz, *J. Poly. Sci.,* **50**, 287 (1961). H. Westlinning and S. Wolff, Paper presented in Vienna, October 1966, at the D.K.G. Meeting, Group S. & S.W., Germany.
(64) S. Ross and J. P. Olivier, *On Physical Adsorption*, John Wiley & Sons, New York–London (1964); J. H. deBoer, *The Dynamical Character of Adsorption,* Clarendon Press, Oxford (1953).
(65) W. C. Wake, *Proc. Int. Conf. on Rubber,* Washington, D.C., November 1959, p. 406.
(66) J. A. C. Harwood, L. Mullins, and A. R. Payne, *Polymer Letters B,* **3**, 1119 (1965); *Trans. IRI,* **42**, T14 (1966).
(67) J. J. Brennan, E. M. Dannenberg, and Z. Rigbi, *Proc. Inter. Rubber Conf.,* Brighton, 123 (1967). E. M. Dannenberg and J. J. Brennan, *Rubber Chem. & Tech.,* **39** 577 (1966).
(68) R. Peremsky, *Kaucuk. Plast. Hmoty.,* 37 (2) (1963).
(69) W. H. Hess, C. E. Scott, and J. E. Callan, *Rubber Chem. & Tech.,* **40**, 371 (1967).
(70) P. J. Corish and M. J. Palmer, "Some New Aspects of Polymer Blends," paper presented at the Loughborough Conference of the IRI, September 1969.
(71) P. A. Marsh and A. Voet, *Rubber Chem. & Tech.,* **41**, 344 (1968).

4

PROCESSING AND VULCANIZATION TESTS

R. W. WISE
Monsanto Co.
Akron, Ohio

INTRODUCTION

In this chapter we will direct our attention to the test procedures and instruments used to measure two of the most critical properties of rubber and rubber compounds – processibility and vulcanization. These characteristics are vitally important because they affect the success of all the basic steps in the manufacturing process used to convert raw rubber into final usable products.

Doubtless previous chapters have outlined the fundamental principles involved in the rubber manufacturing process. However, for the sake of clarity and a better understanding of the technology, some of the basics will be reviewed here. The rubber industry, in common with other industries, has developed a terminology of its own, the meaning of which is not always self-evident to the newcomer.

In this context, it is interesting to recall the apparent paradox under which the industry labors. In general, it takes a tough, sometimes elastic material and spends large amounts of money and energy converting this to a soft, pliable, plastic, workable material, incorporating additives (fillers, oils, chemicals, etc.) along the way. This process is called "breaking-down" the rubber, "compounding" it and "mixing," "milling," and generally "masticating" the resulting rubber mixture or "stock."

In this state the compound may be formed by the application of force, and since it is predominantly plastic, it will retain the shape imposed upon it. This can be accomplished by squeezing it between rolls (calendering), pushing it through an orifice having the desired shape (tubing or extruding), or by confining it under pressure in a mold or cavity of the required dimensions. This procedure is called "processing" and the measure of the performance of a compound in this phase is called the "processibility." The plastic flow (plasticity) of a stock is dependent upon temperature, force, and the rate at which that force is applied, and it follows that these three variables are significant in measuring processibility.

After forming the stock to the desired shape, the compound needs to be converted to a strong elastic material. This is accomplished by the process of "curing" or "vulcanizing" the unvulcanized or "green" stock. The basis for vulcanization is chemical bonding (crosslinking) accomplished usually by means of sulfur and an accelerator under pressure at elevated temperatures. During vulcanization the stock changes from an essentially plastic body to one that is predominantly elastic. Its inherent resistance to deformation increases as does its strength, bounciness, and toughness. At the end of the vulcanization process, the rubber is in its final form with which we are most familiar. Vulcanization tests measure the performance of a rubber compound during this curing process.

The remaining terms need clarification before getting on to description of test methods and apparatus.

Scorch Scorch is premature vulcanization in which the stock becomes partly vulcanized before the product is in its final form and ready for vulcanization. It reduces the plastic properties of the compound so that it can no longer be processed. Scorching is the result of both the temperature reached during processing and the amount of time the compound is exposed to elevated temperatures. This period of time before vulcanization starts is generally referred to as "scorch time." Since scorching ruins the stock, it is important that vulcanization does not start until processing is complete.

Rate of Cure The "rate of cure" is the rate at which crosslinking and the development of the stiffness (modulus) of the compound occur after the scorch point. As the compound is heated past the scorch point, the properties of the compound change from a soft plastic to a tough elastic material required for use. During the curing step crosslinks are introduced, which connect the long polymer chains of the rubber together. As more crosslinks are introduced, the polymer chains become more firmly connected and the stiffness or modulus of the compound increases. The rate of cure is an important vulcanization parameter since it in part determines the time the compound must be cured, i.e., the "cure time."

State of Cure In general, "state of cure" is a term used to indicate the development of a property of the rubber as cure progresses. As the crosslinking or vulcanization proceeds, the modulus of the compound increases to various "states of cure." Technically, the most important state of cure is the so-called "optimum." Since all properties imparted by vulcanization do not occur at the same level of cure, the state for optimizing a particular property may not be the best for other properties.

Cure Time Cure time is the time required during the vulcanization step for the compounded rubber to reach the desired state of cure.

Overcure A cure which is longer than optimum is an "overcure." Overcures may be of two types. In one type, the stock continues to harden, the modulus rises, and tensile and elongation fall. In other cases, including most natural rubber compounds, reversion occurs with overcure and the modulus and tensile strength decrease.

To clarify the test procedures needed to best define the various steps involved in the conversion of raw rubber to a final usable product, the remainder of this chapter will be subdivided into tests used to characterize each succeeding step of the process. The first section is devoted to measurement of processibility, i.e., flow and scorch characteristics, followed by a section which covers tests used to define the behavior of the processed compound during the final step of the process, i.e., vulcanization.

PROCESSIBILITY

Processibility is dependent on the viscosity or plasticity of the rubber mixture, i.e., its resistance to flow, and the time required for crosslinking of the polymer to occur under processing conditions. Plasticity determines the energy required to extrude or form the rubber, while the time to initial crosslinking, i.e., scorch, indicates the amount of heat history which can be tolerated before the polymer is converted from a plastic to a rubbery state in which processing is virtually impossible.

Plasticity The rate at which the rubber compound is sheared or deformed has a marked effect on its apparent plasticity. The various rubber processes – mixing, extrusion, calendering, the molding – involve different levels of shear and, therefore, can be expected to affect each rubber stock differently. Figure 4.1 shows the influence of shear rate on shear stress, i.e., its resistance to flow, for two compounds. These two compounds have the same flow characteristics and behave

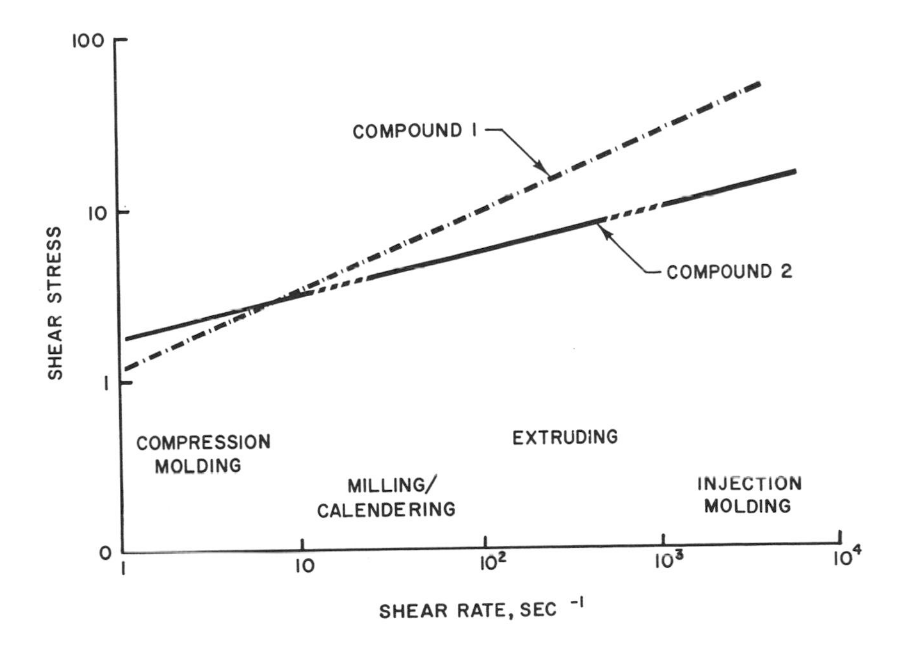

Fig. 4.1 Effect of shear rate on shear stress.

identically in the low shear rate steps of processing, such as compression molding, but would exhibit markedly different flow behavior under the high shear rates encountered in extruding. Therefore, tests to ascertain the processing characteristics of a particular stock should ideally expose the test sample to shear rates identical to those encountered in the actual process step. Unfortunately, many of the tests used to predict the manner in which stocks will process in the plant do not impose shear rates on the test sample equivalent to those involved in the actual process step. As a consequence, poor correlation between laboratory test findings and the behavior of the rubber stock in the plant results. Therefore, whenever practicable, a processibility test should be chosen which simulates as nearly as possible the conditions which prevail in the plant process.

Plasticity Tests Plasticity tests determine the deformability or flow characteristics of a rubber or rubber compound which normally has little elastic recovery. Rubber, before vulcanization, is in between a plastic and an elastic state; when warm it becomes more plastic and less elastic. Commercially there is little use for rubber in its plastic state· nevertheless, plasticity tests are important because their results assist in determining the processibility of the material.

Due to the nature of the mixing process, consecutive batches of supposedly identical compounds differ from one another. One characteristic which is sensitive to these minor types of variation is plasticity. Because this is so, plasticity tests are frequently used for process control purposes. Although a whole family of testing equipment exists, there are two criteria essentially involved. One class of tests forcibly deforms the sample and measures the force required to produce a known displacement, or conversely, measures the displacement caused by a known force. A second class of test is based on the appearance and dimensions of the rubber after deformation. For example, in extrusion, the smoothness and dimensions of the extrudate often determine whether it is suitable for use in building the final product. Due to the elasticity present even in unvulcanized rubber, energy will be stored in the compound during deformation and then released when the force which caused the deformation is removed. The net result is that the compound will swell after extrusion, i.e., "die swell."

Test Methods In general, there are three accepted test methods. One is the parallel-plate type such as the Williams plastometer in which a pellet of standard dimensions is squeezed between two plates by applying a known force and measuring the deformation after a given length of time at a prescribed temperature. The second involves extruding or mixing devices in which a quantity of material is extruded through a small orifice, or masticated in a small mixing chamber. These devices have the advantage that they closely simulate the conditions present in plant extruders and mixers. The last and most widely used is the rotating disk viscometer, in which a knurled rotor is rotated in a mass of rubber contained in a mold under pressure, and the torque required to rotate the rotor is measured.

Parallel Plate Plastomer The Williams Parallel Plate Plastomer has been a widely used type of plastomer and still enjoys fairly wide but diminishing usage. In this device, a cylindrical preheated rubber sample 2 cc in volume is placed between two parallel plates and a 5-kg load is applied on it for a standard period of time, commonly 3, 5, or 10 minutes. The resulting thickness of the sample, in hundreds of a millimeter multiplied by one thousand, is called the "plasticity" number. The load is then removed and the increase in thickness after 1 minute is called the "recovery" value. The plasticity number and the recovery, being related to flow properties and the elastic component, respectively, are useful in predicting the processibility characteristics such as ease of forming and die swell. The method is described in detail in ASTM Standard Method D926.

Extrusion and Mixing Tests Extrusion, possibly the most direct method for evaluating the processing characteristics of vulcanized rubber compounds, is described in ASTM Standard Method D2230. A conventional laboratory screw type extruder is equipped with a special extrusion die widely referred to as a Garvey die. The rubber compound is extruded through the die under closely controlled conditions and the appearance of extruded rubber is rated visually against a series of standards. Additionally, the length of extrudate per unit of time is indicative of extrusion rate, i.e., displacement of the rubber upon application of a given force.

A laboratory instrument which is ideally suited for extrusion testing is the C. W. Brabender Plasti-Corder®. The heart of the instrument is a recording torque dynamometer combined with variable speed drive which operates over the range of 2 to 200 rpm. A variety of both mixing and extruder heads, simulating the various plant processes, can be attached to the instrument. The advantage of the Plasti-Corder is that the torque can be recorded throughout the extrusion process. The torque values give a good indication of the plasticity of the compound under conditions closely approximating those encountered in plant extruders. By varying the screw speeds, a wide range of shear rates can be generated, ranging from approximately 2 to 10^3 sec^{-1}. These cover shear rates encountered in milling, calendering, and extrusion.

Another attachment which can be added to the Plasti-Corder is a mixing head or test chamber very similar to a conventional Banbury mixer used commonly in the rubber industry. The mixing or shearing action is accomplished by rotating two masticating blades in opposite directions within the heated test chamber. As with the extruder head, the rate at which the blades are rotated can be varied and the resistance of the test compound to mastication continuously recorded.

Breakdown energies for various rubbers can be identified in terms of the area under the torque-time curve. Effects of reinforcing agents, process oils, plasticizers, peptizers, stabilizers, heat promoters, and other additives can be observed as corresponding increases or decreases in torque. Data obtained by employing this "mixing-in-miniature" principle can be used to estimate breakdown times and energies as well as mixing rates on production equipment. It should be kept in mind, however, that attempts to "scale-up" results obtained on small laboratory extruders and mixers to large plant equipment is difficult. This is because it is almost impossible to duplicate exactly in a laboratory test the complex shear rates

®Trademark of C. W. Brabender Co., Hackensack, N. J.

and conditions which prevail in plant size equipment. This is particularly true of the heat build-up in the compound and is one of the main reasons for the lack of success in scale-up procedures.

Rotating Disk Viscometer A number of plastometers of this type were developed during the rapid growth of the rubber industry in the 1920–1940 period. However, the shearing disk viscometer developed by the late Melvin Mooney has become the "work-horse" of the rubber industry. The Mooney viscometer is widely used as a laboratory control instrument. All synthetic and natural rubbers are graded in Mooney viscosity units which are included in their specifications. It is also widely used to determine the scorch of compounded rubbers. These methods are described in detail in ASTM standard method D–1646, titled "Viscosity and Curing Characteristics of Rubber by the Shearing Disk Viscometer."

The shearing action is performed by a disk rotating in a shallow cylindrical cavity filled with the rubber under test, as shown in Fig. 4.2. The rubber is squeezed into the cavity under considerable pressure. The surface of the disk and the dies which form the cavity are grooved to avoid slippage. The test specimen consists of two pieces which completely fill the test chamber. One piece is placed above the rotor and the other beneath it. In the older models the rotor was connected through a vertical shaft, worm gear, and horizontal floating shaft to an electric motor as shown in Fig. 4.3. The rotor is rotated at 2 rpm by an electric motor. The resistance of the rubber to the shearing action develops a thrust on the shaft which presses against and deflects a calibrated U-spring. The deflection of the U-spring is proportional to the torque required to rotate the rotor and is read on a dial gage graduated in thousands of an inch, each thousand corresponding to one Mooney unit. The top platen and die is moved downward to engage the lower die and form the test chamber by a pair of double acting levers. The upper and lower platens are heated electrically or by steam, and maintained at the desired temperature. In more recent models, the double lever system has been replaced with a pneumatic cylinder which presses the dies together with 2500 pounds of force. The Mooney viscometer was subsequently further modernized by replacing the dial gage with strain gages

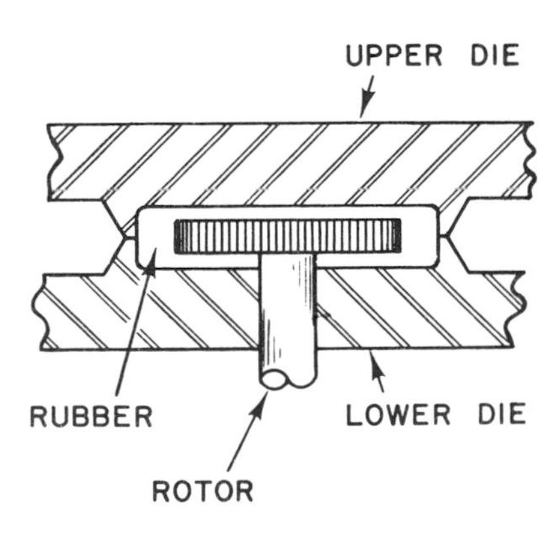

Fig. 4.2 Mooney chamber and rotor.

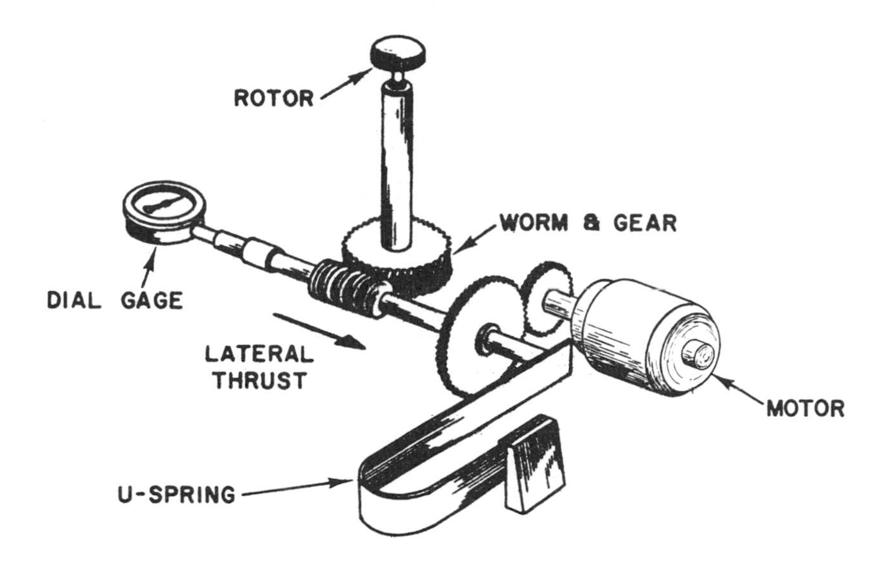

Fig. 4.3 Mooney torque measuring system.

bonded to the U-spring to provide, when connected to an autographic recorder, a continuous record of viscosity in Mooney units versus test time. The U-spring technique has recently been supplanted with other more sophisticated types of electronic torque measuring transducers. Mooney Viscometers ranging from the older lever/dial gage units to the more modern pneumatic cylinder/electronic recording instruments are now found in most laboratories.

In viscosity measurements, the sample is allowed to warm up for one minute after the platens are closed. The motor is then started and a reading taken after an appropriate time. For example, a 4-minute reading is usually taken for SBR and most other rubbers while 8 minutes is used for butyl rubbers. Figure 4.4 shows typical viscosity time curves for two rubbers. The viscosity taken from the dial or recorder chart is not constant with time but varies considerably depending on the particular type of rubber. Of importance is the fact that the specimen does not attain temperature equilibrium for approximately ten minutes after the start of the test. A sharp drop in viscosity is usually noted at the beginning of the test due to the thixotropy exhibited by most rubbers. Finally, most rubbers show a steady decrease in viscosity with continuing shearing, which is the result of rubber breakdown occurring during the test.

The rotor of the Mooney Viscometer is normally rotated 2 rpm which corresponds to a shear rate of approximately 2 reciprocal seconds (sec^{-1}). This corresponds to shear rates encountered in molding but is substantially below those encountered in extrusion and injection molding. Therefore, some caution should be used in relating Mooney viscosity to the behavior of compounds in these high shear rate stages of the process.

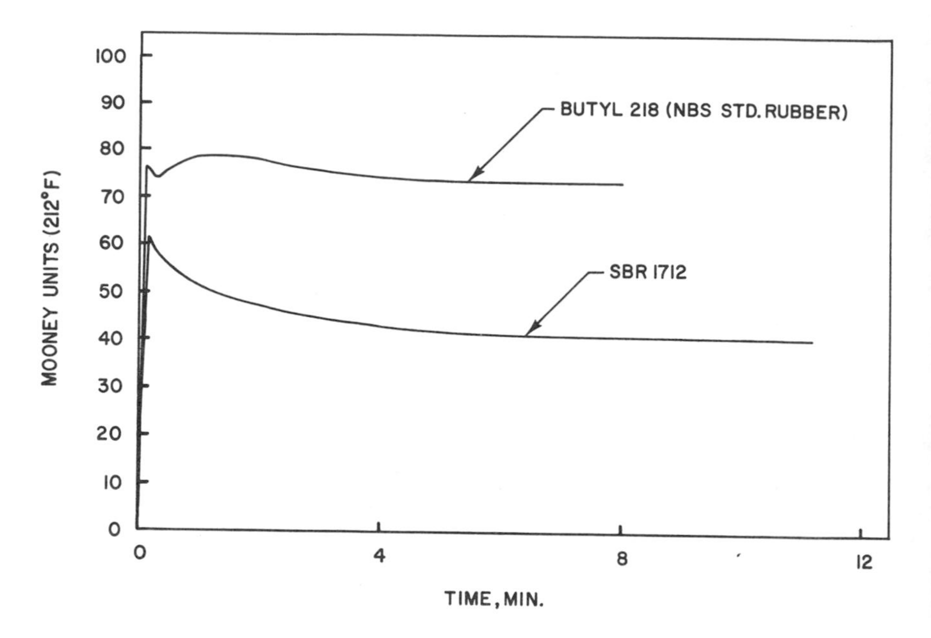

Fig. 4.4 Typical Mooney viscosity curves for different rubbers.

The results for viscosity measurements are reported in the following form:

50-ML 1 + 4 (100C)
Where 50-M is the Mooney viscosity number, L indicates the use of the large 1.5 in. rotor (S would indicate the small 1.25 in. rotor), 1 is the time in minutes that the specimen was permitted to warm in the machine before starting the motor, 4 is the time in minutes after starting the motor at which the reading is taken, and 100°C is the temperature of the test.

Scorch As mentioned before, scorch is premature vulcanization which may occur during the processing of the rubber compound due to accumulated effects of heat and time. Therefore, the time for the compound to scorch will slowly decrease as the compound moves through each stage of the process. Thus samples taken from the same batch at different stages in the process will have progressively shorter scorch time. The scorch time of a batch at a particular point in the process is called its "residual scorch time." The effect of "heat history" on decreasing the scorch time of a compound is illustrated in Fig. 4.5. (The cure curves presented in Fig. 4.5 are obtained with a Curemeter which will be described in a subsequent section of the chapter.) Any good factory stock will have a scorch time slightly longer than the equivalent of the maximum heat history it may accumulate during processing. If all the scorch time is depleted during processing, it will no longer process and must be scrapped.

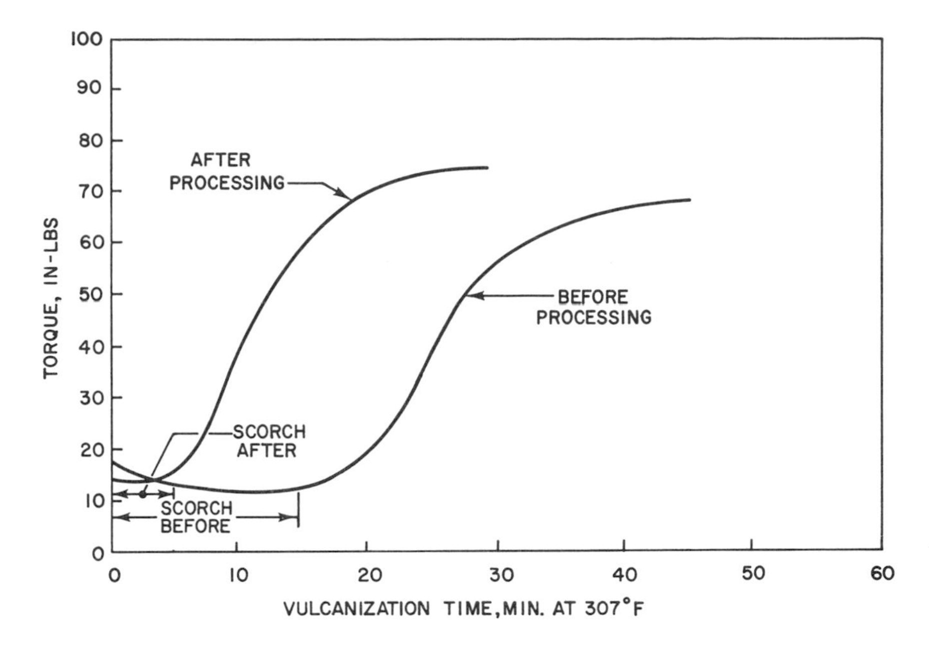

Fig. 4.5 Effect of heat history on scorch time.

The Mooney viscometer described in the previous section is widely used to determine scorch characteristics of compounded rubber. Normally the test is conducted at temperatures encountered during the processing of the rubber. Typically this is in the temperature range 250 to 275°F. From a chart of Mooney units versus test time, the time required for the compound to scorch (for vulcanization to start) can be easily determined by noting when the cure curve turns upward as shown in Fig. 4.6. The most common method of measurement is to run the compound in the Mooney viscometer until the viscosity shows a 5-point rise above the minimum. The viscosity of the compounded rubber at processing temperatures can also be obtained from the minimum of the curve. The values normally taken from a Mooney cure curve are:

MV= minimum viscosity
t_5 = time to scorch at MV + 5 units
t_{35} = time to cure at MV + 35 units
ΔT_L = cure index which = $t_{35} - t_5$

VULCANIZATION

After the rubber has been compounded by the addition of the appropriate curing agents, processed and formed, it is then vulcanized. The vulcanization process occurs in three stages: (1) an induction period; (2) a curing or crosslinking stage; and (3) a reversion or overcure stage.

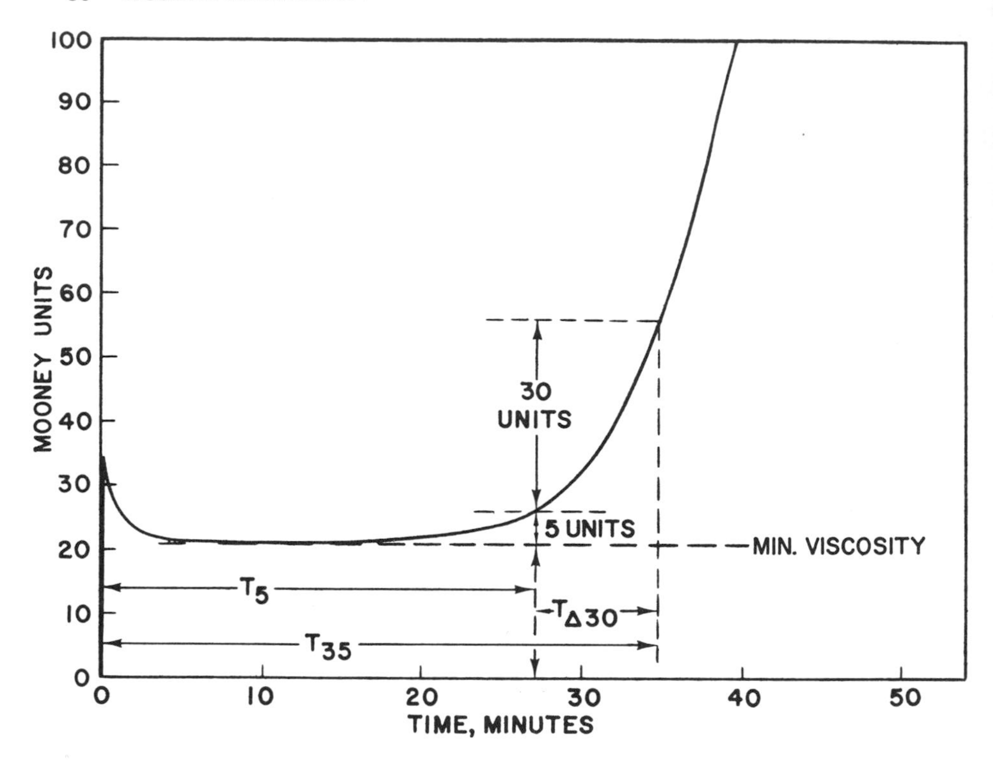

Fig. 4.6 Mooney scorch curve.

The location of these three stages in the vulcanization cycle is shown in Fig. 4.7. The induction period represents the time at vulcanization temperature during which no measurable crosslinking has occurred. It is of practical importance since its duration determies the safety of the stock against "scorching" during the various processing steps which precede the final vulcanization. As mentioned in the previous section, the Mooney viscometer test conducted at processing temperatures gives an excellent indication of "scorch" time. Following the induction period, crosslinking proceeds at a rate which is dependent on the temperature and the composition of the rubber compound. When crosslinking proceeds to full cure continued heating produces an overcure which may result either in a further stiffening or softening of the compound. In the development or production of rubber compounds, the rubber technologist strives to arrive at a balance between a tendency to scorch and a vulcanization rate which best fits the processing and cure requirements of his final product.

Vulcanization Test Vulcanization tests are all aimed at determining the behavior of the compound during these three phases of the vulcanization cycle. There are three techniques currently used for measuring and following the development of the properties during the vulcanization process: (1) chemical methods; (2) physical test methods; (3) continuous measurements (Curemeters).

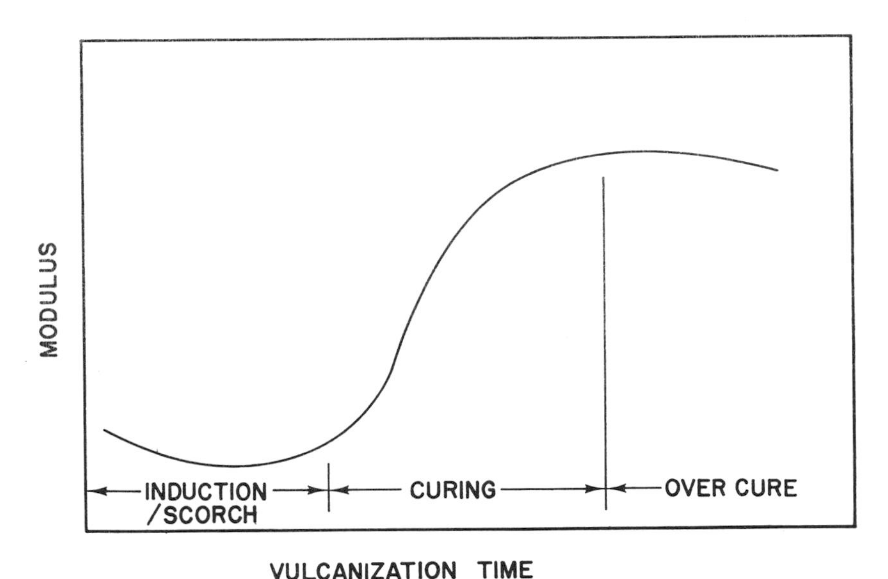

Fig. 4.7 Steps in the vulcanization process.

Chemical Methods Since in a normal vulcanization system at least part of the sulfur combines with the rubber during vulcanization, an obvious way of following vulcanization is to measure the decrease in free sulfur. This method is not used extensively since it is well known that the combination of free sulfur does not correlate well with the development of crosslinks or other physical properties. In addition, the analysis is lengthy and costly. However, free sulfur determinations are often made on finished products as a means for checking for uniformity of the product and to estimate the degree of cure. Figure 4.8 shows the rate of sulfur combination at different vulcanization temperatures for a typical natural gum rubber compound. They exhibit the typical sigmoid shape which typifies all cure curves.

Physical Test Methods The classical method for following vulcanization by physical methods is to vulcanize a series of sheets for increasing time intervals and then measure the stress-strain properties of each and plot the results as a function of vulcanization time. Typically, sheets are vulcanized for five or ten minute intervals and then the modulus at 300% elongation and tensile strength at break are measured on dumbbell specimens cut from each sheet. Figure 4.9 shows a typical of data for two natural rubber stocks having different accelerators. This method is satisfactory for discerning large differences in vulcanization but is inadequate for detecting small differences due to large testing errors. It is also inadequate for measuring the scorch period. Combining the Mooney scorch test with stress-strain tests on partial cures can overcome this objection to some extent. It does have the advantage that the vulcanization time required for development of a maximum physical property, be it ultimate tensile strength, modulus, or some other property,

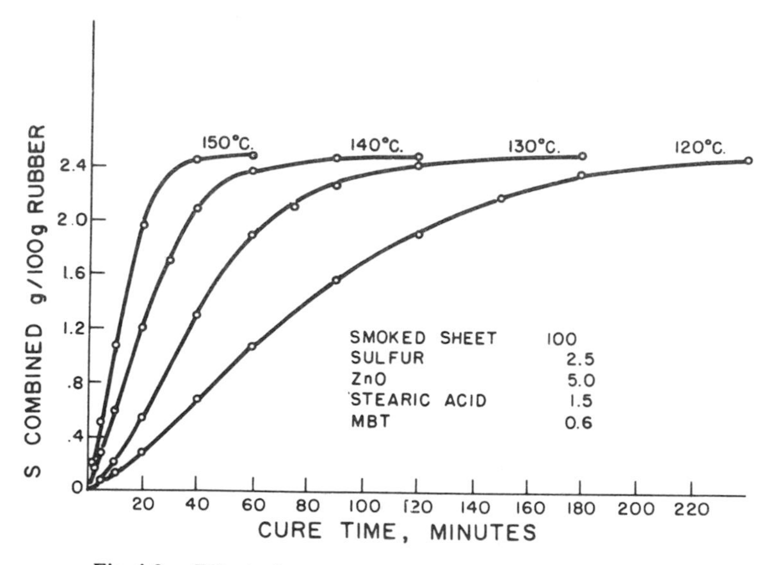

Fig. 4.8 Effect of temperature on rate of sulfur combination.

can be determined. This is significant because the optimum vulcanization time often depends on which property of the rubber it is desired to optimize. For example, maximum tensile strength usually occurs prior to the development of maximum modulus. The compounder is often faced with choosing a cure time which will optimize the property or properties which are most critical in the performance of the final product, while not reducing other important properties below acceptable limits. For example, crack growth decreases with cure while hysteresis is improved. Thus the optimum state of cure is often a compromise and the best cure for a particular application is often referred to as the "technical cure," thus signifying it is the best compromise cure that can be technically obtained.

A modification of this test generally called a "rapid modulus test" is widely used in the industry as a production control test. A single sample taken from a production batch of compounded rubber is vulcanized at a high temperature and its tensile modulus measured. Temperatures as high as 380°F are used to reduce the vulcanization time to only a few minutes. Any modulus value outside predetermined acceptance limits indicates that the batch is defective and is rejected.

Continuous Measurement of Vulcanization In all the methods previously described for measuring state of cure, it is necessary to cure specimens of the test sample for each of a series of curing times and then perform the desired test on the vulcanizate. It occurred to a number of investigators that a savings in time and more complete information could be obtained if a test could be made continuously on a single specimen during vulcanization. The Mooney viscometer test approaches this objective. However, a weakness of the Mooney viscometer test is that the test is completed before a measurable modulus value passed the scorch point has been

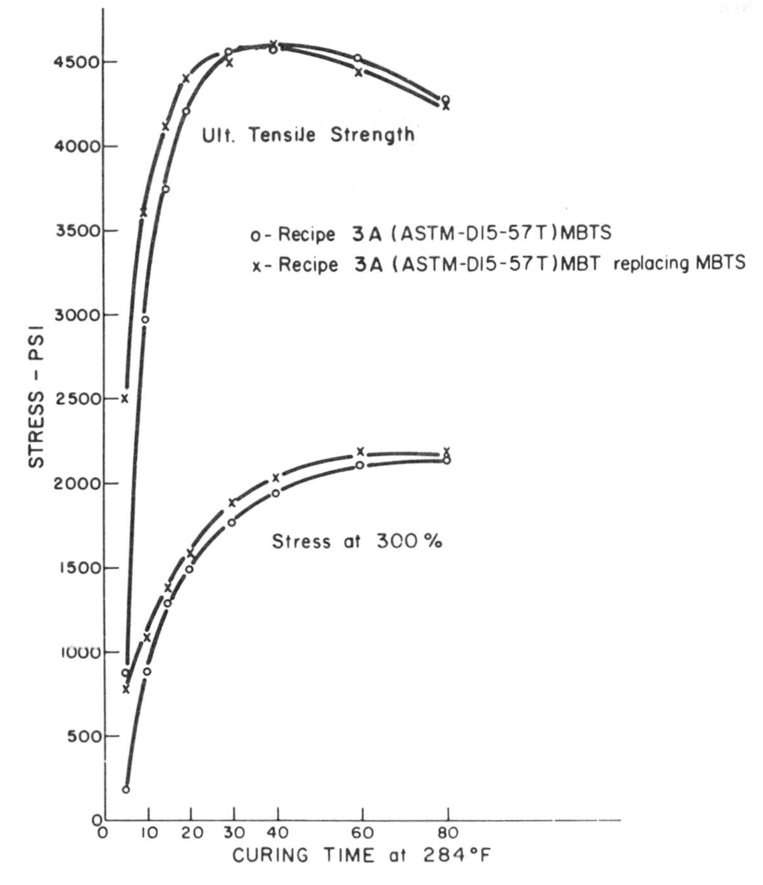

Fig. 4.9 Comparison of cure curves for natural rubber tread type stocks accelerated with MBTS and MBT.

obtained. This is because the test sample is destroyed after the induction period is passed due to tearing by the continuous rotation of the rotor. To overcome this deficiency and provide a total cure curve for the entire vulcanization cycle, a series of instruments called *curemeters* were developed. In each of the test instruments that have been developed, the compound stiffness or modulus was chosen as the parameter to measure continuously.

A variety of different types of curemeters were developed during the early 1960's. The Vulkameter, developed at Bayer in Germany, was the first of the curemeters to be developed. This apparatus continuously measures the dynamic shear modulus of the heated test specimen against vulcanization time, using the type of test specimen and apparatus shown in Fig. 4.10. The paddle located between the two test specimens is oscillated back and forth through a small distance to continuously shear the specimen. The force required to move the paddle is measured to yield the desired cure curve of modulus versus vulcanization time. The

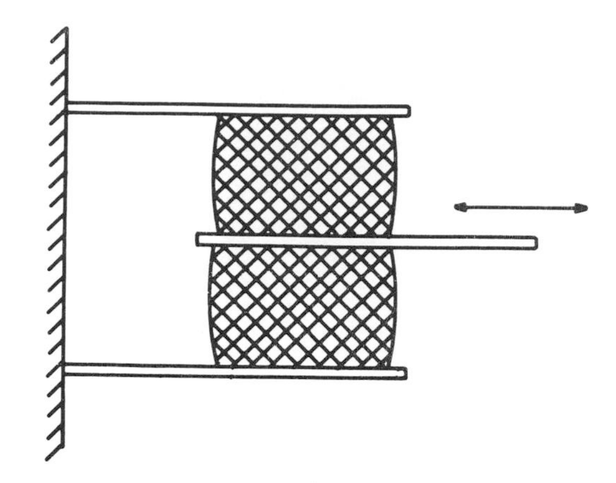

Fig. 4.10 Vulkameter and curometer apparatus and test specimen.

Wallace-Shawbury curemeter is also commercially available* and it operates on the same principle as the Vulkameter. Both these devices suffer from the fact that the sample is not maintained under the pressures similar to those used in plant vulcanization equipment. This has the consequence that the test sample develops porosity and often degrades during the test.

As a result of these problems and to overcome these deficiencies, oscillating disk type curemeters have emerged as the preferred type of curemeter. They are a modification of the Mooney viscometer, but unlike the Mooney, the rotor is oscillated through a small arc rather than continuously rotated. They have the decided advantage that the sample is maintained under high pressure throughout the test. In this way the curemeters measure the torque required to oscillate the rotor, which is embedded in the rubber sample confined in a die cavity under pressure and controlled at the desired vulcanization temperature. The sample is thus subjected to an oscillatory shearing action of constant amplitude. Oscillation of the rotor does not result in the destruction of the sample such as occurs in the Mooney viscometer. As vulcanization proceeds the torque required to shear the rubber increases and a curve of torque versus cure time is generated. Since the rotor is straining the rubber, the torque value is directly related to the shear modulus of the rubber.

Figure 4.11 shows a typical cure curve obtained with an oscillating-disk curemeter.* From this curve of torque versus cure time, all of the vulcanization characteristics of the test compound can be determined directly, as shown on the chart.

A photograph of one of the several commercially available curemeters is shown in Fig. 4.12. The use of this type of curemeter is described in detail in ASTM Tentative Method D2084-71-T, "Measurement of Curing Characteristics with the Oscillating Disk Rheometer Curemeter." This method recommends that the

* H. W. Wallace & Co. Ltd. Croydon, England.
* Rheometer Model TM-100, Monsanto Co., Akron, Ohio.

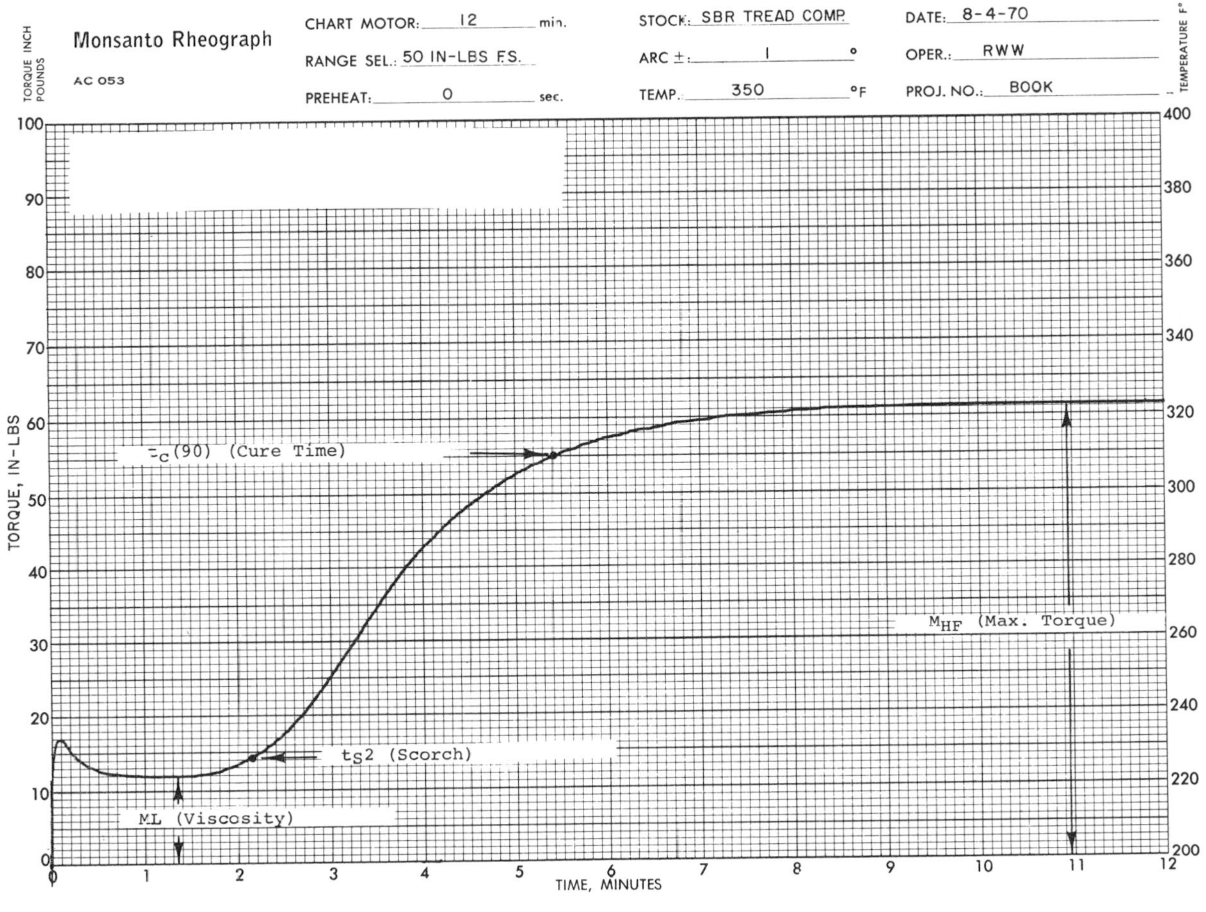

Fig. 4.11 Typical curemeter curve.

Fig. 4.12 Photograph of a commercial oscillating disk type curemeter. (*Courtesy Monsanto Company*)

following values be taken from the curve:

M_L = Minimum torque

M_{HF} = Equilibrium torque

M_{HR} = Maximum torque obtained for curve which exhibits reversion

M_H = highest torque value attained where no constant or maximum value is obtained

t_{sx} = scorch time to x units of torque increase above minimum torque

$t_{c(x)}$ = cure time to (x) percent of maximum torque development.

Figure 4.13 shows the three different types of cure curves which can be obtained with different types of rubber compounds. For example, some synthetic rubber compounds attain a constant or equilibrium torque level (M_{HF}) while most natural rubber compounds exhibit reversion (M_{HR}). The rubber technologist normally strives to develop a compound which neither reverts nor increases in modulus (torque) with overcure. Thus by using the above nomenclature recommended in the

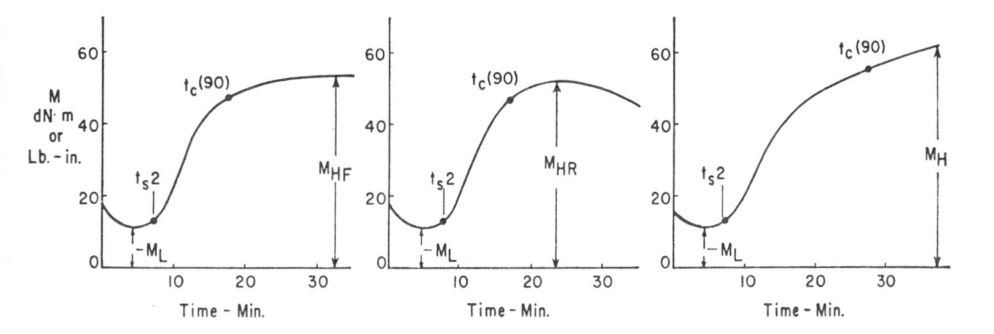

Fig. 4.13 Type of cure curves. Left curve: Cure to equilibrium torque. Middle curve: Cure to a maximum torque with reversion. Right curve: Cure to no equilibrium or maximum torque. (Source: ASTM D2705–68-T, Fig. 2, p. 38.)

ASTM method, the type of cure curve as well as all the other vulcanization characteristics of the compound is defined.

Uses of Curemeters The effects of compound variations on curing characteristics are important in compound development studies or production control. Curemeter tests are ideally suited for use in both these areas. In compound development, the composition of the compound can be varied until the desired vulcanization characteristics are obtained. The effect of compound changes on viscosity and scorch can be determined from the early portion of the cure curve, while from the latter portion, the effect on rate of vulcanization and the modulus of the cured compound can be measured. Figure 4.14 shows how relatively minor changes in the concentration of a compounding ingredient can be detected, and illustrates the immense value of curemeters in this area. The cure curve obtained with a curemeter is a "fingerprint" of the compound's vulcanization and processing character.

The ability of a curemeter to detect minor changes in the composition of the rubber compound has made it a widely accepted production control test. Another advantage is that the test can be run very rapidly. For example. by operating at test temperatures in the 350 to 400°F range, a test can be completed in approximately five minutes, which coincides closely with a typical Banbury mixing cycle. The procedure usually followed is to establish specification limits at several points along the cure curve as shown in Fig. 4.15. A cure curve is obtained on each batch of production stock prior to its use in fabricating the final product. Batches that yield cure curves which fall outside these acceptance limits are rejected.

Limits of acceptability for quality control purposes using a curemeter are normally set by running 30–40 consecutive batches of a compound on a single chart. This gives a representative "picture" of the batch-to-batch variation occurring. By correlating this information with experience, it is possible to set up limits of acceptability. By introducing deliberate variations in the mix in the laboratory and comparing the resulting rate curves, reasons for batch rejection can be determined.

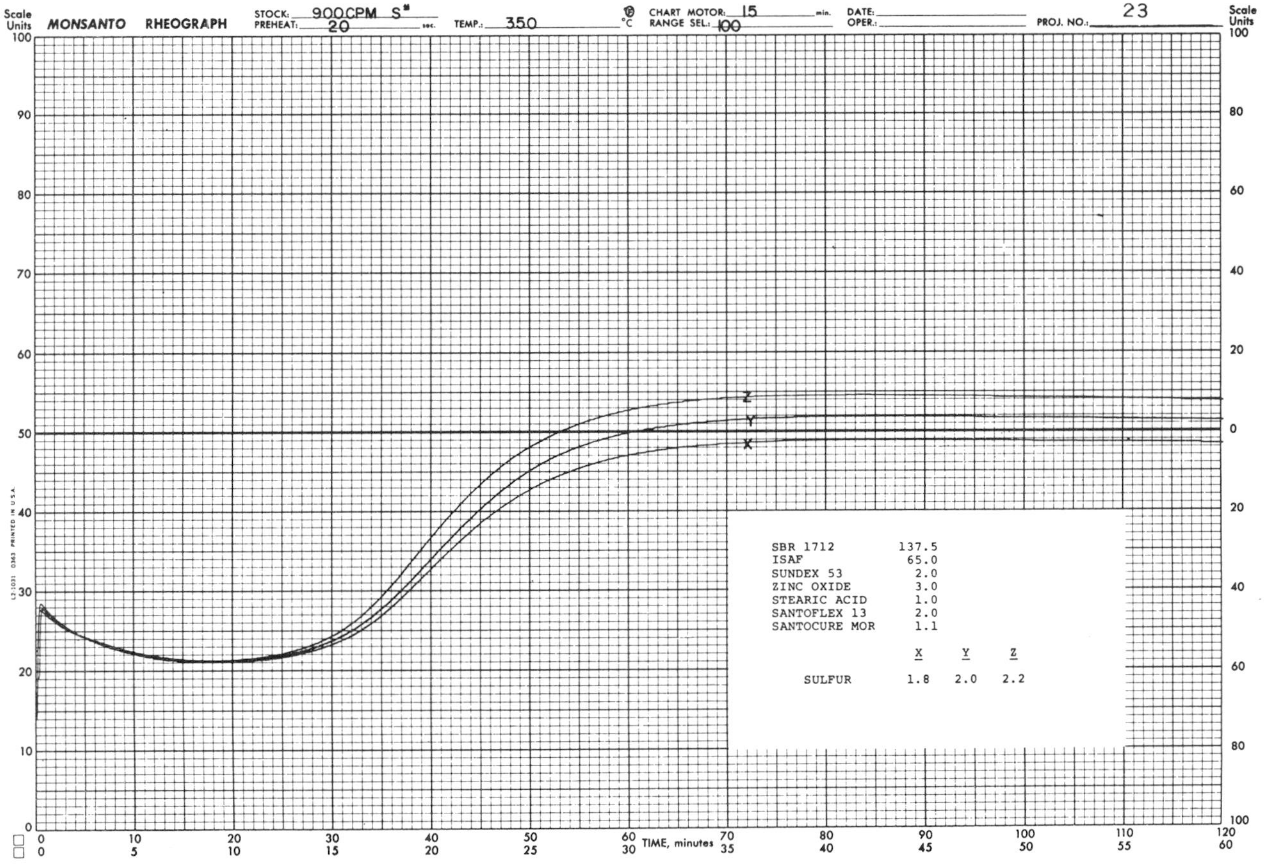

SBR 1712	137.5
ISAF	65.0
SUNDEX 53	2.0
ZINC OXIDE	3.0
STEARIC ACID	1.0
SANTOFLEX 13	2.0
SANTOCURE MOR	1.1

	X	Y	Z
SULFUR	1.8	2.0	2.2

Fig. 4.14 Sensitivity of curemeter cure curves to minor changes in sulfur content. (Monsanto Technical Bulletin O/R C-3.)

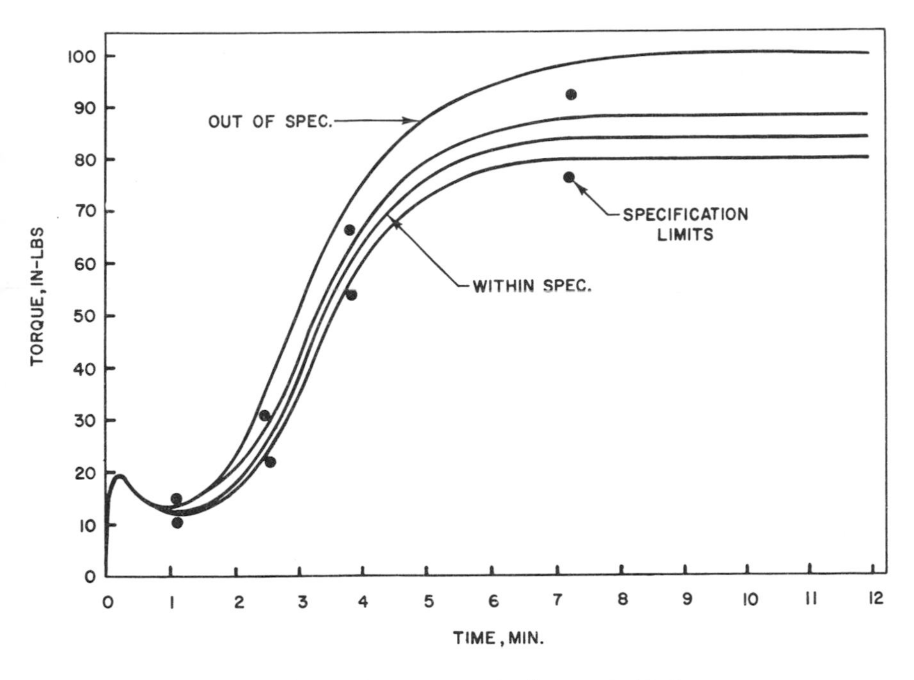

Fig. 4.15 Curemeter production control test.

The two types of cure curves which are typical of most natural and neoprene rubber based compounds are shown in Figs. 4.16 and 4.17, respectively. Few compounds have a completely flat "plateau" although, by judiciously selecting the proper types and amounts of curing ingredients, a compound can be obtained which exhibits neither reversion nor a "marching modulus." Curemeter cure curves are also used to aid in selecting the proper cure time for vulcanizing laboratory specimens for use in subsequent physical tests. As previously mentioned, the classical procedure for ascertaining the optimum cure time is to vulcanize individual test specimens for incremental times and then conduct the desired physical test on each specimen. A typical example is the preparation sheet specimens for stress-strain tests. Since good agreement exists between the cure curve obtained with a curemeter and tensile modulus measurements made on partial cures as illustrated in Fig. 4.18, a practice which is being widely used is to determine the optimum cure time directly from the cure curve and prepare only a single specimen for physical test. This has the advantage that only a single rather than a multiple number of test samples need be prepared, with a commensurate saving in testing time and cost.

The good agreement between curemeter results and tensile modulus measurements makes curemeters an attractive alternate procedure for controlling the production step and testing the quality of raw materials such as polymers and fillers. Curemeter tests have many advantages over the traditional stress-strain (tensile modulus and strength) procedures which have been historically used as

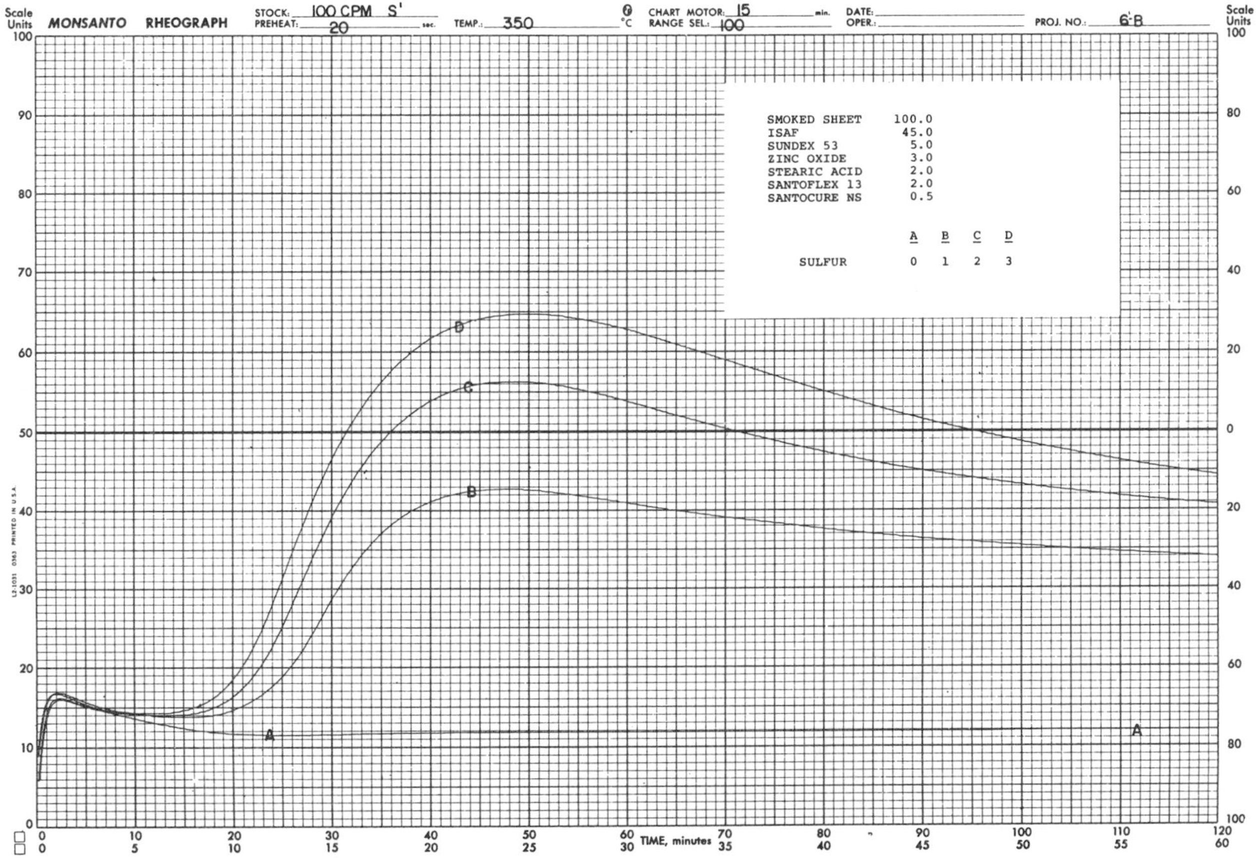

Scale Units
MONSANTO RHEOGRAPH
STOCK: 100 CPM S'
PREHEAT: 20 sec.
TEMP.: 350
°C
CHART MOTOR: 15 min.
RANGE SEL.: 100
DATE:
OPER.:
PROJ. NO.: 6-B
SMOKED SHEET 100.0
ISAF 45.0
SUNDEX 53 5.0
ZINC OXIDE 3.0
STEARIC ACID 2.0
SANTOFLEX 13 2.0
SANTOCURE NS 0.5
A B C D
SULFUR 0 1 2 3
TIME, minutes

Fig. 4.16 Illustration of "reverting" type cure curve (ASTM type M_{HR}).

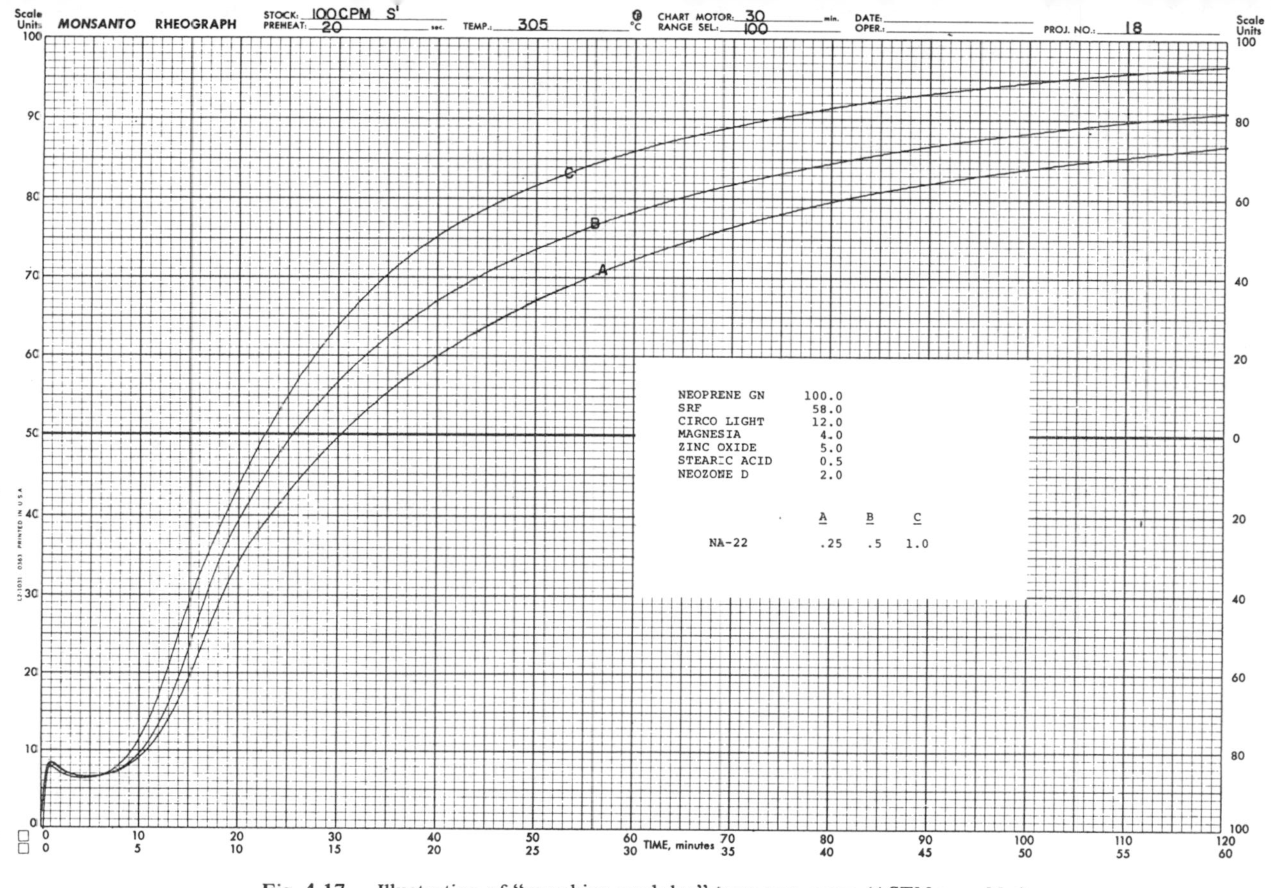

Fig. 4.17 Illustration of "marching modulus" type cure curve (ASTM type M_H).

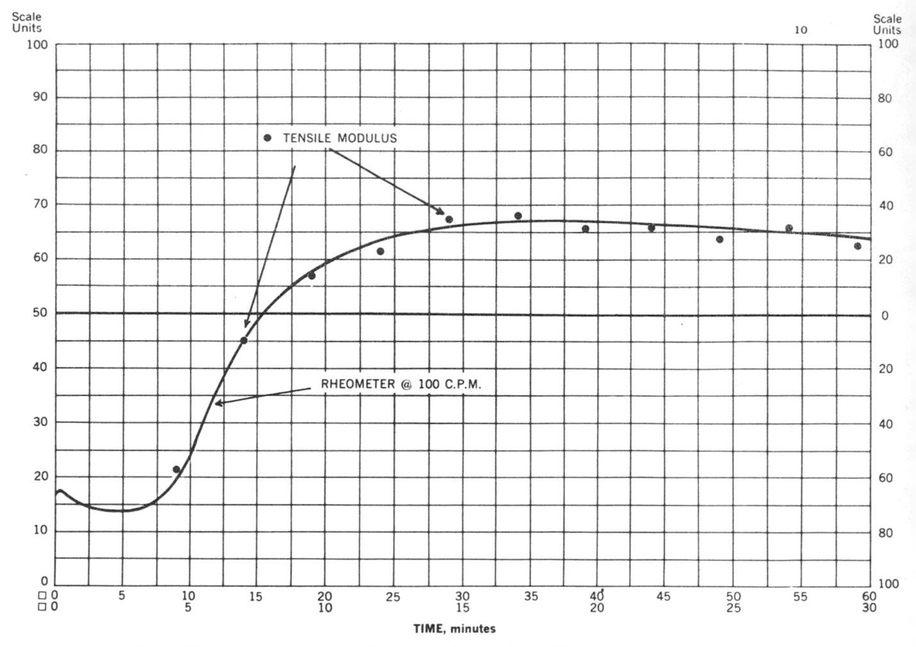

Fig. 4.18 Comparison of curemeter and tensile modulus cure curves.

acceptance tests. Among these advantages is the fact that curing and testing are combined in a single test so that a more comprehensive delineation of vulcanization behavior is obtained.

Effect of Temperature on Vulcanization Rate The rate of vulcanization of rubber compounds is greatly influenced by temperature, as is typical of other chemical reactions. A rule of thumb widely used in the industry is that the rate of cure is doubled for every 10°C (18°F) increase in temperature. While this is sufficiently accurate for most practical work it becomes more and more in error the further one moves in either direction from the referenced temperature. This error is reflected in Fig. 4.19, which shows a plot of cure time vs. temperature. The dashed line is based on a temperature coefficient mentioned above, i.e., 2.0/10°C, while the solid line represents actual values. The deviation of the dotted line from the straight line is a reflection of the change in the temperature coefficient of vulcanization at different curing temperatures. Figure 4.20 shows a nomograph for transposing cures from one temperature to another in which a temperature coefficient of 2.0/10°C was used in its construction. This type of nomograph is widely used in the industry for correlating cure times for a given compound at different temperatures. The fact that the rate of vulcanization doubles for every 10°C indicates that the temperature of any test designed to determine cure times and rates of cure must be closely controlled if meaningful results are to be obtained.

Curemeters normally expose the test specimen to a constant temperature.

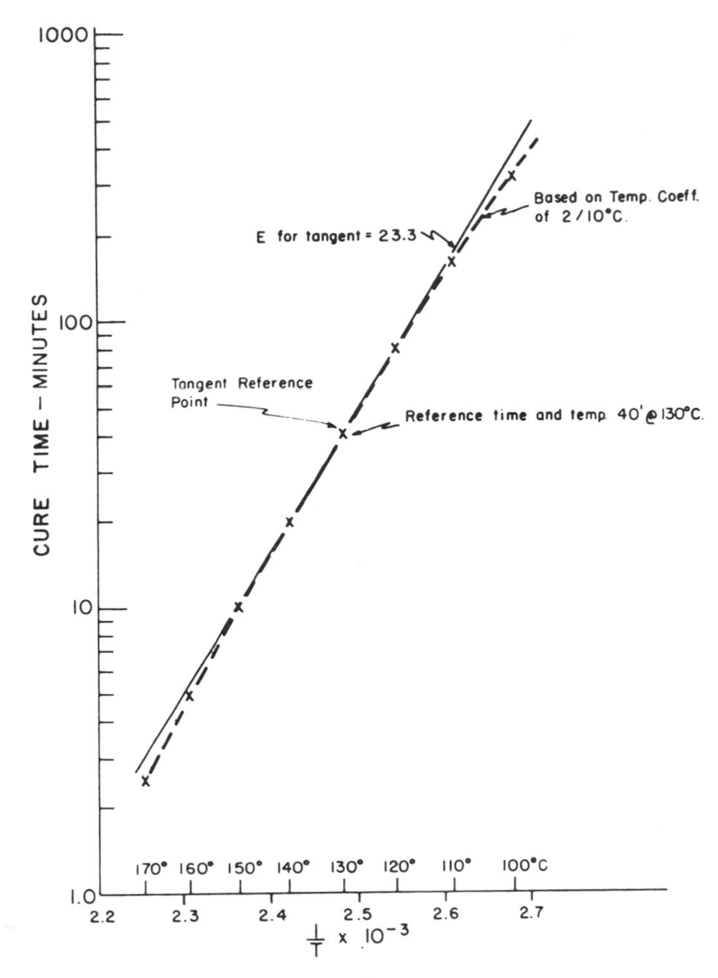

Fig. 4.19 Error involved in use of a temperature coefficient.

However, in practice, especially in thick rubber articles, rubber is never cured under constant temperature conditions. The best example of this is in thick sections of tires where, with rubber being such a poor heat conductor, the inner sections of the tire are slow to heat up, and conversely, slow to cool down, as shown in Fig. 4.21. The various sections of the tire must be compounded to take this into account. The standard practice in the rubber industry has been to first determine the actual temperature to which the rubber is exposed as a function of time, usually by inserting thermocouples in the article and recording temperature versus time. Exhaustive calculations are then employed using curing characteristics determined under constant curing temperatures to calculate equivalent states of cure under the variable time-temperature conditions which actually exist.

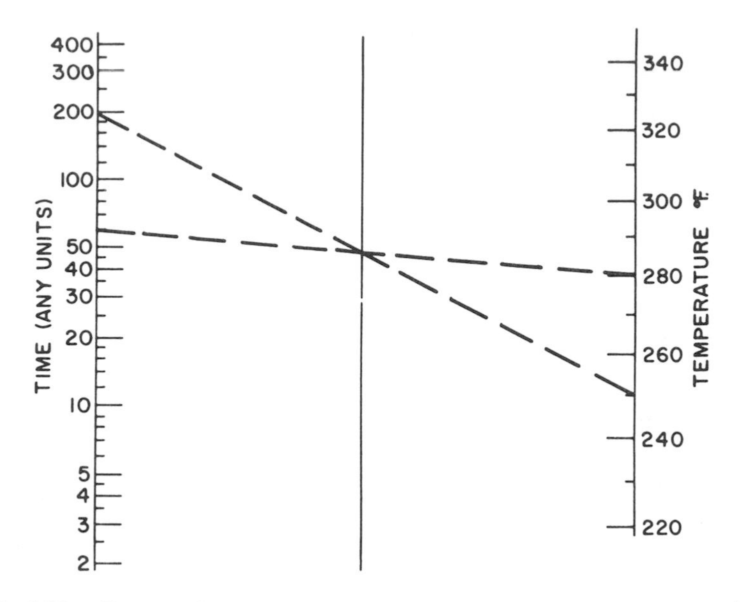

Fig. 4.20 Nomograph for transposing cure time from one temperature to another.

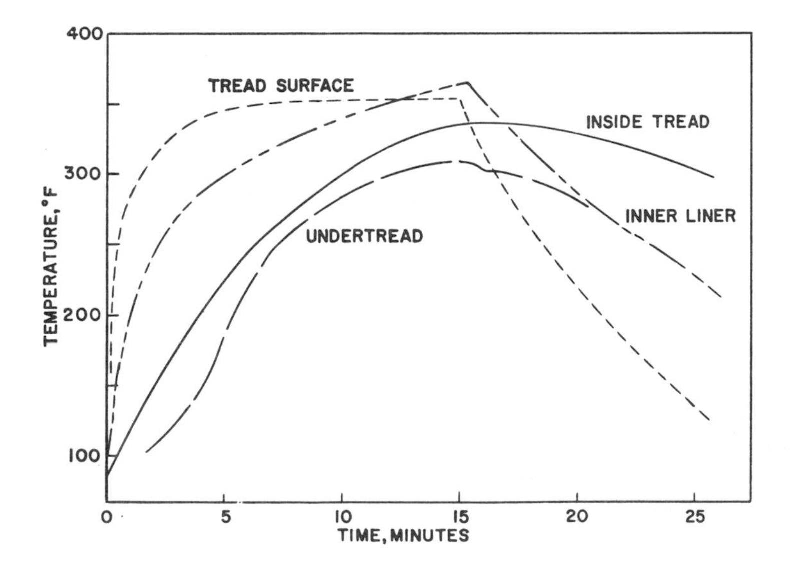

Fig. 4.21 Typical cure-time temperature records of various sections of passenger car tires.

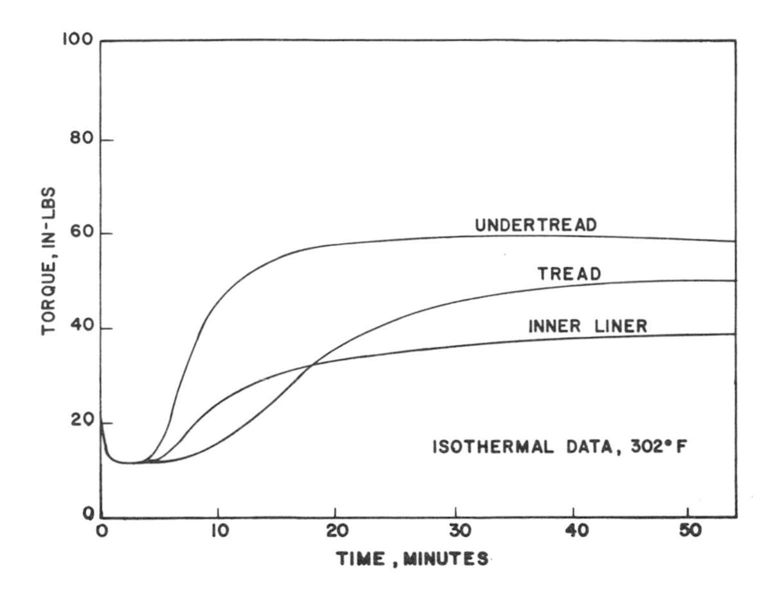

Fig. 4.22 Isothermal cure curves of typical tire compounds.

A modified version of an oscillating-disk type curemeter* has provisions for continually varying the temperature of the test speciment in accordance with any function of time versus temperature. The modification involves the use of a low mass directly-heated die system in order to allow the temperature of the die to be changed rapidly. The temperature of the die and thus the sample is varied by a curve-follower temperature controller system. The time versus temperature program is plotted on the chart of the curve follower which then tracks the curve and varies the temperature of the sample accordingly. A typical cure curve of torque versus cure time is thus generated.

Cure curves obtained at constant temperature on an oscillating disk type curemeter using compounds from each section of the tire on which the temperature was measured during vulcanization are shown in Fig. 4.22. When the same compounds were subjected to the varying time-temperature conditions which prevailed at each section, a set of markedly different cure curves were obtained as shown in Fig. 4.23. This modification of the curemeter shows great promise in following the vulcanization step under those nonisothermal conditions which are widely encountered in the vulcanization of thick articles.

Selection of the "Best Test" The selection of the test most applicable to a given problem is, at best, difficult. Ideally, the chosen test should correlate well with performance of the rubber or compound in the corresponding plant process. Obviously, there are many other considerations such as whether the test is to be used for control purposes where test speed is important or where a single test is

* Cure-Simulator, Monsanto Company.

TABLE 4.1 SELECTION OF PROCESSING AND VULCANIZATION TEST

	Processibility Test					*Vulcanization Test*		
Test Method	*Mixing*	*Extrusion*	*Calendering*	*Compression Molding*	*Injection Molding*	*Scorch*	*Rate of Cure*	*State of Cure*
Parallel plate plastomer (ASTM method D926)	poor	poor	poor	fair	poor	–	–	–
Extrusion test (ASTM method D-2230 or torque measuring extruder)	good	excellent	good	good	good	good	–	–
Mixing test (torque measuring internal mixer)	excellent	fair	good	fair	poor	good	–	–
Mooney viscometer (ASTM D-1646)	fair	fair	fair	good	poor	excellent	poor	–
Physical properties test	–	–	–	–	–	poor	fair	excellent
Curemeter (ASTM D2084)	fair	poor	poor	good	poor	good	excellent	excellent

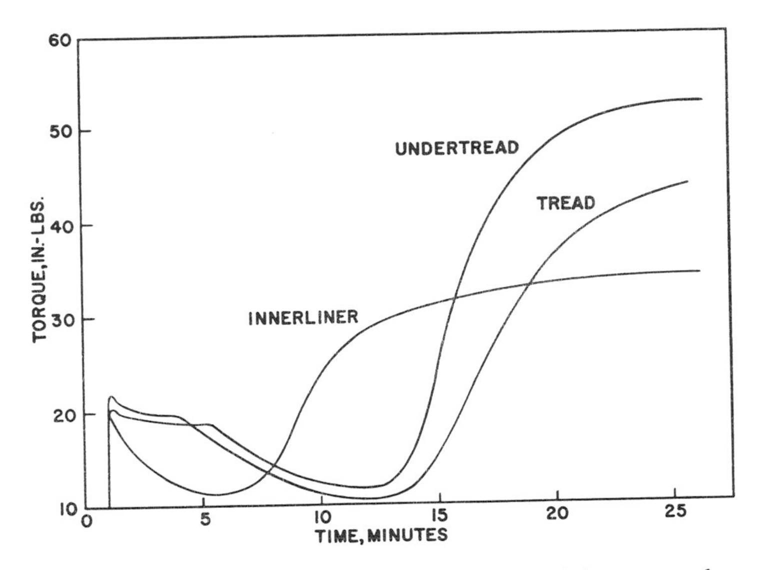

Fig. 4.23 Programmed temperature cure curves of typical tire compounds.

needed to define the behavior of the compound during several process steps. The choice of the test, therefore, is often compromise in which all of these factors must be carefully assessed. Table 4.1 is intended to provide rubber technologists with only a general guide to the applicability of each test method to the various process steps used in the conversion of raw rubber to the final product.

REFERENCES

(1) M. Morton, ed., *Introduction to Rubber Technology*, Van Nostrand Reinhold, New York, 1959.
(2) L. Bateman, ed., *The Chemistry and Physics of Rubber-Like Substances*, Maclaren & Sons, Ltd., London, 1963.
(3) J. R. Van Wazer, J. W. Lyons, K. Y. Kim, and R. E. Colwell, *Viscosity and Flow Measurement; A Laboratory Handbook of Rheology*, Interscience Publishers, New York, 1963.
(4) G. Alliger and I. J. Sjothun, *Vulcanization of Elastomers,* Van Nostrand Reinhold, New York, 1964.
(5) G. G. Winspear, ed., *The Vanderbilt Rubber Handbook*, R. T. Vanderbilt Co., New York, 1968.
(6) Am. Soc. Test. Matl., *1971 Book of ASTM Standards,* Part 28, Philadelphia, Pa., 1968.
(7) Am. Soc. Test. Mat., "Continuous Measurement of the Cure Rate of Rubber," ASTM Special Tech. Publication No. 383, Philadelphia, Pa., 1965.

5

PHYSICAL TESTING OF VULCANIZATES

F. S. Conant
The Firestone Tire and Rubber Company
Akron, Ohio

INTRODUCTION

Scope

The purpose of this chapter is to introduce the subject of physical testing of rubber from the viewpoints of its rationale, methods, trends, and limitations. As the name implies, physical testing involves measurement and evaluation of physical properties. This could include a wide diversity of characteristics but, by experience, certain categories of tests have emerged as being satisfactory adjuncts to economical development, production, and acceptance of rubber products.

Brief discussions of physical properties of rubber are included to demonstrate the necessity for certain tests, to aid in selecting the most appropriate test in a given situation, and to explain the meaning of the test results. References are included for those who wish to pursue this aspect of rubber technology.

The methods chosen for discussion are mostly those that have been standardized by The American Society for Testing and Materials (ASTM). This does not imply that these are necessarily superior to others but that they are widely used and representative. References are given to ASTM standards, where these exist, for detailed requirements on apparatus and method and to literature discussions on nonstandardized tests. Operating instructions for a particular apparatus are normally provided by the manufacturer.

Reasons for Physical Testing

As stated by Jones[1] a rubber laboratory may perform any or all of the following functions:

(1) Quality control of incoming material.
(2) Quality control of in-process material.
(3) Quality control of finished goods.

(4) Factory technical service.
(5) Customer technical service.
(6) Preparation of samples for sales department.
(7) Development compounding.
(8) Research.
(9) Advertising assistance.
(10) Product performance testing.

These are often condensed into three principal categories[2]: compliance with specifications, quality control, and research and development.

Specifications are requirements, usually physical rather than chemical, imposed on a material or an end product. They dictate the tests to be made and acceptable test results. Specifications may be made by a customer to ensure a uniform product of adequate quality or by the manufacturer to maintain processibility. Duplicate tests are sometimes made by the final inspection department of a producer and by the acceptance laboratory of a consumer, each to ensure that stated specifications have been met. Fortunately, the current trend in specification writing is to require only those properties which have a demonstrable value to further processing or to end use rather than the formerly common "general quality indices," such as tensile strength or compression set whether or not these qualities were actually necessary. Systems for classifying rubbers by "line call-out" given in ASTM D2000 and D1207 are useful in specifications for rubbers to be used in automotive applications.

Control tests are made by the manufacturer, at any stage in the fabrication process, for the purpose of maintaining processibility or quality of a finished product. Such tests do not have as much need for standardization as do specification tests which may require that two laboratories get the same results. Cooperation between laboratory and production is essential for efficient utilization of control testing results. Unless some use is made of the data, it is completely nonproductive.

In both specification and control testing the subject either passes or fails; these are often called "go-no-go" tests. In research and development tests, on the other hand, much more information is desired; single point data is no longer sufficient. Results are often obtained in a series of situations in which a component or a test condition is varied and the results are plotted to show trends. Special instruments devised for such tests often serve as the basis for later development of standardized tests.

In summary, physical testing is always done for economic reasons. Unfortunately, however, the programs and evaluations are not always carried out in the most efficient manner. Specific instances of inefficiency are pointed out in some of the following sections.

GENERAL CONSIDERATIONS

Standardization

Standardization of a test means that many of its users have agreed on all of the critical requirements and that these have been written up by a standardizing group. Furthermore, the method will have been tried in an interlaboratory program so that its repeatability within a laboratory, and reproducibility between laboratories, are known. Each developed country has a national body with overall responsibility for such standardization. In the U.S.A. this is the American National Standards Institute (ANSI), which coordinates the various groups that actually establish standards. Committee D-11 of ASTM has the principal responsibility for standards which affect producer-consumer relationships in the American rubber industry. These are published in Part 28 of the ASTM Book of Standards. Other American standards include Federal Test Method Standard 601,[3] military and other federal standards,[4] motor vehicles standards (U.S. Dept. of Transportation), and methods written into specifications such as Federal Specifications 501 and Society for Automotive Engineers (SAE) Handbook. Specifications for rubber products are written by many other groups such as Underwriters Laboratories, AMS Aeronautical Board, National Hospital Association, and International Electrotechnical Commission. The predominant international standards group is the International Organization for Standardization (ISO), of which Technical Committee 45 is concerned with rubber products.[5] Activities of this latter group have included many physical tests on rubber which are performed in a standard manner worldwide.

Standard test methods are not static documents which are perpetuated without change. The ASTM methods, in particular, are continuously updated as new instrumentation, new materials, and new knowledge become available. Much cooperative research often precedes writing of new methods or revising of older methods. A new *Book of Standards* is published every year. Reference to an ASTM method should include the date of its latest revision. For example, ASTM D1415-68, "International Hardness of Vulcanized Natural and Synthetic Rubbers," is the write-up as revised and adopted in 1968. It differs slightly from the previous version D1415-62T.

In addition to standardized test methods and specification, there exists a standard reference material program conducted by the National Bureau of Standards.[6] In 1969 the catalog listed 670 materials. A major purpose for standard materials is to calibrate test systems, including sample preparation, instrumentation, and method. Stocked materials of concern to rubber technologists include natural rubber, SBR 1500 and 1503, butyl rubber, zinc oxide, sulfur, stearic acid, accelerators, carbon blacks, and age resistors. This storage has been established to provide the rubber industry with standard materials for rubber compounding. They are useful for the testing of rubber compounding materials in connection with quality control of raw materials and the standardization of rubber testing. Each material has been statistically evaluated for uniformity by mixing rubber compounds and vulcanizing them in accordance with ASTM D3100 through D3109 and determining the stress-strain properties of the resulting vulcanizates.

Precision, Accuracy, and Validity

A test is *valid* if the results are actually a measure of the desired property. If a valid wear test for tire tread rubber existed, for example, there would be little need for expensive fleet testing for wear resistance. Even though an apparatus and method may be *valid*, however, the *accuracy* can be impaired by either instrumental or procedural defects. An oven aging test, for example, is inaccurate if the temperature measurement is incorrect. A test is *precise* if results are closely reproducible. A low temperature brittle point test, for example, may be both valid and accurate if results on many specimens are averaged, yet have a wide scatter of results, and hence be imprecise, on certain types of materials.

Problems of precision, accuracy, and validity each exist in varying degrees in all physical tests on rubber. Juve[7] has pointed out many weaknesses in rubber testing and concluded that much of it is nonproductive. Examples of nonvalid testing include unnecessary specifications by the consumer. High tensile strength may be specified routinely when, in fact, it is totally unnecessary to the proposed use. In this case it is not a valid test. Another example cited is use of the "peanut" or "angle" specimen for tear testing. Since results are unduly influenced by stock modulus, they do not truly represent tear resistance. While tests may be valid in a limited application, such as the simple hysteresis tests (rebound, Yerzley oscillograph), they may become invalid because of faulty *interpretation* of data. "The inadequacy of the simple device lies in the acceptance of the data as an unqualified measure of hysteresis, regardless of the service conditions for which the stock is intended."[7] The validity of a method for a particular application should be indicated in its scope. A further caution is that tests should be chosen which can be interpreted. Some tests measure such complex combinations of basic factors that it is nearly impossible to understand the results.

Interlaboratory tests on rubber usually give a wide scatter of data. Invariably some laboratories report results which are so far from the mean of the others that some instrumental or procedural problem is indicated. In other words, the data is *inaccurate* to a degree unsuspected by personnel of that laboratory. Surprisingly often the problem is temperature measurement or calibration of force measuring equipment. These problems can be avoided by testing standard materials periodically or by continual reciprocal testing with other laboratories.

Precision or reproducibility problems may be caused by variability in test material, improper sampling, inadequate control of test conditions, or instrument problems such as friction or electronic drift. Certainly the precision of a test should be known, either from previous records or by testing many specimens, so that confidence levels of test results can be established statistically.

Specimen Preparation

Most physical tests on vulcanizates are performed on specially prepared samples. Some variations in properties must be expected even on supposedly identical samples because of nonuniformities in raw materials and processing techniques. For this reason a *control* stock is often included in each group of *experimental* stocks. Dispersion of test results on the control must then be considered in determining a degree of confidence in evaluations of results on the experimental stock.

Some standard methods for stock mixing, sample curing, and specimen forming are given in ASTM methods D3100 through D3109. Many properties of rubber important to good physical testing practice are implicit in these directions. For example, different aging periods after mixing, after remilling, and after curing are required for compounds based on different polymers. In general, no tests should be made until at least 16 hours after vulcanizing a sample because significant post-vulcanization changes in structure of the material occur during this period. Of course, in control tests such delays are intolerable, so the change in properties between actual testing time and 16 hours after curing should be ascertained and taken into consideration.

Some test results are influenced by flow of rubber in the mold, so a good general practice is to use no specimens cut from material near a mold edge. Also the *grain* in rubber caused by the action of the mill in the final pass may affect certain properties. Tensile tests, for example, are always made along the grain.

Specimens are sometimes required from a finished product rather than from laboratory prepared samples. In such cases they are often buffed to obtain uniform thickness, to remove fabric or fabric impressions, or to remove glaze or skin coats from the rubber. Abrasive grinding wheels of about 30 grit, 5-inch diameter, running at 2500 to 3500 rpm are satisfactory for this purpose. The grinder should have a slow feed so that very little material is removed at each cut, to avoid both high heat and coarse abrasive pattern at the surface.

Standard Temperature

ASTM D1349 gives a list of standard test temperatures which agrees with the ISO list. An advantage of using only the standard temperatures is that many unnecessary temperature changes in environmental chambers are eliminated. A standard humidity is also provided although it is doubtful that humidity has much effect on physical properties of rubber. Properties of fabrics or rubber-fabric composites, however, are very sensitive to moisture.

Break-in

For some tests a rubber specimen should be *broken-in* before recording any results. This means that it should be deformed several times to the same extent that it is in the recorded test. This usually gives a softer material, but one on which results are reproducible. Upon resting the specimen properties gradually return toward the initial values. The rebound test is an example of one which requires a break-in.

Fundamental Constants of Rubbers

Many physical properties of rubber are of the *basic* type, that is they are independent of the measuring instrument, e.g., thermal conductivity, specific heat, thermal expansion, density, refractive index, etc. Typical values for such properties have been collected by Wood.[8]

STRESS-STRAIN TESTS

Stress-Strain Terminology

Physical testing of rubber often involves application of a force to a specimen and measurement of the resultant deformation or, conversely, application of a deformation and measurement of the required force. Two common modes of deformation, tensile and shear, are illustrated in Fig. 5.1. Since we are interested in material property, test results should be expressed in a manner than is independent of specimen geometry. For this purpose the concepts of *stress* and *strain* are used. Stress is the force per unit cross sectional area, i.e., F/A for either tensile or shear deformations. Strain is the deformation per unit original length ($\Delta L/L_0$) in tensile tests or deformation per unit distance between the contacting surfaces (S/D) in shear tests.

Stress is usually expressed in units of pounds per square inch (psi). In the SI system of international units the newton per square meter (N/m^2) is used. Strain is

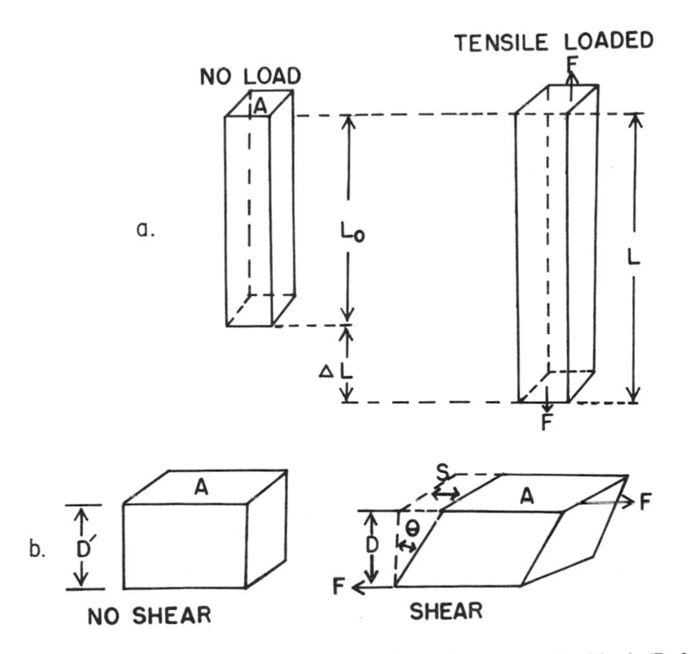

Fig. 5.1 a. Tensile stretching of a bar; b. Shear of a rectangular block (Ref. 23).

usually expressed in per cent. Since it is the ratio of two lengths, it is dimensionless. An elongation of 300%, for example, means that the specimen has been stretched to four times its original length.

In the common parlance of rubber technology the stress required for a given elongation is used to represent the material stiffness. This quantity is called the modulus. A 300% modulus, for example, means the stress required to produce a 300% elongation. In mechanical engineering usage, however, the term *modulus* is defined as the ratio of stress to strain. If this ratio is a constant the material is said to obey Hooke's law and the constant is called *Young's modulus.* In practice, the term *Young's modulus* is often used to represent the ratio of stress to strain even in situations where it may vary with change in elongation. In the terms used in Fig. 5.1A, Young's modulus E is given by:

$$E = \frac{F/A}{\Delta L/L_0}$$

The same equation applies when the bar is decreased in length by a compressive force. Compressive tests, at low deflection, give moduli equal to or slightly greater than do tensile tests at comparable deflection.

Shearing of a rectangular block is illustrated in Fig. 5.1B. The rigidity or shear modulus G is defined as the ratio of shearing stress to shearing strain:

$$G = \frac{F/A}{S/D}$$

Shear is also involved in torsional deformation of beams but calculation of modulus from torsional tests involves a shape factor which, for rectangular beams, depends upon the ratio of thickness to width.[9]

A third type of modulus is the *bulk modulus B*. It is defined as the ratio of the hydrostatic pressure to the volume strain:

$$B = \frac{\text{hydrostatic pressure}}{\text{volume change per unit volume}}$$

Rubber is highly incompressible, so its bulk modulus is much higher than its Young's modulus.

When a material is stretched, its cross-sectional area changes as well as its length. *Poisson's ratio, ν*, is the constant relating these changes in dimension:

$$\nu = \frac{\text{change in width per unit of width}}{\text{change in length per unit of length}}$$

If the volume of a material remains constant when it is stretched, Poisson's ratio is 0.50. This value is approached for rubbers and liquids.

The indices discussed are interrelated by the following equations, which are strictly valid only in the small strain regime:

$$E = 2G(1 + \nu) = 3B(1 - 2\nu)$$

Young's modulus is seen to be three times the shear modulus if Poisson's ratio is 0.50.

Tensile Tests

The stress-strain test in tension, including ultimate tensile and elongation, is probably still the most widely used test in the rubber industry. Among the purposes for such tests are: to ensure that all compounding ingredients have been added in the proper proportions, to determine rate of cure and optimum cure for experimental polymers and compounds, for specification purposes, and to obtain an over-all quality check on the compound. Proportionately fewer tensile tests are run now than in previous years, for two principal reasons: the cure meters have been given a large share of the tests for state of cure, and more emphasis is being given to testing for the properties desired in a particular stock than to tests for general quality. High tensile strength is seldom required in service and, by itself, does not guarantee the level of any other property. However, since a single test can yield modulus at specified elongations, ultimate elongation, and ultimate tensile strength in well standardized tests and with short testing time, tensile tests are far from outdated.

The standard method for tension testing of vulcanized rubber is given in ASTM D412. Three shapes of test specimens are permitted, each usually stamped out of a flat sheet by impact with a metal die of specified contour and dimensions. The straight-sided rectangular specimen is least preferred for ultimate strength and elongation tests because of its tendency to break in the clamps. It may be required, however, because of the size or shape of the sample, e.g., from tubing or electrical insulation. Also the clamp separation can be used as a fair measure of the elongation.

In the U.S. the most common tensile specimen is the dumbbell, so-named for its shape, which has tabbed ends for gripping in the test machine and tapering to a central constricted section of uniform width. *Bench marks*, often one inch apart, may be stamped on the constricted section to facilitate manual following of the elongation during test. Six different shapes of dies for *clicking out* dumbbell specimens are currently permitted by ASTM D412. Advantages of the dumbbell specimen include its easy preparation, breakage in a predetermined area (usually), means for following the elongation, and the immense accumulated background and specifications for such data. Many styles of grips are available so that specimens of widely varying properties may be held properly. These include air clamps which maintain constant clamping pressure.

The circular ring specimen is also normally died out from a flat sheet. For testing, the ring is held by rollers in the machine so that the stress is equalized.

Advantages of this specimen include the possibility of using clamp separation as a good measure of elongation and distribution of stress over a greater length than in the dumbbell so as to include more possible defects. This method for measuring elongation becomes important when autographic recorders are to be used and in tests at high speed or at other-than-normal room temperature or atmosphere. A disadvantage, especially for low elongation tests, is that the inside of the ring is stretched more than the outside.

Tensile stress is calculated as the ratio of observed force to the cross-sectional area of the unstretched specimen. Elongation for straight and dumbbell specimens is given by

$$\text{Elongation, percent} = \frac{L - L_0}{L_0} \times 100$$

where L = observed distance between bench marks on the stretched specimen and L_0 = original distance between bench marks.

In order that stress can be read directly in pounds per square inch, many tensile machines have provision for adjustments to compensate for the varying thickness (gage) of different specimens. In the pendulum type machines this is done mechanically by positioning a weight along the calibrated pendulum arm. The load cell instruments have electronic compensators.

Tensile stress at rupture is usually higher for a specimen having a small cross-sectional area than for one having a larger cross-sectional area and higher for a specimen having a short test section than for one having a long section.[10] The reason is the same for both cases. Rubber in tension normally fails at a flaw, which may be caused by the die used to cut the specimen, by a region of poor dispersion, by porosity, by inclusion of foreign matter, or by any of a number of other reasons. Obviously, an increase in the amount of rubber being strained increases the chance of a flaw being in the strained area. For this reason, comparison of results from different types of specimens should be made with caution.

The standard speed for tension testing of rubber is at a machine jaw separation rate of 20 inches per minute. A standard speed is necessary because all tensile properties vary with change in elongation rate. Of course, if a vulcanizate is to be used at a high deformation rate, it should be tested at a high rate. As the rate is increased the modulus increases, the ultimate elongation decreases, and the tensile strength may either increase or decrease. The rate of change of each of these properties with change in rate of elongation depends upon both the test temperature and the glass transition temperature (T_g) of the material; the effect becomes less as the test temperature is moved farther above T_g. As discussed by Juve[11] the effect of rate of elongation is related to stress relaxation in the specimen. At low rates there is time for more stress to relax during the elongation than is the case at high rates. As the glass transition temperature is approached, stress relaxation becomes progressively slower, so the test piece registers higher stress (at a given rate of elongation) than it would at higher temperatures. This

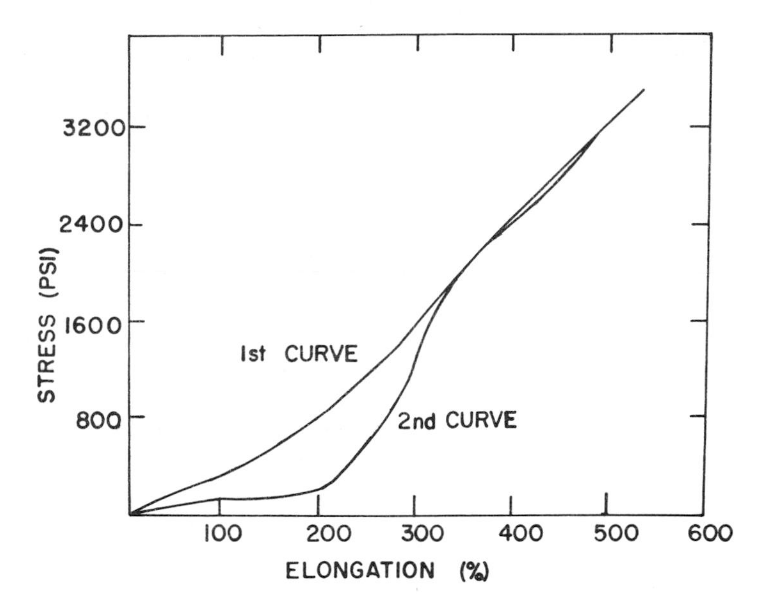

Fig. 5.2 Stress-strain curves showing effect of prestretching (Ref. 25).

point is discussed in more detail in the section entitled "Glass Transition Temperature."

An illustrative stress-strain diagram for rubber is given in the 1st curve of Fig. 5.2.[11] The small hump in the curve near the origin is characteristic of most rubber compounds. The slope at the origin, if accurately measured, is directly related to the hardness of the material. Stress, in this case, is load per unit original cross-sectional area. If the attained cross-sectional area were used, the indicated stress would be much higher. Rubber does not have a yield point in the conventional sense as associated with metals, so that its stress continues to increase as the strain increases until rupture. If the material crystallizes under strain, as exemplified by gum natural rubber, the slope of the stress-strain curve becomes very high near the breaking point.

The slope of a curve, such as that in Fig. 5.2, is a modulus in the engineering sense, since it is the ratio of stress to strain. Because the slope is not constant, however, two kinds of moduli should be distinguished. The slope of a straight line from the origin to a point on the curve at a particular elongation may be called a *secant modulus* at that elongation, whereas the slope of a straight line tangent to the curve at any point may be called a *tangent modulus* at that point.

Although tensile data, like all physical testing data, are usually shown as plotted through exact points, they should, in reality, be plotted as bands or belts to allow for error and statistical variation. This implies that in obtaining these values we should always use random and multiple sampling. Distribution of tensile data is nearly normal, only slightly skewed in the direction of low values. This means that a standard deviation is a good measure of variation. Since it is impractical, however, to test enough specimens of each sample to obtain a reliable standard deviation, recourse may be had to average standard deviations for the type of material and type of test being used. The ASTM method specifies that the middle value obtained from tests on three specimens should be used. This eliminates any effect of abnormally low values associated with defects but gives no information on precision of test or reproducibility of samples. The compounder should have some knowledge of these factors in order to assess a degree of confidence in his results.

The stress-strain curve is almost always obtained on a specimen which has not been stretched previously. This is a unique curve which cannot be repeated on subsequent stretchings because of material changes which are only partially reversible. These changes are more evident in compounds containing a reinforcing filler than in gum compounds. Figure 5.2 shows an original stress-strain curve and a second curve obtained by prestretching a duplicate specimen three times to 300%. Up to 300% elongation the second curve shows lower stress but beyond that elongation it almost coincides with the first curve. Neither the ultimate tensile strength nor elongation was much affected. This behavior has been called the *Mullins effect*, after an early investigator, or *stress softening*. Many of the bonds which are broken in the first stretch will reform if the specimen is rested, more rapidly if it is heated.

EVALUATION OF RATE AND STATE OF CURE

Tensile Methods

After an *induction time* at curing temperatures a vulcanizable compound starts to stiffen because of crosslinking. The crosslinking proceeds at a rate determined by the stock composition and the temperature. Every physical property of the stock is affected as the cure progresses, but at different rates, as is illustrated by the tensile properties in Fig. 5.3. The average *rate* of cure might be considered as the inverse of the time to reach a full cure after the induction period. A fast cure is desirable but this must be balanced against safety from *scorching* (premature stiffening from accumulation of heat history).

The *optimum cure* for a rubber product is the one which optimizes the properties desired in that product. This has traditionally been judged by use of tensile strength, modulus, and ultimate elongation data to estimate a best *technical* cure. Standard methods for performing these tests are given in ASTM D412, but interpretation of the results to establish optimum cure is not so well standardized.[1,2]

In a series of cures the time which gives the maximum tensile strength has often been taken as optimum. On this basis the best cure for the SBR compound in Fig. 5.3 might be chosen at about 65 minutes. Many prefer to select a somewhat lower time since product aging in effect causes an increase in cure, so about 45 minutes might be chosen.

The time at which a modulus curve bends sharply away from the modulus axis is often selected as optimum, especially for compounds such as those based on SBR in

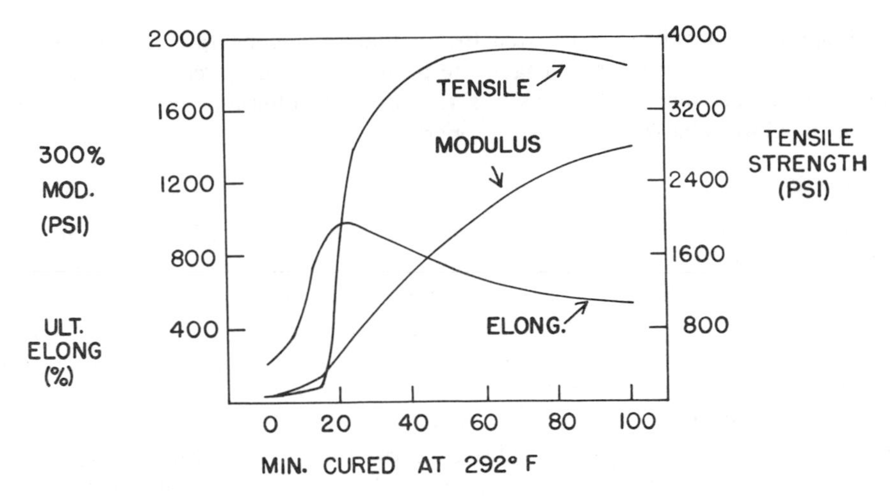

Fig. 5.3 Effect of state of cure on tensile properties of a butadiene/styrene compound tested at 77°F (Ref. 26).

which the modulus does not reach a maximum but continues to increase as the cure increases. For the case shown in Fig. 5.3 a cure time of about 70 minutes might be selected from the modulus curve. Another method[13] is to draw a straight line from the origin to a point on the modulus curve at a long cure time, such as 100 minutes in the figure. A second line is then drawn parallel to the first and tangent to the modulus curve. The time indicated at the point of tangency, again about 70 minutes in the example, is taken as optimum. As in most tensile testing, first-stretch modulus is normally used. In reinforced compounds this contains contributions from filler-polymer interactions – a fugitive structure which breaks down on repeated stretching (Mullins effect). Perhaps, then, modulus measurements for determining state of cure should be made on specimens which have been stretched repeatedly beyond the elongation used in the modulus test.

When the modulus shows either a peak or a plateau at its maximum value, the cure time required to reach 90% of that value is often taken as optimum. If the modulus continues to increase over the normal range of cure times, however, the 90% index cannot be read directly. In such a case it has been suggested that another series of cures be made at a temperature high enough to get a modulus curve showing a peak or a plateau. The time for 90% of full modulus on this curve can be translated to time at the lower temperature by the approximate rule that curing time is doubled by a 10°C (18°F) temperature decrease.

Ultimate elongation is usually considered along with tensile strength and modulus in selecting an optimum cure time. For most compounds the breaking elongation reaches a well-defined maximum at quite low curing times, about 25 minutes in the illustration. If it drops off sharply beyond the peak, this may be a strong factor in selecting an optimum cure since a vulcanizate which breaks at low elongation is usually undesirable.

Properties other than tensile strength, modulus, and elongation can also be used to determine optimum cure. In fact, the property of greatest consequence in the proposed application should be used, since different optimum cures will generally be obtained when based on different properties.

Cure Meters

Instead of curing each test compound at separate ranges of temperatures and making separate tensile tests, many laboratories use a cure meter, of which several types have been developed.[14] In these instruments modulus change is monitored during the cure. Many properties are obtainable from the data; the most common being minimum and maximum stiffness, scorch time, cure time to 90% or 95% of maximum stiffness, and a cure rate index.

The most widely used cure meter is the Rheometer (ASTM D2705), in which the specimen is contained in a sealed test cavity under a positive pressure and maintained at an elevated temperature. A biconical disk, embedded in the specimen, is oscillated through a small arc. The autographically recorded force is proportional to the shear modulus of the rubber. An envelope of a typical cure curve is given in Fig. 5.4. The minimum torque (M_L), maximum torque (M_{HF}), scorch time (t_{s2}) and time to 90% cure ($t_c(90)$) are indicated. Also obtainable[15] are initial viscosity at zero time, minimum viscosity, and reversion (time after M_{HF} to reach 98% of M_{HF}).

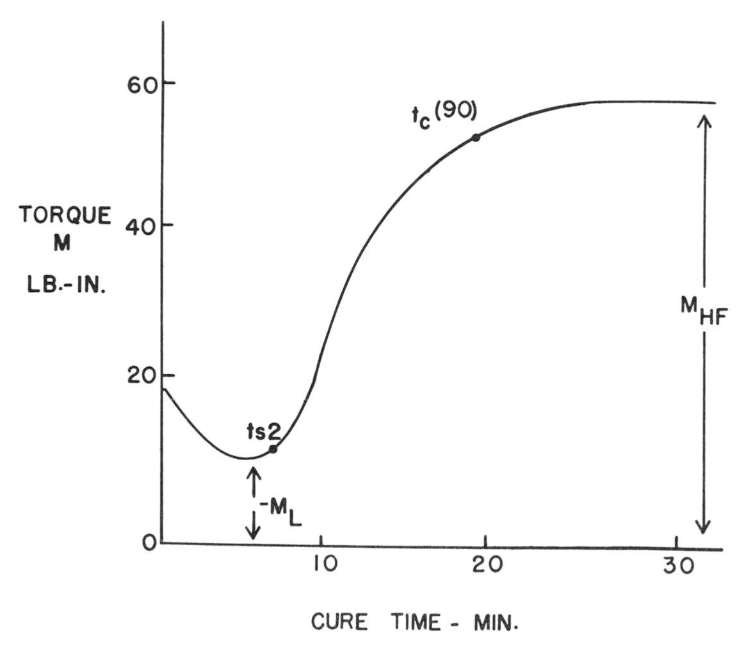

Fig. 5.4 Envelope of rheometer cure curve which attains an equilibrium torque (ASTM D2705).

Many illustrations of rheometer use have been described[16,17] as well as of the viscurometer,[18] which works on a similar principle. The biconical rotor used in each instrument produces a shear rate which is uniform over the entire rotor surface. In instruments such as the Mooney viscometer, which use a flat disk, the shearing rate is near zero at the central shaft and a maximum at the disk periphery.

Other Methods

Any property of a rubber which changes in a regular manner with time or temperature of cure may be used to evaluate the state of cure. It should be emphasized, however, that methods based on different properties will not in general give results which agree with each other. For example, a sample which is overcured for best tear resistance might be grossly undercured for optimum hysteretic properties. Indices which have been used[12] include free and combined sulfur, tensile product (maximum tensile strength multiplied by the ultimate elongation), permanent set, equilibrium modulus, swelling in solvents, crosslink density, and hardness.

HARDNESS

Hardness, as applied to rubber, may be defined as the resistance to indentation under conditions which do not puncture the rubber. Different instruments designed to measure hardness do not usually agree well with each other for any of several reasons: definition of scale end points, shape and size of indentor point, total load applied, rate and time of load application. Hardness is a property of rubber which must be expressed in terms of instrument parameters rather than in basic units. A high modulus rubber is, of course, also hard, but the relationship is not easy to quantify.

The spring-loaded pocket durometer is the most common instrument for measuring hardness of elastomers. The Shore durometer (ASTM D2240) in particular is known and used world-wide. In this instrument the scale runs from zero hardness for a liquid to 100 for a hard plane surface such as glass. The Type A durometer is used for soft stocks, up to a reading of 90. Above 90 the Type D durometer, having a different indentor shape and different stiffness spring, is used. One reason for the popularity of this instrument is its portability for field use and adaptability to fairly irregular surfaces. Difficulties arise, however, in reproducibility of results by different operators – differences of five units and practical tolerances and ten to twelve points are not rare.

Better reproducibility is obtained by dead weight loading, as in ASTM D314 or D531, than in the spring loaded, hand-held durometer. These particular instruments, however, are not as widely used as is D1415, in which results are expressed as International Rubber Hardness Degrees. These units are directly related to Young's modulus while still being approximately equivalent to Shore durometer readings. The test consists of measuring the difference between the depths of penetration of a ball point into the rubber specimen under a small initial load and a large final load. This is essentially the same as the ISO R48 method.

The initial reading of a durometer depends on the rate of loading and the final reading on the duration of loading, both because of creep in the rubber. Some portable instruments, such as the Rex or some Shore models, register the maximum reading. Most bench models, however, are designed to be read after some loading period such as 30 seconds. Since different materials have different creep characteristics it is very important to precision that the rate of loading be standardized and the duration of loading be specified.

Since so many hardness meters exist, the choice of which to use is difficult. The Shore is undoubtedly the most widely used and understood but is not as precise as bench models. For referee purposes the ASTM committee on physical testing of rubber recommends D1415 for ordinary soft rubber compounds and D530 for hard rubber products.

In ordinary SBR rubber compounds the hardness increases with increased cure. In natural rubber compounds the hardness increases to a maximum and then decreases because of reversion as the cure time is increased.

TRANSITIONS IN ELASTOMERS

Types of Transitions

Two types of reversible, temperature-dependent transitions are important concepts in rubber technology: first order and second order. In a first order transition there is an abrupt change in level of a property, such as volume, specific heat, or modulus, within a small temperature range. In a second order transition there is no abrupt change of level, but rather a change in slope of the line representing that property plotted against temperature. For example, there is no change in volume at a second order transition temperature, but a definite change in coefficient of thermal expansion. First order changes are usually attributed to crystallization and second order changes to vitrification, i.e., to becoming glass-like. More than one second-order transition temperature may exist for a given polymer, however, only one of which is truly a glass transition (T_g). The main difference between a rubber and a rigid plastic is that rubber has a glass transition below room temperature while a plastic has a glass transition above room temperature.

As is detailed in ASTM D832, crystallization, vitrification, and simple temperature effects have many distinguishing characteristics, each of which could lead to a definitive test. Some of these tests are described briefly in the following sections. First, however, let us consider some of the principles involved.

Effect of Rate of Deformation

According to the kinetic-molecular theory of rubber elasticity, the modulus of a rubber increases as its temperature is increased. This presumes, however, an equilibrium deformation, which is probably unattainable; some creep inevitably persists. Modulus tests with a finite deformation usually, but not always, show a decrease in modulus with increase in temperature. This occurs because the stress relaxation is a time-dependent process and more time is required for a given amount of stress relaxation as the temperature is lowered toward the glass transition temperature. An effect of this principle is that if a deformation is forced at a greater rate than can be accommodated by the elasticity of the specimen, it breaks. A low temperature brittle point is, therefore, very sensitive to rate of deformation. Another consequence is the difficulty of getting the same results from different types of low temperature testing equipment, which usually have different time rates of sample deformation. An illustration of this difficulty was demonstrated in a comprehensive study of stiffness testing at low temperatures,[19] which included the following reasons for variation in results from different instruments and different laboratories in an interlaboratory program:

(1) Rate of testing.
(2) Geometry of the apparatus.
(3) Nonlinear characteristics of the elastomers.
(4) Variance in the way the materials reacted to temperature.
(5) Variations in testing conditions.
(6) Variations in methods.

Differential Thermal Analysis (DTA)

Differential thermal analysis is a technique for studying the thermal behavior of materials as they undergo physical and chemical changes during heating and cooling. The name is derived from the differential thermocouple arrangement, consisting of two thermocouples wired in opposition as shown in Fig. 5.5.[20] Thermocouple A is placed in a sample of the material to be analyzed. Thermocouple B is placed in an inert reference material, which has been selected so that it will undergo no thermal transformations over the temperature range being studied. When the temperature of the sample equals the temperature of the reference material, the two thermocouples produce identical voltages and the net voltage output is zero. When sample and reference temperatures differ, the resultant net voltage differential reflects the difference in temperature between sample and reference at any point in time.

Physical properties of elastomers which may be studied by DTA include first order transitions (crystallization) and second order transitions (e.g., glass transition). In either case the sample and a material of comparable heat capacity and

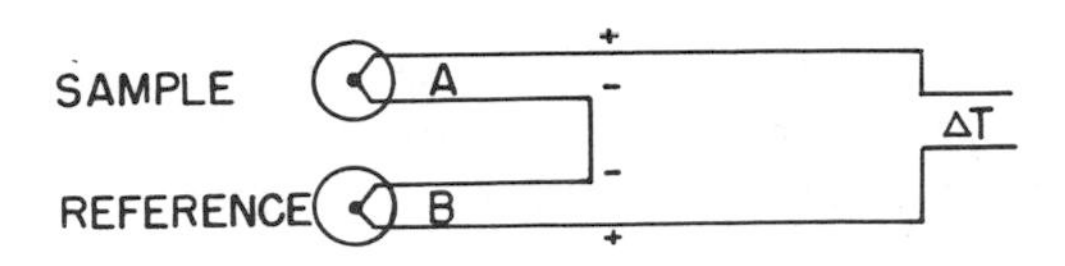

Fig. 5.5 Differential thermocouple (Ref. 34).

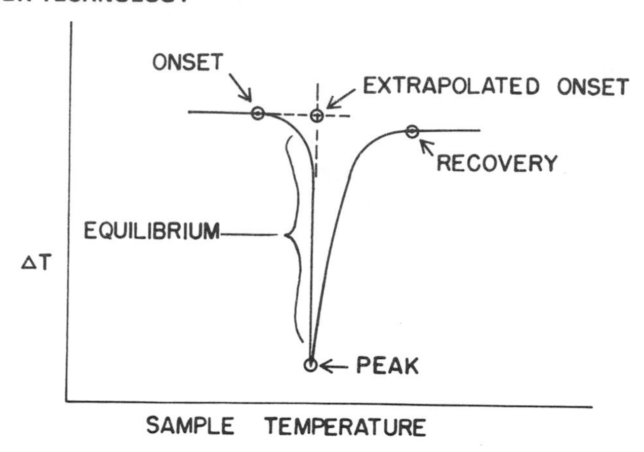

Fig. 5.6 Thermogram for melting of crystals in an elastomer (Ref. 34).

thermal conductivity, each containing a thermocouple, are cooled rapidly in the DTA chamber to below the suspected transition temperature. The chamber temperature is then raised at a given rate and a plot obtained of ΔT against sample temperature. At transition points the sample will interchange heat with the chamber without a change in its own temperature until the transition is complete, giving a plot of the nature shown in Fig. 5.6 to illustrate crystal melting. In crystal formation the peak would be above the base line while at a second order transition point there would normally be a change in base line to a lower level at the higher temperature.

In addition to its use in locating transition temperatures, DTA has many applications to rubber testing which might be considered chemical rather than physical. These include identification, composition, solvent retention, thermal stability, oxidative stability, polymerization, curing, and thermochemical constants.

Tests for Crystallization

The DTA test discussed in the previous section is usually satisfactory for measuring the temperature at which a polymer crystallizes most rapidly. Sometimes, however, crystallization rates are too slow to fit well into the normal time scale of the test as, for example, unstretched natural rubber. In such cases the specimen may be held for long times in the temperature range of interest with periodic measurements of some property such as modulus. ASTM D797 is an example of such test – one in which simple beam stiffness in bending is the measured property. The Young's modulus increases, of course, with increase in degree of crystallinity.

Crystallization rate of an elastomer is increased by stretching – a fact which is used in the T-R (temperature-retraction) test for crystallization (ASTM D1329). In this test the specimen is first elongated to 250–350%, and then frozen to a practically nonelastic state and released. As the temperature is raised at a uniform rate the recovery attained is measured at regular temperature intervals. The difference between the temperature of 10% retraction (TR10) and that of 70% retraction (TR70) increases as the tendency to crystallize increases.

The DTA method for evaluating crystallinity is based on a thermal property: heat of crystallization. The ASTM methods are based on a mechanical property: modulus. An optical property, the polarizing effect of crystals, is used in the photomicroscope method. A density change is the basis for dilatometer methods and X-ray reflection properties for X-ray methods; diffuse crystals give rings in the diffraction patterns and aligned crystallites give spots. This roster of tests is a good illustration of the principle that a measurement of any material property which changes in a regular manner with change in test conditions may serve as the basis for a test.

Low Temperature Stiffness

Rubber stiffness as measured by nonequilibrium deformation increases with a decrease in temperature, but the relationship between modulus and temperature is by no means linear. It is rather characterized by plateaus such as shown in Fig. 5.7. Outside the "rubbery" plateau a rubber cannot serve its intended purpose. Various tests have been devised to measure the temperature at which an elastomer becomes inserviceable. For some uses a "leathery" response is still satisfactory but for other uses a lower limiting modulus must be specified. One proposed value[21] is a Young's modulus of 10,000 psi as measured with a ten-second loading time. Since a value in this region occurs at about 30°C above the glass transition temperature, T_g is a possible serviceability index. More direct measurements are usually chosen, however.

Perhaps the most widely used low temperature test for elastomers is that based on the Gehman torsional wire apparatus, ASTM D1053. A schematic diagram of the working parts of the instrument is shown in Fig. 5.8. Multiple specimens F on the rack D can be rotated to clamp individually to the base of the calibrated torsion

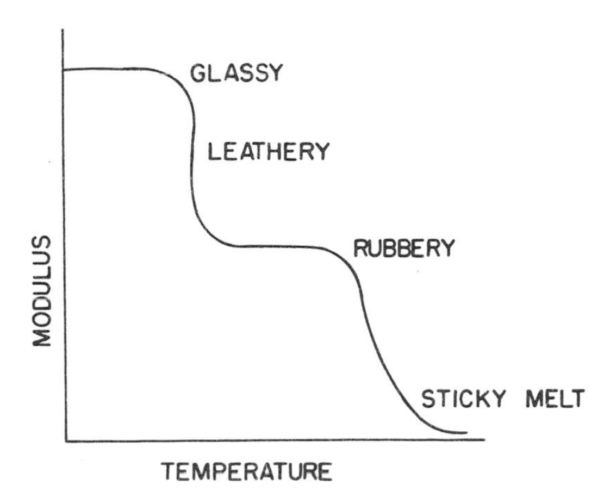

Fig. 5.7 Characteristic temperature-modulus curve for polymers (Ref. 33).

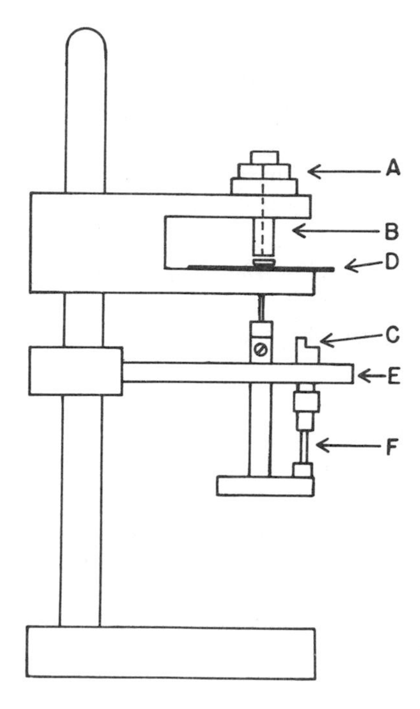

Fig. 5.8 Gehman low temperature apparatus (ASTM D1053): A-Torsion head, B-Torsion wire, C-Clamp stud, D-Movable protractor, E-Specimen rack, F-Specimen.

wire B. A 180° rotation of the torsion head A rotates the top of the specimen by an amount measured on the protractor. Since the torque is known, the apparent modulus of rigidity is calculable. If tests are made at a series of temperatures, plots of relative modulus vs. temperature are readily obtained. Commonly used indices are temperatures at which the modulus reaches 2, 5, 10 and 100 times its value at room temperature. The apparatus is also usable for tests of continued stiffening at a given temperature, e.g., crystallization or plasticizer incompatibility.

Another ASTM test for low temperature stiffening is D797, Young's modulus in flexure. The specimen is supported as a simple beam and a known load applied centrally. Young's modulus is calculable from the specimen and support geometry, the added load, and the resultant deflection. Data are normally plotted as modulus vs. temperature. Indices such as the temperature at which the modulus reaches 10,000 psi can be read from the graph. This apparatus is also suitable for crystallization or low temperature creep studies.

Other ASTM tests for low temperature stiffening include the Temperature-Retraction (TR) test, D1329 and a bend test for rubber-coated fabrics, D2136. Many other tests have also been used commercially, ten of which were included in the interlaboratory study of reference 19. No "best" test resulted from this study. Poor agreement among results from different methods forced the conclusion that the term "modulus" as applied to rubberlike materials is rather vague unless the method of obtaining that modulus is also given.

Low Temperature Brittleness

When rubber is forced to deform faster than its relaxation time permits, it breaks. At room temperature extremely high speed would be required to produce a brittle fracture but as the temperature is lowered a point is reached at which a specimen breaks under a given deformation at a given speed. In ASTM D746 specimens 0.075 in. thick are bent sharply by impact with a striker arm moving at six to seven feet per second. When using multiple specimens care must be taken that the breaking energy is not high enough to slow the striker arm below six feet per second *after* the impact. Otherwise too low a brittle temperature would be indicated.

As with any low temperature test, either a gaseous or liquid heat transfer medium may be used if they have been shown to give equivalent results on specimens of a material having similar composition to that of the test material. This means that the liquid medium should not corrode or swell the rubber and that the rubber should not crystallize in the range of test temperatures. The longer times required for thermal equilibrium in the gaseous medium would promote more crystallization than would the liquid medium.

Low temperature stiffening and low temperature embrittlement do not correlate well enough with each other to permit either index to be inferred from the other. The choice of test to be run must be made on the basis of anticipated service conditions.

Brittle fracture is visually distinguishable from tensile failure in that the failure surfaces have a glassy rather than a ragged appearance.

Other Low Temperature Tests

The principal "rubbery" characteristics which an elastomer must maintain to remain serviceable at low temperatures are a low modulus and freedom from embrittlement. Another property which sometimes becomes important, however, is that of recovering original dimensions after removal of a deflecting force at low temperatures. This requires a test such as ASTM D1229 "Low Temperature Compression Set of Vulcanized Elastomers." In this test the specimen, in the form of a cylinder, is compressed 25% at room temperature, cooled to the test temperature, and kept there for 22 or 94 hr. It is then released and the thickness measured after 10 sec and 30 min recovery at the test temperature.

Compression set measurement is often used to evaluate seal retention at low temperature, although a more direct test would be a stress relaxation measurement. This is analogous to the set test except that the compression is maintained and decay of restoring force with time is measured. Crystallization is especially dangerous for seal retention but low temperature stiffening and second order transition are also important.

Tension set measurement at low temperature is not as common as compression set because rubber is seldom used in tension, but the TR test (ASTM D1329) could be used for this purpose. The room temperature tension set test of ASTM D412 could also be adapted to low temperature use.

Most static rubber properties measurable at elevated or room temperatures could also be made at low temperatures with proper regard for the time scale of deformation. In a hardness test, for example, the loading time would be much more critical than it would be at higher temperatures. Dynamic properties are hard to measure at low temperatures because of the complications introduced by heat generation. A necessary decision is whether to measure the property during the first few oscillations before appreciable heating has occurred or after thermal equilibrium has been reached.

DYNAMIC MECHANICAL TESTS

In most of its uses rubber is subjected to relatively large deformations during which it absorbs mechanical energy, transforming it into heat energy. The deforming force is thus resisted by both *elastic* and *damping* forces. The principal purpose of dynamic mechanical tests is to evaluate these forces. A bewildering variety of tests has been developed for this purpose, none of which can be regarded as universally used. The examples cited here are some for which operating procedures have been standardized, but this does not necessarily imply widespread usage. Extensive reviews of this subject have been written[22,23,24,25] which include much theoretical basis for the types of tests that have been developed.

Terminology for dynamic properties testing has been fairly well standarized. The definitions given in ASTM D2231, for example, are in essential agreement with those adopted by ISO TC/45. Four more general terms have also been defined.[25]

Resilience. In a rubber-like body subjected to and relieved of stress, resilience is the ratio of energy given up on recovery from deformation to the energy required to produce the deformation. Resilience for these materials is usually expressed in percent.

Hysteresis is the percent energy lost per cycle, or 100 minus the resilience percentage.

Dynamic modulus is the ratio of stress to strain under vibratory conditions. It is usually expressed in pounds per square inch (psi) for unit strain.

Damping refers to the progressive reduction of vibrational amplitude in a free vibration system. Damping is a result of hysteresis, and the two terms are frequently used interchangeably.

Measured values of dynamic properties of rubber may be influenced by the particular apparatus and method used, so it is not sufficient to specify a certain dynamic modulus, for example, unless the details of its measurement are also specified. The types of tests used can be classified generally as: impact rebound, forced vibration in both resonance and nonresonance conditions, and free vibration.

Rebound Tests

The simplest test for resilience is the falling ball rebound.[26] If the drop height of the steel ball is divided into 100 equal parts, the rebound height is equal to the resilience. The same principle is used in the Bashore Resiliometer (ASTM D2632) – a compact instrument in which the plunger follows a guide rod in its fall and rebound.

In a pendulum rebound test the specimen, held at the rest position of the pendulum (zero degrees), is impacted by the center of percussion of the arm. The angle of rebound is followed on a scale and resilience calculated by the formula:

$$R = \frac{1 - \cos(\text{angle of rebound})}{1 - \cos(\text{angle of fall})} \times 100$$

Many designs for the pendulum rebound apparatus have been used. In the Goodyear-Healy design (ASTM D1054) provision is also made for measuring the depth of penetration of the pendulum head.

European laboratories prefer the Lupke pendulum, adopted by ISO TC/45, or the Schob pendulum. The Lupke design is interesting in that the impacting weight is held by four flexible suspension lines (Fig. 5.9). This minimizes absorption of energy in the instrument. As with any rebound test, however, care must be taken that the specimen support is very rigid.

Interlaboratory tests have shown that high correlation should not be expected between results from two different types of rebound testers. Results are influenced by plunger weight, design of the impacting head, drop height, penetration, and energy absorbed in the apparatus. In any rebound test the specimen should be preconditioned by about six impacts before a reading is taken. After this the rebound height is usually stabilized.

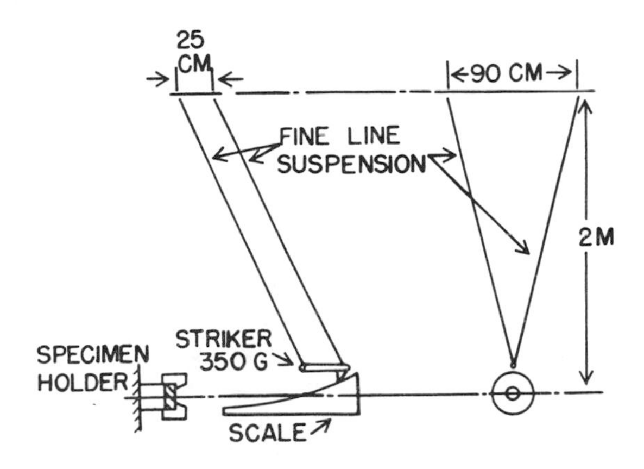

Fig. 5.9 Lüpke pendulum.

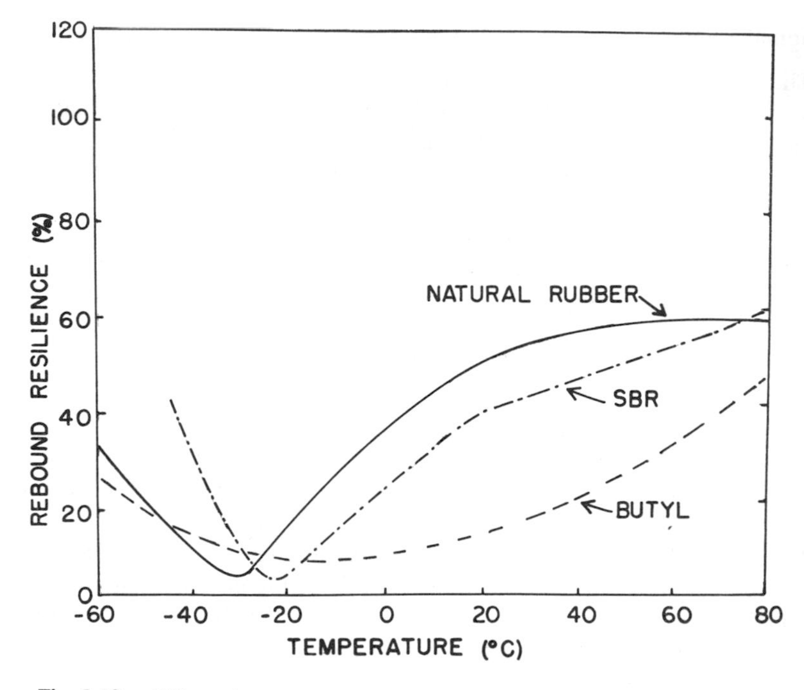

Fig. 5.10 Effect of temperature on rebound of various rubbers (Ref. 36).

Rebound tests are very sensitive to bulk temperature of the specimen but, fortunately, not to surface temperature. This allows tests to be made in the open laboratory of specimens conditioned at either low or elevated temperatures, with very little error. As shown in Fig. 5.10, as the specimen temperature is lowered the resilience decreases to a minimum, then increases again. This minimum occurs when the rubber is in the "leathery" state of Fig. 5.7.

Rebound is increased by an increase in modulus or by a decrease in internal friction. If these properties are to be measured separately another type of test must be used – such as free or forced vibration.

Free Vibration Tests

In free vibration tests the rubber specimen forms the spring in a mechanical system with inertia chosen so that a damped oscillation of the desired frequency results from release of a deformation. Deformation in compression, shear, torsion, tension, or torsion plus extension have been used. Perhaps the most common instrument of this type is the Yerzley Oscillograph, ASTM D945, of which a sketch is shown in Fig. 5.11. A balanced beam is supported at its center and designed so that its motion is controlled by a rubber specimen strained in either compression or shear. A pen mounted at one end of the beam draws a trace on a recorder drum rotating at a constant speed. The natural frequency of free vibration is measured from this trace and used to calculate the effective dynamic modulus K of the

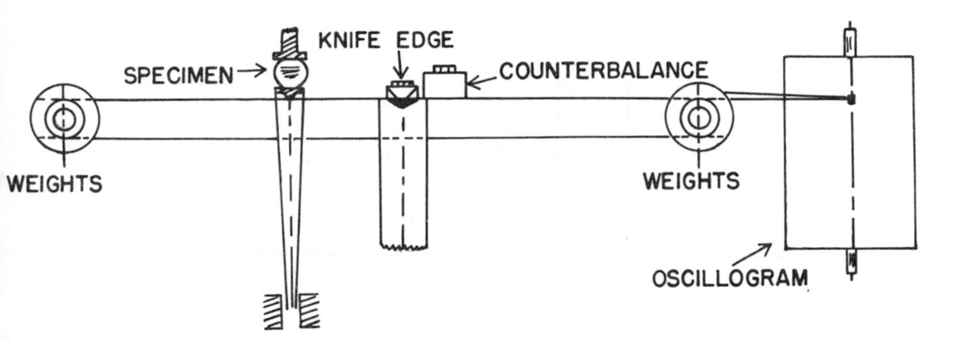

Fig. 5.11 Yerzley oscillograph (ASTM D945).

rubber from the formulae

$$K_c = 210\ If^2 \text{ (compression)}$$

$$K_s = 105\ If^2 \text{ (shear)}$$

where f equals frequency and I equals moment of inertia of the system.

A number of other properties can be obtained from the trace. For example, the average ratio of heights of successive oscillations gives an artibrary measure of resilience of the rubber which, expressed as a percentage, is termed the *Yerzley resilience.* In a trace such as that of Fig. 5.12 the Yerzley resilience equals

$$\left(\frac{A_3}{A_2} + \frac{A_4}{A_3}\right) 50$$

Another type of free vibration instrument is the torsion pendulum,[23,24] a classical method for studying dynamic properties of rubber. Many designs of

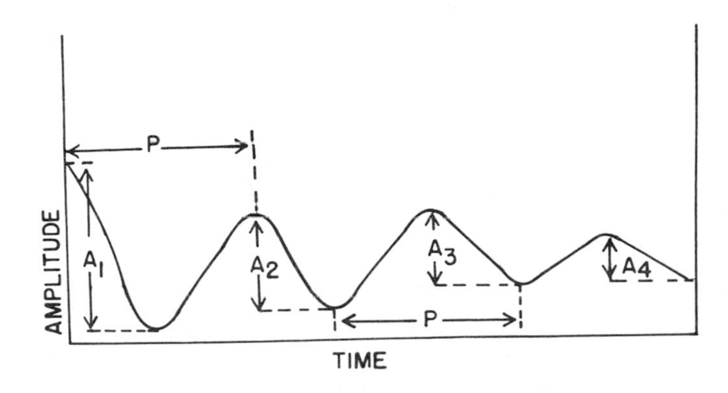

Fig. 5.12 A typical damped oscillation curve (Ref. 37).

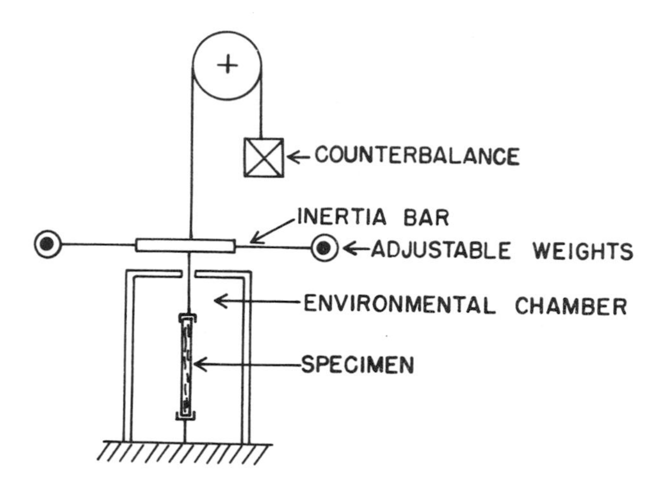

Fig. 5.13 A torsion pendulum device.

apparatus have been used; that in Fig. 5.13 will illustrate the principles. The bottom of the ribbon shaped specimen is rigidly attached while the top is fastened, either solidly or through a torsion wire, to a structure having adjustable moment of inertia. The weight of this device is counterbalanced through a cord having negligible torsional stiffness.

In performing the test the inertial system is displaced through a small angle (about 1.5°) and released. The shear modulus may be calculated from the period[23]; the shorter the period the greater the modulus. A measure of damping is given by a quantity called the logarithmic decrement (see Fig. 5.12):

$$\Delta = \log_e \frac{A_3}{A_2} = \log_e \frac{A_2}{A_3} = \ldots$$

An objection to any free vibration method is that amplitude is not maintained so that a calculated modulus, which depends on amount of deformation in a polymer, is not consistent from one material to another. This can be overcome in the torsional pendulum by exerting a mechanical moment (by magnetic means) which is equal in magnitude but reversed in sign to that caused by internal friction in the specimen. The required force is a measure of damping. In effect, however, this then becomes a nonresonance forced vibration test at low frequency.

Forced Vibration Tests

A rubber is placed in forced vibration when the disturbing force is periodic and continuing. The system is in a *steady state* condition when the hysteresis energy exactly balances the impressed energy and in a *resonance* condition when the exciting frequency equals the natural frequency of the spring and mass system. Many forced vibration methods have been developed[22,23,24,26] with perhaps the greater number being resonant. These include forced vibrators in both shear and compression, reed vibrators where several modes may be excited during a single test, and resonant beam tests for low frequency studies. Calculations are simpler in resonance than in nonresonance conditions.

The theory of forced vibration testing is too extensive to enter into here, and the reader is referred to ASTM D2231 for pertinent definitions, concepts, and factors affecting dynamic measurements. Each type of dynamic mechanical test yields a different damping term. In order to compare data obtained by different methods, it is necessary to convert all to a common basis. This is done by using complex moduli, G^* and E^*:

$$G^* = G' + i\,G'' \text{ (shear)}$$

$$E^* = E' + i\,E'' \text{ (Young's)}$$

G' and E' are the "real" parts of the shear modulus and Young's modulus respectively, i.e., the elastic parts. The quantity i is equal to $\sqrt{-1}$, so G'' and E'' are the "imaginary parts" of the moduli, i.e., those due to damping forces, often called the *loss moduli.* A very useful damping term, called the *dissipation factor* or *loss tangent*, is defined as G''/G' or E''/E'. The dissipation factor is proportional to the ratio of energy dissipated per cycle to the maximum potential energy stored during a cycle.

Dynamic properties testing provides a good example of the modern systems approach to physical testing in which a single apparatus performs the tests and computes the results, or different modules can be used in a single basic unit to extend its capability over a number of different tests. Commercial units are available, for example, which both test and compute spring rate and damping coefficient for selected values of mean load, cyclic amplitude, frequency, and temperature. This is based on the fundamental differential equation for a spring-mass system:

$$F = Kx + C\frac{dx}{dt} + M\frac{d^2x}{dt^2}$$

where x is the instantaneous displacement of the mass from its rest position, F is the displacing force at that time, K is the spring rate, C is the damping coefficient, M is the vibrating mass, and t is the time at which the displacement is measured. When a load cell is used, only its platen, the specimen, and its clamp constitute the mass, so M can be ignored in the above equation. The fundamental material properties K and C are then easily calculable in the nonresonant electrohydraulic closed loop system.

Other systems are available which can perform an amazing variety of tests and basically require only changing of clamps. The commercial MTS electro-hydraulic system, for example, includes the following tests on a single basic apparatus:

(1) Conventional stress-strain relationships; tension or compression tests, etc.
(2) High strain rate testing (tension or compression).

(3) Short term, long term creep testing.
(4) Fatigue testing – constant amplitude; programmed amplitude, mean level and/or frequency; and random amplitude/frequency.
(5) Service simulation tests (sample trace testing from service records).
(6) Component testing.
(7) Low cycle fatigue using constant strain, constant deflection, constant load.
(8) Testing in other loading modes (e.g., bending, torsion).
(9) Environmental tests.
(10) Biaxial and triaxial testing.

In tests for dynamic properties either the amplitude or frequency or both are kept low so that heat build-up in the specimen is minimized. In many European methods the test results are obtained during the first few cycles before the specimen heats up much; the argument being that the specimen is then at a known uniform temperature. American methods, however, usually specify attainment of thermal equilibrium before taking readings, believing that a slight temperature gradient is not as important as being sure that stress softening (Mullin's effect) has reached an equilibrium.

Heat Build-up Tests

Damping properties as measured in rebound or vibration tests provide a measure of heat generating potential under the usually mild conditions of those tests. More severe tests are used to measure actual heat build-up and resistance of the sample to deterioration at the attained temperature.

Perhaps the best-known of these tests are the Goodrich flexometer and the Firestone flexometer (ASTM D623). In the Goodrich apparatus a definite compressive load is applied to the cylindrical specimen while superimposing a high-frequency cyclic compression of definite amplitude. The increase in temperature at the base of the specimen is measured with a thermocouple. Specimens may be tested under a constant applied load, or a constant initial compression. Change in specimen height can be measured continuously during flexure. By comparing this change in height with the observed set after test, the degree of stiffening (or softening) may be estimated.

In the Firestone flexometer a rotary motion is applied to one end of the specimen, which is shaped like the frustrum of a rectangular pyramid, while it is held under a constant compressive load. The time required for a definite change in height of the specimen is determined. At the end of the test the temperature may be measured by inserting a needle thermocouple into the center of the specimen. This is perhaps the more common use for the apparatus. It can also be used as a *blow-out* test, however, by using a heavy load and large throw and running to failure. Either flexometer can be used at elevated temperatures as well as at room temperature.

Flexometers are not recommended for use in purchase specifications because both correlation with service life and reproducibility between different laboratories are uncertain. As with so many other physical tests, comparisons between compounds are more valid when they have similar composition than when they are widely different as, for example, being based on different polymers.

Flex Resistance

In compression fatigue tests by the flexometer methods, discussed in the preceding section, rubber failure is mostly due to heat. In fact, the "blow-out" condition is accompanied by pyrolitic decomposition with gaseous products. Tests for fatigue failure caused principally by mechanical action are usually made in the bending mode. The term *flexure* generally means deformation in bending although *flexometers* compress or shear a specimen and *flexible foams* are those which are easily compressed. The mechanical action in flexure is, of course, extension on the outside of the bend and compression on the inside. The *neutral plane*, near the center of an isotropic specimen, is neither extended nor compressed.

Flexural fatigue failure in a rubber-fabric composite is usually manifested by delamination. The ASTM test for this type of durability is the Scott Flexing Machine (D430 Method A), in which strip specimens are stretched over a rotatable hub and clamped at a bend angle of 135°. After applying a prescribed load on one of the clamps, thus placing the specimen in tension, the test strip is pulled back and forth over the hub at about 160 cycles per minute. Separation between fabric and rubber is indicated by melting of a wax coating caused by the excess heat generated at the failure point. This type of test is applicable to either belts or tire bodies and samples may be specially molded for the test or cut from finished products.

The DuPont flexing machine, also described in ASTM D430, provides an action similar to that of the Scott machine except that a reverse curvature is also included in the flex cycle. A total of 21 test specimens form a test belt which is run over a series of four pulleys arranged in a V-formation. Failure is indicated by visual inspection. Normally the test can be continued until all 21 specimens have failed.

Flexural fatigue failure in rubber or in the rubber part of composites is usually divided into two categories: crack initiation and crack growth. The DuPont apparatus can be used for crack initiation tests, in which case failure is indicated by nicks or pinholes in the corrugations which are molded on the rubber face of the specimen.

Crack Initiation Tests

Perhaps the most widely used test for both crack initiation and crack growth, however, is the De Mattia (National Flexing Machine) described in ASTM D430 for initiation and in D813 for growth tests. Either tensile or flexure deformation may be used for crack initiation tests. In the tensile type tests normal dumbbell specimens are alternately stretched and completely relaxed so that a slight bend is induced. This is an important point because flex life in any type of deformation is sometimes much shorter if zero strain is included in the cycle than if all the flexing is entirely on one side of zero, especially for crystallizable materials.

The De Mattia specimen most commonly used is 6 inches long, 1 inch wide, and

0.25 inch thick with a groove molded across its width. The flex cycle is such that the specimen is bent almost double at the groove, then almost straightened out; no gross extension occurs. This type of bending, without a mandrel, can allow progressively greater strain concentration as the bending line becomes hotter and weaker. The energy absorbed in the critical area is thus maintained or even increased as the test proceeds. This should lead to more reproducible end points than occur in mandrel tests such as the DuPont or Scott where bending energy may actually decrease at a hot or weak point.

Results of crack initiation tests may be expressed in any of several ways: (a) a severity comparison of the various samples at a definite number of flexing cycles; (b) the number of cycles required to attain a definite severity rating; (c) comparison of the number of cycles required to attain progressive degrees of severity ratings; (d) after a suitable number of cycles, depending on the compound, the specimens are examined and rated according to the degree of cracking by comparison with a set of standard specimens graded 0 (no cracking) to 10 (completely cracked through).[15] The latter two methods give combinations of crack initiation and crack growth ratings.

Crack Growth Tests

The ranking of a group of compounds for crack initiation is often quite different from the ranking for crack growth, so separate evaluations are needed. Compounds of SBR, for example, normally resist crack initiation much better than do compounds of natural rubber but once a crack is started in each it proceeds much faster in the SBR sample. Some types of apparatus may be used for either test, however. The De Mattia method for crack growth is described in ASTM D813. The grooved specimen is used and a crack is started by piercing with a specified tool at the center of the groove. Crack length is measured at regular intervals with the end point being the number of cycles required to extend the crack to 0.5 or 0.75 inches. Alternatively the average rate of crack growth over the entire period or a certain portion of it may be reported.

Another standardized test for cut growth is the Ross flexing machine, ASTM D1052, in which the pierced section of the specimen is bent over a mandrel through a 90° angle. The number of cycles for the cut length to increase 100, 200, 300, 400, and 500% of its original length is reported.

Effect of Test Conditions

Both crack initiation and crack growth are speeded in oxygen, especially in ozone, and slowed in a nitrogen atmosphere. Crack growth of SBR is speeded by increase in temperature but that of natural rubber may be either speeded or slowed. All cracking tests are constant amplitude, so sample modulus and thickness variations affect actual stresses and hence the rate of cracking.

Flexing tests do not necessarily correlate well with service performance, a situation which occurs with many accelerated tests. For this reason flexing tests should not be specified unless close correlation can be shown with service.

AGING TESTS

Aging tests are performed to evaluate, within a relatively short period of time, the susceptibility of a rubber to deterioration of physical properties because of environmental effects. As with any test in which service conditions are exaggerated for economy of testing time, care is required to ensure that the degree of acceleration is uniform. For example, the temperature of an aging test should not be increased beyond the point where a sudden change in tensile strength occurs unless it can be shown that the same change occurs over a longer time at lower temperatures.

All ASTM aging methods carry the warning that no correlation with natural aging should be implied. The user must determine for himself whether or not useful correlation exists.

Heat Aging

Perhaps the most common aging test is that of comparing tensile and hardness properties before and after heating in air. Standard methods include the Geer oven (ASTM D573) in which all the specimens being tested at a given time are held in a common air space. This is satisfactory if it can be shown that air contamination does not influence the results. Careful cleaning between tests is also needed to avoid contamination.

Cross-contamination among specimens having different compositions may be prevented by using the test tube method (ASTM D865). Test tubes, holding no more than three specimens each, are heated in an oil bath or an aluminum block. Convective air circulation is permitted through glass tubes inserted in the tube stoppers. Self contamination is effectively prevented by use of a tubular oven (ASTM D1870) in which heated fresh air enters one end of the tube and is exhausted from the system without being recirculated.

A more severe test is the air pressure heat test (ASTM D454) in which the specimens are held at 260°F and 80 psi air pressure. This was originally intended for use on innertube compounds and is not now widely used. The community aging is an objectionable feature of this test also.

Aging tests for specific products include ASTM D622 for automotive air brake and vacuum hose, D1055 for latex foam rubbers, D1056 for expanded rubbers, and D296 for fire hose.

Oxygen aging

In a method analogous to the air pressure heat test, oxygen may be used instead of air (ASTM D572). The pressure is 300 psi and temperature 158°F. Tensile properties are again used to judge the aging effects.

Many instruments have been devised to measure directly the rate of combination of oxygen with rubber: gravimetric (measures increase in weight of rubber), manometric (measures drop in oxygen pressure), and volumetric (measures drop in oxygen volume). These are used more in research and development than in control or specification testing. Oxygen absorption, although usually associated with degradative processes, may, especially in the early stages of cure, contribute to the development of desirable properties. In fact, oxygen may compete with sulfur as a vulcanizing agent. Thus the interpretation of oxygen absorption curves requires study in terms of the experimental conditions.

Ozone is a form of oxygen which has a particularly severe effect on rubber, especially if it is stretched. In ASTM D1149 either bent specimens or stretched, straight, or tapered specimens are exposed to ozone concentration of 50 parts per hundred million in air. Effects are judged by visual examination and comparisons with standards. Other tests often provide for flexing or stretching of the specimen during at least part of its exposure time. Other ASTM tests exist for evaluation of specific products: D1373 for insulating tape, D1352, D574, D2526, and D470 for wire and cable insulation.

Weather Resistance

Laboratory tests for ozone resistance do not necessarily correlate well with outdoor weathering tests. The outdoor tests introduce such additional variables as light-catalyzed oxidation, water leaching, and changing temperature and ozone concentration. In ASTM D750 the specimens are exposed, either continuously or intermittently, to carbon arc lights and water spray. Ozone may also be present. Exposure effect is judged by tensile tests and visual examination.

In ASTM D518 detailed instructions are given for mounting specimens for either laboratory or outdoor weather exposure. Effects are judged by visual examination. In the D1171 test for weathering of automotive compounds, triangular specimens are stretched around a mandrel. Specimen cracks are rated against a set of standard photographs. Cracks appear perpendicular to the direction of stretch.

Discoloration due to sunlight and heat is the subject of ASTM D1148. An important feature of this type of test is the emphasis on control of light intensity. Wavelengths between 2000 and 2500 angstroms are especially critical. Sunlight produces crazing in either stretched or unstretched rubbers, whereas ozone produces cracks.

Exposure to Liquids

Another deteriorating factor for rubber products is exposure to various liquids such as fuels, oils, chemicals, and even water. Any rubber may be swollen by certain chemicals. Although an "oil-resisting" rubber may resist the swelling effect of certain oils, there are other oils or solvents which will cause it to swell, thus causing a deterioration in physical properties. Detailed procedures for evaluating change in properties of elastomeric vulcanizates resulting from immersion in liquids is given in ASTM D471.

CREEP, STRESS RELAXATION, AND SET TESTS

The term *creep* refers to increase in deformation under a constant force, while *stress relaxation* means the decrease in retractive force of a specimen held at constant deformation. Although the two can be mathematically related they are usually evaluated separately. An example of a need for a creep test is a use requiring dimensional stability, as in a motor mounting. Stress relaxation tests might be indicated in a use requiring continued sealing ability.

Any type of deformation might be used in creep or stress relaxation tests. ASTM D1390 describes a stress relaxation test in compression. Although there is at present no ASTM test for creep of rubber, Subcommittee D11.14 is considering a shear creep method similar to the modified ICI test.[27] This is suitable for tests at either normal or subnormal temperatures.

The term *set* refers to the strain remaining after complete release of the load producing a deformation. ASTM methods D412 and D395 describe methods for evaluating tension and compression set, respectively. The usual measurement of set is intended to evaluate delayed elastic recovery. If chemical changes have occurred, however, the residual deformation may be permanent. Low temperature set tests are discussed in the section on "Transitions in Elastomers."

TEAR TESTS

Tear test results are strongly dependent on the type of specimen used, the rate of tearing, and the temperature. As described in the scope of ASTM D624, "The method is useful, therefore, only for laboratory comparisons and is not applicable for service evaluations, except when supplemented by additional tests, nor for use in purchase specifications." Nevertheless, many such tests are run, perhaps because they seem to be logical extensions of the hand tear evaluation which was so useful to the old-time compounder.

Three types of tear specimens are classified by Buist[28]: indirect tearing as in the trousers specimen of Fig. 5.14, tearing perpendicular to the direction of stretching as

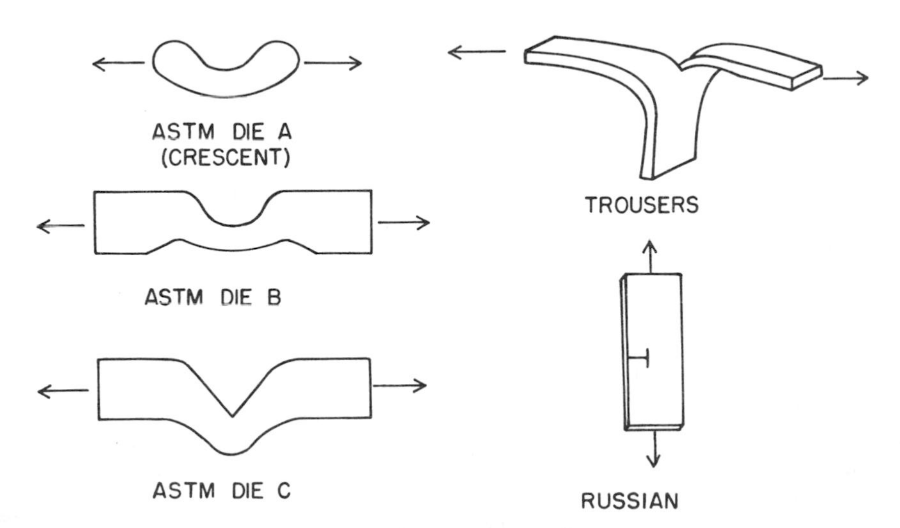

Fig. 5.14 Types of tear specimens.

in the ASTM methods, and tearing in the direction of stretching as in the Russian test piece. Except in the ASTM Die C specimen nicks of prescribed lengths are cut into the region of desired stress concentration.

Rate of stretching in the ASTM method is 20 inches per minute. An increase in rate would normally decrease the tearing energy for SBR rubbers but give a more complicated effect in natural rubber. On the theory that a portion at least of tire wear is a result of high speed tear, such a test has been considered for evaluation of abrasion resistance.

Tear test results are usually expressed as the pull in pounds required to tear a specimen one inch thick, but the required pull on specimens of the ASTM type is dependent on width as well as thickness. It has been suggested[28] that results on such specimens should be called *tear strength*, whereas those on direct tearing methods such as the trousers specimen should be called *tear resistance*.

Tear tests sometimes yield smooth plots of force against strain but other times, notably with natural rubber, the curve is oscillatory in nature. In such cases, called *knotty tear*, the high force points on the oscillatory curve are usually recorded as the required pull.

Of the three types of specimens the lowest force values are given by the trousers specimen, next by ASTM Die C, next by Die A or B, and highest by the Russian specimen. Low correlation of results from different specimens should be expected.

Tear resistance of black-loaded general purpose elastomers is quite sensitive to state of cure, often showing a sharp "optimum" curing time which is usually lower than the optimum as judged by modulus and tensile properties. Tear in black-loaded stocks often progresses from the smooth type at low states of cure to a jagged fibrous type at tighter cures. Low vulcanization temperatures and short cures are desirable for high tearing resistance.

ADHESION TESTS

ASTM D413 presents two standard tests for rubber adhesion to fabric: (1) the dead weight method measures the rate of separation per unit width under constant load, and (2) the machine method measures the required force for constant rate of separation. Since the resulting plots are often jagged, average results are given. As in tear strength, resistance to separation is decreased sharply by an increase in test temperature. Constant temperature rooms are suggested.

ASTM D429, for testing adhesion of rubber to metal, also provides two methods: (1) rubber part assembled between two parallel metal plates and separated by straight pull, and (2) rubber part assembled to one metal plate and stripped at 90°. In the first case results are expressed in force per unit area of adhered surfaces and in the second case as force per unit specimen width.

In addition to these general tests, special methods for products such as friction tape, hose, insulating tape, and belting have been standardized by ASTM. A common problem in adhesion testing is failure in the rubber part rather than at the bond. In such cases the bond can be presumed to be satisfactory if product stress concentration at the bond is no greater than that of the test specimen. Otherwise a specimen must be devised with sufficient concentration of stress to provide bond failures.

ABRASION TESTS

Abrasion resistance is a property of rubber that can be only incompletely evaluated by laboratory tests. Three principal reasons may be cited for the well-known lack of validity of such tests, especially as applied to tire tread wear: (1) any acceleration of the normal wear process introduces new wear mechanisms, so that the original mechanism is not being truly evaluated; (2) no satisfactory method has been found to maintain a constant sharpness of a laboratory abrasive surface and simultaneously remove abraded rubber from the working surface; and (3) relative wear rating depends upon severity of test conditions. Road service includes a greater range of conditions than could reasonably be duplicated in a laboratory situation.

Wear involves removal of surface material which, for tires at least, includes such diverse mechanisms as removal of chunks by cutting, high speed tear, degradation of surface to low molecular weight material which can transfer to opposing surface, and even degradation to gaseous products. Temperature is, of course, critical to each of these mechanisms and the temperature cycle at the tread surface during a tire's rotation on the pavement is not well known. In spite of these problems, abrasion resistance tests can be useful if the range of validity is recognized. For example, correlation with service rating is often high if tests are restricted to compounds based on a single polymer. This is often adequate, e.g., in evaluation of carbon black.

The Pico Abrader (ASTM D2228) shows good agreement both within and among laboratories, i.e., its precision and reproducibility are high, but no claim is made for validity. The method involves the use of a pair of tungsten carbide knives of specified geometry and controlled sharpness which are rubbed over the surface of the specimen in a rotary fashion under controlled conditions of load, speed, and time. A dust is used on the specimen and at the interface between the knives and the specimen to engulf the particules removed and to maintain the cutting knives free from contamination. Weight loss of the specimen is determined.

FRICTION TESTS

Friction requirements for rubber range from the very low coefficients in antifriction bearings to the very high desired in tire treads or rubber heels. Several extensive reviews of this subject have been made,[29] in which it was pointed out that there are two components to rubber friction – adhesive and hysteretic. Except in very well lubricated sliding the adhesive component dominates but in tire traction, for example, wet pavement often presents a critical situation where the hysteretic properties of rubber become important.

Perhaps the most common rubber friction apparatus is the British Portable Skid

Tester (ASTM E-303 Part 11, which was actually developed as a pavement evaluation instrument. This is a portable device in which the specimen is attached to the base of a pendulum arm and contacts the opposing surface during a swing of the pendulum. The weighted pendulum head is free to move longitudinally on the pendulum arm so that the swing amplitude is determined by the friction of the rubber against the pavement surface. This instrument is very well adapted to testing on a wet surface.

NONDESTRUCTIVE TESTING

Nondestructive testing of rubber products is a growing field which is at the present time mostly limited to specific items. With their proven capabilities, however, more general tests are likely to appear. A summary of their range and potentialities has been given by Halsey.[30] These methods include holography, ultrasonics, infrared radiation, microwaves, X-rays, and gages based on nuclear, capacitive, optical, or magnetic principles.

Ultrasonics have been used for thickness measurements, including that of a coating of one material on another, and for locating voids. Infrared radiation permits noncontacting surface temperature measurements to be made on stationary or moving objects. Microwaves can be used for noncontacting thickness measurements without radiation hazard. Both the loss tangent and the dielectric constant of the material can be measured independently. X-rays, including both xeroradiography and fluoroscopy, are sensitive to flaws of various kinds – especially to inclusion of foreign material that is more opaque to X-rays than is the test material.

COMPUTER USAGE

Any test instrument which produces an electrical signal that can be calibrated in terms of the measured property level is a potential candidate for tie-in with a computer. This may be an *on-line* tie-in, in which the coded results are fed directly to the computer for immediate processing, or the results may be stored and introduced to the computer at a later time. Depending at least partly on the speed at which data must be logged, storage can be by punched card, paper tape, or magnetic tape.

An on-line (real time) computerized system of four tensile testers as used at DuPont has been described by Stanton.[31] This used a *dedicated* computer (for that purpose only) which was programmed to accommodate nine tests performed on the tensile testing machine: stress-strain, Finch tear, ASTM Die B tear (Winkleman), ASTM Die C tear (Grains), flexural modulus of elasticity, compression-deflection, paper tensile, peel adhesion, and trouser tear. Four of these programs had alternate routines, making 13 tests in all. The computer not only processed the data and printed out the calculated results but also controlled a segment of the electronic circuit in each unit. This application illustrates the principal reasons for computerizing physical tests: saving employee's time, accurate logging and calculating of large amounts of data, consistent test control, and rapid operation.

Test Design

Apart from logging of data, the use of a computer can make certain types of testing programs practical that would otherwise require excessive computation time. An example is the so-called *designed experiment*. By this we mean a system of tests in which the factor levels of each variable are distributed according to a design table that permits calculation of main effects and interaction effects of each variable. A great deal more information is available from this type of program than from the one-variable-at-a-time traditional test, especially in research or development testing. An example has been given by Derringer[32] for evaluating effect of type and amount of accelerator in a silica-filled natural rubber. Control plots were used to analyze trends in 300% modulus, 500% modulus, tensile, elongation at break, Mooney scorch, and 95% Rheometer cure time.

Another example of the combined use of designed experiment and computer, in this case a special purpose analog computer, has been given by Claxton et al.[33-35] Means are described for maximizing a chosen response (e.g., economy) while maintaining other responses (e.g., durometer hardness, cure rate constant, cure induction time) at acceptable levels in a compounding study involving various levels of sulfur, Bismate, Santocure NS, Vultrol, and Paraflux. The method and computer are adaptable to any compounding study involving up to five factors and eight responses.

Trends in Physical Testing

With the development of new polymers and more versatile tests the rubber compounder is more nearly able than formerly to tailor-make an elastomer for each application. Consequently less reliance is placed on meeting arbitrary specifications and more on achieving properties which demonstrably relate to the applications. A new polymer usually requires a new interpretation of test results since its compounds will have a new combination of properties which may meet a specific requirement as well as would previously recognized combinations.

The compounder's responsibility for product performance is increasing which, again, requires extra care in selecting tests and evaluating the results. Since composition and design of a rubber product are often interrelated in achieving optimum performance and reliability at minimum cost, more tests are being made on actual products under operating or accelerated conditions. There is, consequently, a decline in testing for fundamental properties, such as tensile strength, ultimate elongation, or hysteresis, unless the need for certain levels of such properties can be demonstrated. Both static and dynamic modulus tests, however, are assuming even greater roles.

The new tensile testers now available commercially reflect the increased demand for versatility in physical testing for research and development purposes. Such items as a large range of crosshead speeds, environmental chambers, high speed recording, computer interfacing, or even mini-computer components are more important for such applications than is high volume output. The versatility of basic instruments is further increased by use of modules for particular applications and by provisions for cycling and by choice of constant rate of deflection or constant rate of load application. Often either stress relaxation or creep tests are also possible on the same apparatus. Principles proven on such multi-use instruments may then be applied to development of specialized high volume testers for control applications.

As Murtland[36] states, a big push is on to expand quality control testing into the production line. This will require computerized control to amass and analyze the flood of acquired data. Instruments will continue the trend toward programmed cycles, since shortage of skilled operators will be further compounded by the great increase in the number of instruments installed. End product testing will see a great increase of nondestructive testing applications developed for use of the production line.

Each new consumer protection law requires additional testing for quality control – sometimes duplicate or replicate testing because each processor of a material must assume responsibility for each previous processor. The amount of physical testing per unit of production is thus continually increasing and it is more important than ever that tests be used as effectively as possible.

REFERENCES

(1) R. W. Jones, "Planning the Rubber Laboratory," Southern Rubber Group Meeting, Nov. 9, 1962 at New Orleans, La.
(2) Rolla H. Taylor, "Testing Equipment for the Rubber Laboratory," Southern Rubber Group Meeting, November 9, 1962 at New Orleans, La.
(3) Anon. "Rubber: Sampling and Testing," Federal Test Method Standard No. 601. General Services Administration, Business Service Center, Region 3, Seventh and D Streets, S.W., Washington, D.C.
(4) Anon. "Index of Federal Specifications and Standards," Superintendent of Documents, U.S. Gov't Printing Office, Washington, D.C. 20402.
(5) ISO Publications available from American National Standards Institute, 10 East 40th St., New York, N.Y. 10016.
(6) Anon. "Standard Reference Materials: Catalog and Price List," NBS Spec. Publ. 260 July, 1969 Ed. U.S. Dept. of Commerce, Jan. 1970 Supplement.
(7) A. E. Juve, "On Testing of Rubber" (Goodyear Medal Award Lecture), *Rubber Chem. Technol.,* **37** xxiv April–June (1964).
(8) L. A. Wood, "Tables of Physical Constants of Rubber," *Polymers Handbook,* J. Brandrup and E. H. Immergut, eds., Interscience Publishers, Division of John Wiley & Sons, New York, 1965; *Rubber Chem. Technol.,* **39**, 132–142 (1966).
(9) Lawrence E. Nielsen, *Mechanical Properties of Polymers*, Reinhold Publishing Corp., New York 1962, Chapt. 1.
(10) Takeru Higuchi, H. M. Leeper, and D. S. Davis, "Determination of Tensile Strength of Natural Rubber and GR-S. Effect of Specimen Size," *Anal. Chem.,* **20** 1029 (1948); *Rubber Chem. Technol.,* **22**, 1125 (1949).
(11) A. E. Juve, "Physical Testing," Chapt. 19 in *Introduction to Rubber Technology*, Maurice Morton, ed., Van Nostrand Reinhold, New York, 1959.

(12) F. S. Conant, "The Effect of State of Cure on Vulcanizate Properties," Chapt. 3 in *Vulcanization of Elastomers*, G. Alliger and I. J. Sjothun, eds., Reinhold Publ. Co., New York, 1964.
(13) Anon. "Lectures from a Course in Basic Rubber Technology," given by Philadelphia Rubber Group with Villanova Univ. 1955. Lecture IV.
(14) "Continuous Measurement of the Cure Rate of Rubber," ASTM Special Technical Pub. No. 383, 1965.
(15) Anon. "Test Methods," Chapt. 3 in *The Vanderbilt Rubber Handbook*, George G. Winspear, ed., R. T. Vanderbilt Co., Inc., New York 1968.
(16) Joseph R. Weber and Hector R. Espinol, "Cure State Analysis," *Rubber Age,* **100** (3), 55 (1968).
(17) G. E. Decker, R. W. Wise, and D. Guerry, Jr., "Oscillating Disc Rheometer," *Rubber World,* **147** (3), 68 (1962); *Rubber Chem. Technol.*, **36**, 451 (1963).
(18) A. E. Juve, P. W. Karper, L. O. Schroyer, and A. G. Veith, "The Viscurometer – An Instrument to Assess Processing Characteristics," *Rubber World,* **149** (3), 43 (1963).
(19) F. S. Conant, "A Study of Stiffness Testing of Elastomers at Low Temperatures," ASTM Bulletin July, 1954, pp. 67–73 (TR145).
(20) Anon. "DuPont 900 Differential Thermal Analyzer," Instruction Manual, E. I. du Pont de Nemours & Co., Instrument Products Division, Wilmington, Del.
(21) F. S. Conant and J. W. Liska, "Some Low Temperature Properties of Elastomers," *J. Appl. Phys.*, **15**, 767 (1944).
(22) A. C. Edwards and G. N. S. Farrand, "Elasticity and Dynamic Properties of Rubber," Chap. 8 in *The Applied Science of Rubber,* W. J. S. Nauton, ed., Edward Arnold Ltd., London, 1961.
(23) Nielsen, op. cit., Chap. 7.
(24) S. D. Gehman, "Dynamic Properties of Elastomers," *Rubber Chem. Technol.*, **30**, 1202 (1957).
(25) Anon. "Handbook of Molded and Extruded Rubber," The Goodyear Tire and Rubber Co., Akron, Ohio, 1969.
(26) J. H. Dillon, I. B. Prettyman, and G. L. Hall, "Hysteretic and Elastic Properties of Rubberlike Materials Under Dynamic Shear Stresses," *J. Appl. Phys.*, **15**, 309 (1944).
(27) J. M. Buist and R. L. Stafford, "Yield Stress in Frozen Rubbers," *Trans. IRI,* **29**, 238 (1953).
(28) J. M. Buist, "Physical Testing of Rubber," Chapt. 9 in *The Applied Science of Rubber*, W. J. S. Nauton, ed., Edward Arnold Ltd., London, 1961.
(29) F. S. Conant and J. W. Liska, "Friction Studies on Rubberlike Materials," *Rubber Chem. Technol.*, **33**, 1218 (1960).
(30) G. H. Halsey, "Non-Destructive Testing," *Rubber Age,* **100** (2), 62 (1968).
(31) J. L. Stanton, "Computerized Physical Testing of Elastomers," ACS Div. of Rubber Chemistry, Los Angeles, April 30, 1969.
(32) George C. Derringer, "Predicting Rubber Properties with Accuracy," ACS Div. of Rubber Chemistry, Los Angeles, April 30, 1969.
(33) W. E. Claxton, "Basics for Blend Optimization," *Rubber World,* **159** (3), 42 (1968).
(34) W. E. Claxton, H. C. Holden, and J. W. Liska, "A Blend Optimizer for the Rubber Industry," *Rubber World,* **159** (3), 47 (1968).
(35) W. E. Claxton, "Use of Computers in Formulating Rubber Compounds," International Rubber Conference, Paris, June 1, 1970.
(36) W. Murtland, "Rubber Testing Sparks Instrument Boom," *Rubber World,* **162** (1), 53 (1970).

6

NATURAL RUBBER

STEPHEN T. SEMEGEN
Natural Rubber Bureau,
Hudson, Ohio

Natural rubber has rightfully been called "the supreme agricultural colonist of all times." Although indigenous to forests in the Amazon Valley, natural rubber has been cultivated principally in Southeast Asia, especially the countries of Malaysia and Indonesia. Where once only dense jungles and rain forests stood, more than fourteen million acres of land have been cleared and planted with rubber trees.

Natural rubber can be isolated from more than 200 different species of plants, including even such surprising examples as dandelions or golderod.[19,20] However, only one tree source, *Hevea Brasiliensis*, is commercially significant. At one time, early in the century, a large tree *Funtumia elastica* was exploited in tropical Africa, as was the *Landolphia species*, a climbing shrub. In Mexico, the large tree *Castilloa elastica* also provided rubber. In South America, the Ceara Rubber tree, *Manihot glaziovii*, was another source. None could compete with *Hevea Brasiliensis* in yield, frequency of tapping, or longevity.

All contain more resin, i.e., acetone soluble material, than does *Hevea.* The Russian dandelion *Taraxacum koksaghys* contains 10% of rubber in its roots. It achieved some popularity and use during World War II as an indigenous source of supply. Similarly, the Guayule plant *Parthenium argentatum* was cultivated in the state of California. It has one unusual captive use, in its ability to retain high amounts of sulfur, and thereby is an effective masterbatching base for factory compounding. Also, the shrub *Cryptostegia grandiflora* was grown in Florida.

EARLY HISTORY

Of all the wonderful tales reported by Christopher Columbus after his second voyage to the New World in 1496, none was stranger than the story of natives in Haiti, who played with a ball made from the gum of a tree – a ball which *bounced.* In South America, such trees were called "Cau-uchu," or "weeping wood." In fact, games were played with such a rubber ball, known as far back as the eleventh century, throwing light on the advanced Mayan civilization.

Natives were known to waterproof articles of clothing and footwear with the gum, dried over smoke fires. Although the Spanish and Portuguese made a few experiments to explore possible usefulness, it remained for the French scientists, La Condamine and Fresneau, to demonstrate practical uses. However, such rubber articles suffered from stickiness in hot weather and brittleness in cold.

The English chemist, Priestley, gave the name "rubber" to the raw material in 1770, when he found it would "rub off" pencil marks. Ironically, in so doing, he may have unawaredly become the first one to add carbon black to rubber. In the early 1800's, a Scotsman, MacIntosh, patented the first raincoat, consisting of a layer of rubber sandwiched between two thicknesses of cloth. A London coachman, Thomas Hancock, then cut rubber into long strips, and invented the "rubber band" in 1823. In 1839, the tremendous discovery of *vulcanization* occurred, which ultimately made possible the multibillion dollar industry of today. Credit is generally shared by Hancock and Charles Goodyear, who found that the combination of sulfur dust and heat added to raw rubber resulted in remarkable resistance of the "vulcanized" rubber to extreme temperature changes.

In 1846, Thomas Hancock made solid rubber tires for the carriage of Queen Victoria, perhaps thereby launching our present vast tire industry. Even so, the discovery of vulcanization timed with the industrial revolution caused a great demand for raw rubber. At that time, known wild rubber areas in South America and Africa, coupled with exploitation of both vine and tree types, were able to keep up with demand. However, it soon became obvious that more rubber would have to be grown, and in regions where transportation and dense jungle growth did not pose obstacles.

While not supported in fact, legend and story romanticize the expeditions of Sir Henry Wickham to Brazil is search of rubber seeds for growth experiments. Wickham "smuggled" 70,000 Hevea seeds from the Tapajoz region of Brazil, shipped them to England, where the seeds were planted in greenhouses at the Kew Gardens outside London. Only a few percent of the seeds germinated. These young plants were then shipped to Ceylon, where they grew vigorously. In fact, some of these original trees are still standing in Ceylon, and even yielding latex upon tapping. Ironically, nowhere in the Kew Gardens does a marker even exist to commemorate this momentous event.

A subsequent consignment of 22 seedlings was shipped to Singapore in 1877 and planted in the Economic Gardens. From these 22 trees, 75% of the total cultivated rubber in the world has sprung. By 1888, about one thousand trees were scattered through the Malay peninsula. At this point, the most significant step in the rubber growing industry took place. Henry Ridley took over the Singapore Gardens in 1888. At that time, coffee was the main crop of Malaya, and his suggestion to plant the slow-growing rubber trees resulted only in earning him the title of "Mad Ridley."

It is to his credit that Ridley anticipated that the newly invented pneumatic tire, credited to John Dunlop in England, would revolutionize transportation and create an unprecedented demand for rubber. It was further apparent that sporadic collection of hosts of widely different rubbers from wild jungle areas would not answer the needs of the fledgeling industry. Hence grew what we now know as the rubber plantation.

PLANTATION RUBBER

Ridley set about to demonstrate that rubber trees could be grown in cleared areas in regular order, tapped frequently and economically, and the liquid latex converted to dry, clean raw rubber ready for use. In so doing, Ridley rapidly discovered the phenomenon of "wound response," wherein latex flowed more freely after the initial tapping, that the tree could be tapped at intervals of a few days, that the excised bark would renew itself after a few years for subsequent tapping, thereby permitting trees to yield latex economically and efficiently for at least 30 years (as we now know).[21] He was also first to use acetic acid to coagulate the latex, permitting the coagulum to be rolled out into flat sheets for drying, rather than the wet balls and lumps formerly dried on the ends of poles over a fire. Ridley even experimented with the control of diseases known to attack the rubber tree. By the time of his retirement in 1912, Ridley had firmly established the position of the rubber plantation. The rapid growth of rubber production is shown in Fig. 6.1.[30]

Problems of transporting rubber from the interiors of South American and African jungles to world shipping centers, coupled with the killing onslaught of the

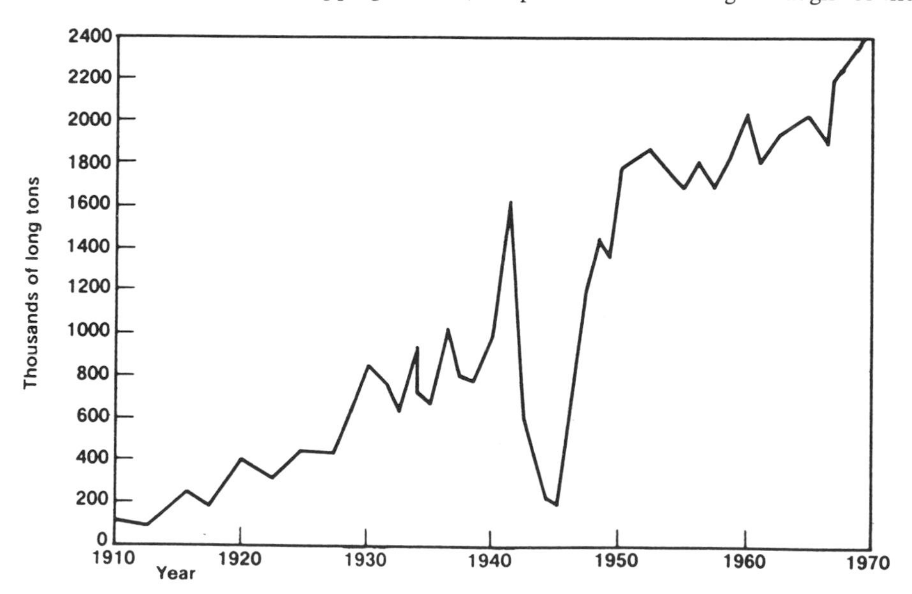

Fig. 6.1 World production of natural rubber since 1910.

"yellow leaf" blight, left the door open for the countries of Southeast Asia to become dominant in production of natural rubber.

TABLE 6.1. ORIGIN OF NATURAL RUBBER BY COUNTRY

Country		*Production, thousand long tons (1969)*
Malaysia		1,258
Indonesia		750
Thailand		278
Ceylon		149
Africa		142
India		69
Cambodia		51
Vietnam		24
Brazil		23
Other Asia		26
Other Latin America		7
	Total	2,777

Production of dry rubber is performed approximately equally between *estates*, i.e., plantations of more than 100 acres in size, and small holdings, i.e., farms of less than 100 acres in size, and usually only three to eight acres. Due to differences in yield per acre, actual acreages under cultivation total ten million small holdings and less than five million acres in estates. Some of the larger estates include thousands of acres, tending to specialize in concentration of bulk latex and premium rubbers.

BOTANY

It has been stated earlier that the bulk of trees now producing rubber came from seeds of the descendants of the original plantings in Singapore and Ceylon. It has also been estimated that more than half of the cost of producing rubber is involved in tapping of trees and collecting the latex. Accordingly, it is vital to obtain the maximum yield per tapping operation. Early efforts to increase yields were centered around selection of seeds from known high-yielding trees. These had certain limitations.

The next step was *bud grafting*. By grafting buds from high-yielding trees onto the stems of seedling stock, a large number of high-yielding trees were obtained. All trees derived by vegetative reproduction from a single mother tree are known as a *clone*.[22] The vast majority of plantation acreage has now been replanted with such high-yielding clones. In fact, these clones do not number more than about two dozen.

Further improvements were made in the genetic strain by cross-fertilization where both parents are high-yielders. Of the best known clones, each has its own set of properties and characteristics. One may be best for yield, but poor in resistance to wind damage. Others grow best in certain soils, climate, and topography. Among the more common are the Tjipandji 1 clone, popular in Indonesia, the PB series, and especially the RRIM series. The latter were developed by the Rubber Research Institute of Malaya, the largest research institute in the world devoted to one single agricultural commodity. The RRIM 500 series and the RRIM 600 series have shown increases in yield from 300 pounds per year per acre for unselected seedlings to as much as 1100 pounds per year per acre. In fact, even 2,000–3,000 pounds per year per acre have been obtained under careful conditions.[23] The second and third generation of the RRIM 600 series have even been reported to yield up to 6,000 pounds per year per acre in small, experimental plantings.

AGRICULTURE

Rubber trees normally grow to a height of 60 feet. However, the long range objective is to produce a more stumpy tree with less top growth and sturdier trunks. Although soil culture is important, it is surprising how well the Hevea tree can grow in relatively poor soil. A well-drained, acid and sandy clay is most common, to which mineral nutrients and organic matter contribute much. A hot, damp climate with equable distribution of rainfall and temperature is essential.

An annual rainfall of 100 inches is preferable, with temperature ranges of 70–95°F. A flat terrain is desirable, although terraced hillsides are not uncommon. Generally, rubber is not grown at elevations exceeding 1,000 feet. While the rubber tree goes through a wintering season, with yellowing and loss of leaves, quality latex is produced and tapped year-round. The bark of the tree is smooth and a mixture of light browns. The fruit of the tree is a pod of three cells, popping open when ripe to fling the smooth, shiny seeds many feet away. The seeds are slightly larger than acorns.

Young plants are raised in a nursery for the first year. These are then transplanted to the permanent field. The trees are planted in regular rows from 14 to 20 feet apart. After thinning, usually 100–150 trees per acre are left.[25] Originally, about seven years were required before trees could be tapped properly. With new nursery techniques, the time to maturity is now only five years.[31] In addition, latex yield is markedly improved by the use of plant hormones, such as 2,4D and copper sulfate. As a rule of thumb, a girth of 22 inches when measured at two feet above ground is sufficient for tapping. At such time, trees can be 30 feet tall. Legumes, such as soybean, are common for ground cover plants, providing fixed nitrogen. Phosphate fertilizers are also used extensively.[26,27]

Yellow leaf blight,[28] caused by the fungus *Dothidella ulei*, ultimately destroyed the rubber tree in South America, its birthplace. Surprisingly, Africa and Southeast Asia have not suffered from this disease, although drastic measures must be taken at times. If detected, hundreds of acres of trees are burned out to isolate the spread. No effective fungicide is known.

Brown bast, which attacks the trunk, is another of the worst diseases. It develops a canker of the bark, due to some physiological disturbance. Other diseases attack the stem and even the roots by moldy rot.

LATEX TAPPING

Figure 6.2 illustrates the delicate nature of slashing the bark to reach the latex vessels, without touching the cambium layer. The latex vessels are a network of capillary tubes throughout all parts of the tree. In the trunk, these vertical bundles are inclined from right to left at about a five degree angle.[30] The latex vessels are in a layer only two to three millimeters thick.

Ridley originally developed a herringbone design on the tapping panel, shaving about one or two millimeters thickness of bark with each cut. Tapping is usually done in the early morning hours, although a rainfall at such time drastically curtails latex flow. The latex flows for several hours, followed by plugging of the latex vessel ends, and a subsequent coagulation of latex skin over the cut. Parkin in 1900 in Ceylon discovered the phenomenon of wound response. If the cut is reopened the next day, even more latex is obtained, indicating the rapid regeneration of latex by the tree.

The cut is made with a special knife or gouge, sloping from left to right at about 20–30° from the horizontal. Several systems of tapping are used, including half circumference alternate days, full circumference every fourth day, etc.[24] An unusual system is the double tap, usually on opposite sides of the trunk, with one cut about four feet above the other. Latex from the tree consists of 30–40% dry rubber by weight. The tapper is paid by hydrometer measurements of the "d.r.c." (dry rubber content) of his volume of latex collected each day. The contents of

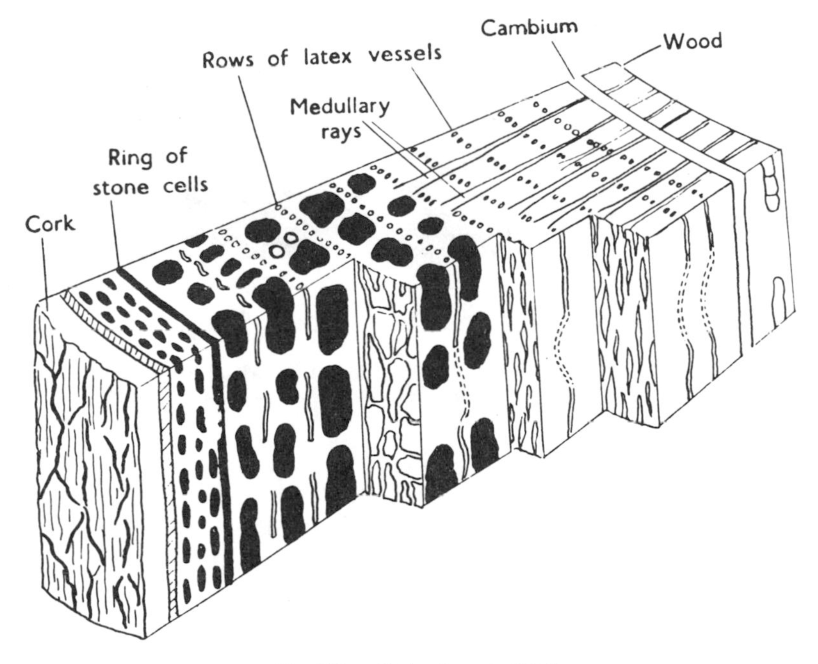

Fig. 6.2 Outer layers of tree.

Fig. 6.3 Removing tree lace (dried rubber) before tapping. (*Courtesy Natural Rubber Bureau*)

each latex cup mounted on each tree are collected in five gallon containers, such as those for milk, then taken to pickup trucks for transport to storage tanks at bulking stations.

Premature coagulation of the latex is prevented by the addition of small amounts of stabilizers, such as ammonia (0.01%), sodium sulfite (0.05%), or formaldehyde (0.02%).[32] Fresh latex from the tree has a normal pH of 7.

DRY RUBBER PREPARATION

Depending on the producing country, 10–20% of the tree latex is concentrated by creaming or centrifuging, and shipped to consumer countries, where it is used as such to make finished articles, including foam for upholstery and bedding, a variety of tubing and dipped goods, etc. The remainder is processed into dry rubber, as sheets, crepes, remills, etc. These are graded according to International Standards of Quality and Packing (The Green Book) established by an international body of endorsing organizations. The latter include rubber manufacturers' associations, rubber trade associations, rubber growers' associations, etc. These standard grades may number as many as thirty-six types.[39] More recently, a growing innovation among producing countries is the classification of rubber to *technical specifications*, not visual inspection.

Smoked Sheets. The largest single type of dry rubber is the ribbed smoked sheet, and recently air dried sheet. Field latex is strained into large bulking and blending tanks, diluted with an equal volume of water to a dry rubber content of about 15%, then coagulated. To 1,000 parts of the diluted latex, 50 parts of a 1% formic acid solution are added. Acetic acid can also be used.

The latex is treated in aluminum tanks about 16 inches deep and slotted for aluminum partitions. Thereby, overnight storage results in a series of soft, thick gelatinous slabs. The coagulum is then passed as separate sheets or a continuous slab through a series of squeeze and wash rolls at even speed, with continuous water spraying. After going through the five or six rolls set progressively tighter, the serum is washed away, and the rubber has been squeezed down from about 1 1/4 to about 1/8 inch thickness.[34] The characteristic rib pattern is embossed primarily for increased surface area for drying.

The wet sheets are mounted on poles or racks set on trolleys, then passed into a "smoke-house." The smoke-house ranges in temperature from about 100°F at the entry to 140°F at the exit, the entire period covering about four days. Too fast drying results in blistering and tackiness of the rubber. Many estates generate 85% of their total crop as RSS.

Pale Crepe. Probably only 5–10% of latex is converted into pale crepe. This is categorized as a premium grade of rubber, for use where lightness of color is important, as in white sidewalls of tires, drug sundries, surgical goods, baby nipples, bathing accessories, etc.

After dilution, latex is treated with about 0.5% sodium bisulfite on the d.r.c. to prevent darkening of the rubber by enzymes. In some cases, a chemical bleaching agent may be added, such as 0.1% of a mercaptan, like xylyl mercaptan.[36] If extremely pale rubber is desired, fractional precipitation may be utilized. In such case, about 0–15% of the normal amount of coagulating acid is added as a 2.5% solution (acetic acid preferred). After about three hours of stirring, approximately 10% of the latex has coagulated and is skimmed off. This contains the large bulk of beta-carotene, the yellow coloring pigment in natural rubber.[35]

The remainder of the latex is coagulated by the addition of more acetic acid, in partitioned troughs, and allowed to set into slabs. The wet slabs are passed through a series of rollers at differential speeds under a water spray. After about eight sheeting passes in total, the sheet has been creped and marked with a macerating roll to its final, characteristic crinkled surface. Drying is done in sheds heated only to 90–100°F for about six days, and in the absence of any discoloring smoke.

Brown Crepe. Whereas the sheets and pale crepes constitute the great majority of the various rubber grades, and are prepared from fresh field latex, the brown crepes constitute the larger part of the remaining rubber grades. The brown crepes are prepared from field coagulum, coagulum formed during the handling of latex, cup lump (the dried latex film left behind in the cup after pouring off the latex), and even tree lace (the thin latex skin which forms over the cut on the tree after latex flow ceases).

Due to the heterogeneous composition of the raw material, the various coagula are soaked overnight in a dilute solution of sodium bisulfite to lighten the color and to remove surface dirt. After maceration, the crumbs are washed, dried, and creped similarly to the pale crepes. The brown crepes comprise 5–10% of total production.

Miscellany. Other commercial types include *flat bark crepe*, made primarily from latex drippings on the bark or ground; *blanket crepe* (ambers), made from RSS clippings and remilled cup lump; and skim rubber, either sheet or crepe, recovered from the serum or mother liquor, after centrifugal concentration of field latex.

The proportionate amount of each rubber grade produced varies depending on the origin (smallholder or estate) and also by country. For example, Malaysia specializes in the top quality RSS and pale crepes from fresh field latex. Indonesia produces a large amount of the brown crepes and ambers from fresh coagulum. Smoked sheets and pale crepes account for about 65% of the total crop, while the browns and ambers represent about 30%, with 5% for the flat bark crepes. Paradoxically, modern techniques are producing higher amounts of the top grade rubbers, whereas demand, primarily in passenger tires, continues to increase for the lower grades of natural rubber. A balance must be attained between economics and quality.

MARKET PACKING

Originally, natural rubber was packed for shipping in wooden cases, following the practice of the tea industry. Splinters in the rubber were sometimes a troublesome problem. In the 1930's, burlap or jute bags were used, but stray fibers sticking to the rubber often required burning off. Finally, "bare back" baling became the most standard practice. Rubber sheets were plied together under a hydraulic press to form a bale of five cubic feet, weighing 250 pounds. The outer wrap is also comprised of sheets coated with a very light slurry of soapstone, talc, or whiting in water (to prevent bales from sticking in transit).[37]

Pale Crepe is normally shipped in 75–160 pound bales, within multi-wall treated paper bags or wooden boxes.[38]

In recent years, new processes for making natural rubber to *technical* specifications (to be discussed later) have led to small rectangular bales of 70–75 pounds, sealed in polyethylene bags and packed on wooden pallets containing usually one long ton.

Latex is transported in the 60–62% concentrated form. About 0.7% of ammonia is added for mechanical stability, although other stabilizers can be used for special purposes. On land, bulk transport is by tank car, whereas large tankers are used for sea transport. Many consumers prefer to receive latex in 50 gallon drums coated internally to prevent corrosion or contamination.

PROPERTIES OF NATURAL RUBBER

First recorded analysis of natural rubber was performed by Faraday.[51] He reported it to be composed of hydrogen and carbon in the ratio expressed by C_5H_8 as early as 1826. Williams[52] found in 1860 by destructive distillation that isoprene was the building unit of natural rubber. Tilden[53] also reported that the probable structure of rubber was 2-methyl-1-3 butadiene. That is, one double bond unit existed for each C_5H_8 group. Bouchardat[54] recognized in 1879 that isoprene could be polymerized into natural rubber.

Composition. Commercial raw natural rubber has a small, but highly important number of nonrubber constituents. These may compromise as much as 5–8% of the total composition. Most important are the natural-occurring antioxidants and activators of cure, represented by the proteins, sugars, and fatty acids. Typical composition is as shown in Table 6.2.

TABLE 6.2. ANALYSIS OF REPRESENTATIVE NATURAL RUBBER

Ingredient	*Ave. %*	*Range %*
Moisture	0.5	0.3–1.0
Acetone Extract	2.5	1.5–4.5
Protein	2.5	2.0–3.0
Ash	0.3	0.2–0.5
Rubber Hydrocarbon	94.2	
Total	100.0	

TABLE 6.3. APPROXIMATE CONCENTRATIONS OF NONRUBBER CONSTITUENTS IN CENTRIFUGED LATEX CONCENTRATES

Constituent	*Percentage by weight of latex*
Fatty acid soaps (e.g., ammonium oleate)	0.5
Sterols and sterol esters	0.5
Proteins	0.8
Quebrachitol	0.3
Choline	0.1
Glycerophosphate	0.1
Water-soluble carboxylic acid salts (acetate, citrate, etc.)	0.3
Amino acids and polypeptides	0.2
Inorganic salts (ammonium and potassium carbonate and phosphate, etc.)	0.2

Trace elements present include potassium, magnesium, and phosphorus, as well as copper, manganese, and iron. Although usually present only to the extent of 2–3 parts per million, the latter are important as catalysts to promote the oxidation of vulcanized rubber. Amounts higher than 8–10 parts per million are not tolerated.

Table 6.3 outlines in detail the complex variety of chemical substances present in the dry rubber and latex.

Physical Properties. Physical properties of natural rubber may vary slightly due to the nonrubber constituents present and to the degree of crystallinity. When natural rubber is held below 10°C, crystallization occurs, resulting in a change of density from 0.92 to about 0.95. Listed in Table 6.4 are some average physical properties.

TABLE 6.4. SOME PHYSICAL PROPERTIES OF NATURAL RUBBER

Density	0.92
Refractive index (20°C)	1.52
Coefficient of cubical expansion	0.00062/°C
Cohesive energy density	63.7 cal./c.c.
Heat of combustion	10,700 cal./g
Thermal conductivity	0.00032 cal./sec/cm^2/°C
Dielectric constant	2.37
Power factor (1,000 cycles)	0.15–0.2*
Volume resistivity	10^{15} ohms/c.c.
Dielectric strength	1,000 volts/mil

* The power factor is reduced to 0.0015 and the resistivity substantially increased in deproteinized rubber.

Natural rubber exhibits two phases, sol and gel. The differences are exhibited in degree of solubility in certain organic solvents due to different amounts of highly branched and lightly crosslinked components intertwined in the rubber. Mechanical shear, as in milling, or oxidative breakdown with heat and oxygen tends to disaggregate the gel phase, thereby increasing solubility.[64] Effective solvents for masticated natural rubber include aliphatic and aromatic hydrocarbons, chlorinated hydrocarbons, ethers, and carbon disulfide. Nonsolvents include the lower ketones, alcohols, and lower esters.

Molecular Weight. Since natural rubber has a high molecular weight, the classical methods of measurement based on depression of freezing point or elevation of boiling point cannot be used. Instead, osmotic pressure and viscosity measurements are used. Fractionation of solution of natural rubber reveals that a range of polymers of varying molecular weight is present. Thereby, only an average molecular weight can be assigned. Even so, such average molecular weight can range from 200,000–500,000.

It is possible to deposit simple molecules from solution on to suitable substrates, then to measure the size of the "molecular particles" in the electron microscope. By bromine addition to the rubber hydrocarbon, deposition of the

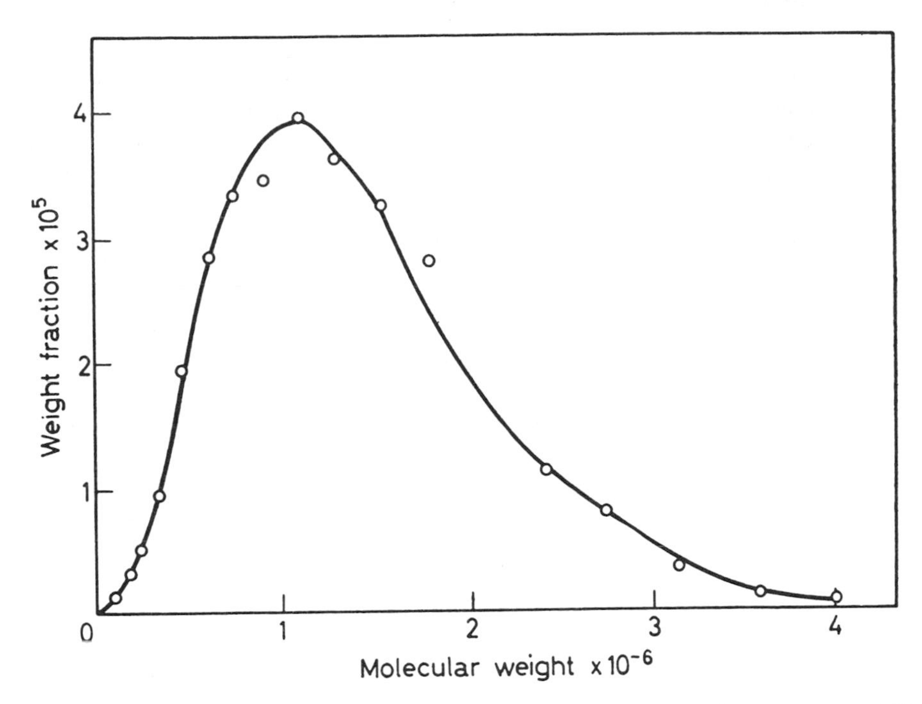

Fig. 6.4 Molecular weight distribution of natural rubber by electron microscope method.[55]

molecular particles is facilitated, and can be shadowed by standard methods. The size/frequency data can then be converted into a molecular weight distribution curve as above.

Crystallinity.[61,62,63] Natural rubber exists in two geometric configurations, i.e., *cis* and *trans* isomers.[56] The standard rubber, such as sheets and crepes, is the cis-polyisoprene form, while *gutta-percha* or *balata* is the trans-polyisoprene form. The latter is small in commercial importance. Originally used in underwater cable insulation, it is now utilized only in golf ball covers and special adhesives. Balata, or gutta-percha, is an opaque, hard and horny substance, with superior resistance to impact and to embrittlement. When warmed over 180°F, the material softens and sheets out on a mill, undistinguishable from milled sheets. However, upon cooling, the original hardness (or high crystallinity) returns.

Gutta comes from a substantial tree, a third of whose twigs, leaves, etc., are harvested each year. Extraction is by maceration and hot water treatment. Plantations of gutta trees are in Malaysia.

Balata is obtained from a small bush, primarily in the Surinam and Guiana regions on the northeastern coast of South America, and also in certain areas of Indonesia. Volume use does not justify plantation growing, hence the plant grows wild. The entire bush is cut down and macerated to squeeze out the latex, which when coagulated and dried, is extremely high in dirt content and resins. The latter comprise about 75% of the total. Balata may be purified by dissolving the unrefined form in a light naphtha solution, then precipitating the pure material by freezing out or adding a nonsolvent, such as acetone. When subsequently dried and milled into slabs, the refined balata has less than 2% residual resin and nonrubbers.

X-ray diffraction studies of raw rubber in an unstretched state reveal a pattern of two concentric broad bands, indicating an *amorphous* state, at room temperature. However, when the rubber is stretched, preferably to at least 700% elongation, a "fiber pattern" is obtained.[60] This consists of well defined interference spots or crystallites. The regular arrangement permits calculations of the unit cell of the crystal lattice, indicating a repeat distance along the chain structure of 8.1A.

Raw rubber also exhibits a curious phenomenon called "racking." If the rubber specimen is stretched rapidly and repeatedly near its breaking elongation, then cooled quickly, a highly fibrous structure is developed, which is quite stiff and extremely strong. In fact, such fibers can be drawn out to elongations as high as 10,000%.

Related to the elastic structure of rubber are two effects, thermal in nature, which comprise the Gough-Joule effect: (1) when stretched rapidly, rubber heats up, contrary to most materials; and (2) if the rubber sample is held at one and stretched under a given load, it will *retract* as the temperature is raised.

The importance of crystallization in natural rubber vulcanizates stems from its effect on strength of the rubber. For example, tensile strength is governed by the degree of crystallization attained as the breaking point is approached. This is particularly true in gum or nonfilled rubber.

RUBBER DERIVATIVES

Over the years, a wide variety of chemical and physical reactions of natural rubber has been studied. In so doing, some rather unusual products were obtained. Some of these laboratory curiosities have attained varying degrees of commercial utility.

Depolymerized rubber. When raw rubber is exposed for several hours at 140°C or higher, degradation of the polymer occurs, as evidenced by liquefaction of the rubber to a pourable state, and considerable reduction of molecular weight. There are commercial grades which are used for casting articles and can be vulcanized to a soft or hard rubber state. Potting compounds, binders for abrasive wheels, casting molds, etc., are examples.

Cyclized rubber.[48] In the presence of heat and acidic catalysts, rubber is transformed into a resinous material of greatly reduced unsaturation. The substances increase in density to 0.99, and exhibit a softening point of 90–120°F. (a) *Thermoprenes* are the types formed with acidic catalysts, including sulfuric acid, p-toluenesulfonic acid and its chloride, phosphoric acid, or trichloracetic acid, etc. By heating such a mixture for 1–4 hours at 125–145°C, a series of products ranging from a balata-like state to a shellac-like form can be obtained. These are used primarily for binders and adhesives, especially for protecting metals against corrosion and to adhere metals to rubber. The cyclizing reaction can be carried out in latex and the latter coagulated with a normal latex, for use in rigid moldings and soling.

(b) *Amphoteric halides.* Heavy metal chlorides, such as those of tin, antimony, iron, aluminum, or titanium, when added to a rubber solution, form a complex substance. The latter is recovered by precipitation with alcohol or by distilling off the solvent. The white, amorphous powder can be used as a molding material or as a vehicle for coating compounds.

Chlorinated Rubber. Natural rubber readily undergoes both substitutive and additive reactions with chlorine.[4] Although it is possible to chlorinate in the rubber latex system,[5] the best process is the direct chlorination of a rubber solution in an inert solvent, such as carbon tetrachloride at about 80°C. The fine, white powder of chlorinated rubber is obtained by steam distillation to remove the solvent.

Fully chlorinated rubber (65% chlorine) has a specific gravity of 1.63–1.66, and is readily soluble in aromatic hydrocarbons and chlorinated solvents.[6] Since it is inert, chlorinated rubber is used as a base in chemically resistant paints. It is also useful as an adhesive for rubber to metal bondings.

Rubber Hydrochloride. As with the halogens above, only the hydrogen chloride addition product of rubber is feasible.[11] At one time, rubber hydrochloride was made in a rubber solution by passing in dry hydrogen chloride gas. The product was obtained either by precipitation with a nonsolvent or by steam distillation.

A commercial product, called "Pliofilm," had importance for packaging because of low permeability to water vapor and good transparency. It was also used in rubber to metal adhesives. An antiacid stabilizer had to be incorporated to prevent decomposition and corrosion.

Rubber hydrochloride (28–30% chlorine) has a specific gravity of 1.16 and a refractive index of 1.533. It is also highly crystalline, orienting upon stretching, and melts at about 115°C.[12,13]

MODIFIED RUBBERS

A variety of other modifications of natural rubber has been accomplished, several with rather important commercial results, as follows.

(a) AC rubber. Better known as "anticrystallizing" natural rubber, AC rubber is an equilibrium mixture of *cis* and *trans* isomers, which interfere with the orientation, and hence crystallization of the rubber when exposed to low temperatures. It is obtained most readily in two ways: (1) addition of thiols or thiol acids to the rubber latex at room temperatures, overnight; or (2) treatment of dry rubber with sulfur dioxide for 10–15 minutes at 170–180°C. In this case, a sulfur dioxide source, such as butadiene sulfone, is most commonly used.[14,15,16,49] Ingredients are mill-mixed, while reaction readily occurs in an extruder or internal mixer.

Natural rubber has its maximum *rate* of stiffening at –26°C. If stress-relaxation measurements are made on stretched specimens at such temperature, the crystallization "half-life" of the AC rubber is about 1,000 times that of normal rubber.[57] Product applications under such arctic conditions include brake boot hydraulic cups, gaskets, tubing, and even adhesives.

(b) Grafted rubber. Starting in 1941, the Institut Francais du Caoutchouc performed considerable work on chemically attaching polymerizable monomers, such as vinyl benzene (styrene), vinyl cyanide (acrylonitrile), acrylic esters, etc., to natural rubber.[17] These have been called *grafted rubbers.* Measurements by differential thermal analysis and by selective solvency confirm that a truly homogeneous material can be formed.

Heveaplus is the generic name for a family of graft rubbers, wherein methyl methacrylate monomer has been polymerized in natural rubber latex, thereby attaching side chains of methyl methacrylate on the linear backbone of natural rubber.[18] These materials were developed by the Natural Rubber Producers' Research Association in England, and are now commercially available, either in the latex form or as dry rubbers. Compositions ranging from a weight content of 30%–50% methyl methacrylate are available.

The Heveaplus materials are self-reinforcing, and can be added to natural or synthetic rubbers to confer high hardness and impact strength to nonblack compounds. However, the most useful applications are as adhesives, either as a latex, organic solution, or dry rubber film. Heveaplus has the ability to adhere many "like" and "unlike" surfaces together. This stems from the fact that the grafted rubber combines both polar and nonpolar groupings in the same material, thereby effectively acting as a chemical bridge.

(c) Superior processing rubbers.[44] These are rubbers coagulated from a *latex* blend of 80% normal rubber and 20% lightly vulcanized rubber. The latter is prepared by adding a dispersion of a special curative system to the latex, heating with live steam for several hours at 82–85°C, cooling down overnight, and then blending with the normal latex.[43] The SP rubbers are finished and packed as with normal rubber.

A more concentrated form of SP rubber is available as PA-80.[45] The latter contains 80% of the above vulcanized rubber and 20% of normal rubber. It is useful for blending with either natural or synthetic rubbers to confer the superior processing properties. These include better dimensional stability of extrusions or calendered sheets, improved smoothness, less die swell, faster tubing rate, reduced porosity and pitting, and less water marking in open steam or hot water vulcanization.

TECHNICALLY SPECIFIED RUBBER

In recent years, rubber producers have examined new processes for making natural rubber, as well as new presentation methods for packing, grading, and shipping the rubber. The general approach has been to prepare these rubbers in such way that the present 36 grades of standard rubber can be reduced to a few grades. In so doing, current grading and marketing by *visual* inspection is superseded by strict *technical specifications.*

The rubber producers of Malaysia, working with the Rubber Research Institute of Malaya, introduced the new system commercially in 1965 as SMR (Standard Malaysian Rubber). Initially, cleanliness and freedom from other contaminants were emphasized. Current SMR technical specifications are shown in Table 6.5.

TABLE 6.5. STANDARD MALAYSIAN RUBBER SPECIFICATIONS

Standard Malaysian Rubber specification limits	*SMR 5*	*SMR 20*	*SMR 50*
Dirt content %	0.05	0.20	0.50
Ash content, %	0.5	1.0	1.5
Copper content, p.p.m.	8	8	8
Manganese content, p.p.m.	10	10	20
Nitrogen content, %	0.7	0.7	0.7
Volatile matter, %	1.0	1.0	1.0

Most SMR rubber is processed in a crumb form by one of several mechanical or chemical processes. After washing and drying, the crumb rubber is compressed hydraulically into 70–75 pound bales (not over 112 lb), wrapped and sealed in polyethylene bags, then packed on one-ton wooden pallets for shipment. Bales are either 28 x 14 x 6 1/2 inches, or 22 1/2 x 15 x 7 inch measurements. The pallet units to handle the various size bales are 59 x 45 x 36 inches for the former, and 48 x 40 x 46 inches for the latter.

Advantages of the new process rubbers include increased uniformity, greatly improved cleanliness, and better appearance, as well as ease of handling from the producing area, the loading dock, or ship storage for dock unloading, trucking to the user, and finally for processing and storing the rubber in the consuming factory.

Production of SMR rubbers rose from a few thousand tons in 1966 to more than 380,000 tons in 1972. By 1975, it is anticipated that 50% of total Malaysian rubber will be produced as SMR. In addition, a similar growth rate is estimated for other major producing areas: notably Indonesia, Ceylon, Cambodia, Nigeria, and Liberia in the field of similar technically specified rubber.

Heveacrumb process. The Rubber Research Institute of Malaya pioneered a mechano-chemical process, using crumbling agents such as castor oil, to form a fine crumb. Small amounts (0.2–0.7%) of such incompatible oil are added to the latex, followed by conventional coagulation with acid to the usual gelled slab form. The wet slab is then passed through a series of several crepeing rolls, finally dropping into a drying tray as fine, rice-like particles. The crumb can be dried in a few hours in a batch oven, or in a tunnel dryer, easily adapted to a conveyor system.

The Heveacrumb process can also be easily adapted for masterbatching with high amounts of carbon black and/or extender oils. In these cases, the dispersion of black and emulsion of oil are simply added to the latex, prior to the addition of the crumbling agent. The rest of the procedure stays the same.

Comminution process. A strictly mechanical process, using no additives, has been developed by several of the larger estate producers. This simply consists of reducing slab coagulum to the size of pea-like crumbs, with either a rotary knife cutter or a granulator, such as the Cumberland type. Other means of shredding or chopping rubber particles can also be utilized.

Washing, drying, nd packing of thc comminuted rubber follow the same pattern as for Heveacrumb. Further, the whole raw material crop, whether latex, cup lump, tree lace, etc., can be utilized readily. The proportion of each category of crop used will determine the SMR grade into which the finished rubber will fall.

Extruder-dryer process. Extrusion drying equipment, similar to that used in making general purpose synthetic rubber (SBR), has been designed to handle natural rubber. Wet coagulum is first shredded, then washed, put through a dewatering screw press which reduces the water content to 10–15%, and then through a second screw extruder which essentially dries the rubber. Baling and packing go on as before.

To justify the cost of the equipment, large capacity is required (1–3 tons per hour) for continuous operation. Accordingly, only the largest producers can consider the installation.

Miscellaneous processes. There are other new processes for making rubber to technical specifications. A few speciality rubbers which are noteworthy in their own right, are as follows.

(a) *Technically classified rubber.* Technically classified (TC) rubber was developed to reduce the variability in cure rate of natural rubber. The latter may vary depending upon differences in technique in converting from the latex to the dry rubber, and usually is related to the nonrubber ingredients present.

The classification is simply a sorting of the rubber into one of three designations: red, yellow, and blue circle. The three groups are respectively: slow, medium, and fast curing. Cure rate designation[41] is based on limits of percent strain of a pure gum vulcanizate (ACS test recipe No. 1, cured 40 minutes at 140°C) under a load of five kg/cm^2. For example, producers' limits for red circle rubber are 103 to 85%, yellow between 85 and 73%, and blue ranges from 73 to 55%.[42] On the average, world production of TC rubber is about 60% yellow, 30% blue, and 10% red. Addition of very small amounts of a quaternary ammonium salt to the latex can speed up the cure rate, while formaldehyde will slow down the cure rate.

(b) *Constant viscosity rubber.* The most recent development in technically specified rubbers is the constant viscosity (CV) rubber. Natural rubber can vary in plasticity due to variations in molecular size and structural arrangement of the hydrocarbon, as shown by Bloomfield in 1951.[40] Moreover, the raw rubber stiffens in storage progressively from the time of fresh preparation, during shipment from the Far East, and even in the warehouses of the rubber factories. The extent of increased plasticity can be as much as 20–30 units of Mooney value.

Work by Sekhar and associates at RRIM demonstrated that stiffening was due to crosslinking between linear chains at the site of aldehyde groups occurring in very small amounts along the chain. Stabilization of viscosity can be obtained in either of two ways: (a) By addition of a monofunctional amine, especially the water soluble hydrochloride salt of hydroxylamine, which inactivates the aldehyde linkage. Original normal Mooney viscosity of 55–65 (ML,4′@212°F) is maintained. (b) By addition of a difunctional amine, such as hydrazine hydrate, which acts as a bridge between aldehyde groups on adjacent chains. In so doing, the raw rubber is slightly stiffer, such as 70–85 Mooney viscosity.

Retention of original viscosity of CV rubber within a few points has been verified for a period of several years. Such a phenomenon has great economic potential in uniformity of processing during product fabrication and in possible elimination of the *premastication* step, hitherto indispensable and costly.

PRODUCT USES

Prior to World War II, natural rubber accounted for practically 100% of the total rubber usage in all products. Since then, SBR has become the largest volume general purpose rubber in use. However, natural rubber is still the preferred polymer in many areas for the following reasons: superior building tack, green stock strength, better processing, high strength in nonblack formulations, hot tear resistance, retention of strength at elevated temperatures, high resilience, low hysteresis (heat build-up), excellent dynamic properties, and general fatigue resistance.

A distribution of percent usage of natural rubber is shown in Table 6.6.

TABLE 6.6. DISTRIBUTION OF NATURAL RUBBER BY PRODUCT USAGE

Product	*Per Cent*
Tires and Tire Products	68.0
Mechanical Goods	13.5
Latex Products	9.5
Footwear	5.5
Adhesives	1.0
Miscellaneous	2.5
Total	100.

It is generally believed that natural rubber usage in tires has diminished. However, statistics for the past ten years do not shown any change in percentage. In passenger tires, primarily for carcasses and white sidewalls, 28% of the natural rubber is used. The remainder of the tire usage is in racing cars, airplanes, heavy duty trucks and buses, and off-the-road tractor and farm vehicles.

Natural Rubber Latex. Natural rubber latex is still in strong demand, especially for blending with synthetic rubber latex, because of superior wet gel strength, tear resistance, strength properties, and wet tack. Main product areas are in latex foam for upholstery and bedding, carpet and rug backing, dipped goods, surgical goods and drug sundries, adhesives, and textile thread.

PRODUCT COMPOUNDING

The start of all natural rubber compounding is with a standard test recipe for the rubber as it arrives at the factory. Most widely used has been the so-called ACS-I recipe, recommended by the crude rubber committee of the Division of Rubber Chemistry of the American Chemical Society, and subsequently adopted by the D-11 Committee of the American Society for Testing and Materials. It is as follows:

ACS-I Test Recipe

Natural Rubber (test sample)	100.
Zinc Oxide (Amer. Process-low lead)	6.
Stearic Acid (Comm. double pressed)	0.5
Mercaptobenzothiazole (Captax)	0.5
Sulfur (Rubber makers' grade)	3.5
Total	110.5

Cure: 20, 30, 40, 60, and 80 minutes at 260°F.
Tests: Modulus at 500, 600, and 700% elongation; tensile strength and elongation at break on all cures.

For wild rubbers, or rubber shown to be slow in cure, primarily due to low fatty acid content, an ACS-II test recipe has been suggested by the same organizations. It is the same as for ACS-I, except stearic acid is increased to *four* parts. Cure conditions are changed to 5, 10, 15, 20, 30, 45, 65, 100, 150, and 225 minutes at 286°F.

Due to the inherent difficulties in testing *pure gum* vulcanizates and objections to the use of MBT as not typical of factory formulations, a new *black-loaded* test recipe has very recently been recommended by the Crude Rubber Committee of ASTM, as follows:

Test Recipe 1-I

Natural Rubber (or *cis*-polyisoprene)	100.
Zinc Oxide	5.
Stearic Acid	2.
N-tert-butyl-2-benzothiazole sulfenamide	0.7
Oil Furnace Black (HAF)	35.
Sulfur	2.25
Total	144.95

It is anticipated that both the pure gum and black-loaded recipes will be used, depending upon the nature of the product usage intended.

Tires. Typical natural rubber formulations for use of tire products would be as shown in Tables 6.7–6.9.

TABLE 6.7. TREAD

Ingredient	Truck Tread Normal	Truck Tread Oil-Extended
Natural Rubber (RSS or SMR-5)	100.	53.6
Cis-Polybutadiene	–	13.4
Process Oil (naphthenic or paraffinic)	–	36.
Stearic Acid	2.5	2.
Zinc Oxide	3.5	5.
ISAF Black	50.	–
HAF Black	–	55.
Nonox ZA	2.	2.
CBS (Santocure)	0.8	0.8
Sulfur	2.	2.
	160.8	169.8
Cure: 15 minutes at 316°F	4200.	2530.
Tensile Strength, psi	620.	460.
300% Modulus, psi	1440.	1575.
Shore A Hardness	59.	62.
Crescent Tear, lb/in.	650.	

TABLE 6.8. WHITE SIDEWALL

Ingredient	Normal	High Ozone Resist.[a]
Natural Rubber (Pale Crepe or SMR-5L)	100.	70.
Neoprene W	–	20.
Hypalon 20	–	10.
Stearic Acid	2.	1.
Zinc Oxide	55.	25.
Titanium Dioxide	15.	35.
Antioxidant (nonstaining)	1.5	1.5
Ultramarine Blue	0.3	0.3
Light Process Oil	–	15.
Anti-Sunchecking Wax	3.	–
Altax	1.5	1.2
Tuads	0.2	0.4
Sulfur	2.	2.
Total	180.5	181.4
Cure: 20 minutes at 307°F		
Tensile Strength, psi	3575.	2950.
% Elongation	600.	650.
300% Modulus, psi	900.	700.
Shore A Hardness	55.	52.
Crescent Tear, lb/in.	280.	235.

[a] More recently, vulcanizable ethylene-propylene terpolymers blended with natural rubber and SBR (about 25/40/35 ratio) have also shown considerable merit.

TABLE 6.9. CARCASSES

Ingredient	*Passenger*	*Truck*
Natural Rubber (RSS or SMR-5)	–	100.
Natural Rubber (Br. Crepe or SMR-20)	35.	–
SBR (1500 type)	35.	–
Whole Tire Reclaim (nonstaining)	60.	–
Stearic Acid	1.5	2.
Zinc Oxide	3.	5.
Antioxidant	2.	2.
SRF Black	15.	15.
FEF Black	25.	10.
Light Process Oil	3.	–
Pine Tar	–	3.
Amax	0.7	–
Santocure	–	0.5
Sulfur (insoluble)	2.	2.5
Total	182.2	140.
Cure:	25 minutes at 307°F	
Tensile Strength, psi	1950.	3800.
% Elongation	500.	600.
300% Modulus, psi	1100.	900.
Shore A Hardness	55.	50.
Crescent Tear, lb/in.	250.	350.

Mechanical Goods (Industrial Products). The largest volume articles in this area have the typical formulations shown in Tables 6.10–6.14.

While rubber compounding has long been regarded as an art, rather than a science, in recent years the formulation of rubber compounds has lent itself to computerization. Punch-card formulary services are also commercially available.

NATURAL RUBBER RESEARCH

The natural rubber industry has grown and thrived on the fruits of wide-ranging research and its application. In the producing areas, the Dutch pioneered Ceylon and Malaya. The Rubber Research Institute of Malaya began in 1925 and is now the largest institute in the tropics devoted to research on a single crop. Malaya also supports the Natural Rubber Producers' Research Association in England, the prime unit for consumption research on natural rubber.

The rubber manufacturing companies, as well as government laboratories such as the National Bureau of Standards, continue to extend the technical knowledge of natural rubber, especially in product usage and service performance. Contributions from the laboratories of the rubber industry are recognized to have been of fundamental significance in the many essential aspects of the science of elastomers.

TABLE 6.10. CONVEYOR BELTS

Ingredient	*Cover*	*Friction*
Natural Rubber (RSS or SMR-5)	100.	–
Natural Rubber (Br. Crepe or SMR-20)	–	100.
Stearic Acid	2.	2.
Zinc Oxide	5.	5.
Process Oil	4.	–
Pine Tar	–	3.
Rosin Oil	–	3.
HAF Black (low structure)	45.	20.
Whiting	–	70.
Flectol H	2.	1.
Age Rite White	–	1.
Sunproof Wax	1.	–
Santocure	0.5	1.
Sulfur	2.5	3.
Total	162.	209.
Cure:	20 minutes at 307°F	
Tensile Strength, psi	4575.	2500.
% Elongation	575.	560.
300% Modulus, psi	1650.	950.
Shore A Hardness	60.	50.
Crescent Tear, lb/in.	600.	520.

TABLE 6.11. HOSE (TUBE AND COVER)

Ingredient	*Acid-Resistant*	*Sandblast*
Natural Rubber (RSS or SMR-5)	100.	100.
Stearic Acid	2.	2.
Zinc Oxide	5.	10.
Age Rite Resin D	1.	–
Age Rite Powder	–	1.
Dixie Clay	20.	–
Blanc Fixe	–	5.
Methyl Tuads	0.25	–
Altax	–	0.5
Methyl Zimate	–	0.1
Sulfur	2.	2.5
Total	130.25	121.1
Cure:	10 minutes @ 290°F	
Tensile Strength, psi	2970.	2800.
% Elongation	720.	730.
300% Modulus, psi	320.	220.
Shore A Hardness	38.	37.

TABLE 6.12. EXTRUSIONS

Ingredient	*Transparent Tubing*	*Sealing Gasket*[a]
Natural Rubber (Pale Crepe or SMR-5L)	50.	–
PA-80	50.	–
SP Natural Rubber	–	100.
Stearic Acid	0.75	2.
Zinc Oxide	0.75	5.
Antioxidant 2246	1.	–
Flectol H	–	1.
Altax	1.	–
Tetrone A	0.4	–
Sulfasan R	0.4	–
MBT (Captax)	–	0.8
Tuads	–	0.2
Sulfur	–	1.2
Total	104.3	110.2
Cure:	20 minutes at 285°F	15 at 307°F
Tensile Strength, psi	3050.	2100.
Tensile Strength (aged 7 days @ 158°F)	3200.	1950.
% Elongation	520.	480.
% Elongation (aged 7 days @ 158°F)	480.	430.
Shore A Hardness	60.	70.
Shore A Hardness (aged 7 days @ 158°F)	62.	–

[a]Fillers may be added as deemed advisable for other physical properties.

TABLE 6.13. ENGINEERING USES

Ingredient	*Bridge Bearings*	*Rail Pads*
Natural Rubber (RSS or SMR-5)	100.	100.
Stearic Acid	1.	2.
Zinc Oxide	10.	5.
Dutrex R	2.	3.
SRF Black	35.	–
MT Black	–	60.
China Clay	–	20.
UOP-88 (Antiozonant)	4.	–
Age Rite Powder (or A.O. MB)	1.	1.
Paraffin Wax	–	1.
Santocure (CBS)	0.7	1.
Sulfur	2.5	2.5
Total	156.2	195.2

TABLE 6.13 (*continued*)

Ingredient	*Bridge Bearings*	*Rail Pads*
Cure:	20 minutes at 285°F	15 at 307°F
Tensile Strength, psi	3050.	2880.
Tensile Strength (aged 7 days @ 158°F)	3200.	2590.
% Elongation	520.	540.
% Elongation (aged 7 days @ 158°F)	480.	510.
Shore A Hardness	60.	66.
Shore A Hardness (aged 7 days @ 158°F)	62.	–

TABLE 6.14. MISCELLANEOUS

Ingredient	*Thread Rubber*	*Shoe Foxing*
Natural Rubber (Air Dried Sheets or SMR-5L)	100.	100.
Stearic Acid	1.	0.5
Zinc Oxide	5.	5.
Age Rite White	0.5	–
Age Rite Resin D	0.5	–
Age Rite Superlite	–	0.5
Anti-Sunchecking Wax	–	0.5
Process Oil	–	10.
Cumar Resin	–	5.
Clay	–	50.
Whiting	–	0.5
DOTG	0.2	0.5
Tuads	0.1	–
Altax	–	1.
Sulfur	1.2	2.5
Total	108.5	235.5
Cure:	5 minutes at 287°F	10 minutes @ 287°F
Tensile Strength, psi	4000.	2100.
Tensile Strength (aged 14 days @ 158°F)	4000.	–
% Elongation	750.	700.
% Elongation (aged 14 days @ 158°F)	600.	–
300% Modulus		625.
Shore A Hardness	39.	48.

REFERENCES

(1) G. F. Bloomfield, *Communications,* 271, 272, **13** (1951).
(2) W. H. Stevens and H. P. Stevens, *I.R.I. Trans.,* **11**, 182 (1935).
(3) H. Staudinger and W. Feisst, *Helv. Chim. Acta,* **13**, 1361 (1930).
(4) G. Kraus and W. B. Reynolds, *Journal Amer. Chem. Soc.,* **72**, 5621 (1950).

(5) G. F. Bloomfield and E. H. Farmer, *J. Soc. Chem. Ind.*, **53**, 43T, 47T (1934).
(6) P. Schidrowitz and C. A. Redfarn, *J. Soc. Chem. Ind.*, **54**, 263T (1935).
(7) G. F. Bloomfield, *J. Chem. Soc.*, 114 1944.
(8) B.P., *456, 536.*
(9) C. O. Weber, *J. Soc. Chem. Ind.*, **19**, 218 (1900).
(10) W. A. Caspari, *J. Soc. Chem. Ind.*, **24**, 1275 (1905).
(11) G. J. van Veersen, *Proc. Second Rubber Tech. Conf.*, 87 1948.
(12) S. D. Gehman, J. E. Field, and R. P. Dinsmore, *Proc. First Rubber Tech. Conf.*, 961 1938.
(13) C. W. Bunn and E. V. Garner, *J. Chem. Soc.*, 654 1942.
(14) J. I. Cunneen, *J. Chem. Soc.*, 36 1947.
(15) J. I. Cunneen, *B.P.* 580, 514, *B.P.* 820, 261, *B.P.* 820, 262.
(16) J. I. Cunneen, *Rubber Age*, **85** 650 (1959).
(17) J. Le Bras and P. Compagnon, *Comptes rendus*, **212**, 616 (1941); *Bull. Soc. Chim.*, **11**, 553 (1944); *Trans. Faraday Soc.*, **50** 759.
(18) G. F. Bloomfield, F. M. Merrett, F. J. Popham, and P. McL. Swift, *Proc. Third Rubber Tech. Conf.*, 185, Heffer & Sons, Cambridge 1954.
(19) H. Brown, *Rubber: Its Sources, Cultivation, and Preparation,* John Murray, London 1918.
(20) G. Martin, *I.R.I. Trans.*, **19** 38 (1943).
(21) Anon, *Rubber Res. Inst. Malaya Planters' Bulletin*, **21**, 95 (1955); **23**, 39 (1956).
(22) Anon, *ibid*, **40**, 3 (1959).
(23) Anon, *ibid*, **28**, 2 (1957).
(24) Anon, *ibid*, **22**, 11 (1956).
(25) Anon, *ibid*, **22**, 14 (1956); **34**, 5, (1958).
(26) G. Owen, D. R. Westzarth, and G. C. Iyer, *J. Rubber Res. Instit. Malaya*, **15**, 29 (1957).
(27) W. B. Haines, *Commonwealth Bureau of Soil Science, Tech. Comm.*, **46**, 2117, (1949).
(28) Anon, *Rubber Res. Instit. Malaya, Planters' Bulletin*, **3**, 54 (1952).
(29) Anon, *Rubber Res. Instit. Malaya, Planters' Bulletin*, **21**, 95 (1955); **23**, 39 (1956).
(30) W. Bobiloff, *Anatomy of Hevea Brasiliensis*, Zurich (1923).
(31) G. W. Chapman, *India Rubber World*, **125**, 94 (1951).
(32) Anon, *Rubber Res. Instit. Malaya, Planters' Bulletin*, **6**, 61 (1953).
(33) Anon, *Bull. d'information concernant l'Institut des Recherches sur le caoutchouc en Indochine*, **15**, 17 (1956).
(34) Anon, *Rubber Res. Instit. Malaya Planters' Bulletin*, **37**, 74 (1959).
(35) B. J. Eaton and R. G. Fullerton, *Rubber Res. Instit. Malaya, Planters' Bulletin Quarterly J.*, **1**, 135 (1929).
(36) M. W. Philpott, *B.P.*, 681, 486.
(37) F. J. Paton, B. J. Newey, T. H. Barnwell, and Dunlop Rubber Co., Ltd., *B.P.*, 695, 813.
(38) Anon, *Rubber Res. Instit. Malaya, Planters' Bulletin*, **22**, 2 (1956).
(39) *Type Description and Packing Specifications for Natural Rubber Grades Used in International Trade*, Rubber Manufacturers' Association, N.Y. 1957.
(40) G. F. Bloomfield, *J. Rubber Res. Instit. Malaya*, **13**, Part I. 273 (1951).
(41) G. Martin, *I.R.I. Trans.*, **19**, 38 (1943).
(42) Anon, *Rubber Res. Instit. Malaya, Planters' Bulletin*, **35**, 28 (1958).
(43) Anon, *Rubber Res. Instit. Malaya, Planters' Bulletin*, **32**, 83 (1957).
(44) B.R.P.R.A. and M. W. Philpott, *B.P.*, 803, 013.
(45) B. C. Sekhar and G. W. Drake, *J. Rubb. Res. Instit. Malaya*, **15**, 205 (1958).

(46) Anon, *Rubber Res. Instit. Malaya, Planters' Bulletin*, **33**, 108 (1957).
(47) Anon, *B.R.P.R.A. Tech. Bulletin No. 1, Heveaplus M.*
(48) B. L. Davies and J. Glazer, *Plastics Derived from Natural Rubber*, Monograph, Plastics Instit., London 1955.
(49) L. Bateman, *British Rubber Producers' Research Association, Annual Report*, 27 1958.
(50) Anon, *Vanderbilt Rubber Handbook*, 6 1968.
(51) M. Faraday, *Quarterly J. of Science and Arts*, **21**, 19 (1826).
(52) C. G. Williams, *Phil. Trans.*, 241 1860; *Proc. Royal Soc.*, **10**, 516 (1860); *J. Chem. Soc.*, **15**, 110 (1862).
(53) W. A. Tilden, *J. Chem. Soc.*, **45**, 410 (1884).
(54) G. Bouchardat, *Comptes rendues*, **89**, 1117 (1879).
(55) G. V. Schulz and A. Muls, *Proc. Natural Rub. Res. Conf. (Kuala Lumpur)*, 602 1960; *Makromol. Chem.*, **44**, 479 (1961).
(56) C. C. Davis and J. T. Blake, *The Chemistry and Technology of Rubber*, Reinhold, N.Y., 119 1937.
(57) J. I. Cunneen, *Proc. Int. Rubber Conf.*, Amer. Chem. Soc., Wash., D.C., 514 1959.
(58) P. B. Dickenson, *J. Cell Biology* 1963.
(59) A. Frey-Wyssling, *Arch. Rubbercultuur*, **13**, 392 (1929).
(60) E. H. Andrews, *Proc. Roy. Soc.*, **A270**, 232 (1962).
(61) L. A. Wood and H. Bekkedahl, *J. App. Phy.*, **17**, 362 (1946).
(62) N. Bekkedahl and L. A. Wood, *Ind. Eng. Chem.*, **33**, 381 (1941).
(63) P. J. Flory, *J. Chem. Phy.*, **15**, 397, (1947); **17**, 223 (1949).
(64) W. F. Busse, *Proc. Second Rubb. Tech. Conf.*, London, 288 1938.

7

STYRENE-BUTADIENE RUBBERS

WILLIAM M. SALTMAN
Research Division,
Goodyear Tire & Rubber Co.
Akron, Ohio

The most important synthetic rubber and the most widely used rubber in the entire world is SBR, a copolymer of styrene and butadiene. The history of its development from an interesting research material into the dominant commercial rubber of our time is a reflection of the political and social turmoil of our century.

In the 1920's the ability of emulsion systems using free radical catalysts to produce both high polymerization rates and high molecular weight products was first recognized. In the next decade the Nazi German Government stimulated research into synthetic rubbers in an effort to free itself of dependence on foreign sources of raw materials. The first butadiene-styrene copolymer from an emulsion system (Buna S) was prepared at the research laboratories of I. G. Farbenindustrie by Bock and Tschunker and was followed shortly thereafter by the analogous butadiene-acrylonitrile copolymer (Buna N).

Although these products were of poor quality compared to natural rubber, the technology, with considerable improvement and modification, formed the basis for synthetic rubber production in the United States.

The U.S. Government foresaw shortages of natural rubber due to spreading of war in the Far East and established a government corporation (The Rubber Reserve Company) in 1940 to accumulate a stockpile of natural rubber and start a synthetic rubber research and development program. The program was expanded when the U.S. entered World War II; arrangements were made for exchange of technical knowledge and for coordinated research programs amongst the interested rubber, oil, and chemical firms who dropped their traditional competitive activities. Emulsion polybutadiene was rejected as unsatisfactory and a styrene-butadiene copolymer made in emulsion with a charge ratio of 25% styrene and 75% butadiene was chosen as the best possible general purpose rubber for development during the emergency.

In mid 1942 production of GR-S (as SBR was then called) in a government plant began; in 1945 production was over 820,000 long tons. To achieve this goal, the

U.S. Government financed construction of fifteen SBR plants, two butyl rubber plants, sixteen butadiene production facilities, and five styrene plants. Between 1946 and 1955 these plants were sold to various private companies that have maintained and improved them ever since.

In recent years solution methods of polymerization have become popular. The resulting new rubbers are competing in cost and quality with SBR. Among the new polymers are styrene-butadiene copolymers, similar in properties to the emulsion type, but with some differences and advantages.

Raw Materials. Butadiene and styrene are the chief raw materials required to manufacture SBR. Others required in smaller amounts are the various emulsifiers, modifiers (e.g., thiols), catalysts, short-stops, coagulating agents, antioxidants, and antiozonants.

Butadiene

Present butadiene consumption in the United States is about 3.4 billion pounds per year, and about 80% of this is used to make synthetic rubber. Over half of the butadiene is derived from butenes, about 37% from butanes, and the remainder from miscellaneous refinery streams. World wide, butadiene demand should be about six billion pounds in 1970 and is expected to rise at a rate of about 5% per year. In both Europe and Japan, the C_4 by-product stream from naphtha crackers is a major source of butadiene. This source is expected to grow even more important in the future as the demand for ethylene increases. By the mid-seventies, Europe and Japan are expected to be exporting substantial amounts of this butadiene into the U.S.

In the process for production from butenes (Phillips process), the butenes are obtained from a C_3–C_4 stream.* The propane and isobutane are removed by fractionation and the isobutylene by absorption in sulfuric acid. The remainder is a mixture of C_4 straight chain hydrocarbons including about 30% of a mixture of 1-butene and 2-butene. This is fed to a catalytic dehydrogenator after preheating to about 550°C and diluting with superheated steam. The crude butadiene product must be purified before becoming suitable for polymerization use. Butane may be dehydrogenated in a single-step process to form butadiene using a chromia-alumina catalyst (Houdry process). Naphtha cracking yields a complex mixture containing ethylene as a major component.

Purification and concentration of the butadiene from all these streams is required. Emulsion SBR may tolerate the presence of as much as 5000 parts per million of acetylenic impurities in the polymerization recipe but solution BR or SBR cannot allow more than about 50 ppm. Purification of butadiene may be effected by complex formation with cuprous ammonium acetate followed by distillation or by extractive distillation with a solvent which preferentially absorbs butadiene. Several solvents such as furfural, acetonitrile, dimethyl formamide,

*Mixture of petroleum refinery gases containing hydrocarbons having 3 or 4 carbon atoms per molecule.

dimethyl acetamide, and N-methyl pyrrolidone are in commercial use. After a final fractionation the purified butadiene is inhibited against polymerization during storage by the addition of 25–200 ppm of *t*-butyl catechol or a similar substance. Although rubber or technical grade purity specifications call for a 98% minimum butadiene content, most commercial suppliers furnish monomer that is more than 99% pure. Absolute purity is less important than the presence (or absence) of polymerization poisons. Cost is about 9 1/2 to 11 1/2¢/lb depending on availability of refinery streams, purity requirements, etc. Base costs as low as 7¢/lb have been projected.

Styrene

U.S. production capacity of styrene is currently over 4.6 billion pounds per year, of which about 22% is used in SBR manufacture. Substantially all styrene is made by dehydrogenating ethyl benzene, the reaction product of benzene plus ethylene, though a small portion is made from ethyl benzene recovered directly from refinery streams. Current cost of their product is approximately 8¢/lb.

Ethyl benzene is prepared by alkylating benzene with ethylene in the presence of aluminum chloride in a typical Friedel-Crafts alkylation reaction. In some cases ethyl chloride is added as an activator for the reaction. The reaction mixture after neutralization with sodium hydroxide is fed to atmospheric distillation towers where the ethyl benzene is separated from the tars and heavy ends. In the direct recovery process the ethyl benzene is separated by super-distillation from the slightly higher boiling mixed xylenes.

In either case the ethyl benzene is mixed with superheated steam and dehydrogenated at high temperature, low pressure and a high ratio of steam to ethyl benzene. Chromia-alumina or iron oxide catalysts are used in the same way as in the preparation of butadiene from butane. Conversions to styrene are about 35–40% per pass. The product is distilled at low pressure to remove benzene and toluene light ends and tarry bottoms. The product, over 99% pure, is generally inhibited with a small amount of *t*-butyl catechol to prevent premature polymerization.

Polymerization

Polymerization of butadiene ($CH_2{=}CH{-}CH{=}CH_2$) occurs by addition of one butadiene unit to another repeated several thousand times over. Because there are two double bonds in the monomer, additions may be of three varieties, known as *cis*-1,4; *trans*-1,4; or vinyl (also called 1,2). In a polymer chain these may be all joined together randomly.

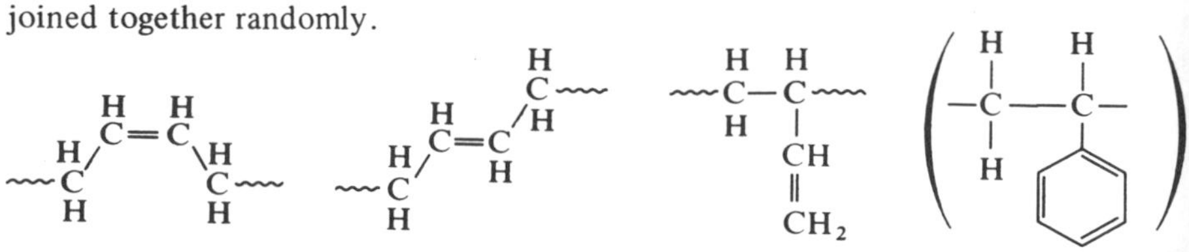

cis-1, 4　　*trans*-1, 4　　vinyl　　styrene

In SBR there are styrene units in the chain as well as the three butadiene forms above. These copolymers of styrene and butadiene may be randomly dispersed mixtures of the two monomers or blocks (where large segments of each kind follow one another ~~~~ SSSSBBBBBSSSBBBBB~~~~) or grafts (where the segments dangle from the main chain).

SBR made in emulsion usually contains about 23% styrene (but see Tables 7.3 and 7.5 for other levels of styrene) randomly dispersed with butadiene in the polymer chains. The structure of the butadiene units is about 18% *cis*, 65% *trans*, and 17% vinyl (percent of the butadiene portion only). SBR made in solution contains about the same amount of styrene but both random and block copolymers have been made. They have lower *trans*, lower vinyl, and higher *cis* contents than emulsion SBR. The major differences between emulsion and solution SBR appears to be in the linearity and molecular weight distribution of the molecular chains rather than in their microstructures.

Emulsion Polymerization

Polymerization in emulsion was investigated by early workers in order to imitate the physiological conditions which they conjectured existed during the formation of natural rubber latex. The preparation of stable latices with soaps and the sulfonate emulsifiers, and the subsequent discovery of water-soluble free radical polymerization initiators such as hydrogen peroxide and potassium peroxydisulfate, gave great impetus to emulsion research, since high polymers were formed in comparatively swift reactions.

In a homogeneous (solution or bulk) system with free radical initiators the maximum degree of polymerization, i.e., the chain length, attainable is too low (molecular weights below 10^4) for elastomer use unless the rate of polymerization is exceedingly slow. If the free radical concentration is high enough to give appreciable reaction rates, the competition for monomer is so keen that termination occurs before growth to a high degree of polymerization. This limitation does not apply to emulsion systems for which both high propagation rate and high molecular weight are simultaneously possible because the emulsion physically isolates each growing radical and prevents them from terminating one another. Other advantages for emulsion systems lie in the high rate of transfer of the heat of polymerization through the aqueous phase. Because the heat of polymerization amounts to about 18 kcal/mole for dienes, good heat transfer is necessary if temperature control is to be maintained. The emulsion systems also allow ready removal of unreacted monomers and remain fluid even though high concentrations of high polymer are present.

A typical emulsion system contains water, monomer(s), initiator, and an emulsifier (soap). Although early investigators believed that the locus of polymerization was at the monomer-water interface of the emulsion droplet, later work showed that polymerization begins in the monomer solubilized in the much smaller soap "micelle" of the aqueous phase and not in the aqueous solution nor in the monomer droplets.

In free radical polymerizations in emulsion the chain reaction is initiated by radicals generated, e.g., by the decomposition of a peroxide or a peroxydisulfate (persulfate). Because the rate of formation of free radicals by the initiator is temperature dependent, the earliest free radical emulsion polymerizations were carried out at about 50°C or higher in order to obtain reasonable polymerization rates. Subsequent research led to the discovery of other oxidation reduction reactions (redox) capable of generating radicals in sufficient numbers for adequate polymerization rates at temperatures as low as −40°C.

Commonly employed initiators include potassium peroxydisulfate ($K_2S_2O_8$), benzoyl peroxide, cumene hydroperoxide, and azobisisobutyronitrile (AIBN). These compounds are thermally unstable and decompose at a moderate rate to release free radicals. The combination of potassium peroxydisulfate with a mercaptan such as dodecyl mercaptan is used to polymerize butadiene and SBR. In hot recipes, the mercaptan has the dual function of furnishing free radicals through reaction with the peroxydisulfate, and also of limiting the molecular weight of polymer by reacting with one growing chain to terminate it, and to initiate growth of another chain. This use of mercaptan as a chain transfer agent or modifier is of great commercial importance in the manufacture of SBR and of polybutadiene in emulsion, since it allows control of the toughness of the product which otherwise may limit the processibility in the factory.

A standard polymerization recipe agreed on for industrial use became known as the "mutual," "standard," "GR-S," or "hot" recipe and was as follows:

Butadiene	75.0	Potassium Peroxydisulfate	0.3
Styrene	25.0	Soap Flakes	5.0
n-Dodecyl Mercaptan	0.5	Water	180.0

At 50°C conversion to polymer occurred at 5–6% per hour. Polymerization was stopped at 70–75% conversion since higher conversions led to polymers with inferior physical properties, presumably because of crosslinking in the latex particle to form microgel or highly branched structures. The termination of the reaction was effected by the addition of a "shortstop" such as hydroquinone (about 0.1 part by weight), which reacted rapidly with radicals and oxidizing agents. Thus the shortstop destroyed any remaining initiator and also reacted with polymer free radicals to prevent formation of new chains. The unreacted monomers were then removed, first the butadiene by flash distillation at atmospheric, followed by reduced pressure, and then the styrene by steam stripping in a column. A dispersion of antioxidant, such as N-phenyl-β-naphthylamine (PBNA), was added (1.25 parts) to protect the product from oxidation. The latex was partially coagulated (creamed) by the addition of brine and then fully coagulated with dilute sulfuric acid or aluminum sulfate. The coagulated crumb was washed, dried, and baled for shipment. This general procedure is still basic for the present-day production of all emulsion polymers. Many variations and refinements have been developed which make possible improved products with distinct and unique properties, but the underlying procedure is the one described above. One of the first major improvements on the basic process was the adoption of continuous processing shown schematically in Fig. 7.1. The styrene, butadiene, soap, initiator, and activator (an auxiliary initiating agent) are pumped continuously from storage tanks into and through a series of agitated reactors maintained at the proper temperature at a rate such that the desired degree of conversion is reached at the exit of the last reactor. Shortstop is then added, the latex is warmed by the addition of steam, and the unreacted butadiene is flashed off. Excess styrene is then steam stripped off and the latex is finished, often by blending with oil, creaming, coagulating, drying, and baling as shown in the figure.

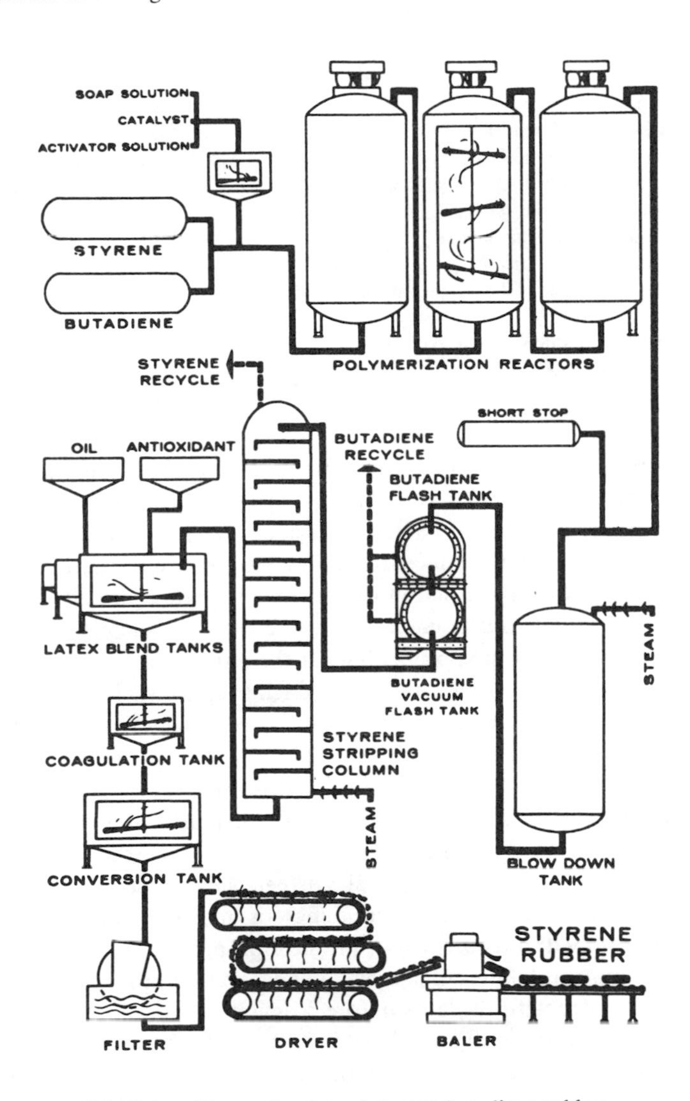

Fig. 7.1 The production of styrene-butadiene rubber.

Cold Rubber. Another major improvement in SBR production arose from the use (about 1947) of more active radical initiating systems which permitted polymerization at 5°C with high rates of conversion. The "cold" SBR polymers produced at the lower temperature, but stopped at 60% conversion, were found to have properties superior to those of "hot" SBR. Two typical recipes for a "cold" SBR are given in Table 7.1.

TABLE 7.1. TYPICAL FORMULATIONS FOR "COLD" SBR

	1	*2*
Butadiene	72	71
Styrene	28	29
Tert-Dodecyl Mercaptan	0.2	0.18
Diisopropylbenzene Monohydroperoxide	0.08	–
P-Menthane Hydroperoxide	–	0.08
Ferrous Sulfate ($FeSO_4 \cdot 7H_2O$)	0.14	0.03
Potassium Pyrophosphate ($K_4P_2O_7$)	0.18	–
Trisodium Phosphate ($Na_3PO_4 \cdot 10H_2O$)	–	0.50
Tetrasodium Salt of Ethylenediamine-tetracetic Acid (EDTA)	–	0.035
Sodium Formaldehyde Sulfoxylate	–	0.08
Resin Acid Soap	4.0	4.5
Water	180	200

At 5°C 60% conversion to polymer occurs in about 12 hr. The mercaptan and soap perform the same functions as in "hot" recipes. The main difference lies in the initiator systems. The phosphates and EDTA act as buffers and also complex with ferrous ion and thus limit the concentration of the free ion. Initiation results from the radicals generated by the reaction of the iron and hydroperoxide. The reaction is very rapid even at 0°C and in emulsion, where the components are primarily in separate phases and the reaction occurs only at the interface.

In many "cold" recipes, an auxiliary reducing agent such as glucose or the sulfoxylates has been used. The sugars are no longer in widespread use because of their cost and susceptibility to bacterial attack during storage.

A major improvement in SBR quality came with the availability of cold rubbers. It was found that gel-free elastomers with a higher than usual molecular weight, and consequently difficult to process with ordinary factory equipment, could be modified by the addition of up to 50 phr of petroleum base oils, to permit easy factory handling. Not only do these extending oils improve processing characteristics, but they do so without sacrifice in physical properties (which depend on the high molecular weight). In the commercial polymerization process the oil is usually emulsified and blended with the latex before coagulation.

Recent trends in SBR production have been toward grades designed for specific uses. Improvements have been made in the color of SBR, which is important in many nontire uses. This has resulted from the use of lighter colored soaps, shortstops, antioxidants, and extending oils. An example is the substitution of dithiocarbamates for hydroquinone as shortstop; the latter is now rarely used and only in some hot SBR where dark color is not objectionable. A shortstop such as sodium dimethyldithiocarbamate is more effective in terminating radicals and destroying peroxides at the lower temperatures encountered in preparing the cold rubbers.

Masterbatches. Another improvement directed toward specific end uses has been the preparation of carbon black masterbatches of regular and oil-extended cold SBR. These are of interest to rubber manufacturers having limited mixing capacity and those who wish to avoid factory handling of loose blacks.

Addition of carbon black in masterbatch manufacture is accomplished by slurrying the pigment in water which may contain an anionic surface-active agent. The SBR latex and slurry are then mixed, along with the emulsified extending oil and antioxidant. The resulting intimate blend is then coagulated rapidly, with vigorous agitation, by addition of an excess of acid or salt/acid. The creaming step is usually omitted since it results in preferential precipitation of an unredispersible black with resultant poor qualities of finished product. Other methods for black masterbatching include blending the components in a steam jet and coagulation by addition of the latex-black blend to the salt-acid mixture.

Commercial Grades. The International Institute of Synthetic Rubber Producers, Inc., is now responsible for assigning numbers to various commercial grades of SBR and butadiene polymers and latices. The numbering system instituted under the government synthetic rubber program is still, in the main, adhered to by private industry, although there is a recent tendency for the companies to assign an ASTM code number or their own codes or trade names for newer products. The Institute numbering system is arranged as follows:

Series:
1000 hot polymers
1100 hot black masterbatch with 14 or less parts of oil per hundred parts SBR
1500 cold polymers
1600 cold black masterbatch with 14 or less phr oil
1700 cold oil masterbatch
1800 cold oil-black masterbatch with more than 14 phr oil
1900 miscellaneous dry polymer masterbatches
2000 hot latices
2100 cold latices

There are many types of SBR, but only a portion of these are available from any one manufacturer. Each producer will generally prefix the code number with his distinguishing trade name. Table 7.2 lists the trade names of the major American manufacturers and the ASTM code (semi-commercial) numbers assigned to each. The variety of products obtainable from these producers are listed in *The Elastomers Manual* (see References).

TABLE 7.2A. SBR AND BR PRODUCERS

Producer	*Abbreviation*	*Trade Name*	*Semi-Commercial Numbers*[a]
American Rubber and Chemical Company	A	CISDENE	
American Synthetic Rubber Corporation	AS	ASRC	3000–3499
Copolymer Rubber and Chemical Company	C	COPO, CARBOMIX	3500–3999
Dewey and Almy Chemical Division – W. R. Grace & Company	DA	DAREX	
Firestone Synthetic Rubber & Latex Company	F	FR-S	4000–4499
General Tire and Rubber Company	G	GENTRO, GENTRO-JET, JETRON	9000–9499
Goodrich-Gulf Chemicals, Inc.	GG	AMERIPOL	4500–4999
Goodyear Tire and Rubber Company	GT	PLIOFLEX PLIOLITE	5000–5499
International Latex Corporation	IL	TYLAC	
Phillips Petroleum Company	PP	PHILPRENE	6500–6999
Polymer Corporation Ltd. (Canada)	P	POLYSAR, KRYLENE, KRYNOL, KRYMIX, KRYFLEX	7000–7499
Shell Chemical Company	S	S.	7500–7999
Texas-U.S. Chemical Company	TU	SYNPOL	8000–8499
United Carbon Company	U	BAYTOWN	8500–8999
Uniroyal, Chemical Division	US	NAUGAPOL, NAUGATEX	6000–6499

[a] For those producers with blocks of 500 numbers, each block should be further divided by type according to the following tabulation.

Solution SBR. Several solution SBR's are offered commercially. The random copolymers are rubbery and like emulsion SBR but with several improved properties. The block polymers tend to be thermoplastic and are not recommended for tire use. Random products have narrower molecular weight distribution, less chain branching, higher *cis* content, lighter color, and less nonrubber constituents than the emulsion SBR's. As a result, they are reported to have better abrasion resistance, better flex, higher resilience, and lower heat build-up than the emulsion rubbers. Tensile, modulus, elongation, and cost are comparable.

At present, successful solution copolymerization of styrene and butadiene has been carried out only with lithium and alkyl lithium catalysts. Coordination catalysts such as are used for stereospecific polybutadiene and polyisoprene have not been satisfactory for styrene-butadiene copolymers. Polymerization plants for preparation of the solution polymers have many points of similarity. The

TABLE 7.2B. SBR AND BR PRODUCERS

Producer's Code No.	*Product Type*
0 to 49	Hot nonpigmented polymers
50 to 99	Hot black masterbatch with 14 or less parts of oil per 100 parts SBR
100 to 149	Cold nonpigmented polymers
150 to 199	Cold black masterbatch with 14 or less parts of oil per 100 parts SBR
200 to 249	Cold oil masterbatch
250 to 299	Cold oil black masterbatch with more than 14 parts of oil per 100 parts SBR
300 to 349	Hot latex
350 to 399	Cold latex
400 to 499	Unassigned

The American Society for Testing and Materials has recently published a "Recommended Practice for DESCRIPTION OF TYPES OF PETROLEUM EXTENDER OILS," ASTM Designation: D2226 which is shown below:

Classification of Oil Types

Types	*Asphaltenes max., per cent*	*Polar Compounds max., per cent*	*Saturated Hydro-carbons, per cent*
101	0.75	25	20 max
102	0.5	12	20.1 to 35
103	0.3	6	35.1 to 65
104	0.1	1	65 min

The classifications highly aromatic, aromatic, naphthenic, and paraffinic correspond to the ASTM classifications of Types 101, 102, 103, and 104, respectively.

polymerization catalysts are all attacked by air and moisture so that strict exclusion, down to parts per million, of these and other catalyst poisons is imperative. In general, continuous reactor systems are used.

The catalyst is added to the thoroughly dried mixture of monomer(s) and hydrocarbon solvent. As with the emulsion SBR system there may be several reactors in a chain. As the polymerized solution (cement) leaves the last reactor, stopper and stabilizer are added. The cement is steam stripped to form a rubber crumb and to recover the solvent and unreacted monomers for recycling. The drying of the rubber crumb is completed on tray or extruder driers. If oil extended polymer is desired, the oil is added to the cement after the stabilizer addition but before the steam stripping operation. Molecular weight of the rubber is controlled during polymerization by control of catalyst level and reaction temperature. To make copolymers, such as SBR, careful control of monomer charge rate and total conversion is necessary.

Polymerization of butadiene by lithium has been known for many years but was not exploited commercially until the 1960's. Butyl lithium, being soluble in hydrocarbons, is widely used as a catalyst. Propagation occurs as butadiene units are inserted in between the lithium and butyl group. The most striking effect of this system is the absence of a termination mechanism. Such a system leads to polymers with very narrow molecular weight distributions since all the growing chains have equal probability of growth. In actual practice some termination and broadening of the distribution occur because of the presence of impurities in the system. In many cases, reagents may be added to the polymerized cement to deliberately introduce branching or broadened molecular weight distributions into the polymer to enhance its storage stability or factory processing behavior.

When styrene and butadiene are copolymerized with alkyl lithium in a *batch* reactor the reactivity ratios are such that the butadiene units polymerize first and with almost total exclusion of any styrene present. Only when all the butadiene monomer is consumed does the bulk of the styrene enter the polymer chain. Thus, a block polymer, BBBB . . . . BBSS . . . SSSS, is formed. Since there is no termination step, tri-block (SBS) polymers may be made by adding in succession, styrene, butadiene, and then styrene again to the catalyst solution.

Since the styrene blocks behave like polystyrene, these polymers are not suitable for tire uses although they have many interesting properties and uses. Random copolymers, suitable for tire uses, may be made, as indicated above, by careful control of the charge to continuous reactors. Random copolymers may also be made by addition of modifiers (such as an ether) to the catalyst. The presence of the modifier not only results in increased random incorporation of styrene units in the chain but also increases the vinyl microstructure of the butadiene segments from about 8 to 20%.

The variety of solution SBR's available is also indicated in *The Elastomers Manual* Philips' Solprene 1204 is a block SB polymer and not recommended for tire use, but the random Solprenes 1205 and 375 are. Firestone's Stereon 750, recommended for tire and other uses, has very low vinyl content in the butadiene moiety. Shell Chemical makes several SBS tri-block copolymers under the tradename "Kraton." Although not general purpose rubbers suitable for tires, they are finding many other uses. Because the two rigid polystyrene blocks tend to anchor the central rubbery polybutadiene block, the raw polymer behaves like a cured elastomer at room temperature. When heated above the glass transition temperature of polystyrene (~100°C), the styrene segments lose their rigidity and the polymer behaves like an uncured elastomer. Thus, products, such as shoe soles, may be injection molded at high temperatures without requiring a cure step. Scrap, or excess from such operations may be recovered and reused. These block polymers are discussed in greater detail in Chap. 20.

Physical Properties

There are a host of physical properties which are of importance to the technologist for determining how suitable a particular rubber is for a particular use. Only a few of the more fundamental properties are discussed here. The single most important property for classifying a material as an elastomer is the glass transition temperature (or temperature range). This is the region where a viscous or flexible (rubbery) material above the transition changes to a stiff and brittle material below. The transition temperature (T_g) is characterized by small changes in the rate of change of specific volume with temperature (thermal expansion coefficient) and similar variations for specific heat, sound velocity, refractive index, etc. In many ways it resembles a second-order thermodynamic transition.

Theory has connected the transition with the temperature below which free rotation of polymer chain segments is not possible. It may be conceived as a "free-volume" related to polymer molecular weight, degree of crosslinking, copolymer composition, bulkiness of side groups, and chain stiffness.

The glass transition temperature may be readily measured by instruments designed for differential thermal analysis or differential scanning calorimetry. In addition, a simple relation was found between the glass-transition temperature and the Gehman low-temperature torsion flex test (ASTM D 1053).

For SBR copolymers prepared by emulsion polymerization at 50°C, the glass-transition temperature is calculable from the styrene content (S = weight fraction of styrene) by:

$$T_g - (85 + 135S)/(1-0.5S)$$

For a similar copolymer prepared at 5°C, T_g is given by:

$$T_g = (-78 + 128S)/(1-0.5S)$$

No equations have been proposed for the solution SBR's but qualitatively the T_g will be higher, the higher the styrene content. For a given styrene content a solution SBR will generally have a T_g slightly lower than emulsion SBR with the same styrene level because the butadiene portions of solution SBR are lower in vinyl structure and higher in *cis* structure.

Molecular weight and its distribution are also important in determining polymer properties. In addition to the traditional molecular weight (the "number average," $\overline{M}_n$), polymer molecular weights may be estimated from light scattering, ultracentrifuge, ($\overline{M}_w$), solution viscosity, and other measurements. Viscosity measurements are very convenient and can be empirically related to molecular weight, using the Kuhn-Mark relation:

$$[\eta] = KM^a$$

where $[\eta]$ is the intrinsic viscosity (in dl/g), M is the polymer molecular weight, and K and a are experimentally determined constants. Values of K and a for SBR prepared at 50°C were 5.4×10^{-4} and 0.66 respectively when the viscosity measurements were made in toluene at 30°C.

Although a high molecular weight is desirable to obtain better physical properties, low molecular weight polymer processes more readily. It is thus necessary to compromise to obtain optimum quality rubber. Thus not only is the average molecular weight important but also the portion of polymer above and below the average. The distribution of molecular weight can be measured by gel permeation chromatography (GPC). The heterogeneity index ($HI = \overline{M}_w/\overline{M}_n$) is also often used to indicate the breadth of the molecular weight distribution.

For emulsion SBR the distribution varies with reaction conversion, type of mercaptan modifier used, reaction temperature and the commercial source. A type 1500 SBR typically might have an intrinsic viscosity of about 2.0, an $\overline{M}_w$ of 320,000 to 400,000, and an $\overline{M}_n$ of 80,000 to 110,000. Heterogeneity indices might range from 3 to 5. A type 1712 SBR might have molecular weight about 30% higher but with the same HI.

Solution SBR's tend to have very narrow molecular weight distributions, and to improve processing it is often desirable to deliberately broaden the distribution by adding some branching reagent immediately after polymerization. A typical solution SBR may have an $\overline{M}_n$ of above 100,000 and a heterogeneity index of 2.0 after broadening.

Compounding and Processing

Compounding recipes in enormous variety are available from various manufacturers for a wide range of end uses. The inexperienced compounder should regard such recipes as good points of departure for determining an optimum compound for use with his own ingredients in his own plant. Unfortunately, minor variations in the nature of the carbon black, oil, or other compounding ingredients can effect major variations in rubber physical properties. Minor differences in processing equipment, such as a die configuration, can make one recipe extrude smoothly and another raggedly. Any production run should be preceded by trial until experience justifies otherwise.

Modern styrene-butadiene rubber no longer requires the repeated mixing cycles of the earlier products. Its extrusion properties are superior to those of natural rubber, and its stocks have less tendency to scorch in processing. Although cold SBR is often preferable to hot for optimum physical properties, use of hot SBR types, when possible, can contribute to both processing and product improvements. They break down more readily, develop less heat, and accept more filler in processing. For many uses, blends of SBR and other rubbers such as natural rubber or *cis*-polybutadiene are made. Compounding recipes should be proportioned to balance the requirements for each type of rubber used.

All types of SBR use the same basic compounding recipes as do other unsaturated hydrocarbon polymers. They need sulfur, accelerators, antioxidants (and antiozonants), activators, fillers, and softeners or extenders. SBR requires less sulfur than natural rubber for curing. The usual range is about 1.5–2.0 phr of sulfur; for oil-extended SBR this should be based on the rubber hydrocarbon only. All styrene-butadiene rubbers, because of their lower unsaturation, are slower curing than natural rubber and require more acceleration.

Compounding recipes with low sulfur or containing only organically bound sulfur (as in a thiazole) improve aging properties but are slower curing. Zinc stearate (or zinc oxide plus stearic acid) is the most common activator for SBR. There are many accelerators which are useful to speed up slow-curing stocks and retarders for slowing down the cure rate of scorchy stocks. Recipes may also contain plasticizers, tackifiers, softeners, waxes, reclaim, etc. It is not unusual to find a recipe with 15 or more ingredients.

A fairly typical tread recipe which might be suitable for a passenger tire and containing SBR and polybutadiene has been given before (Table 2.1 on p. 20).

Other recipes directed towards nontire uses are given in Tables 7.3 to 7.6. Table 7.3 lists two recipes which may be used for general purpose mechanical goods. The

TABLE 7.3. SBR COMPOUNDS FOR GENERAL PURPOSE MECHANICAL GOODS

Compound	*phr*	*phr*
SBR 1712C[a]	100	100
Zinc Oxide	5.00	5.00
Stearic Acid	1.00	1.00
Mineral Rubber	10.00	20.00
FEF Black	20.00	20.00
SRF Black	75.00	30.00
Hard Clay	–	100.00
Sun Proof Improved	5.00	3.00
Paraffin Wax	1.00	–
Wing Stay T	1.00	–
PBNA	–	1.00
Para Flux	10.00	15.00
Altax	1.75	2.00
DPG	0.50	–
Tuads	–	0.50
Sulfur	2.25	2.00
Total	232.50	299.50
Specific Gravity	1.241	1.398
Mooney Plasticity, ML4' @ 212°F	47	42
Mooney Scorch, MS @ 270°F		
Minimum	18	18
Minutes to 10 Point Rise	14	15
Best Cure, Minutes @ 310°F	20	20
Physical Properties		
Tensile Strength, psi	1625	1175
Elongation, %	400	620
300% Modulus, psi	1425	600
Hardness, Shore A	65	62
Crescent Tear (Die B), lb/in.	198	152
NBS Abrasion Index	123.7	22.8
% Rebound (Goodyear-Healy)	43.7	31.5
Compression Set, "B," Aged 22 hr @ 158°F	13.0	27.0

[a] Plioflex 1712C manufactured by the Goodyear Tire and Rubber Company.

TABLE 7.4. SBR COMPOUNDS FOR WIRE AND CABLE JACKETING

Compound	*phr*	*phr*
SBR 1503[a]	100	–
SBR 1708[a]		100
Zinc Oxide	10.00	10.00
Mineral Rubber	40.00	30.00
Hard Clay	125.00	125.00
Atomite	75.00	75.00
Wax Paraffin	3.00	3.00
Sulfur	2.00	2.00
Altax	2.50	2.50
Methyl Zimate	1.25	1.00
	358.75	348.50
Mooney Viscosity, ML 4' @ 212°F	60	41
Specific Gravity	1.558	1.578
Tensile Strength, psi		
5'/310°F	1250	1050
10	1310	1020
15	1250	1040
20	1300	1060
Elongation, %		
5'/310°F	620	730
10	590	620
15	580	605
20	590	640
300% Modulus, psi		
5'/310°F	600	320
10	650	380
15	630	410
20	660	370
Hardness, Shore A		
5'/310°F	69	58
10	70	61
15	71	62
20	71	62
Remilled Data-Best Cures, 10' @ 310°F		
Tensile Strength, psi	1250	1010
Elongation, %	570	625
300% Modulus, psi	640	390
Hardness, Shore A	70	60
Crescent Tear; Die C, Lb/Inch		
10'/310°F	149	90
Permanent Set – 2 Minutes Hold; 2 Minutes Rest		
150% Elongation		
10'/310°F, % Set	13	14.5

TABLE 7.4 (*continued*)

Compound	*phr*	*phr*
Mooney Scorch, MS @ 250°F		
Minimum MV	24	23
Minutes to 3 Pt Rise	13.6	15.4
Minutes to 5 Pt Rise	14.2	16.2
Minutes to 10 Pt Rise	15.4	16.7
Aged 96 Hours in Oxygen Bomb @ 158°F, 300 psi (10′ @ 310°F Cures)		
Tensile psi	1220	950
Retained %	97	89
Elongation %	510	535
Retained %	89	86
Hardness, Shore A	75	68
Points Change	+5	+8
Volume Resistivity; Ohm-Cm × 10^{-14}		
10′/310°F	3.5	3.6
Volume Resistivity; Aged 7 Days in H_2O @ 158°F, Ohm-Cm x 10^{-14}		
10′/310°F	2.15	1.1
Water Absorption; Aged 7 Days in H_2O @ 158°F, Mg/cm^2		
10′/310°F	1.19	1.26

[a]Plioflex 1053 and 1708 manufactured by the Goodyear Tire and Rubber Company.

use of clay in the second of these poses the usual dilemma for the compounder since it gives both lower cost and lower physical properties.

Table 7.4 lists recipes and some physical properties for electrical grade rubbers. Special requirements for such rubbers are low water absorption and low ash content to maintain high dielectric properties.

Table 7.5 shows a recipe for open cell sponge rug underlay. Because such rubbers are open cell and are blown with sodium bicarbonate, the recipes usually call for low Mooney and fast cures. SBR 1905, used here, is a rubber/resin masterbatch containing 25 parts of a high styrene resin per 100 parts of rubbery SBR.

Table 7.6 is for a recipe for a closed-cell sponge compound suitable for shoe soling applications. The data in the table show the effect of cure time on the hardness and specific gravity of the sponge. Properties and cure rates of sponge are usually very sensitive to changes in accelerator type and level in the recipe.

Processing of SBR compounds is similar to that of natural (or other) rubber. The ingredients are mixed in internal mixers or on mills, and may then be extruded, calendered, molded, and cured in conventional equipment. Mixing procedures vary with the compound. In general, the rubber, zinc oxide, antioxidant, and stearic acid are mixed; then the carbon black is added in portions with the softener or oil. (This

TABLE 7.5. SBR COMPOUND FOR RUG UNDERLAY

Compound	*phr*
SBR 1905[a]	125.00
Zinc Oxide	6.25
Whiting	200.00
Naphthenic Oil	85.00
Sulfur	4.25
Wing Stay L	1.00
Methyl Tuads	2.50
Micronized Sodium Bicarbonate	18.00
Methyl Zimate	.15
Petrolatum	12.50
ML4 – 212°F	5
Grab Tear Tests, lb	
Room Temperature	7.10
220°F	.80
Compression Deflection, 25% psi	.50
Compression Set B, 22 hr at 158°F %	18.0
Cure, 340°F, min	10

[a] Plioflex 1905 manufactured by the Goodyear Tire & Rubber Company.

may be considered a black masterbatch.) It may be desirable at this point to dump, sheet out, and cool the batch. The second phase now includes mixing in all the other ingredients with the accelerator and sulfur being added last. Mixing is then continued until the sulfur is well dispersed.

The earlier shortcomings of poor cut-growth resistance and the requirements for repeated mixing cycles have been overcome in the styrene-butadiene rubbers now in use. Tread-wear and aging properties are superior to those of natural rubber. Building tack is still poor and dynamic properties are such that heavy-duty tires become too hot in use for satisfactory life or service. When reinforcing fillers such as carbon black or silica are not used, physical properties are poor. For maximum tensile properties and abrasion resistance cold SBR is preferable to hot. Hot SBR is no longer used for tire-tread compounds. In applications where lower physical properties can be tolerated hot SBR types may contribute processing advantages.

Although the styrene-butadiene elastomers were originally manufactured during the war years as a rather mediocre replacement for natural rubber, which was unavailable, they now stand on their own merits. A major reason for their popularity is cost (advantages realized with oil extended elastomers). Quality, however, is also important since present-day styrene-butadiene rubber often has better abrasion characteristics and better crack initiation resistance than natural rubber. With their lower unsaturation styrene-butadiene rubbers also have better heat resistance and better heat-aging qualities. SBR extrusions are smoother and maintain their form better than natural rubber does.

TABLE 7.6. CLOSED CELL SPONGE FOR SHOE SOLING – EFFECT OF CURE ON PERCENT BLOW

Compound	*phr*
SBR 1507[a]	100.00
Pliolite S6B[b]	25.00
Zinc Oxide	3.00
Stearic Acid	2.00
Zeolex 23	55.00
Hard Clay	40.00
Atomite	10.00
Coumarone Indene MH	10.00
Medium Process Oil	10.00
Sulfur	2.80
Amax	1.00
Triethanol Amine	1.00
Cellogen	7.00
	266.80

6" x 6" x 1/4", beveled mold – Cured the following times under pressure, plus 4 hours in 250°F oven:

Mooney Plasticity ML 4' @ 212°F	46
Specific Gravity	
8'/310°F	0.288
10'/310°F	0.367
12'/310°F	0.435
15'/310°F	0.525
20'/310°F	0.645
Hardness, Shore A – Buffed/Unbuffed Surfaces	
8'/310°F	22/32
10'/310°F	38/41
12'/310°F	39/50
15'/310°F	48/58
20'/310°F	54/63
Split Tear (Cured 15' @ 310°F)	
Tear, lb	15

[a]Plioflex 1507 manufactured by the Goodyear Tire and Rubber Company.
[b]Pliolite S6B is a high styrene resin manufactured by the Goodyear Tire and Rubber Company.

Mixing procedures for rubber stocks vary with different companies. Some plasticize the rubber in a separate operation before addition of blacks and other ingredients. The procedure may be largely determined by the capacity and mixing ability of the available equipment.

In the case of natural rubber, it is necessary to plasticize the raw rubber mechanically as a separate operation before mixing with other ingredients. With a butadiene-rubber stock, however, some large companies employ a procedure whereby the operator adds carbon black and other ingredients to make a masterbatch, thus eliminating the preliminary plasticizing step. If additional plasticization is required, it may be advantageous to remill the masterbatch before use in the finished batch, rather than to plasticize the rubber by itself. Excessive remilling of SBR compounds may lead to gel formation and poorer extrusions.

The formulations for tread stocks required in the greatest amounts are usually such that remilling of the masterbatch is not necessary for satisfactory extrusion conditions on the tread tubing machine. Common mixing procedures compare as follows (when the tread stock is allowed to cool, and warmed up again on 84″ mills prior to feeding to the tubing machine):

Natural Rubber	*Styrene-Butadiene Rubber*
Plasticize or Peptize	
(By Mechanical Methods)	
Black Masterbatch	Black Masterbatch
Finish	Finish

The masterbatches of certain synthetic tread stocks must be remilled prior to the finishing step in which the vulcanizing ingredients are added. But this is equally necessary for certain natural rubber tread stocks. On an overall basis, the processing of synthetic tread stocks requires less milling time than natural rubber tread stocks, because one step is eliminated.

A typical comparison of the time required per 1000 lb of finished stocks, using a No. 11 Banbury at 30 rpm, would be the following:

Natural Rubber Tread	30 min
SBR Tread	20 min

The power requirements with synthetic stocks are somewhat higher. They vary with the particular formulation, but loads on the Banbury motor may be 20% higher for synthetic than for natural rubber stocks.

In tubing operations, the rates of extrusion of synthetic and natural rubber tread stocks are comparable in terms of output per hour per turn of the screw. Synthetic stocks will break down somewhat more quickly than natural rubber stocks. Thus, if warm-up capacity is the limiting factor, total output of the tubing machine can sometimes be increased for synthetic stocks.

One of the major defects in SBR stocks is the lack of green tack. For this reason one must apply a layer of natural rubber cement to obtain adhesion on all surfaces of the tread which are to be joined to other surfaces when the tire is built. This is done mechanically on the bottom side of the tread tubing unit.

If one uses SBR in stock for coating fabric on a calender, it is necessary to use slightly more crown on the calender roll which forms the gum sheet in order to achieve uniform gauge across the sheet. The greater crown required is .001″ to .002″, depending upon the diameter of the roll.

If the stock being calendered has a synthetic rubber content in excess of one-third of the total rubber hydrocarbon, a thin layer of natural rubber cement must usually be applied to the surface of the coated fabric to obtain satisfactory green adhesion in assembling the tire.

From this point on in the manufacture of passenger tires, there is no essential difference which can be attributed to the type of stock employed, whether it be natural or synthetic.

Uses

While passenger tires and tire products account for the major portion of SBR consumption, other branches of the rubber industry also make extensive use of these synthetics in the fabrication of a wide variety of products. Table 7.7 gives a breakdown of SBR end-uses for 1962 and 1969. Although total consumption has increased, some markets have faltered as a result of competition from other rubbers and plastics. Essentially all this SBR is from emulsion processes. Solution SBR is only now (1970) beginning to be a market factor.

In 1962, a year when the first stereo rubber plants were coming on stream, many prognosticators were predicting sharp drops in future SBR consumption. The product has proved more durable than anticipated and future growth should continue at a slow rate as the use of all rubber products increases with our expanding population and economy. At present, SBR represents about 50% of all new rubber used in the U.S., about 58% of all dry, general purpose rubbers, and about 65% of all synthetic rubber. In comparison about 22% of all new rubber used in the U.S. is natural rubber. For both SBR and natural rubber, the future trend seems to be toward greater tonnage but lesser fractions of the total market.

TABLE 7.7. U.S. CONSUMPTION OF SBR (THOUSANDS OF LONG TONS)

End Use	*1962*	*1969*
Tires and related products	611	890
Wire and cable	13	11
Mechanical goods	43	50
Footwear and shoe	62	62
Foamed products	38	50
Belts, belting	–[a]	8
Hose, tubing	–[a]	17
Sponge	–[a]	10
Proofed goods	6	10
Adhesive	4	10
Miscellaneous, latex uses	–[a]	60
Miscellaneous, solid rubber	123	132
Total	900	1310

[a] Included under miscellaneous.

The growth pattern will undoubtedly change as other nations improve their technology. In 1966 the U.S. had 58% of the Western countries' SBR plant capacity; in 1972 it is estimated it will have only 43% of this capacity. U.S. SBR capacity is growing at a slower rate than that of the rest of the world although this is in part compensated for by a striking growth of plants for production of the stereospecific elastomers.

The polymer scientist of the future will not only have to cope with ever-changing technical problems as new rubbers flourish and fade in the market but also with the economic, social, and political problems of differing costs and qualities desired in foreign lands.

REFERENCES

(1) G. Alliger and I. J. Sjothun, eds., *Vulcanization of Elastomers*, Van Nostrand Reinhold, New York 1964.
(2) L. Bateman, ed., *The Chemistry and Physics of Rubber-Like Substances*, MacLaren & Sons, Ltd., London 1963.
(3) F. Bueche, *Physical Properties of Polymers*, Interscience Publishers, New York 1962.
(4) P. J. Flory, *Principles of Polymer Chemistry*, Cornell University Press, Ithaca, N.Y. 1953.
(5) J. P. Kennedy and E. Tornqvist, eds., *Polymer Chemistry of Synthetic Elastomers*, Interscience Publishers, New York, Part I 1968, Part II 1969.
(6) W. S. Penn, *Synthetic Rubber Technology*, MacLaren & Sons, Ltd., London 1960.
(7) G. S. Whitby, ed., *Synthetic Rubber*, John Wiley & Sons, Inc., New York 1954.
(8) G. G. Winspear, ed., *Vanderbilt Rubber Handbook*, R. T. Vanderbilt Co., Inc., New York 1968.
(9) *The Elastomers Manual*, International Institute of Synthetic Rubber Producers, 45 Rockefeller Plaza, New York 1972.

The journal *Rubber Chemistry and Technology* published by the Division of Rubber Chemistry. One issue each year, since 1957, known as "Rubber Reviews," contains up-to-date review articles on specialized subjects of interest to polymer scientists and technologists.

8

POLYBUTADIENE RUBBER

R. S. HANMER AND H. E. RAILSBACK
Phillips Petroleum Company
Bartlesville, Oklahoma

Although polybutadiene was first produced in Europe in the early 1930's, it was virtually unknown to the rubber industry in the U.S. prior to 1960. The discovery of organometal catalysts for polymerization of butadiene in hydrocarbon solvents brought a remarkable change to the industry. During the last ten years solution-polymerized polybutadiene has found its way into practically all tire compounds and into nontire compounds as well. As a result of this wide acceptance, polybutadiene has taken over the number two position among synthetic rubbers, second only to emulsion-polymerized SBR. In 1969 consumption of polybutadiene in the U.S. amounted to 260,000 long tons.[1] It is used in tire treads in mixtures with other rubbers to improve abrasion resistance and provide crack resistance. In carcass compounds and sidewall stocks up to about 30% of the rubber used may be polybutadiene to improve resistance to heat degradation and blowouts. Nontire applications take advantage of its abrasion resistance, flex-crack resistance, resilience properties, and excellent low-temperature flexibility.

HISTORICAL DEVELOPMENT

For many years polymer chemists tried to duplicate the composition and properties of natural rubber in the laboratory. Although they recognized the building block for natural rubber is isoprene, they were unable to synthesize a high molecular weight polymer starting with this monomer until solution-polymerization systems and organometal catalysts were tried in the 1950's. Isoprene (2-methyl-1,3-butadiene) is a diene and as expected related materials were tested as monomers for synthetic rubber. These studies led to the use of butadiene and the early synthesis of polybutadiene.

In 1958 W. W. Crouch and G. R. Kahle[2] reviewed the developments leading up to the production of solution-polymerized polybutadiene and correctly predicted its ultimate commercial importance. These authors pointed out that following the abandonment of "methyl" rubber produced by Germany during World War I from 2,3-dimethyl-1,3-butadiene, research was done in many countries to develop a practical general purpose rubber having properties at least approaching those of natural rubber. 1,3-butadiene gradually emerged as a monomer of choice.

In 1932 metallic sodium was used by Russian chemists to produce polybutadiene by "rod" polymerization.[3] The process was improved by substituting finely divided sodium for the sodium-coated rods in 1935. This rubber was important in the Russian synthetic rubber industry until the close of World War II.[4,5] Also in the 1930's potassium-catalyzed polybutadienes were developed in Germany. Polybutadiene rubber did not become as important in Germany as it did in Russia because of the German discovery and development of emulsion polymerization and the shift thereafter to copolymers of butadiene and styrene.

The synthetic rubber industry in the United States started in the 1930's with specialty rubbers such as neoprene and butyl. The giant SBR segment was developed as a war time necessity in the early 1940's and was based on free-radical polymerization of butadiene and styrene in emulsion systems. Polybutadiene made in emulsion systems was not considered to be a satisfactory rubber at that time.

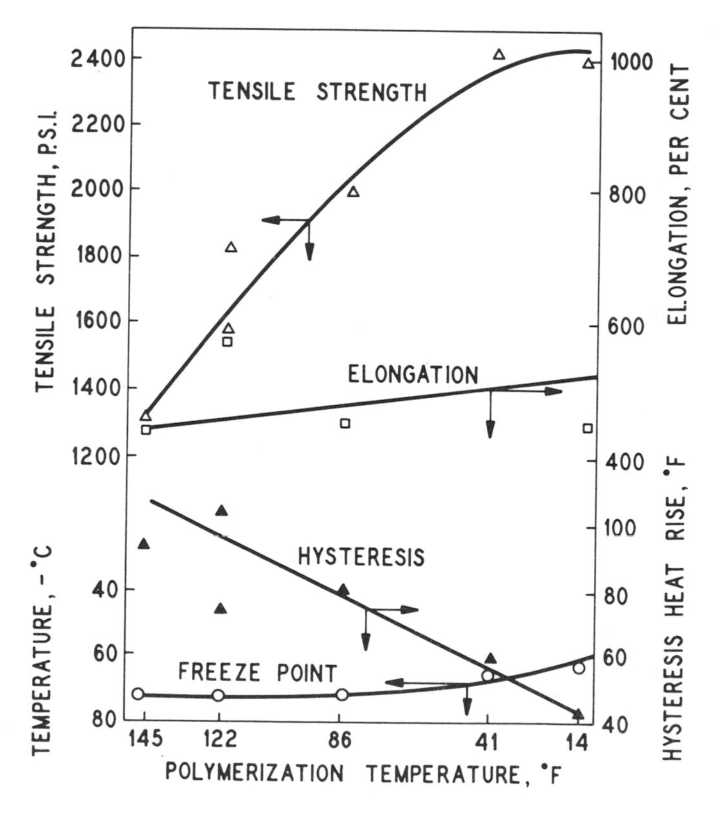

Fig. 8.1 Polymerization temperature as it affects emulsion polybutadiene rubbers.

However, work was continued with these systems, encouraged by the low cost of butadiene and by periodic shortages of raw material for styrene manufacture.[6,7]

Emulsion polybutadienes were prepared in accordance with the GR-S recipe at 122°F to conversions of 72% and about 50 Mooney viscosity. These rubbers exhibited poor processing characteristics. Dispersion of carbon black reinforcing fillers was difficult, compounds would not adhere to the mill rolls, and extrusion appearance was usually poor. Although processing was improved for polybutadienes produced at lower temperatures (down to 14°F), these polymers showed a tendency to crystallize after prolonged exposure to low temperature. Figure 8.1 shows some properties of emulsion polybutadienes as a function of polymerization temperature.[2]

In cooperation with the Office of Rubber Reserve, Phillips Petroleum Company developed a promising emulsion polybutadiene-carbon black masterbatch on a pilot plant scale in 1949. A successful production run was made in the Borger, Texas, Copolymer Plant operated by U.S. Rubber Company in 1950.[6,7] A polymerization temperature of 86°F was chosen as a compromise between physical properties and processability and the polymer was made to 25-Mooney viscosity to further enhance processing. Government tire tests[8] indicate the polybutadiene-HAF carbon black masterbatch was equal to 41°F SBR in treadwear and showed no cracking in 25,000 miles of testing.[2] In spite of these encouraging results polybutadiene was not commercialized at that time.

A variation on emulsion polybutadiene also developed under the Government Synthetic Rubber Program was alfin polybutadiene by A. A. Morton and coworkers.[9] Because of the characteristic very high molecular weight, alfin polybutadiene requires special treatment to render it suitable for compounding. It was not produced commercially. Recently alfin-type rubbers have been manufactured in Japan[10]; however, these are actually copolymers of butadiene containing minor amounts of styrene or isoprene.

One distinguishing feature of polybutadiene is its microstructure; i.e., the ratio of *cis, trans,* and vinyl configurations. Sodium and potassium give high 1-2 addition while alfin and emulsion polymerization systems yield polybutadienes high in *trans* content. The solution systems with various organometal catalysts are capable of producing high *cis*, high *trans*, high vinyl-polybutadienes, or polymers with different combinations of *cis, trans,* and vinyl groups. Two types of solution polymerized polybutadienes have emerged as important commercial rubbers. These are *cis*-polybutadiene, first disclosed to the industry by Phillips Petroleum Company in April 1956,[11] and medium-*cis*-polybutadiene, first disclosed by Firestone Tire and Rubber Company in 1958.[12]

Variations in the Phillips catalyst have been made so that today we have three classes of *cis*-polybutadiene made from (1) titanium, (2) cobalt, and (3) nickel-based catalysts. The medium-*cis*-polybutadiene produced by Firestone is made with alkyl lithium catalyst. Phillips Petroleum Company also has developed a medium-*cis*-polybutadiene as a part of its Solprene® rubber line. Today commercial polybutadienes are made almost exclusively by solution polymerization processes

TABLE 8.1. COMMERCIAL SOLUTION-POLYMERIZED POLYBUTADIENE RUBBERS (Ex. EASTERN BLOC COUNTRIES)[a]

Producer	*Location*	*Capacity, Metric Tons/Yr*	*Trade Name*
	cis-Polybutadiene		
Phillips Petroleum Company; General Tire & Rubber Company; Armstrong Tire & Rubber Company	Borger, Texas	47,500	*Cis*-4; Duragen; –
Goodyear Tire & Rubber Company	Beaumont, Texas	100,000	Budene
Am. Rubber & Chem. Company	Louisville, Ky	60,000	Cisdene
Ameripol, Inc.	Orange, Texas	54,000	Ameripol CB
Michelin & Cie	Bassens, France	35,000	
Compagnie Francais des Produits Chimiques Shell	Marseille, France	25,000	Cariflex BR
ANIC SpA.	Ravenna, Italy	20,000	Europrene Cis
Stereokautschuk-Werk GmbH. & Co. KG	Huls, Dormagen, Germany	55,000	Buna CB
Polymer Corp., Ltd.	Sarnia, Ontario	20,000	Taktene
Australian Synthetic Rub. Co.	Melbourne, Australia	8,000	Austrapol CB
Japanese Geon Co. Ltd.	Tokuyama, Japan	20,000	Nipol CB
Japanese Syn. Rub. Co. Ltd.	Yokkaichi, Japan	40,000	JSR BR-01
	Subtotal	484,500	
	medium-cis Polybutadiene		
Firestone Tire & Rubber Co.	Lake Charles La. Orange, Texas	70,000	Diene
Phillips Petroleum Company	Borger, Texas	8,000[b]	Solprene
Coperbo	Recife, Brazil	27,500	Coperflex
Negromex S.A.	Salamanca, Mexico	8,000[b]	Solprene
Petrochim N.V.	Antwerp, Belgium	10,000[b]	Solprene
Firestone-France S.A.	Port Jerome, France	18,000	Diene
Calatrava, S.A.	Santander, Spain	8,000[b]	Solprene
International Syn. Rub. Ltd.	Grangemouth, Scotland	50,000	Intene
Phillips Imperial Chem. Ltd	Kurnell, Australia	4,000[b]	Solprene
Asahi Chem. Industry Co. Ltd.	Kawasaki, Japan	50,000	Asadene
AA Chemical Company	Oita, Japan	4,000[b]	Solprene
	Subtotal	257,500	
	Total	742,000	

[a] Published capacities.
[b] Plant makes SBR as well as polybutadiene – capacity for polybutadiene is author's estimate.

employing organometal catalysts capable of control of microstructure, molecular weight distribution, and branching. Solution-polymerized polybutadienes available commercially are listed in Table 8.1 by trade name and type. Published capacity for *cis*-polybutadiene amounts to 484,500 and medium-*cis*-polybutadiene to 257,500 metric tons annually.

Using modern analytical procedures it is relatively easy to measure polymer microstructure and it is therefore natural to relate polymer differences to *cis*, *trans*, and vinyl configuration.[13] Polymer characteristics such as molecular weight distribution and branching are not easy to measure; however, rheological studies indicate these parameters also play an important role in polybutadiene behavior. Solution polymers are characterized by fairly narrow molecular weight distribution and less branching than emulsion polybutadiene, which may account for some of the major differences in processing and performance.

Present solution-polymerized polybutadienes represent a compromise between processability and performance. Narrow molecular weight distribution and linearity favor low hysteresis loss, high resilience, and high abrasion resistance. However, such polymers do not mix well, do not band on the mill, and usually require special packaging because of high cold flow. Alterations in the polymer architecture can be made in solution-polymerized rubbers to the extent necessary to solve these problems and yet preserve the desirable performance characteristics.

PROCESSING

As used in this discussion processing refers to the handling characteristics of the polymer with regard to the preparation of useful compounded stocks. This includes breakdown characteristics, mixing, mill banding, and extrusion (feed, rate, appearance, and shrinkage). Most polybutadiene rubbers possess inherently high resistance to breakdown, poor mill banding characteristics, and rough extrusion appearance when made to a Mooney viscosity considered normal for SBR rubber. Although solution polybutadiene rubbers may display poor processing characteristics, these polymerization systems have the advantage of permitting modification of molecular weight distribution, degree of branching, microstructure and breakdown to alleviate the most serious processing problems.

Very early in the work with *cis*-polybutadiene rubber it was observed that a transition in milling behavior occurs with a change in the temperature of the stock.[14,15] At temperatures below 100 to 110°F the rubber is continuous on the mill rolls, glossy and smooth in appearance, and bands tightly. As the temperature of the stock is increased the polymer becomes rough and loose on the mill, loses cohesion, and bags so that milling is poor. Smith and Willis[16] have discussed the processing behavior of medium-*cis*-polybutadiene.

Polybutadiene rubbers normally display very little breakdown as a result of intensive mixing as shown by the data in Table 8.2.

TABLE 8.2. EFFECT OF MIXING (10 MINUTES AT 300°F)

	cis-Polybutadiene	*SBR 1500*	*Premasticated Natural Rubber*
ML-4 at 212°F			
Initial	40	54	90
Final	38	45	53

However, polybutadiene can be broken down with certain chemical peptizers to obtain some improvement in processing. Not all commercial peptizers are effective; some that are effective, however, are listed in Table 8.3.

TABLE 8.3
PEPTIZERS FOR *CIS*-POLYBUTADIENE (5 MINUTE MIX TO 300°F)

Additive	*None*	*A**	*B**	*C**
ML-4 at 212°F				
Initial	41	41	41	41
Final	40	27	37	26
	50 phr Carbon Black, 5 phr Resin, 5 phr Oil			
Compounded Mooney, ML-4	73	68	67	58
Mill Banding	Poor	Fair	Fair	Fair
Extrusion at 250°F, Garvey Die				
grams/minute	95	85	89	93
appearance rating (12 best)	7	7	7	11

* 2 phr
A – oil soluble sulfonic acid with paraffin oil
B – di-ortho-benzamidophenyl disulfide
C – modified zinc salt of pentachlorothiophenol

Extensive mixing studies comparing the Banbury and roll mill have shown that more uniform dispersion and improved processing compounded stocks having better properties are obtained if an internal mixer is used. Only on stocks that can be kept banding while milling does the roll mill appear to be satisfactory. A recommended cycle for Banbury mixing follows:

	Mix Time (minutes)
Polybutadiene Rubber	0
Softener	0.5
One-half black and chemicals	1.0
Balance of black and chemicals	2.5
Discharge	300°F

Cooling water is regulated to obtain stock temperatures of 300 to 320°F in 4 to 6 minutes. Contrary to the usual procedure, addition of some plasticizing oil first appears to cause the polybutadiene to mass quickly and facilitates black incorporation. If all of the oil is withheld until black incorporation is complete, excessive mixing times may be required and more "sleepers" are encountered.

Rubber containing 50 parts of N-330 carbon black and 10 phr plasticizer must be peptized to a compound Mooney viscosity of less than 60 to obtain significant improvement in extrusion appearance. Mill banding is only fair even when raw viscosity of the polybutadiene is decreased 40% by chemical peptizing. Manufacture of the polybutadiene to low Mooney viscosity also has a beneficial effect on processing as indicated in Table 8.4.

TABLE 8.4. EFFECT OF MOONEY VISCOSITY ON PROCESSING *CIS*-POLYBUTADIENE

ML-4 at 212°F				
Raw	*Compounded*[a]	*Mill Banding*	*Garvey Die Extrusion grams/minute*	*Rating 12-best*
26	53	Fair	96	12
43	75	Poor	98	7

[a] 50 phr HAF carbon black, 10 phr plasticizer.

Reduction in Mooney viscosity of the raw polymer is effective for improving processing, but this procedure leads to high cold flow which creates packaging and storage problems, increased curative requirements, and a tendency to higher heat build-up in the vulcanizates. Some manufacturers have resorted to preprocessing,[17] broadening of the molecular weight distribution, and increasing the degree of long chain branching to improve milling and extrusion. One producer has introduced "weak links" in the polymer during manufacture, sensitive to chemical scission, as a means of improving the mill handling and extrusion of the rubber. An example of the latter is a medium *cis*-polybutadiene that is sensitive to breakdown when mixed with organic acids.[18,19] Without additives these polymers show only a small decrease in Mooney viscosity (4 points) after mixing for 10 minutes at 250°F. With 2 parts stearic acid the same polymer decreases 6 points in viscosity during the first minute of mixing and 19 units after 5 minutes mixing (Fig. 8.2).

The following organic acids have also been demonstrated to be effective in reducing Mooney viscosity in this particular polybutadiene: benzoic, lauric, salicylic, sorbic, and abietic acid. Since breakdown is chemical in nature, as little as 0.5 part per hundred of rubber is effective with a minimum of mixing. Commercial peptizers are not as effective in inducing breakdown under these moderate conditions.

Polybutadiene rubbers reported to have greater long chain branching band better on the roll mill and display lower compounded Mooney viscosity and better extrusion appearance than polybutadienes with more linear molecular structure as indicated in Table 8.5.

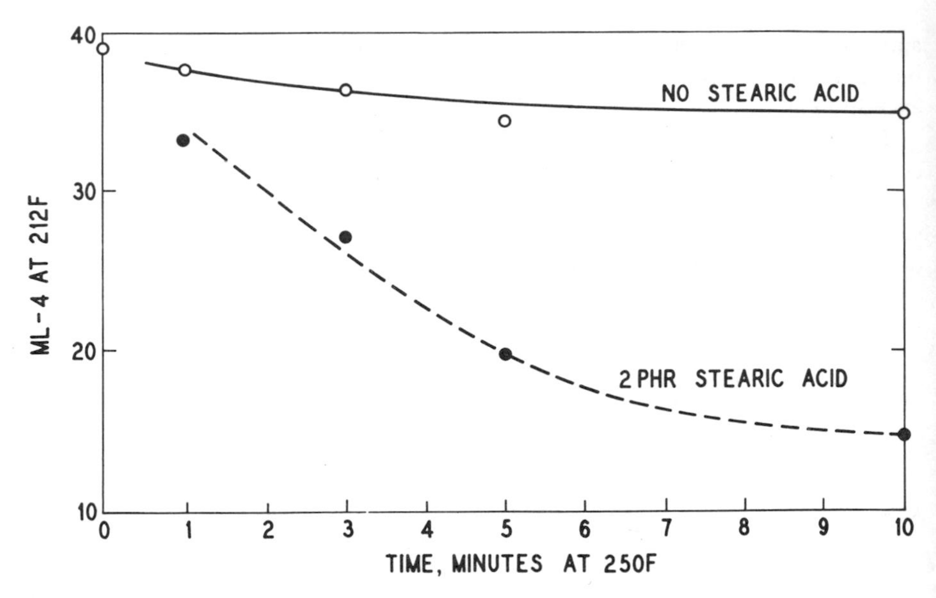

Fig. 8.2 Medium *cis*-polybutadiene rubber breakdown.

TABLE 8.5. EFFECT OF BRANCHING ON PROCESSING

	Moderate	*Slight*
Cold Flow, mg/min	0.5	1.0
Raw ML-4 at 212°F	41	45
Compound ML-4 at 212°F	60	75
Mill Banding	Good	Poor
Extrusion at 250°F, Garvey Die		
grams/minute	82	84
appearance rating	10	7

The large difference in Mooney viscosity between that of the raw polymer and that of the compound for the more linear polymer is typical. Because of the high viscosity and lack of shrinkage, the more linear polymer bands poorly.

Maximum extension of rubber to obtain minimum cost and still retain quality is a well established practice. Polybutadiene is admirably suited to this cost reducing process yielding desirably good properties even when highly extended and, in fact, exhibiting improved processability under these conditions. Formulations have been studied wherein polybutadienes have been extended with various levels of black and oil to obtain stocks which would be considered to have satisfactory milling and extrusion characteristics. Compounds of this type, of course, may only be useful for certain applications because of the high degree of extension. The data in Fig. 8.3 show the effects of oil and black levels on the Mooney viscosity, extrusion rate, and appearance of polybutadiene. An extrusion rate of 40 inches per minute and an appearance rating of 10 or higher is considered satisfactory for most applications. If

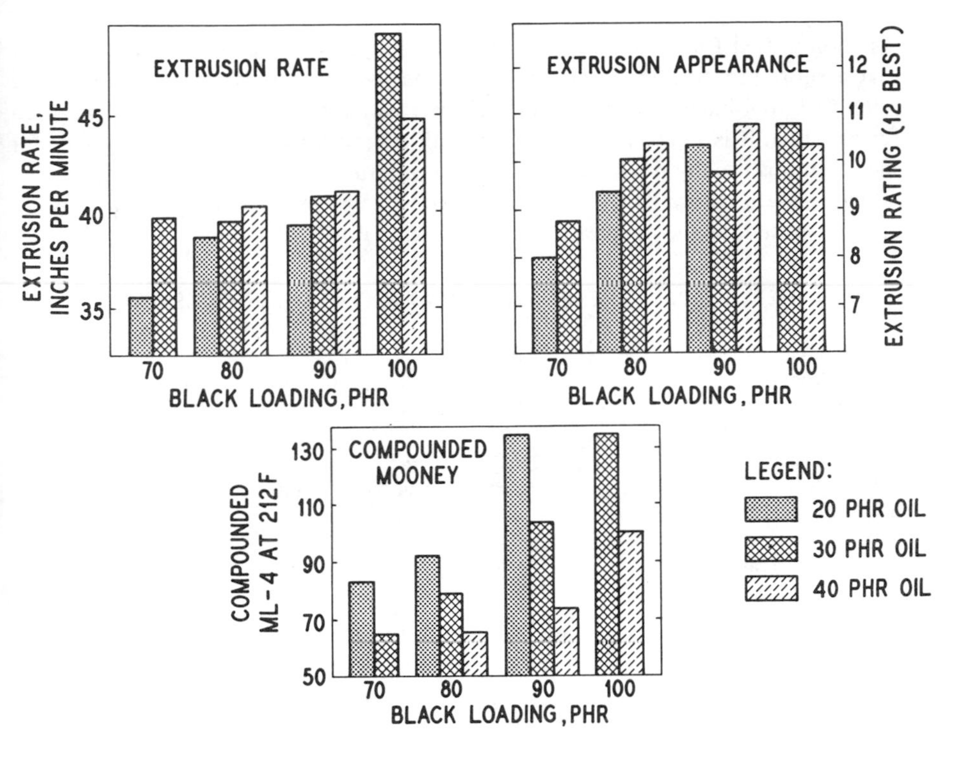

Fig. 8.3 Processing of highly extended *cis*-polybutadiene.

these criteria are used, several of the compounds in Fig. 8.3 have adequate processing characteristics for many commercial plants.

Blends with Natural Rubber. Early in the development of solution polybutadiene rubbers it was noted that these polymers had certain desirable properties when compared to natural rubber and SBR.[20,21] These included a high tolerance for extender oil, excellent abrasion resistance, and outstanding hysteresis properties.[22] Blends of *cis*-polybutadiene and natural rubber were made initially as a means of obtaining improved processing; it was then noted that polybutadiene rubber conferred many of its desirable properties to the blends. Mix cycles

TABLE 8.6. MIX CYCLE FOR POLYBUTADIENE RUBBER BLENDS

	Minutes
Polybutadiene and Blend Rubber	0
1/2 Filler and powders	0.25 to 0.5
Balance of filler and powders	1.5
Plasticizer	250°F
Discharge	300°F

commonly used for natural rubber may be employed to prepare such blends except that a short (15 to 30 second) premix of the two rubbers is considered advisable prior to initial black addition. Although premasticated natural rubber is preferred for blending, it is not mandatory. Even with a short premix the total mix time for such blends is usually the same or less than for natural rubber.

Typical effects on the compounded Mooney viscosity, milling characteristics, and extrusion rate and appearance are shown in Table 8.7.

TABLE 8.7. EFFECT OF BLEND RATIO ON PROCESSING

cis-Polybutadiene	100	75	50	25	–
Natural Rubber	–	25	50	75	100
Carbon Black, phr	50	50	50	50	50
Extender, phr	10	10	10	10	10
ML-4 at 212°F	72	68	60	60	60
Mill Banding	Poor	Fair	Good	Good	Excellent
Extrusion at 250°F, Garvey Die					
grams/minute	85	96	100	101	102
appearance rating	8	9	11	11	12

In general, greater changes are effected in the range between 100/0 and 50/50 ratios of polybutadiene to natural rubber, and processing appears to change less rapidly after the proportion of natural rubber exceeds 50 per cent.

Blends with SBR. Likewise, blends of polybutadiene rubber with clear and oil-extended SBR can be made to obtain easy processing compounds. That blends with oil-extended SBR or oil-black masterbatch are easily prepared is apparent from the data in Table 8.8. The masterbatch used as a source of black and/or oil has little effect on processing of the blends. Compounded in the same formulation, all of the blends give low compounded Mooney viscosity, high extrusion rates, and good extrusion appearance.

TABLE 8.8. BLENDS WITH OIL-BLACK OR OIL-EXTENDED SBR RUBBER

	SBR 1808	*SBR 1813*	*SBR 1609*	*SBR 1712*
SBR Masterbatch	112.5	98.75	72.5	68.75
cis-Polybutadiene	50	50	50	50
Black Type[a]	HAF	ISAF	SAF	ISAF
ML-4 at 212°F	46	46	41	43
Mill Banding	Good	Good	Good	Good
Extrusion at 250°F, Garvey Die				
grams/minute	90	120	105	129
appearance rating	10	12	12	12

SBR 1808 – 75 Parts HAF black, 50 parts highly aromatic oil.
SBR 1813 – 60 Parts ISAF black, 37.5 parts highly aromatic oil.
SBR 1609 – 40 Parts SAF black, 5 parts highly aromatic oil.
SBR 1712 – 37.5 Parts highly aromatic oil.

[a]Black-oil adjusted to 70/40 for all stocks.

Polybutadiene rubbers display an unusual tolerance for high carbon black and oil levels.[23] Using black levels of 70 to 100 and oil levels of 50 to 70 phr in polybutadiene/SBR blends, Mooney viscosity changes about 0.9 ML-4 unit for each change of one part of carbon black. Also, 1.2 parts of carbon black appear necessary to compensate for each part of highly aromatic oil if Mooney viscosity is to be kept reasonably constant. At the same oil level, increasing the carbon black loading increases extrusion rate and appearance.

Polybutadiene rubbers are offered in both oil and oil-black masterbatch form. Oil levels range up to 50 parts per hundred of rubber and may be naphthenic, aromatic, or highly aromatic; paraffinic oils have limited compatibility.[24,25] Highly reinforcing carbon blacks of the HAF and ISAF types, both normal and high structure, have been used at levels of 80 and 90 phr, usually with oil levels of 37.5 and 50 parts, respectively. When such polybutadiene masterbatches are blended with other polymers, mix times are shortened because the mechanical incorporation of large amounts of oil or carbon black is avoided. There has been little incentive to use high Mooney base polymer in polybutadiene masterbatches since processing is adversely affected and advantages in physical properties usually consist of slightly lower heat build-up and a marginal improvement in abrasion resistance.

VULCANIZATE PROPERTIES

Polybutadiene rubbers are usually vulcanized with sulfur and accelerator whether used alone or in blends;[26] however, other crosslinking systems (e.g., peroxides) can be used. Selected physical properties at a crosslinking level that appears optimum for each type of rubber using a sulfur accelerator system are shown in Table 8.9.

TABLE 8.9.

	cis-*Polybutadiene*	*SBR 1500*	*Natural Rubber*
Crosslinking, $\upsilon \times 10^4$ moles/cc	2.0	1.5	1.5
300% Modulus, psi	1200	1400	1800
Tensile, psi	2500	3400	4000
Elongation, %	500	580	520
200F Tensile, psi	1400	1500	2800
Heat Build-up, °F	40	67	40
Resilience, %	75	62	72
Blowout, minutes	120	9	9
Shore A Hardness	63	60	62

Basic Recipe: Rubber 100, N330 Carbon Black 50

The lower modulus, even at high crosslinking levels, and lower tensile strength of the polybutadiene tread stock compared to SBR 1500 and natural rubber tread compound are typical. So also is the good retention of tensile strength at elevated temperature. Excellent hysteresis properties fully equivalent to the natural rubber control are obtained, and blowout resistance is superior to that of either control stock.

As discussed previously, blends of polybutadiene with natural rubber or with SBR have been used commercially to obtain adequate processability and take advantage of the unique properties of polybutadiene. Experience has shown that if the Mooney viscosity of the polybutadiene in such blends increases, modulus and resilience increase slightly and heat build-up decreases; ultimate tensile and elongation remain relatively constant. Polybutadiene-natural rubber blend compounds having a useful balance of physical properties can be obtained with a wide range in sulfur levels (1.0 to 2.5 phr). Of course, each sulfur level requires the selection of the proper accelerator level to give the best balance of properties.

TABLE 8.10. EFFECT OF SULFUR AND ACCELERATOR LEVEL IN BLENDS 50/50 *CIS*-POLYBUTADIENE/NATURAL RUBBER[a] (30 MINUTES CURE AT 307°F)

Sulfur	*Accelerator*	*Comp. Set %*	*300% Modulus psi*	*Tensile psi*	*Elongation %*	*Heat Build-up °F*	*Resilience %*	*Shore A Hardness*
1.25	1.2	12	1520	3350	540	41	73	59
1.75	0.8	15	1450	3300	550	42	73	59
2.0	0.7	16	1550	3300	550	40	73	61
2.5	0.6	19	1500	3000	500	40	75	60
			Aged 24 hours at 212°F					
1.25	1.2	–	1950	2700	390	39	75	64
1.75	0.8	–	2020	2450	360	37	76	65
2.0	0.7	–	2080	2650	380	37	77	66
2.5	0.6	–	2250	2450	330	36	78	66

[a] N-330 Carbon Black 50 phr; extender 10 phr.

Higher sulfur levels usually give vulcanizates with somewhat poorer aging characteristics. Delayed action type accelerators give excellent processing safety. Aromatic or naphthenic petroleum base oils or pine tar are satisfactory plasticizers. Highly reinforcing carbon blacks improve stress-strain properties and abrasion resistance and increase hysteresis.

Typical properties for various polybutadiene natural rubber ratios are summarized in Table 8.11. These low sulfur compounds exhibit relatively low modulus and heat build-up. Up to 60 parts of the natural rubber can be replaced before tensile drops below 3000 psi. All of the stocks exhibit the desirable low heat build-up and high resilience given by both polybutadiene and natural rubber.

Typical physical properties displayed by blends of polybutadiene rubber and oil-extended SBR are listed in Table 8.12. At the same curative level the inherently

TABLE 8.11. VARIABLE *CIS*-POLYBUTADIENE/NATURAL RUBBER RATIO[a] (45/5 N-220/OIL)

Polybutadiene	100	75	60	50	25	–
Natural Rubber	–	25	40	50	75	100
Sulfur, phr	1.5	1.5	1.5	1.5	1.5	2.25
Accelerator, phr	1.0	0.9	0.8	0.8	0.7	0.5
		45 Minutes Cure at 292°F				
300% Modulus, psi	800	1000	1100	1150	1200	1650
Tensile, psi	2500	2850	3300	3550	4050	4150
Elongation, %	520	530	580	600	650	570
Heat Build-up, °F	50	47	45	46	46	41
Resilience, %	70	73	73	71	70	67
Shore A Hardness	58	60	60	62	59	64

[a] Zinc oxide 3, stearic acid 3, antioxidant/antiozonant 1/2.

lower modulus and tensile strength and better hysteresis properties of polybutadiene rubber are reflected in the properties of the blend compounds.

The response of a 40/60 blend of medium-*cis* polybutadiene/SBR 1712 to various types of sulfenamide accelerators and also to the use of secondary accelerators is shown in Table 8.13. At equal levels of accelerator, NOBS Special imparts the greatest scorch resistance; modulus and heat build-up are very similar for the three primary accelerators. Replacement of 0.3 part of NOBS Special or Santocure NS with various secondary accelerators decreases Mooney scorch time significantly compared to compounds accelerated with the single system. In the NOBS Special system Butyl Zimate and Monex kickers are both more active than

TABLE 8.12. BLENDS WITH OIL EXTENDED SBR[a]

SBR 1714	75	–	–
SBR 1712	–	68.25	137.5
cis-Polybutadiene	50	50	–
N220 Carbon Black, phr	68	68	68
Highly Aromatic Oil, Total phr	40	40	40
		Cured 30 Minutes at 307°F	
300% Modulus, psi	750	850	970
Tensile, psi	2500	2650	3220
Elongation, %	640	610	650
Heat Build-up, °F	56	60	59
Resilience, %	62	58	54
Shore A Hardness	54	59	59

[a] Basic recipes also include zinc oxide 3, stearic acid 2, antioxidant 4, wax 3, sulfur 2, and accelerator 1.2.

TABLE 8.13. ACCELERATOR SYSTEMS IN MEDIUM-*CIS* POLYBUTADIENE/SBR 1712 BLENDS[a]

	Mooney Scorch at 280°F	*300% Modulus psi*	*Heat Build-up °F*
	Primary Accelerator 1.3 phr		
NOBS Special	19	1230	72
Santocure	14	1180	73
Santocure NS	14	1210	71
	Primary 1.0; Secondary 0.3		
NOBS Special/DPG	10	1140	74
/Altax	9	1180	74
/Captax	8	1190	73
/Butyl Zimate	9	1330	72
/Monex	9	1550	66
Santocure NS/DPG	8	1160	75
/Butyl Zimate	8	1230	69
/Monex	7	1510	67

[a]Basic recipe: N220 carbon black 70, highly aromatic oil 40, sulfur 2.0. 30 Minutes cure at 307°F.

diphenylguanidine, Altax or Captax if the modulus and heat build-up of the vulcanizates are used as criteria of state of cure.

APPLICATIONS

Tires. Large volume use of polybutadiene rubbers has been primarily in blends with other polymers where advantages of the inherently good hysteresis properties and abrasion resistance can be realized. It was also found that blending polybutadiene rubber with either SBR or natural rubber conferred improved crack growth resistance to the blend. Large scale usage therefore has been concentrated in the tire tread field, both passenger and heavy duty truck types. Other end use applications almost always consist of blends, e.g., modification of plastics, admixture with rubbery polymers to improve molding characteristics, low temperature properties, resistance to heat degradation, or resistance to abrasion or cracking.

Tire service performance data for polybutadiene/natural rubber blends (40/60) show how polybutadiene plus increased black and oil levels can be used to obtain better abrasion and chipping resistance than with natural rubber alone.

Increasing the amount of *cis*-polybutadiene in the blend while keeping the carbon black level constant results in an improvement in abrasion resistance. With higher levels of polybutadiene, increasing carbon black and oil levels maintains or gives improved abrasion performance. Cracking and cutting resistance of poly-

TABLE 8.14. TIRE PERFORMANCE (10:00 x 20 SIZE)

	cis *Polybutadiene-Natural Rubber Blend*			*Natural Rubber*
N220 Carbon Black	45	55	60	45
Highly Aromatic Oil	5	13	18	5
Sulfur, phr	1.7	1.7	1.7	2.25
Accelerator, phr	0.9	1.0	1.1	0.5
Abrasion Index	116	115	114	100
Chipping Resistance	Fair	Good	Excellent	Good

butadiene/natural rubber tread compounds are consistently good compared to natural rubber when low sulfur levels, e.g., 1.5 phr, are used. Polybutadiene/natural rubber blends have been observed to perform even better in abrasion resistance in comparison to natural rubber when test conditions are severe.

TABLE 8.15. 10:00 x 20 TIRES – ROAD TEST RESULTS[a] (*CIS*-POLYBUTADIENE/NATURAL RUBBER)

Ratio	50/50	60/40	60/40	60/40	100/0
Black type	ISAF	ISAF	ISAF	HAF	HAF
level, phr	45	45	60	60	70
Oil, phr	5	5	18	18	20
Abrasion Index	114	120	140	125	160

[a] Abrasion indices are in relation to factory first line natural rubber tread – usually 40–45 parts ISAF carbon black and 5 phr oil.

Resistance to heat degradation is desirable in carcass as well as in tread compounds where long cure times, retreading operations, or high temperatures developed in service can degrade the vulcanizate. An abbreviated carcass formulation for truck tires utilizing polybutadiene/natural rubber blend or natural rubber alone is shown below:

TABLE 8.16. TRUCK TIRE CARCASS STOCK

Rubber	100
N550 Carbon Black	35
Sulfur, phr	2.7
Accelerator	1.0

	40/60 Medium-cis Polybutadiene/Natural Rubber	**Natural Rubber**
Hand Tack (10 best)		
2 hours	5	9
24 hours	8	8
ML-4 at 212°F	38	33

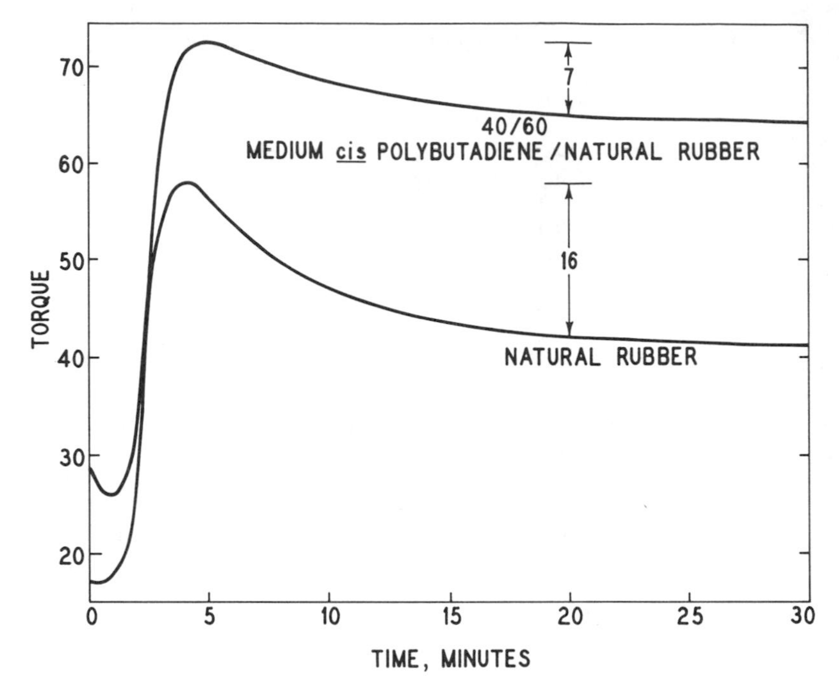

Fig. 8.4 Resistance to reversion.

Resistance to reversion during vulcanization is depicted for these compounds by the Rheograph in Fig. 8.4. The natural rubber carcass stock reaches a peak in torque after 4 minutes at 356°F and shows sharp reversion (16 units at 20 minutes). The 40/60 polybutadiene/natural rubber blend requires slightly longer to reach a maximum in torque and displays much less degradation (7 units) with time.

Physical properties of carcass stocks vulcanized 45 minutes at 293° F show that similar modulus and hysteresis properties are realized with the blends compared to all natural rubber and that tensile strength is low; however, blowout resistance is much better (20% longer) when polybutadiene is used to replace a part of the natural rubber.

In passenger tire tread formulations the effect of sulfur and accelerator ratio has been studied extensively using polybutadiene/SBR blends. If modulus is held constant, abrasion resistance in passenger tire treads will not be affected significantly. A marginal improvement in abrasion resistance may be realized when modulus is increased by changing curative levels. Modulus can also be controlled by the type of carbon black used or the oil level. If the abrasion index of a tread containing N220 carbon black is pegged at 100 then N242 (high structure) black gives a 5% improvement in treadwear. Reducing oil level also improves abrasion resistance as would be expected.

Abrasion performance of polybutadiene/SBR 1712 blends with changes in the level of polybutadiene, carbon black, and oil levels is shown in Table 8.17. It is apparent that a significant improvement in abrasion resistance over SBR 1712 is obtained with 75 parts carbon black and 50 parts oil at either 25 or 50 part levels of polybutadiene. At black/oil levels of 95/75 the greater tolerance for oil extension inherent in the polybutadiene rubber permits improved abrasion resistance to be realized if a 50/50 ratio of polybutadiene to SBR 1712 is used.

TABLE 8.17. EFFECT OF POLYBUTADIENE/SBR RATIO AND EXTENSION

	Medium-cis Polybutadiene/SBR 1712				*SBR 1712*
Rubber Ratio	25/75	50/50	25/75	50/50	0/100
N220 Carbon Black	75	75	95	95	70
Highly Aromatic Oil, phr	50	50	75	75	40
Abrasion Index	108	114	95	112	100

The improved crack growth resistance imparted by the admixture of polybutadiene with oil-extended SBR is illustrated by the data shown in Table 8.18. Treads containing the highest level of polybutadiene gave only one-third the crack growth of oil-extended SBR treads.

TABLE 8.18
TIRE PERFORMANCE – PRECUT GROWTH, TRACTION, AND SKID

cis-Polybutadiene	50	40	25	–	–
SBR 1712	68.75	82.5	103	137.5	–
SBR 1500/1712	–	–	–	–	100
N220 Carbon Black, phr	80	80	80	70	–
N330 Carbon Black, phr	–	–	–	–	60
Highly Aromatic Oil, Total phr	40	47.5	50	45	20
Abrasion Index	121	111	113	109	100
Precut Growth, %	53	85	131	135	395
Static Traction Index					
Wet asphaltic concrete	103	102	102	104	100
Wet concrete	104	99	101	107	100
Skid Resistance Index					
Wet asphaltic concrete	108	105	110	114	100

Not only are advantages shown in abrasion resistance and crack-growth but the inherent poor wet skid resistance of polybutadiene rubber is improved by blending with SBR and extending the compounds with high levels of carbon black and oil. Draw bar pull in pounds required to spin the rear wheels of an automobile held stationary on wet concrete or asphalt paving serves as a measure of static traction on wet surfaces. Skid resistance is measured by determining the distance an automobile will skid on wet asphaltic concrete at 15 to 35 miles per hour with its wheels locked. Both skid and traction results are reported as indices, and numbers over 100 represent improved performance relative to control.

Traction and skid resistance results show that treads with up to 50 per cent polybutadiene, when highly extended with carbon black and oil, give road holding performance equal to SBR 1500/SBR 1712 and very close to SBR 1712.

Nontire. Comparison of the vulcanizate properties of polybutadiene rubber to those of SBR 1500 and natural rubber for nontire application shows there are numerous differences. Polybutadiene, in a recipe containing similar filler level, displays lower tear strength, water absorption, and brittle point than either SBR 1500 or natural rubber. Polybutadiene rubber vulcanizates have higher dynamic modulus and greater permeability to gases than natural rubber and are similar to this rubber in electrical resistivity and dielectric strength. The excellent low temperature behavior of polybutadiene rubbers is shown by the small change in resilience with temperature (Fig. 8.5) compared to natural rubber and SBR 1500.

Substantial quantities of polybutadiene rubber are used in blends with specialty polymers to take advantage of the good low temperature properties conferred by the polybutadiene. This is demonstrated by data for 70/30 Neoprene WRT/polybutadiene blends in Table 8.19. Brittle point is much better for the blend. Other advantages are lower compression set and good retention of tensile strength on aging. A polybutadiene/neoprene rubber blend of this type gives lower viscosity, longer scorch time, and much better mold flow than neoprene alone. Some

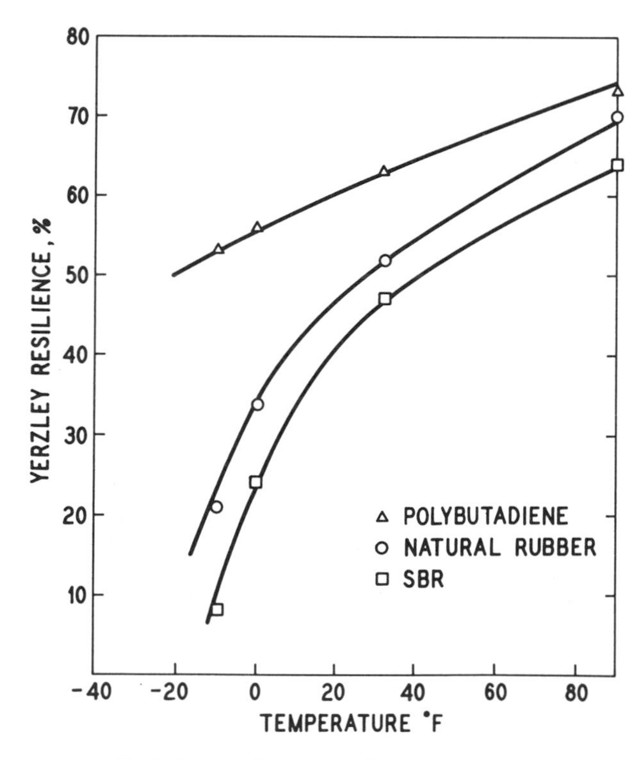

Fig. 8.5 Resilience at various temperatures.

TABLE 8.19. NEOPRENE-POLYBUTADIENE RUBBER BLENDS[a]

Neoprene WRT	70	100
Medium-*cis* Polybutadiene	30	–
N-330 Carbon Black	40	40
Hard Clay	25	25
Naphthenic Oil	25	25
Compound Mooney, ML-4 at 212°F	30	40
Scorch at 280°F (+5), minutes	5	4
20 minutes Cure at 307°F		
300% Modulus, psi	1670	1450
Tensile, psi	2150	2400
Shore A Hardness	61	62
Compression Set, %	15	33
Resilience, %	63	64
Ozone Resistance	Good	Good
Brittle Point, °F	–62	–40
Aged 70 Hours at 212°F		
Tensile Retained, %	94	88
Shore A Hardness Change	+18	+16
ASTM #1 Oil		
Volume Change, %	+9	–5
Hardness Change	–5	+6
ASTM #3 Oil		
Volume Change, %	85	65
Hardness Change	–25	–23

[a] Contains 0.5 phr tetramethylthiuram monosulfide, 1.5 phr sulfur, and 2 phr stearic acid in addition to usual formulation for neoprene vulcanization.

reduction in resistance to ASTM Oil #1 and #3 results, as is indicated by the positive volume change and greater hardness change compared to the control.

Significant amounts of polybutadiene are used in belting compounds as a means of improving abrasion and durability. An example is the blending of polybutadiene with natural rubber or *cis*-polyisoprene in conveyor belt cover stocks to obtain improvements in the cut, tear, and abrasion resistance. It is also necessary that this type of stock display good heat and aging stability and formulations tend to resemble those of high quality tread stocks. It is possible to achieve exceptional flex cracking resistance with a blend of polybutadiene and polyisoprene as shown in Table 8.20. This coupled with the excellent processing characteristics and inherent retention of properties on aging results in the desired high quality cover stock.

Polybutadiene rubbers are in demand for use alone in only a few applications. Usually such uses take advantage of the outstanding resilience or abrasion resistance of the polymer. An example is the manufacture of solid golf balls from polybutadiene rubber. These compounds, when reinforced with silica pigment and

TABLE 8.20. CONVEYOR BELT COVER

cis-Polyisoprene	50
cis-Polybutadiene Masterbatch	120[a]
N-220 Carbon Black	25
45 Minutes Cure at 292°F	
300% Modulus, psi	1500
Tensile, psi	3300
Shore A Hardness	65
Flex Life, M flexures for 1 inch growth	250
Pico Abrasion Index	165

[a] Contains 45 parts carbon black, 25 parts oil; total black/oil 70/40.

very tightly cured, display the desired high rebound and esthetic "click" demanded by the player. Curing may be by either peroxide crosslinking or sulfur vulcanization. Polybutadiene rubbers have also been used to make high rebound toy balls. A typical recipe is: polybutadiene rubber 100, titanium dioxide 10, zinc oxide 3, stearic acid 1, antioxidant 1, MBTS 0.75, DOTG 1.5, sulfur 6.0. Mold cure 2 hours at 200°F to avoid backrinding, and post cure 2 hours at 200°F for high (90%) rebound.

Polybutadiene as well as butadiene-styrene rubbers are used extensively as modifiers of styrene to make impact resistant polystyrene.[27] The rubber may be incorporated into polystyrene by mechanical mixing or into styrene monomer by graft polymerization. The former method successfully raises impact resistance but requires rubber contents as high as 30%. This makes the product opaque and surface characteristics are classified as only fair. The preferred method of incorporation of rubber into styrene is by graft polymerization. This method is efficient and requires only about 5% rubber to impart impact resistance. Polybutadiene rubbers made in solution systems are uniquely suited for this 50 million pound per year market. They can be produced having controlled solution viscosity, free of gel, colorless, odorless, and low in ash and other contaminants. Polystyrenes produced are designated "medium, high, and extra high impact strength" and are used in packaging, appliances, toys, housewares, and numerous other applications.

REFERENCES

(1) *Chem. Week*, March 18 1970, p. 62.
(2) W. W. Crouch and G. R. Kahle, *Petroleum Refiner*, Nov. 1958.
(3) A. Talalay and M. Magat, *Synthetic Rubber from Alcohol*, Interscience Publishers, Inc., New York 1945, p. 187.
(4) J. W. Schade and B. G. Labbe, Office of Rubber Reserve Report CD 850, August 1946 (Library of Congress).
(5) J. W. Schade and N. C. Hill, Office or Rubber Reserve Report CD 958, November, 1946 (Library of Congress).
(6) *Chem. Ind.*, **67**, 29 (1959).

(7) *Chem. Eng.*, **57** (No. 9), 107 (1950).
(8) P. S. Greer and J. S. Blackeney, Government Tire Test Project "BI," Office of Rubber Reserve, July 1950 (Library of Congress).
(9) A. A. Morton, *Rubber Age*, **72**, 473–6 (1953).
(10) *Rubber Age*, **99**, No. 1, p. 15 (1967).
(11) *The Wall Street Journal*, 26th April 1956.
(12) *Rubber World*, **138**, August, p. 761 (1958).
(13) G. Kraus, J. N. Short, and V. Thornton, *Rubber and Plastics Age*, **38**, p. 880 (1957).
(14) H. E. Railsback, C. C. Biard, and J. R. Haws, *Proceedings of International Rubber Conference*, Washington, D.C., p. 502, Nov. 1959.
(15) H. E. Railsback, W. T. Cooper, and N. A. Stumpe, Jr., *Rubber and Plastics Age*, **39**, No. 10, p. 867 (1958).
(16) W. A. Smith and J. M. Willis, *Rubber Age*, **87**, No. 5, p. 815 (1960).
(17) D. V. Sarbach and A. T. Sturrock, *Rubber Age*, **90**, No. 3, p. 423 (1961).
(18) H. E. Railsback and J. R. Haws, *Rubber and Plastics Age*, **48**, No. 10, p. 1063 (1967).
(19) J. N. Short, Fourth International Synthetic Rubber Symposium, London, 30th Sept., 1969 (SRS/4 October 1969).
(20) H. E. Railsback, *Rubber Age*, **84**, No. 6, p. 967 (1959).
(21) F. C. Weissert and R. R. Cundiff, *Rubber Age*, **92**, No. 6, p. 881 (1963).
(22) R. S. Hanmer and W. T. Cooper, *Rubber Age*, **89**, No. 6, p. 963 (1961).
(23) H. E. Railsback, J. R. Haws, W. T. Cooper, and J. H. Tucker, *Rubber World*, **148**, No. 1, p. 40 (1963).
(24) Sun Oil Company, *Rubber World*, **148**, No. 4, p. 66 (1963).
(25) R. W. Hallman, C. A. Brunot, and R. G. Fuller, *Rubber World*, **151**, No. 4, p. 67 (1965).
(26) A. R. Davis and R. A. Naylor, *Rubber World*, **149**, No. 3, p. 54 (1963).
(27) D. L. DeLand, J. R. Purdon, and D. P. Schoneman, *Chemical Engineering Progress*, **63**, No. 7, p. 118 (1967).

9

ETHYLENE/PROPYLENE RUBBER

E. L. BORG
Uniroyal Chemical, Division of Uniroyal, Inc.
Naugatuck, Conn.

INTRODUCTION

Ethylene/propylene rubber was first introduced in the United States, in limited commercial quantities, in 1962. Sample quantities had been available a year or two earlier from laboratories in both Italy and the United States. Though commercial production only began in 1963, ethylene/propylene rubber is now the fastest growing elastomer. There are currently four manufacturers in the United States, three in Europe, and two in Japan.

NOMENCLATURE

Ethylene/propylene rubber is usually called EPDM. This designation, which follows a nomenclature convention endorsed by the American Society for Testing Materials, the International Institute of Synthetic Rubber Producers, Inc., and the International Standards Organization, applies to the more common, sulfur vulcanizable product which includes in the rubber molecule a minor percentage of a diene monomer in addition to ethylene and propylene. The basis for the letter designation, EPDM, is: E for ethylene; P for propylene; D for diene; and M for methylene, which are the repeating units (CH_2), or "vertebrae," in the "spine" of the polymer.

The less common copolymer of ethylene and propylene, containing no diene monomer and no functional unsaturation for sulfur vulcanization, is called EPM. Sometimes this copolymer is called EPR, *e*thylene/*p*ropylene *r*ubber, and this designation is even used to identify the entire category of ethylene/propylene elastomers, including terpolymers as well as copolymers.

HISTORY

In 1951 Professor Karl Ziegler of Germany discovered a new class of polymerization catalysts. These comprised a transition metal halide in association with an organometallic reducing agent, typically an aluminum alkyl. These catalysts induced polymerization by an anionic mechanism, and were found, in most cases, to have a high order of regularity. Ziegler obtained patent protection on his invention.[1] Ziegler's catalysts were first used commercially in the manufacture of linear, low pressure, high density polyethylene.

In Italy, Professor Giulio Natta extended the investigation of such catalysts to show that selected ones were capable of inducing the formation of many polymers frequently differing from one another only in their steric configurations. The catalysts then, at times, came to be called the stereospecific catalysts. Professor Natta also found that polypropylene could be made using these catalysts. This material soon joined polyethylene as the second commercial olefin plastic derived from the new catalyst technology.

Polyethylene and polypropylene are both crystalline, thermoplastic materials. Neither is in any sense rubbery. Extending his investigations still further, Professor Natta found that, using certain selected catalysts of the Ziegler class, he could cause ethylene and propylene to copolymerize in an irregular or random way (*not* stereospecific) to yield an amorphous material with interesting elastic and rubbery properties.[2]

In 1963 Professor Karl Ziegler and Professor Giulio Natta were jointly awarded the Nobel prize for chemistry in recognition of their discoveries which have led to several commercially important new plastics and elastomers including ethylene/propylene rubber.

POLYMER STRUCTURE

The structure of the regular, alternating amorphous copolymer of ethylene

$$H_2C{=}CH_2$$

and propylene

$$H_3C{-}\overset{H}{C}{=}CH_2$$

can be written

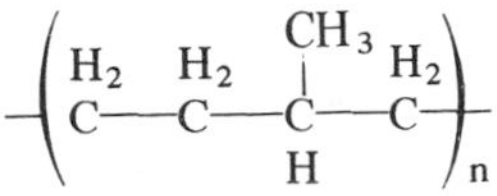

$$-\!\left(\overset{H_2}{C}-\overset{H_2}{C}-\underset{H}{\overset{CH_3}{\overset{|}{C}}}-\overset{H_2}{C}\right)_n\!-$$

ethylene propylene

This structure is remarkably similar to the structure of natural rubber, *cis* 1,4 polyisoprene:

$$-\left(\overset{H_2}{C}-\overset{H}{C}=\overset{CH_3}{\underset{|}{C}}-\overset{H_2}{C}\right)_n-$$

isoprene

It is not surprising, therefore, that the regular, alternating copolymer of ethylene and propylene is a decidedly rubbery material since its classical structure so closely approaches that of the first useful elastomer, natural rubber.

The classical equimolar structure of ethylene and propylene, as shown above, is probably not achieved in the present commercial ethylene/propylene rubbers. The compositions of the commercial materials are generally given on a weight percentage basis. The 50/50 equimolar percentage copolymer would be called a 40/60 ethylene/propylene weight percentage copolymer. Although the rubbery properties of the ethylene/propylene copolymers are exhibited over a broad range of composition, the commercial products are generally in the weight percentage range of 50/50 to 75/25 ethylene/propylene. In addition, even at the equimolar composition, the polymer probably does not follow a regular alternating sequence of ethylene and propylene units but, instead, will contain short chains, or blocks, of both polypropylene and polyethylene interspersed amongst much longer segments of random copolymer.

The ethylene/propylene rubber molecules are not perfectly linear but include, to varying degrees, both short and long-chain branches. This condition, which is influenced to some extent by polymerization conditions and composition, is not as apparent in the ethylene/propylene copolymers (EPM) as it is in the terpolymers (EPDM) where the additional influence of the diene monomer on the tendency to develop chain branches is present. This will be discussed later.

In addition to composition (ethylene/propylene ratio) the average molecular weight of the rubber is controlled by selecting certain catalyst and polymerization variables. While the polymer chemist generally measures average molecular weight in terms of intrinsic viscosity, the rubber compounder uses the more familiar measurement of Mooney viscosity. The ethylene/propylene rubbers are controlled within the range of raw polymer Mooney viscosity which has been found to fit the various processing and applications requirements of the rubber industry and which includes most other commercial synthetic rubbers. Mooney viscosity of EPM and EPDM is generally measured at four minutes after a one-minute warmup with the instrument at a temperature of 250°F. The measurement is expressed as ML 1 + 4 at 250°F.

At any average molecular weight the distribution of the molecular weights of the individual polymer molecules which comprise an elastomer influences both processing and the mechanical properties of the raw, compounded, and cured rubber. In ethylene/propylene rubber the molecular weight distribution is controlled, again by selecting catalysts and polymerization conditions.

An even more sophisticated variable is compositional distribution – different ethylene/propylene ratios in different molecular weight fractions. This, too, can be controlled by catalyst and polymerization conditions.

In summary, then, there are at least six measurable variables in the molecular structure of ethylene/propylene rubber which are controlled individually and in combinations to give a broad selection of commercial materials accommodated to the varying processing and applications requirements of the rubber industry.

Composition. The high propylene rubbers are generally easier mill processing while gum physicals and extrusion improve at higher ethylene.

Crystallinity. Generally zero crystallinity is desirable in the raw elastomer but the development of some crystallinity on stretching is sometimes also desirable.

Branching. Long chain branching should be avoided except where a "dead" compounded stock is desired. Short chain branching is of no significant consequence.

Average Molecular Weight. Cured physicals improve with increasing Mooney viscosity but processing is generally more difficult.

Molecular Weight Distribution. At equal average molecular weight, physicals are better and processing is poorer for the narrow molecular weight distribution rubbers.

Compositional Distribution. Within rather broad ranges this parameter has no measurable effect on practical applications of the rubber.

POLYMER PROPERTIES

The properties of a typical raw ethylene/propylene copolymer (EPM) are shown in Table 9.1. All of these properties are essentially the same for the ethylene/propylene terpolymers (EPDM). It will be immediately apparent to the practical rubber compounder that the low specific gravity of EPM and EPDM can be related to reduced unit costs in the fabrication of finished rubber products.

TABLE 9.1. PROPERTIES OF RAW ETHYLENE/PROPYLENE COPOLYMERS[3]

specific gravity g/cm^3	0.86–0.87
X-ray crystallinity	none
appearance	colorless
Mooney viscosity	varied
heat capacity, cal/g°C	0.52
thermal conductivity, cal/cm sec °C	8.5×10^{-4}
thermal diffusivity, cm/sec	1.9×10^{-3}
thermal coefficient of linear expansion/°C	1.8×10^{-4}
brittle point °C	−95
glass transition temperature[4] °C	−60
relative air permeability cm^2 sec^{-1} atm^{-1}	100[a]

[a]Natural rubber = 100;
IIR = 13[5];
SBR (23% styrene) = 65.

Crude rubbers, natural and synthetics, are bought by weight but used by volume. A low specific gravity rubber compound will yield more finished articles than a high gravity compound. Ethylene/propylene rubber has the lowest specific gravity of all the natural and synthetic elastomers currently available.

The low brittle point and glass transition temperature of ethylene/propylene rubber are important in applications where retention of dynamic properties at low temperatures is desirable. The rebound of EPM (and EPDM) vulcanizates over a range of temperatures is about the same as natural rubber, better than SBR, and much better than IIR.

POLYMER UNSATURATION (EPDM)

The structure of EPM which was illustrated on p. 221 shows this to be a saturated synthetic rubber. There are no double bonds in the polymer chain as there are in the case of natural rubber, the structure of which was illustrated on p. 000, and in most of the common commercial synthetic rubbers (e.g., SBR, CPBR, NBR, etc.). The main chain unsaturation in these latter materials introduces points of weakness. When exposed to the degrading influences of light, heat, oxygen, and ozone the unsaturated rubbers tend to degrade through mechanisms of chain scission and crosslinking involving the points of carbon-carbon unsaturation. Since EPM does not contain any carbon-carbon unsaturation, it is inherently resistant to degradation by heat, light, oxygen, and, in particular, ozone.

The double bonds in natural rubber and the common polydiene synthetics, aside from rendering these elastomers susceptible to environmental degradation, as mentioned above, are essential to their curing into useful rubber products using conventional chemical accelerators and sulfur. The whole world's rubber industry is based on the vulcanization chemistry which was first demonstrated by Charles Goodyear in 1839 and which involves carbon-carbon double bonds in a sulfur crosslinking reaction. EPM, a saturated elastomer, cannot be cured or crosslinked using the long-established manufacturing practices and chemicals pertinent to the unsaturated rubbers.

EPM can be cured into useful rubber articles using special chemicals, and, indeed, this is done every day on a large scale and will be discussed later. A more commercially attractive product would be one which retained the outstanding performance features discussed earlier (e.g., heat, oxygen, ozone resistance) and which included some carbon-carbon unsaturation from a small amount of an appropriate diene monomer to accommodate it to conventional sulfur vulcanization chemistry.

DIENE MONOMERS IN EPDM

The common diene monomers are isprene and butadiene. Efforts to introduce one or the other of these into the EPM molecule were unsuccessful. The catalysts and polymerization conditions which were essential for the manufacture of the random copolymer of ethylene and propylene would not, at the same time, tolerate a conjugated diene and cause it, also, to be included in the molecule. Searches for appropriate nonconjugated dienes at several laboratories resulted in the discovery of, perhaps, fifty such chemicals. The commercially important ones fall into two classes – the nonconjugated straight chain diolefins, and the cyclic and bicyclic dienes.

Both classes have two things in common. They cannot be conjugated and the activities of their two double bonds with respect of polymerization must be substantially different. The lowest molecular weight straight chain diolefin which meets these requirements is 1,4 hexadiene:

$$\overset{H_3}{C}-\overset{H}{C}=\overset{H}{C}-\overset{H_2}{C}-\overset{H}{C}=\overset{H_2}{C}$$

When this chemical is introduced with ethylene and propylene, the terminal double bond is active with respect to polymerization while the internal unsaturation is passive at this stage but remains in the resulting terpolymer as a substituent, or pendant, location for active sulfur vulcanization:

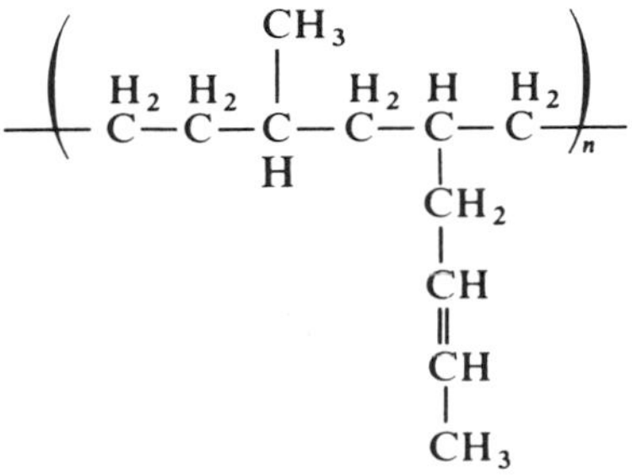

$$\left(\overset{H_2}{C}-\overset{H_2}{C}-\underset{H}{\overset{CH_3}{C}}-\overset{H_2}{C}-\overset{H}{C}(CH_2-CH=CH-CH_3)-\overset{H_2}{C}\right)_n$$

ethylene propylene 1, 4 hexadiene

Ethylene/propylene/1,4 hexadiene terpolymer (EPDM) is an important commercial material.[6]

The bicylic dienes which are used to introduce substituent unsaturation into ethylene/propylene rubber are typified by various derivatives of norbornene. Only three of these chemicals have achieved commercial acceptance in EPDM.

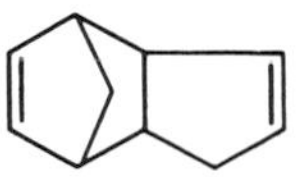

Dicyclopentadiene is used in certain grades of EPDM which are made by a number of manufacturers. It enters the polymer readily with much higher polymerization efficiency than 1,4 hexadiene. The double bond in the bridged, or strained ring, is the more active with respect to polymerization and the five-membered ring is left substituent to the main polymer chain with its double bond then active for sulfur vulcanization. For reasons which are discussed later the activity of this substituent double bond toward sulfur crosslinking is lower than the internal substituent double bond from 1,4 hexadiene. Those grades of EPDM which use dicyclopentadiene as the diene are slower curing than some others.

The high activity of the double bond in the bridged ring extends to other substituted norbornenes. Methylene norbornene

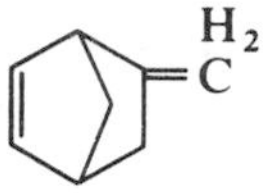

incorporates well into the polymer and yields a fast curing rubber.[7] It was once used in commercial EPDM.

Certainly the most widely used diene in current commercial EPDM is ethylidene norbornene:

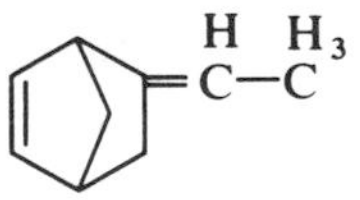

As with the other bridged ring dienes, ENB, as it is commonly called, shows a high rate of polymerization through the double bond in the bridged ring. The substituent internal double bond is also very active with respect to sulfur crosslinking.[8] The structure of EPDM containing ENB is

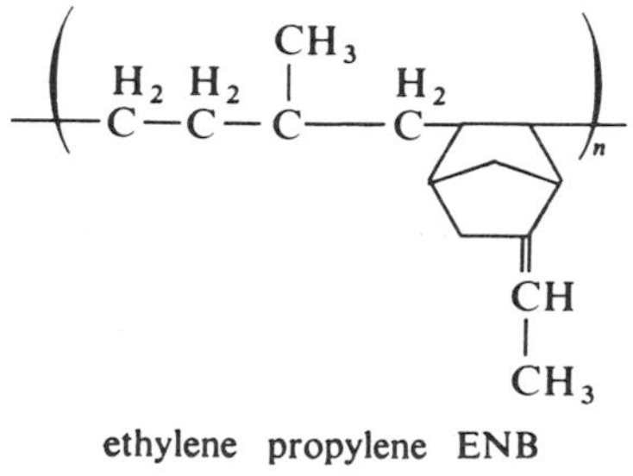

ethylene propylene ENB

The dienes which are used commercially in EPDM are 1,4 hexadiene, dicyclopentadiene (DCPD), and ethylidene norbornene (ENB). These enter the polymer through the double bond which is more active in this function and leave the second double bond substituent or external to the polymer chain where it is then available to function in the classical chemistry of the sulfur vulcanization of rubber.

STRUCTURE AND PROPERTIES OF EPDM

Since one double bond is lost when the diene enters the polymer and the second double bond which remains is not *in* the polymer but pendant and external to it, the outstanding resistance of EPM to degrading attack by heat, light, oxygen, and ozone, which was discussed earlier, is inherent also in EPDM. Indeed, EPDM is still a saturated elastomer insofar as the structure of its spine or backbone is involved. The unsaturation is present only in pendant "handles" or cure-sites and any degradation of these small groups will not affect the integrity of the polymer chain itself.

The amount of diene which is incorporated into the EPDM molecule will affect the rate of cure of the rubber. Generally, from 4% to 5% by weight of diene is sufficient to give a serviceable product which will fit into most rubber factory practices. As much as 10% by weight of diene is incorporated into certain grades of EPDM where ultra-fast cures, or co-cures with polydiene elastomers are desired.

As mentioned earlier, the nature of the diene itself strongly influences the cure rate of EPDM.[9] At equal levels of pendant saturation, dicyclopentadiene EPDM is slowest curing, ethylidene norbornene EPDM is fastest, and 1,4 hexadiene EPDM is intermediate. To some extent, at least, these cure rate differences are related to the availability of allylic hydrogen atoms in the pendant functional groups.

MONOMERS – ETHYLENE AND PROPYLENE

The two principal raw materials for EPM and EPDM, ethylene and propylene, are abundantly available at high purity. In the United States both are produced primarily by cracking natural gas liquids. In Europe, Japan, and elsewhere, they are obtained from cracking a heavier fraction derived from petroleum commonly called naphtha. The production of ethylene and propylene has increased tremendously during the past twenty-five years. Both of these simple olefins are used in the manufacture of plastics, chemicals, and fibers. In the United States the annual production of each of them has reached several billion pounds.

Ethylene and propylene are both gases at atmospheric temperature and pressure. Propylene is commonly stored and transported as a liquid under pressure. Although ethylene can also be handled as a liquid, usually at cryogenic temperatures, it is almost universally transported from production to consumption in pipelines as a gas. An ethylene pipeline manifold or grid has been developed along the Gulf Coast of the United States. Ethylene producers, in addition to supplying "captive" satellites "over the fence," feed into the manifold pipeline. Users of ethylene – for plastics, chemicals, and synthetic rubber – have located their plants astride or adjacent to the ethylene manifold. Subterranean storage in abandoned salt caverns provides the "condenser" capacity to handle surges in production and consumption.

This extensive pipeline distribution of a high purity chemical raw material is typical of the complex and interdependent nature of the petrochemical industry. All of the ethylene/propylene rubber plants are located on the ethylene pipeline manifold; three are in Louisiana and one is in Texas.

MANUFACTURE OF EPDM

The various manufacturers of EPDM have not revealed many details of their processes. One patent[10] describes a continuous process. A suggested process flow sheet has been published.[11] Other patents, while not specifically relating to the process, have, nevertheless revealed some process features.[12-15] From a study of these published references, and of many others which have not been listed in a work of this nature and scope, it is possible to deduce a probable commercial EPDM process.

An important prerequisite is dryness. Ionic polymerizations using Ziegler-type catalysts are extremely sensitive to water and other polar materials. Water cannot be tolerated in any of the feed streams. A few parts per million is a necessary and controllable maximum. The polymerization process is fully continuous. An inert hydrocarbon carrier solvent (e.g., hexane) is used. This is dried in a column by azeotropic distillation and the dry solvent is fed to the polymerization vessels and is used as a carrier for the catalysts to facilitate more precise control of these small streams. High purity ethylene gas (from pipeline supply) is dried by passing it through a molecular sieve. High purity propylene, from liquid tank farm storage, is similarly dried. Diene monomer, a liquid, is also sieve dried.

The catalyst comprises two components. Many combinations of transition metal halides and metal alkyls have been published and patented and shown to be useful in the polymerization of amorphous Ethylene/Propylene Rubber. Typical catalyst compositions are shown in the publications of Kelly.[16] Vanadium oxytrichloride ($VOCl_3$) is a typical transition metal halide catalyst, while ethyl aluminum sesquichloride ($(C_2H_5)_3Al_2Cl_3$), diethyl aluminum chloride ($(C_2H_5)_2AlCl$), and diisobutyl aluminium chloride ($(C_4H_9)_2AlCl$) are examples of typical metal alkyls. The catalyst components are liquids and are transported either in cylinders or tank trucks. In common with all aluminum alkyls, ethyl aluminum sesquichloride is unstable in the presence of water and will spontaneously ignite in moist air. It must be handled in closed vessels and in the total absence of water. Both the vanadium salt and the aluminum alkyl are mixed with dry hexane into the dilute solutions which are more convenient for precise flow control.

Dry hexane, ethylene, propylene, diene, catalyst solution, and cocatalyst solution are continuously and proportionately fed to the first of a series of polymerization vessels. The two catalyst solutions may first enter a premix zone of high intensity agitation with a portion of the monomers to enhance their activity and to ensure their uniform distribution in the polymerization reactor. Polymerization of individual molecules, or chains, is extremely fast – a few seconds, at most, is the average life of a single growing polymer molecule from initiation to termination. Upon termination, the polymer molecule dissociates from the initiating catalyst complex which is regenerated and initiates another polymer molecule. The polymerization is highly exothermic – about 1100 BTU per pound. Since the polymerization temperature (circa 100°F) must be controlled within narrow limits to ensure a product with the desired average molecular weight and distribution, this heat must be removed. This is done through the use of chilled jackets, draft tubes, and internal coils, which are features of the design of the polymerization vessels.

As the polymer molecules form and dissociate from their initiating catalyst, they remain in solution in the carrier hexane. Indeed, this method of ionic polymerization is often called solution polymerization and it is distinguished, in this respect, from the emulsion polymerization processes used in the manufacture of certain other synthetic rubbers such as SBR, NBR, and CR. As the concentration of polymer in the hexane increases, the viscosity of the solution also increases. The practical upper limit of solution viscosity is dictated by considerations of heat transfer, mass transfer, and fluid flow. At a rubber solids concentration of 8% to 10% the solution viscosity is at a level that further increase is inadvisable and the polymerization is stopped.

Polymerization is carried out in a series of two or three vessels. As the polymer solution is displaced from one vessel to the next, additional raw materials, catalysts, and solvent may be added to the succeeding vessels. These added streams, as well as the streams to the first vessel, are carefully controlled in consideration of the desired properties of the finished rubber. On-line process chromatography is essential. Instrumentation, in general, is very sophisticated in the continuous solution polymerization of EPM and EPDM. Polymerization is stopped by adding to the stream leaving the final polymerization vessel a small quantity of a polar material (e.g., water) which destroys the catalyst.

The reactivity of ethylene is high while that of propylene is quite low. Various dienes, as mentioned earlier, have different polymerization reactivities. The viscous rubber solution leaving the last polymerization vessel contains some unpolymerized propylene, more or less unpolymerized diene, and about 10% EPDM – all in homogeneous solution. This solution passes continuously into a flash tank where reduced pressure causes most of the unpolymerized monomers to leave as gases which are collected and reused.

Catalyst residues (particularly vanadium) are undesirable. These are removed as soluble salts in a washing and decantation operation. Vanadium residues are controlled in the finished product to a maximum of a few parts per million. A stabilizer, or antioxidant, is added to the EPDM solution. Certain stabilizers and combinations have been found to be particularly effective in EPDM.[17] If oil-extended EPDM is to be produced, a metered flow of suitable oil is also added at this point.

Next, the rubber is separated from its solvent carrier by steam flocculation. The viscous cement is pumped through orifices into the vapor space of a violently agitated vessel partially full of boiling water. The hexane flashes off and, with water vapor, passes overhead to a condenser and to a decanter for recovery and reuse after drying. Residual unpolymerized ethylene and propylene appear at the hexane condenser as noncondensibles and are recovered for reuse after drying. The polymer, freed from its carrier solvent, falls into the water in the form of floc or crumb which looks very much like white popcorn.

TABLE 9.2. WORLD EPDM MANUFACTURERS

Company Name	*Code*[a]	*Plant Location*	*Plant Start-Up*	*Trade Name*
Copolymer Rubber & Chemical Corp.	C	W. Addis, La. USA	1968	Epsyn
N. I. Nederlandse Staatsmijnen	DS	Geleen, Neth.	1966	Keltan
E. I. duPont deNemours and Co.	DU	Beaumont, Texas, USA	1963	Nordel
Enjay Chemical Co.	EN	Baton Rouge, La., USA	1963	Vistalon
The International Synthetic Rubber Co, Ltd.	IS	Grangemouth, UK	1969	Intolan
Mitsui Petrochemical Industries, Ltd.	MI	Iwakuni-Ohtake, Japan	1969	Mitsui EPT
Montecatini Edison S.pA.	M	Ferarra, Italy	1963	Dutral
Sumitomo Chemical Co., Ltd.	SU	Chiba, Japan	1969	Esprene
Uniroyal, Inc.	US	Geismar, La., USA	1964	Royalene
B. F. Goodrich Chemical Co.		Orange, Texas, USA	1971	Epcar
Chemische Werke Huls		Maarl, W. Germany	1971	
Japan EPR Co.		Japan		
Asahi Chemical Industry Co.		Japan		
Ube Industries		Japan		

[a] For reference to Table IV.

If a diene of low polymerization activity (e.g., 1,4 hexadiene) were used, the hexane recovered from the decanter would contain significant and valuable amounts of this unpolymerized material. Prior to drying, the diene would be separated from the hexane in a fractionation column and both recovered materials would be reused after drying. If a high activity diene were used (e.g., ethylidene norbornene), not enough of it would be present in the recovered hexane to justify its separation and recovery.

The rubber floc or crumb, now a 3%–4% slurry in hot water, is pumped over a shaker screen to remove gross, excess water. The dewatered crumb is fed to the first stage of a mechanical screw dewatering, drying press. Here, in action similar to a rubber extruder, all but 3%–6% of the water is expressed as the rubber is pushed through a perforated plate by the action of a screw. The cohesive, essentially dry rubber then passes into the second stage press. This is similar to the first stage dewatering machine except that the mechanical action of the screw causes the rubber in the barrel to heat up to perhaps 290°F. As this rubber is extruded through the pelletizing perforated die plate at the end of the machine the small amount of remaining water flashes into a vapor and dry rubber pellets are carried off on a cooling conveyor. The EPDM pellets are then continuously weighed, pressed into 75-pound bales, packaged, and palletized for storage and shipment.

MANUFACTURERS

EPM and EPDM are currently manufactured by four companies in the United States and five overseas. In addition, two companies have announced the start of construction of EPDM plants – one in the United States and one in Germany. Finally, an additional three companies in Japan have announced their intention to produce EPDM but their plans are not yet firm enough to assign a start-up time. This information is summarized in Table 9.2.

EPDM MANUFACTURING CAPACITY

The rated capacities of the various EPDM plants have, in some cases, been revealed, usually in advance of start-up, by their owners and builders, and have, in other cases, been estimated by various market analysts, consultants, and editors. There is considerable disparity in these figures and an accurate listing of total domestic and worldwide EPDM capacity is difficult. In his recent annual survey "World Synthetic Rubber – Its Manufacture and Markets" prepared for the Eleventh Annual Meeting of International Institute of Synthetic Rubber Producers, Inc. held in Mexico, D. F., May 19–22, 1970, Clayton F. Ruebensaal of Uniroyal, Inc. gives the capacity figures shown in Table 9.3.

EPDM VARIETIES

The 1970 edition of *The Elastomer Manual* which is published annually by International Institute of Synthetic Rubber Producers, Inc. lists sixty-six numbered

TABLE 9.3. EPDM PLANT CAPACITIES (IN MILLION POUNDS)

	Jan. 1970	*Jan. 1973*
U.S.A.	240	460
Europe	63	235
Japan	31	112
Total Worldwide	334	807

varieties of EPM and EPDM being offered by the nine active manufacturers. That this list has grown so long during the few years that this new synthetic rubber has been in production is witness both to the determination of the manufacturers to tailor their products to specific needs over the whole sweeping range of the rubber fabricating industry's requirements, and to the foresight and skill of the engineers who provided in the new plants the versatility and control to handle a broad range of products.

APPLICATIONS AND CONSUMPTION

The inherent properties of ethylene/propylene rubber, which were discussed earlier, suggest that this material will be useful in fabricating thousands of rubber articles for a great variety of services and environmental exposures. This has, indeed, been the story of EPDM. Aside from applications requiring resistance to hydrocarbon oils, there is scarcely an application for either natural or one of the synthetic rubbers where EPDM has not challenged the incumbent elastomer.

Various authors have published estimates of the current and future consumption of EPDM broken down into its principal areas of application. The most recent of these appeared in the June 28, 1969 issue of *Chemical Week.*[18] Table 9.4. is taken from this publication.

The Sales Department of the Chemical Division of Uniroyal, Inc. prepared an even more recent survey of domestic EPDM markets in October, 1970. A summary of this previously unpublished material appears in Table 9.5.[19]

Chemical Week and Uniroyal are in good agreement with respect to the total domestic consumption of EPDM in 1969 and in 1975; and both foresee an annual growth rate of about 25%.

The "Tires-Related" end-use category which appears in both the *Chemical Week* and Uniroyal surveys reflects the large scale use of EPDM as an additive to the diene rubber (SBR, natural, CPBR) compounds used in tire sidewalls and coverstrips to improve their resistance to ozone and weather cracking while under stress and during flexing. EPDM is now almost universally used in this application, which is discussed in more detail later.

Expressed as percentages of total annual synthetic rubber hydrocarbon consumption in the United States, EPDM has increased from nothing in 1964 to 3.4% in 1970 and will increase further to 8.9% by 1975.[20]

TABLE 9.4. EPDM MARKETS

	(million lb)	
	1969	*1975*
Tires related	28	200
Vehicular	48	150
Appliance	12	27
Hose	8	18
Wire/cable	9	22
Coated fabrics, linings	5	12
Matting, pads, etc.	4	8
Footwear	3	8
Leisure, recreational	1	18
Rug underlay	–	20
Miscellaneous	15	50
Total domestic use	133	533
Export	22	20

Production Capacity for EPDM

	(million lb)	
	1968	*1970*
U.S.		
DuPont	60	75
Enjay	45	55
Uniroyal	40	90
Copolymer	25	55
EUROPE		
Dutch States Mines	26	40
Montedison	12	34
Hoechst/Huels	2	2
Intl. Synthetic Rubber	0	2
JAPAN		
Mitsui	0	6
Sumitomo	0	6

COMPOUNDING

Never before in the history of the synthetic rubber industry has the introduction of a new elastomer been accompanied by the quantity and quality of technical literature which has served EPDM. The EPDM manufacturers have kept their applications and technical service laboratories busy developing techniques and specific recipes to accommodate their products to practically every fabricated rubber article. The manufacturers of rubber chemicals, pigments, carbon blacks,

TABLE 9.5. U.S.A. EPDM CONSUMPTION – 1966-1975

	Millions of Pounds		
End Use Category	*1966*	*1969*	*1975*
Tires & Related Products	8.0	28.8	160.0
Automotive	17.2	33.0	160.0
Wire & Cable	6.5	9.4	50.0
Appliance, Including Hose	4.0	6.6	12.0
Hose	6.0	14.2	25.0
Belting	.8	2.2	12.0
Gaskets	.8	4.4	30.0
Rolls	1.0	1.2	10.0
Proof Goods	0.5	1.1	18.0
Footwear	.5	1.6	2.8
Other	9.7	13.3	25.0
	55.0	115.8	504.8

oils, plasticizers, tackifiers, and resins have also investigated the new EPDM in its association with their products. Many of these investigations have been published in the form of bulletins and manuals and are available from the various manufacturers, all of whom have a history of long and high quality service to the rubber industry.

In addition, EPDM has captured the interest of more theoretical workers in industrial, academic, and government research laboratories all over the world. From their work has come another flood of published literature which has rapidly added to the understanding of the polymerization, structure, properties, and utility of this elastomer.

Finally, the rubber compounders themselves, in the laboratories of hundreds of rubber fabricators in all of the industrialized countries of the world, have made their contributions by learning to use EPDM in applications where its inherent physical properties suggested that it would make a better or lower cost product. While this type of work is not usually published, the rubber industry has long been noted for the candor of its technical personnel, and good ideas spread fast.

Anyone who intends to start a compounding project involving the use of EPDM would be wise to first ask the suppliers of elastomers, chemicals, and pigments for their pertinent literature and help. Next, he might go to the published literature, particularly if his proposed application is unique and requires some understanding of polymer structure and chemistry.

The compounder should be aware of the merits and limitations of the elastomers before making his selection to develop a compound for a new article, application, or replacement. In Table 9.6[21] are listed comparative and typical properties of rubber compounds based on EPDM and the four other elastomers with which it is likely to be directly competitive. NBR, a synthetic rubber with excellent oil resistance, dominates applications in that field and is not included.

In most rubber compounds one finds, in addition to the elastomer, curing chemicals, oils, and pigments.

TABLE 9.6. PROPERTY COMPARISON CHART

	EPDM	*Natural*	*SBR*	*IIR*	*CR*
Specific gravity (polymer)	0.86	0.92	0.94	0.92	1.23
Tensile strength, psi (reinforced) (max.)	3500	4000	3500	3000	4000
Elongation, % (max.) (reinforced)	500	700	500	700	500
Top operating temperature, °F	300 to 350	170 to 250	170 to 250	250 to 350	200 to 300
Brittleness point, °F	–75	–70	–75	–75	–45
Compression set, %[a]	10 to 30	10 to 15	15 to 30	15 to 30	15 to 30
Resilience (Yerzley)	75	80	65	30	75
Tear strength, lb/in.	100 to 250	200 to 250	150 to 200	150 to 200	200 to 250
Dielectric constant	2.5	2.9	2.9	2.5	6.7
Volume resistivity, ohm-cm	10^{14}	10^{15}	10^{15}	10^{15}	10^{12}
Dielectric strength, V/mil	500 to 1400	400 to 600	600 to 800	600 to 900	400 to 600
Resistance to:					
Weathering	E	F–G	F–G	G	G
Ozone	E	P	F	G	G
Acids & alkalis	E	G	G	G	E
Oils & solvents	P	P	P	P–F	G
Abrasion	G–E	G–E	G	G	G–E
Compression set	G–E	E	G	G	G
Tearing	G	E	F	G	G–E
Low Temperature	G–E	G	F–G	G	F
Steam	E	G	G	G	G
Air permeability	P	G	G	E	G

E = Excellent; G = Good; F = Fair; P = Poor.

[a] 22 hr at 212°F ASTM D395 method B.

CHEMICALS

EPM, the saturated copolymer, as was mentioned earlier, requires special curing chemicals. The chemicals, which are usually used in conjunction with a small amount of sulfur, are organic peroxides. The most common, dicumyl peroxide (DiCup), rapidly decomposes in the compound on heating and effects a good, tight cure but develops a decidedly objectionable odor which cannot be tolerated in many rubber articles. Recently, other organic peroxides have been introduced which mitigate the objectionable odor and which should permit EPM to be used in some applications from which it has been excluded. Typical simple EPM compounds using peroxide curatives are shown in Table 9.7.[22]

EPDM, the unsaturated polymer, can be cured using the common rubber accelerators which are found in recipes for most other synthetic rubbers. It is difficult to generalize as to the choice of chemicals used in an EPDM vulcanizate. The particular combination selected is dependent on so many factors – mixing equipment, mechanical properties, cost, safety, compatibility, to name a few, – that it would be futile to identify any one recipe as better than another without also identifying all of the facts which the compounder must evaluate before he makes his final choice.

A cure system using tetramethyl thiuram disulfide (TMTDS) activated with mercaptobenzothiazole (MBT) is basic, active, and safe for a fast curing (ENB or 1,4 hexadiene) EPDM. An even faster curing compound can be achieved by further

TABLE 9.7. COMPARISON OF PEROXIDES

Vistalon 404	100	100	100
SRF black	60	60	60
Hercules-S-890	3.0	–	–
Montecatini PX-60	–	2.7	–
Di-Cip 40	–	–	7.0
Sulfur	0.3	0.3	0.3
Scorch (MS 260°F)			
Minutes to 5 pts rise	27	26	20
Cure: 320°F/20′			
Tensile, psi	2110	2180	2170
Elongation, %	330	380	370
300% Modulus, psi	2000	1800	1840
Tear (cresent) lb/in. R.T.	109	120	118
Compression Set (25′ cure) "B"			
22 hr at 158°F, %	7	8	6
70 hr at 212°F, %	24	22	19
Oven Aging (70 hr at 310°F)			
Tensile, psi	280	250	150
Elongation, %	60	70	50
Odor	Acceptable	Acceptable	Objectionable

TABLE 9.8. LEVEL AT WHICH ACCELERATORS USUALLY WILL NOT BLOOM OUT OF EPDM (ENB)

Accelerator	*Level*
MBT, MBTS, CBS, ZMBT	3.0 phr
ZDBDC, DTDM	2.0 phr
ZDEDC, ZDMDC, TDEDC, TMTDS, DPTTS, TMTMS, TETDS	0.8 phr

activating the basic recipe with a dithiocarbamate such as zinc dibutyl dithiocarbamate (ZDBDC) or ferric dimethyl dithiocarbamate. For the compound to be nonblooming it is essential that the concentrations of the various chemicals in the recipe be kept below their respective limits of solubility in EPDM. The solubilities of the common rubber chemicals used in EPDM compounds are shown in Table 9.8.[23] This reference also includes a very comprehensive study of the accelerator systems for EPDM. Typical cure systems for an ENB EPDM and some physical properties of the vulcanizates are shown in Table 9.9.[24]

TABLE 9.9. TYPICAL CURE SYSTEMS FOR EPDM (ENB)

	A	*B*	*C*
EPDM (ENB)	100	100	100
Zinc Oxide	5	5	5
FEF Black	70	70	70
SRF Black	35	35	35
Naphthenic Oil	100	100	100
Paraffin Wax	1	1	1
Stearic Acid	1	1	1
MBT	0.8	1.0	1.0
TMTDS	0.8	0.8	0.8
ROYLAC 133	0.8	–	–
E. Tellurac	–	–	0.8
Tetrone A	–	0.8	0.8
Sulfur	0.8	0.8	0.8
Compounded Mooney Viscosity			
ML-4 at 212°F	28	27	26
Mooney Scorch at 270°F			
Scorch Time	11′45″	10′30″	9′00″
Cure Rate	4′00″	3′30″	3′00″
Mooney Scorch at 250°F			
Scorch Time	32′00″	18′00″	17′15″

TABLE 9.9. (*continued*)

	A	*B*	*C*
UNAGED PHYSICAL PROPERTIES			
300% Modulus, psi			
3′	220	240	460
5′	320	420	620
7.5′	490	550	780
10′	590	700	830
15′	700	790	930
Tensile Strength, psi			
3′	640	870	1310
5′	1220	1510	1780
7.5′	1650	1790	1800
10′	1810	1870	1840
15′	1910	1970	1900
Elongation, %			
3′	860	900	830
5′	970	940	780
7.5′	860	830	680
10′	810	740	640
15′	760	690	600
Hardness, Shore A			
3′	43	48	50
5′	48	50	51
7.5′	50	52	53
10′	50	53	54
15′	50	55	55
Compression Set, Aged 22 Hours at 212°F			
30′	38.3	46.1	38.5
45′	32.8	33.9	27.4
AGED 1 DAY AIR OVEN AT 300°F			
Tensile Strength, % Retained			
10′	93.9	96.2	97.3
30′	91.9	92.6	102.2
Elongation, % Retained			
10′	43.2	43.2	50.0
30′	61.5	48.3	61.4

Cured at 400°F

TABLE 9.9. (*continued*)

	A	*B*	*C*
UNAGED PHYSICAL PROPERTIES			
300% Modulus, psi			
1′	420	580	780
2′	580	690	790
3′	580	780	830
4′	590	690	830
Tensile Strength, psi			
1′	1420	1610	1700
2′	1600	1790	1820
3′	1650	1810	1900
4′	1690	1800	1900
Elongation, %			
1′	890	820	700
2′	820	730	680
3′	780	700	670
4′	730	670	650
Hardness, Shore A			
1′	44	48	48
2′	45	49	49
3′	45	48	49
4′	43	47	49

OILS

EPDM compounds generally carry fairly high loadings of oil. Indeed, the capacity of the EPDM polymer for extension with oil while retaining a good measure of physical properties is tremendous. Some of the producers have exploited this property of EPDM and add oil during manufacture. Oil-extended EPDM is made by first polymerizing a very high molecular weight, high Mooney viscosity polymer. Then, while the polymer is still in solution the desired amount of oil is added. The oil-extended rubber is then flash flocculated, dried, and baled in the normal manner. The Mooney viscosity of the base polymer is reduced to a practical, workable level in the oil-extended EPDM.

A number of common rubber extending and compounding oils can be used in EPDM. Compatability with the naphthenic and paraffinic types is better and these are preferred to the more aromatic types commonly used in SBR. Viscosity of oil is also important. Naphthenic oils in the medium viscosity range are a good choice for use in EPDM compounds.[25]

The extent to which EPDM can carry high loadings of oil and carbon black is

TABLE 9.10. EPDM IN LOW COST HEATER HOSE

EPDM[a]	100.0	MBT	.5
Zinc Oxide	5.0	ZDMDC	1.0
FEF Black	125.0	TMTDS	1.5
MT Black	120.0	Sulfur	2.0
SRF Black	100.0	Mooney Viscosity	
York Whiting	150.0	ML-4 at 212°F	36
Naphthenic Oil	185.0	Mooney Scorch	
Stearic Acid	1.0	ML at 250°F	23

Unaged Physical Properties, Cured at 320°F

Modulus at 200%	10′	680
	20′	880
Tensile Strength, psi	10′	780
	20′	890
	30′	960
Elongation, %	10′	270
	20′	210
	30′	190
Hardness, Durometer A	10′	74
	20′	76
	30′	78

Heat Aged, D-573, 70 hr at 212°F

Tensile Strength Change %	20′	+12.4
	30′	+4.2
Elongation Change, %	20′	−33.3
	30′	−26.3
Hardness Change, pts.	20′	+4
	30′	+3

Compression Set, D-395 Method B, 22 hr at 212°F

30′ at 320°F, solid	72

[a] ROYALENE® 512

illustrated by the heater hose compound shown in Table 9.10.[26] Its low specific gravity and high oil and filler tolerances are important factors in the selection of low pound volume cost EPDM compounds which are often fully competitive with compounds based on low cost SBR. Oil-extended EPDM is frequently compounded with still additional oil and the necessary fillers in low cost compounds which exhibit the good physical properties associated with the high molecular weight base polymer.

PIGMENTS

Pigments, carbon blacks and nonblacks, are the third class of materials which are found in every EPDM compounds. Almost everything that has been learned and said about the behavior of the various pigments commonly used in general rubber compounding in SBR can be said, also, about EPDM. The reinforcing, semi-reinforcing, easy processing, and high abrasion blacks exhibit their typical behaviors in EPDM compounds. Likewise the soft and hard clays, the reinforcing silica chemicals, and the calcium carbonates whose behaviors are well known to the rubber industry do not show anything unexpected in EPDM.

PROCESSING

The mixing of EPDM compounds sometimes poses some special problems. They can be successfully and economically mixed and processed on all of the machinery commonly found in various rubber fabricating plants. In mill mixing the lower Mooney viscosity EPDM's in lightly loaded stocks are easier to handle. The high viscosity varieties of EPDM are tough and nervy on the mill but their processing improves as they readily accept high loading of oil and fillers.

In Banbury mixing, the cycles and dump temperatures are about the same as would be used for SBR. A typical EPDM compound contains more filler than the total of oil and polymer, so it is desirable to add the filler quickly, near the beginning of the cycle, to ensure adequate dispersion. A Banbury overload of 10%–15% is recommended. "Upside-down" mixing of very highly loaded EPDM compounds is often desirable. Two typical Banbury mixing procedures for EPDM compounds are shown in Table 9.11.[27]

EXTRUSION

EPDM compounds extrude readily on all commercial rubber extruders. Since the mechanical breakdown of EPDM is not as pronounced as with some other

TABLE 9.11. TYPICAL EPDM BANBURY MIXING PROCEDURE (ONE-PASS MIXING)

Right-Side Up *Time, Minutes*	*Addition*
0	EPDM, Zinc Oxide, Stearic Acid
0.5	One-half Filler
1.5	One-Quarter Filler, one-half Oil
2.5	One-Quarter Filler, one-half Oil
3.5	Sweep, Curatives
4.0	Dump below 230°F

Upside-Down **Time, Minutes**	**Addition**
0	Filler, Oil, Zinc Oxide, Stearic Acid
0.5	EPDM
3.0	Sweep, Curatives
3.5	Dump below 230°F

elastomers, but the compounds are readily softened by heat, higher extruder temperatures are generally used for EPDM than for other rubbers. Compound temperatures of 200–230°F at the extruder head have produced good results.[28]

CALENDERING

EPDM compounds can be calendered both as unsupported sheeting and onto a cloth substrate. The former is an important commercial application. EPDM sheeting has been widely used for lining pits, ditches, and reservoirs and, less widely, as a roofing barrier. In both of these applications, and in coated fabrics which are used as truck tarpaulins, the outstanding weather resistance and low temperature flexibility of EPDM contribute to a highly durable product.

SPONGE

Highly durable, weather resistant open and closed cell sponge can be made from EPDM. The various blowing agents commonly used for other elastomers are effective in EPDM. A typical extruded closed cell sponge compound and its physical properties are shown in Table 9.12.[29] Automotive weather stripping and trunk lid gaskets are often made from this type of EPDM compound.

CURING

EPDM compounds can be cured on all of the common rubber factory equipment. Press cures, transfer molding, steam cures, air cures, liquid metal cures, compression molding and injection molding are all practical. Cure cycles can be adjusted by a proper choice of accelerators. Usually the normal unsaturated grades of EPDM will give adequately short cycles. Should a very short cycle be required it is sometimes necessary to use one of the highly unsaturated grades. EPDM, particularly those varieties having high levels of ethylene, is more thermoplastic and more heat resistant than other synthetic rubbers. Higher mold temperatures can frequently be used to achieve both shorter cycles and improved mold flow.

BLENDS

Compared to other polydiene elastomers (viz., natural, SBR, NBR, CPBR) EPDM has a very low level unsaturation. A recipe adjusted to give a rapid, practical cure rate in EPDM will be exceedingly fast and cause overcures in a polydiene. Conversely, a recipe written for SBR or NBR will cause only light, undercures in EPDM. It is, therefore, difficult to achieve a good, tight co-cure of a mixture of EPDM and polydiene rubber. It is, however, often desirable to blend elastomers for applications requiring a combination of properties not found in one or another alone. For instance, a combination of the oil resistance of NBR and the ozone resistance of EPDM would be desirable.

TABLE 9.12. EPDM IN CLOSED CELL EXTRUDED SPONGE

EPDM		100.0
BIK		0.5
Zinc Oxide		5.0
Stearic Acid		1.0
Sunpar 150		70.0
MT Black		100.0
FEF Black		40.0
CELOGEN® AZ		10.0
ZDMDC		3.0
PENZONE B		3.0
MBT		2.0
Tellurac		1.0
Sulfur		2.0
		337.5
Mooney Viscosity, ML-4 at 212°F		39
Mooney Scorch, MS at 250°F		2′45″
Extrusion Data at 150°F and 50 RPM		
Weight/min (g)		64.2
Length/min (inches)		30.5
Comments		Fairly Smooth
Rheometer Data at 400°F		
Viscosity (in. lb)		
Initial		17.4
Minimum		14.8
Maximum		30.8
Viscosity after 1 min		17.5
2 min		27.5
3 min		31.0
4 min		30.0
5 min		28.5
Time to 95% Optimum		2′27″
Density (lb/cu. ft)	4′	19.0
	6′	17.0
	8′	16.0
50% Compression Set 22 Hours at 158°F	4′	40.0
	6′	30.0
	8′	22.0
% Water Absorption	4′	.049
	6′	.046
	8′	.059

The cure rates of EPDM and NBR can be balanced to some extent by using certain delayed action accelerators in the recipe. Compatability with a highly unsaturated elastomer in a blend which is to be co-cured can be improved by introducing a higher level of unsaturation into the EPDM. This is done, as was mentioned earlier, by increasing the diene level (ENB) from the normal 4–5% to 10%. Several EPDM manufacturers now list "ultra fast curing" highly unsaturated varieties and suggest recipes for their co-cure with other elastomers.

A more important application for EPDM in blends with another rubber is to provide ozone resistance without participating significantly in co-cure with the host rubber which comprises the major portion of the blend. Considerable practical use of this technique has been made in enhancing the ozone and weathering resistance of tire sidewalls and coverstrips. When added to an SBR or natural rubber compound as a replacement for 20–30% of the principal elastomer, certain grades of EPDM disperse in the finished compound in the form of extremely small, discrete particles which can be seen only through the electron microscope.

Joseph F. O'Mahoney, Jr.[30] and E. H. Andrews[31] have advanced a theory to explain the behavior of those grades of EPDM which impart ozone resistance to natural rubber (and SBR). It is essential that the two polymers be thoroughly

TABLE 9.13. EPDM IN WHITE SIDEWALLS FOR OZONE PROTECTION

		A	*B*
Natural Rubber		100.00	80.00
ROYALENE® 301T		–	20.00
Titanium Dioxide		35.00	35.00
Zinc Oxide		35.00	35.00
Stearic Acid		2.00	2.00
Circo Light Oil		5.00	5.00
SUNPROOF® IMPROVED WAX		5.00	5.00
DELAC-S®		0.35	0.35
Sulfur		3.50	3.50
Kilocycles Cured 30′ at 320°F			
Outdoor Dynamic Flexing	VVS	744	1671
	VS	1671	
	S	–	
	C	1913	
Ozone Box – 168 Hours Exposure			
12.5% Elongation	VVS	72	OK
50 pphm ozone	VS	168	
	S	–	
	C	–	
Outdoor Static Exposure – Days to Cracking Stage			
Naugatuck	VVS	40	OK
Total Exposure	VS	75	
222 Days	S	–	
	C	97	

mixed before any of the compounding ingredients are added. It is equally important that the compounding ingredients are then added in the proper sequence and at the proper intervals and that mixing be continued for a specified time – a departure from the normal Banbury practice of mixing to a specified temperature. A two-phase mix results and the EPDM, the discontinuous phase, probably does not participate in the subsequent cure of the natural rubber continuous phase except, perhaps, in a narrow zone at the interface.

Upon exposure to ozone, while under stress, cracks develop in the natural rubber phase in a manner to relax the stresses. Before a microscopic crack can propagate to an extent that it becomes visible (macroscopic), it encounters a microscopic particle of EPDM at which the growth of the crack is terminated. Other cracks are initiated through the entire specimen only to suffer the same fate. The practical and visible result is an article or specimen which will pass through a long period of continuous relaxation of stresses while exposed to ozone and yet develop no cracks which are visible to the naked eye. EPDM, in this instance, might be said to be functioning as a polymeric antiozonant.

A typical white sidewall compound which includes EPDM for ozone protection is shown in Table 9.13.[32] In this table the EPDM is identified by its trade name (ROYALENE® 301T) since this is one of the few varieties in which the ozone resistance function in blends is manifest.

TACKIFIERS

EPDM compounds are nontacky. This sometimes causes building and cured adhesion problems in instances where it is necessary to join two stocks as in the lap seams of a pit liner installation using EPDM sheeting, or in the building of an EPDM tire. These problems can be avoided by adding, where necessary, a few parts of a tackifier to the compound. Several tackifiers have been found to be very effective and one or two have been developed specifically for EPDM. There are numerous suggestions in both the EPDM and tackifier manufacturers' technical literature which will guide the compounder to the solution of specific tack problems.

TIRES

At various times since its introduction EPDM has been hailed as the general purpose tire rubber of the future – the next SBR. A tremendous amount of work has been done on the development of special grades of EPDM for the various components of passenger car tires, on the selection of compounding recipes adjusted to established tire fabrication practices, on the solution of building problems largely associated with the tack of EPDM compounds, and on the formulation of special treatments which would ensure an adequate level of adhesion between the cords and the carcass stocks. Probably the most important of these tasks was the development of a polymer and a compound which, in combination, would give a balance of highway and city treadwear and skid resistance equal to present-day first grade passenger tires.

All of these problems have been solved. An all-EPDM passenger tire can be built today on conventional equipment and cured in competitive cycles. Its durability, safety, and treadwear will be adequate. Perhaps its only distinguishing feature discernible to a discriminating motorist would be the freedom from cracking and checking of the ozone resistant EPDM sidewall.

Summaries of their work on all-EPDM tires have been revealed by two manufacturers.[32-34] It is from this work that the foregoing summary was derived. What the future holds for the all-EPDM tire can only be speculated. Work on the all-EPDM tire will certainly continue. The EPDM manufacturers intend to be ready to take their share of the huge tire market when all conditions of quality, performance, and cost have been satisfied.

GLOSSARY OF TRADE-MARKED PRODUCTS

BIK	UNIROYAL Chemical Surface-Coated Urea
CELOGEN® AZ	UNIROYAL Chemical Azodicarbonamide
Circo Light Oil	Sun Oil Co. Naphthenic type rubber extender oil
DELAC® S	UNIROYAL Chemical N-cyclohexyl -2- benzothiazole sulfenamide
DiCup 40	Hercules, Inc. Dicumyl peroxide 40% active
Hercules S-890	(Vul Cup) Hercules, Inc. a,a'-bis (tert-butylperoxy) diisopropyl benzene
Montecatini PX-60	Montecatini Edison S.p.A. A diperoxide of undisclosed composition
Penzone B	Pennwalt Dibutylthiourea
ROYALAC® 133	UNIROYAL Chemical An activated dithio carbamate accelerator
ROYALENE® 301T	UNIROYAL Chemical Ethylene/propylene/DCPD terpolymer (EPDM)
ROYALENE 512	UNIROYAL Chemical Ethylene/propylene/ENB terpolymer (EPDM)
Sunpar 150	Sun Oil Co. Paraffinic type rubber extender oil
SUNPROOF® Improved	UNIROYAL Chemical A controlled blend of selected waxes
Tellurac	R. F. Vanderbilt Tellurium diethyldithiocarbamate
Tetrone A	E. I. duPont de Nemours and Co. Dipentamethylene-thiuram tetrasulfide
Vistalon 404	Enjay Chemical Co. Ethylene/propylene copolymer (EPM)

CURING CHEMICALS CODES

MBT	2-mercapto benzothiazole
ZMBT	Zinc salt 2-mercapto benzothiazole
TMTDS	Tetramethylthiuram disulfide
TMTMS	Tetramethylthiuram monosulfide
TETDS	Tetraethyl thiuram disulfide
DPTTS	Dipentamethylene-thiuram tetrasulfide
MBTS	Benzo-thiazyl disulfide
TDEDC	Tellurium diethyl-dithio carbamate
ZDMDC	Zinc dimethyl-dithio carbamate
ZDEDC	Zinc diethyl-dithio carbamate
ZDBDC	Zinc dibutyl dithio carbamate
DTDM	4,4′ dithio dimorpholine
CBS	N-cyclohexyl -2- benzothiazole sulfenamide

BIBLIOGRAPHY

(1) U.S. Patent 3,113,115, Karl Ziegler, Heinz Martin, and Erhard Holzkamp, assigned to Karl Ziegler, Dec. 3, 1963.
(2) U.S. Patent 3,300,459, Giulio Natta and Giorgio Boschi, assigned to Montecatini Edison S.p.A., Jan. 24, 1967.
(3) G. Natta, G. Crespi, A. Valvassori, and G. Satori, *Rubber Chem. & Technol.*, **36** 1583 (1963).
(4) UNIROYAL Chemical, unpublished work.
(5) W. Hoffman, *Rubber Chem. & Technol.*, **37** (2), 164 (1964).
(6) U.S. Patent 2,933,480, William F. Gresham and Madison Hunt, assigned to E. I. duPont de Nemours and Co., April 19, 1960.
(7) U.S. Patent 3,093,621, Edward K. Gladding, assigned to E. I. duPont de Nemours and Co., June 11, 1963.
(8) U.S. Patent 3,211,709 Stephen Adamek, Edward Allen Dudley, and Raymond Thomas Woodhams (Dunlop Rubber Company, Ltd.), assigned by mesne assignments, to Hercules Powder Co. (now Hercules, Inc.) October 12, 1965.
(9) K. Kujimoto and S. Nakade, *J. Appl. Poly. Sci.*, **13**, 1509 (1969).
(10) U.S. Patent 3,341,503, John L. Paige and Sebastian M. DiPalma, assigned to UNIROYAL, Inc., Sept. 12, 1967.
(11) Peter J. Brennan, *Chemical Engineering*, **72**, 94 (1965).
(12) U.S. Patent 3,496,135, Stanley W. Caywood, Jr., assigned to E. I. duPont de Nemours Co. Feb. 17, 1970.
(13) U.S. Patent 3,000,866, Robert Edward Tarney, assigned to E. I. duPont de Nemours Co., Sept. 19, 1961.
(14) Ital. Patent 587,666, Montecatini Edison S.p.A. *C.A.*, **55**, 1098g (1961).
(15) U.S. Patent 3,153,023, Carl A. Lukach and Harold M. Spurlin, assigned to Hercules Powder Co. (now Hercules, Inc.) Oct. 13, 1964.
(16) R. J. Kelly, *Ind. Eng. Chem. Prod. Res. Devel.*, **1**, 210 (1962).
(17) U.S. Patent 3,361,691, Russell A. Mazzeo, assigned to UNIROYAL, Inc. Jan. 2, 1968.
(18) EPDM Draws a Bead on Markets, *Chem. Week*, **104**, 26 (1969).
(19) Unpublished work, Chemical Division, UNIROYAL, Inc., Naugatuck, Conn.
(20) Ernest L. Carpenter, *Chem. and Eng. News*, **48**, 31 (1970).
(21) Bulletin SYN-67-1273, Enjay Chemical Company, Synthetic Rubber Division (1967).

(22) J. V. Fusco, *Rubber World,* **147**, 48 (1963).
(23) R. P. Mastromatteo, J. M. Mitchell, and R. D. Allen *Rubber Age,* **102**, 64 (1970).
(24) Accelerator Systems for ROYALENE®, 502 Bulletin RR-74, UNIROYAL Chemical.
(25) W. G. Whitehouse, R. R. Barnhart, and J. M. Mitchell, *Hydrocarbon Processing,* **43**, 235 (1964).
(26) Bulletin 570-B166, UNIROYAL Chemical.
(27) Bulletin 16.001, Enjay Chemical Company.
(28) William D. Jones, *Wires and Wire Products,* **41**, 1822 (1966).
(29) Bulletin 570-B137A, UNIROYAL Chemical.
(30) Joseph J. O'Mahoney, Jr., *Rubber Age,* **102**, 47 (1970).
(31) E. H. Andrews, *J. Appl. Poly. Sci.,* **10**, 47–64 (1966).
(32) *ROYALENE in White Sidewalls and Coverstrips*, Technical Bulletin, UNIROYAL Chemical.
(33) Robert W. Kindle and S. van der Burg, *Rubber Age,* **98**, 65 (1966).
(34) "A Research Assessment of Alpha-Olefin Copolymers in Tires," by R. G. Arnold, C. C. McCabe, and J. J. Verbanc, presented at Meeting of Division of Rubber Chemistry, American Chemical Society, Cleveland, Ohio, April 24, 1968.
(35) "Bias-Belted Passenger Tires of Nordel," presented at Akron Rubber Group Meeting, Oct. 24, 1969.

10

BUTYL AND CHLOROBUTYL™ RUBBER

ROBERT L. ZAPP
Enjay Polymer Laboratories
Linden, New Jersey

and PIERRE HOUS
Esso Research, S.A.
Diegem, Belgium

INTRODUCTION

Butyl rubber has been commercially produced since 1942, and at the present time is a well established specialty elastomer used in a wide range of applications. It was first produced by domestic affiliates of the Standard Oil Company (N.J.), but now production has expanded to include three major groups and their associates with production facilities in six countries. In 1960 CHLOROBUTYL was introduced by domestic affiliates of the Standard Oil Company (N.J.). This modified butyl rubber, containing approximately 1.2 weight percent chlorine, possesses greater vulcanization flexibility and enhanced cure compatibility with other general purpose elastomers.

BUTYL RUBBER

Commercial grades of butyl rubber are prepared by copolymerizing small amounts of isoprene, 1–3% of the monomer feed, with isobutylene catalyzed by $AlCl_3$ dissolved in methyl chloride. The reaction is unique in that it is an extremely rapid cationic polymerization[1,2] conducted at the low temperature of –100°C (–140°F). Polymerization is complete in less than a second. The purity of the isobutylene is important to the acquisition of high molecular weight. The n-butene content should be below 0.5%, and the isoprene purity should be 95% or better. The methyl chloride solvent and monomer feed must be carefully dried.

Historical

The origin of butyl rubber springs from early studies on the polymerization of isobutylene, which accounts for about 98% of the butyl molecule. As Table 10.1 outlines, only oily polymers were first produced in the early 1870's.

TABLE 10.1. HISTORY OF ISOBUTYLENE POLYMERIZATION

Workers	*Conditions*	*Nature of Product*
Gorianov and Butlerov (1873)	BF_3 or H_2SO_4; room temp.	Oily polymer
Otto (1927)	BF_3 and its complexes bp of i-C_4 =	Oily polymer
Lebedev *et al.* (1930); Staudinger and Brunner (1930)	Floridin clay; −80° to room temp.; 3 days	Trimers, pentamers + high-mol.-wt polymer
Waterman *et al* (1934)	Solid $AlCl_3$: −78°	Polymer (ca. 1000 mol. wt)
I. G. Farben (1930's)	BF_3, etc.; low temp., hydrocarbon diluents	High-mol. wt (Vistanex)
Std. Oil Dev. (1930's)	$AlCl_3$, BF_3, etc.; low temp. Pure feeds; alkyl halide solvents	High-mol.-wt polymers and copolymers

Subsequent experiments advanced the state of the art so that high molecular weight polymers were produced. These polyisobutylenes are still marketed under the trade names of Oppanol and VISTANEX®. The contributions of R. M. Thomas and W. J. Sparks[3] of Esso Research and Engineering Co. (then known as the Standard Oil Development Co.) produced the first vulcanizable isobutylene-based elastomer by incorporating small amounts of a diolefin and particularly isoprene into the polymer molecule. This produced residual olefinic functionality along the molecular chain sufficient for crosslinking sites.

Butyl Rubber Manufacture

A pictorial diagram of a typical butyl plant is shown in Fig. 10.1. The feed, which is a 25% solution of isobutylene (97–98%) and isoprene (2–3%) in methyl chloride, which is the diluent, is cooled to −140°F in a feed tank. At the same time, aluminum chloride is also being dissolved in methyl chloride. Both of these streams are then continuously injected into the reactor. Because the reaction is exothermic and is practically instantaneous, cooling is very important. In order to remove the heat of reaction, liquid ethylene is boiled continuously through the cooling coils.

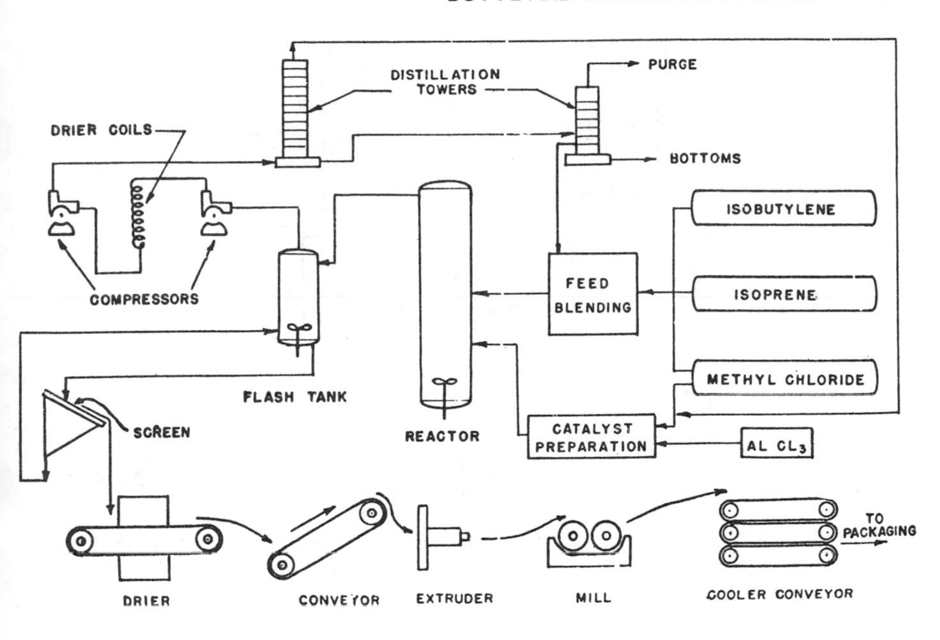

Fig. 10.1 Diagram of a typical butyl plant.

The reaction is thus kept at −140°F. As the polymerization proceeds, a slurry of very small particles is formed in the reactor. This slurry overflows into a flash drum which contains copious quantities of hot water. Here the mixture is vigorously agitated, during which time the diluent and unreacted hydrocarbons are flashed off overhead. It is at this point that an antioxidant and zinc stearate are introduced into the polymer. The antioxidant is added at this point to prevent breakdown of the polymer in the finishing section. Zinc stearate is added to prevent the coagulation, or sticking together, of the wet crumb. The slurry is then vacuum-stripped of residual hydrocarbons. After passing over a screen and filter and through the tunnel dryer, it is extruded and hot-milled. These last operations drive the last traces of residual water from the polymer. Finally, after being sheeted off the mill, the polymer passes along a cooling conveyor to the packaging section. It is then ready for the consumer.

Molecular Characteristics

Commercially available butyl rubbers are copolymers of isobutylene and isoprene and some of the general properties are shown in Table 10.2. The viscosity average molecular weights[4] of these products are in the range 350,000–450,000 and the mole % isoprene in the chain is 0.6–2.5.

In a discussion of the butyl molecule, it is necessary to define the term "mole percent unsaturation." This term is an expression for the number of moles of isoprene per 100 of isobutylene. Thus when we say that butyl contains roughly 2 mole percent unsaturation, we are saying that for every 98 moles of isobutylene,

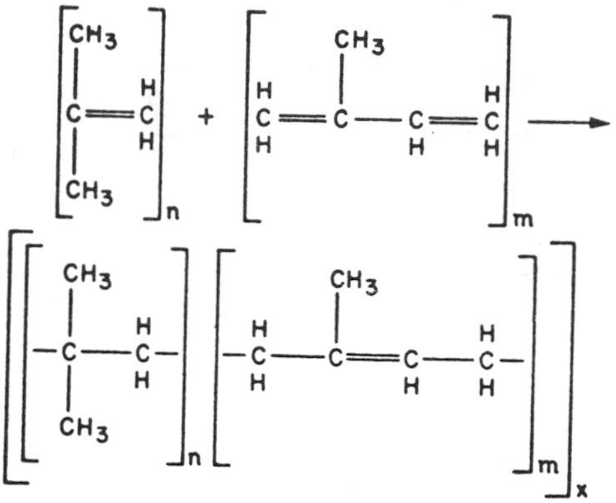

Fig. 10.2

there is present in the molecule 2 moles of isoprene. This is illustrated in Fig. 10.2, in which n represents the moles of isobutylene which are combined with m moles of isoprene. Thus, in our case, n would equal 98, and m would equal 2. In comparison, natural rubber would have n equal to 0 and m equal to 100, since this molecule is 100% polyisoprene. It could be called 100 mole percent unsaturated. Figure 10.3 is an attempt to place the unsaturation of butyl and natural rubber on a spacial basis. For example, a butyl rubber with 1 mole percent of unsaturation would have a molecular weight between points of unsaturation well over 5000. Natural rubber, on the other hand, has a molecular weight between points of unsaturation of only 68. This is the molecular weight of an isoprene molecule. Another way to look at this would be to observe the number of carbon atoms between active sites in the

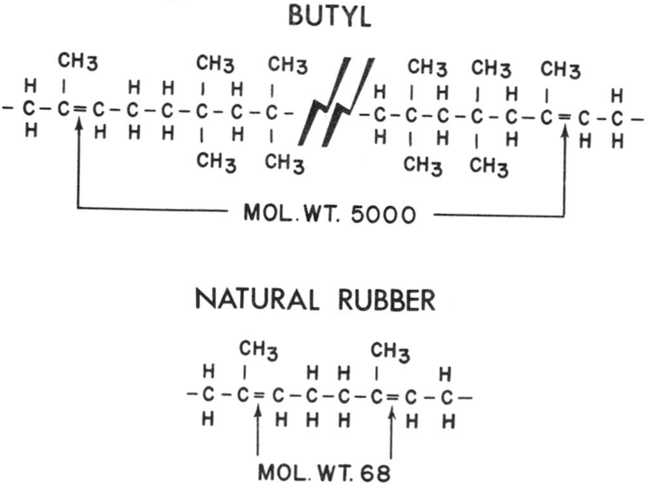

BUTYL

MOL. WT. 5000

NATURAL RUBBER

MOL. WT. 68

Fig. 10.3

two molecules. In natural rubber it would be only 4, while in butyl there would be an unsaturated site every 200 carbon atoms. Unsaturation is determined either by an iodine absorption technique[5] or by ozone degradation of the polymer followed by subsequent viscosity measurements of the degraded fragments.[6] According to Rehner[7] the isoprene unit randomly enters the chain in a *trans* 1,4 configuration, for by chemical analysis there is little evidence for 1,2 and 3,4 modes of entry at unsaturation levels present in commercial grades of butyl:

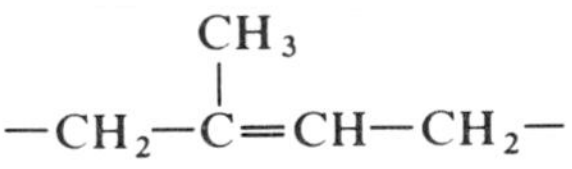

$$-CH_2-\underset{}{\overset{CH_3}{\overset{|}{C}}}=CH-CH_2- \qquad -CH_2-\overset{CH_3}{\overset{|}{\underset{\underset{CH=CH_2}{|}}{C}}}- \qquad -CH_2-\underset{\underset{\underset{CH_3}{|}}{C=CH_2}}{\underset{|}{CH}}-$$

trans-1, 4 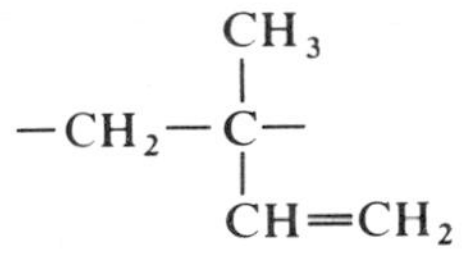   1, 2 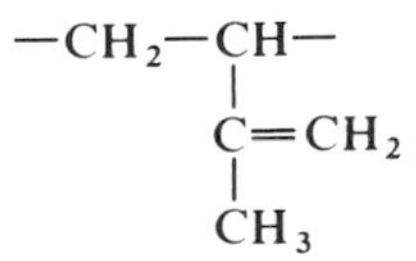   3, 4

The discovery and development of butyl rubber not only furnished rubber technology with a new elastomer, but it also provided a new principle. This was the concept of low functionality. A low functionality elastomer has sufficient chemical unsaturation so that soft vulcanized networks could be produced, but the large inert portions of the chain contribute towards oxidation and ozone resistance. Today the concept of low functionality is well established. The advent of vulcanizable ethylene propylene terpolymers (EPDM) extends the family of low functionality elastomers.

In common with most high polymers, the molecular chains of butyl rubber exist in a wide range of degrees of polymerization or molecular weight. The distribution of the molecular weight has a profound effect upon the rheological behavior of the elastomers.[9] Early methods for the investigation of molecular weight distribution employed fractional precipitation of dilute solutions by controlled additions of nonsolvents.[8] The most recent fractionation technique is gel permeation chromatography, GPC, where molecular size separation from dilute solution is promoted by diffusion through a column packed with porous polystyrene gels.[10] Figure 10.4 presents molecular weight distribution information on a butyl 218 by GPC in the form of weight fraction along the ordinate and species molecular weight on a log scale along the abscissa. From the GPC molecular weight distribution data, the various molecular weight averages can be calculated as indicated in the figure. The ratio of weight average, $\overline{M}_w$, to number average, $\overline{M}_n$,[4] often taken as an index of molecular dispersity, is approximately 3. In a most probable distribution this ratio would be 2.

Properties of Butyl Rubber And Its Applications

The molecular characteristics of low levels of unsaturation between long segments of polyisobutylene produce unique elastomeric qualities that find application in a wide variety of finished rubber articles. These special properties can be listed as: (1) low rates of gas permeability; (2) thermal stability; (3) ozone and

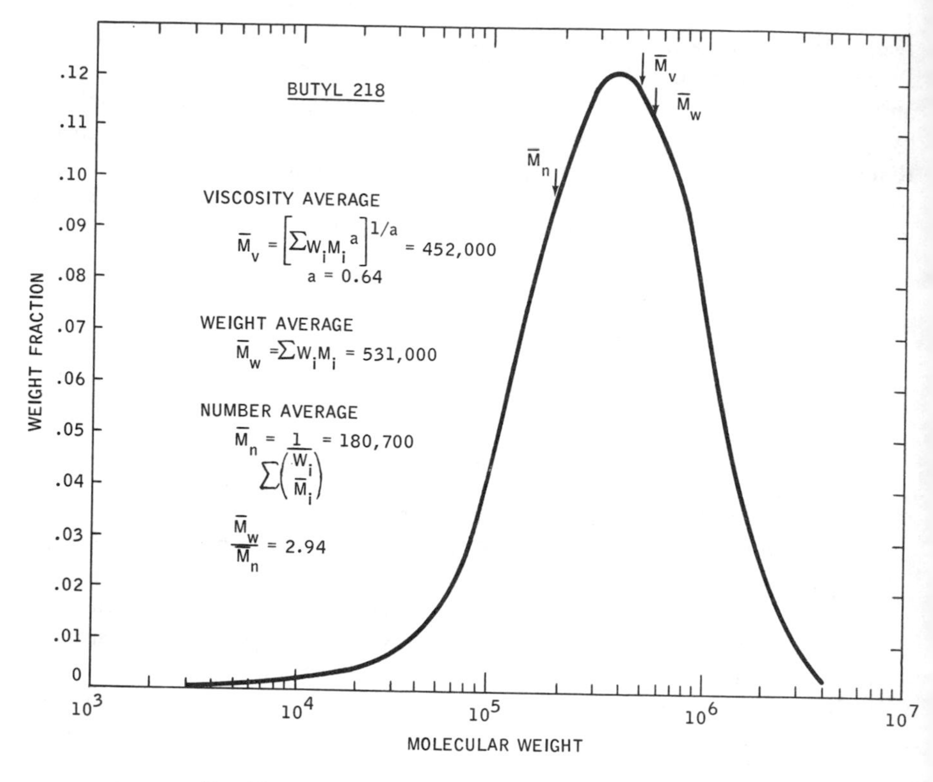

Fig. 10.4 Molecular weight distribution of Butyl 218.

weathering resistance; (4) vibration damping and higher coefficients of friction; (5) chemical and moisture resistance.

The more chemically inert nature of butyl rubber is reflected in the lack of significant molecular weight breakdown during processing. This allows one to perform operations such as heat treatment or high temperature mixing to alter the vulcanizate characteristics of a compound. With carbon black containing compounds, hot mixing techniques promote pigment-polymer interaction,[11] which alters the stress strain behavior of a vulcanizate. The shape of the stress strain curve of the vulcanizate from a heat treated mixture is a reflection of a more elastic network, and heat treatment has resulted in more flexible vulcanized compounds for a given level of a carbon black type. More flexible butyl rubber compositions have also been prepared with certain types of mineral fillers such as reinforcing clays, talcs, and silicas that contain appropriately placed OH groups in the lattice. This enhancement of pigment polymer association has usually been accomplished in the presence of chemical promoters.[12] These promoters can be generally classed as nitroso or dioxime compounds, and are used in quantities from 0.2 to about 1.0% based upon the polymer content of the mix. A partial list of heat treatment

TABLE 10.2. GENERAL PROPERTIES OF ENJAY BUTYL POLYMERS (23)

	Enjay Butyl 035		*Enjay Butyl 150 and 165*		*Enjay Butyl 215*		*Enjay Butyl 217*		*Enjay Butyl 218[a] and 268*		*Enjay Butyl 325 and 365*	
	Min.	*Max.*	*Min.*	*Max.*	*Min.*	*Max.*	*Min.*	*Max.*	*Min.*	*Max.*	*Min.*	*Max.*
Viscosity range												
ML 212°F at 8 min	41	49	41	49	41	49	61	70	71	–	41	49
Viscosity average mol wt[b]	350,000		350,000		350,000		400,000		450,000		350,000	
Mole % unsaturation[b]	0.6–1.0		1.0–1.4		1.5–2.0		1.5–2.0		1.5–2.0		2.1–2.5	
Tensile strength, psi												
40 min cure at 307°F	2600		2500		2400		2400		2400		2300	
Ultimate elongation,												
% minimum	700		650		550		550		550		500	
Modulus, psi at 400% elongation												
20 min cure at 307°F	400	590	590	770	770	950	770	950	770	950	950	1125
40 min cure at 307°F	650	900	900	1125	1125	1375	1125	1375	1125	1375	1375	1600

[a]Scorch recipe for Enjay Butyl 218

Polymer		100
Zinc oxide		2
Sulfur		2
Tetramethylthiuram disulfide		0.6
Scorch time	min.	max.
Minutes at 293°F	6	17

Base recipe for physical properties

Polymer	100
Channel black	50
Zinc oxide	5
Stearic acid	3
Sulfur	2
Benzothiazolyl disulfide	0.5
Tetramethylthiuram disulfide	1

[b] Not a specification; values shown are typical.

Grades: 035, 150, 215, 217, 325 stabilized with 0.1% phenyl beta naphthyl amine
165, 268, 365 (nonstaining grades) stabilized with .1% zinc dibutyl dithiocarbamate.

promoters would include:

Polyac, dinitroso benzene – DuPont
GMF, p-quinone dioxime – Naugatuck
p-nitroso phenol
Nitrol, N-(2-methyl-2-Nitropropyl)-4-nitroso aniline (Monsanto).

The heat treatment of filler-butyl rubber masterbatches can be accomplished at 300–350°F for five minutes in a Banbury mixer with the aid of promoters. In general, the processing of butyl rubber follows accepted factory operations. Banbury mixing in conventional formulations requires no longer times than with other natural and synthetic elastomers. Of course, there is no masticating operation to produce a degree of molecular weight breakdown as in the case of natural rubber, but pre-warming of the butyl rubber prior to mixing reduces mixing times.

Gas Impermeability. The impermeability of elastomeric films to the passage of gas is a function of the diffusion of gas molecules through the membrane and the solubility of the gas in the elastomer.[13] The polyisobutylene portion of the butyl molecule provides a high degree of impermeability to butyl, and is a familiar property leading to an almost exclusive use in inner tubes. For example, the difference in air retention between a natural rubber and a butyl inner tube can be demonstrated by data from controlled road tests on cars driven 60 mph for 100 miles per day. Under these conditions, it was shown (Table 10.3) that butyl is at least 8 times better than natural rubber in air retention.

Other gases such as helium, hydrogen, nitrogen, and carbon dioxide are also well retained by a butyl bladder. While the significance of these properties in inner tubes is waning, it is of importance in air barriers for tubeless tires, air cushions, pneumatic springs, accumulator bags, air bellows, and the like. A typical formulation for a butyl rubber passenger tire inner tube is given below:

Enjay Butyl 218	100
GPF Carbon Black	70
Paraffinic Process Oil	25
Zinc Oxide	5
Sulfur	2
TMTDS[a]	1
MBT[b]	0.5

Cure Range 5′ at 350°F – 8′ at 330°F

[a]TMTDS = tetramethyl thiuram disulfide
[b]MBT = mercaptobenzothiazole

TABLE 10.3. AIR LOSS OF INNER TUBES DURING DRIVING TESTS

Inner Tube	*Original Pressure (psi)*	*Air Pressure Loss (psi)* 1 Week	2 Weeks	1 Month
Natural rubber	28	4.0	8.0	16.5
Butyl	28	0.5	1.0	2.0

Thermal Stability. Butyl rubber sulfur vulcanized formulations tend to soften during prolonged exposure to elevated temperatures of 300–400°F. This deficiency is largely the fault of the sulfur crosslink coupled with low polymeric unsaturation which allows no compensating oxidative hardening. However, certain crosslinking systems and specifically the resin cure[14] of butyl provide vulcanized networks of outstanding heat resistance. This has found widespread use in the expandable bladders of automatic tire curing presses. Two tire curing bladder formulations are given below:

	1	2
Enjay Butyl 218	100	100
Neoprene GN	5	–
HAF Carbon Black	50	50
Process Oil	5	5
Zinc Oxide	5	5
Reactive Phenol Formaldehyde Resin[a]	10	–
Brominated Phenol Formaldehyde Resin[b]	–	10

[a]Amberol ST 137 – Rohm and Haas
[b]SP-1055 – Schenectady Varnish Co.

In compound 1, the neoprene serves as a halogen-containing activator, while compound 2 uses a partially brominated resin that does not require an external source of halogen. Butyl rubber tire curing bladders have a life of 300–700 curing cycles at steam temperatures of 350°F or higher for approximately 20 minutes per cycle. Another application would be conveyor belting for hot materials handling.

Ozone and Weathering Resistance. The low level of chemical unsaturation in the polymer chain produces an elastomer with greatly improved resistance to ozone when compared to polydiene rubbers. Butyl with the lowest level of unsaturation (Enjay Butyl 035) produces high levels of ozone resistances which are also influenced by the type and concentration of vulcanizing crosslinks.[15] For maximum ozone resistance, as in electrical insulation, and for weather resistance, as in rubber sheeting, for roofs and water management application, the least unsaturated butyl is advantageously used. A typical butyl rubber sheeting compound is given below:

Enjay Butyl 035	100
HAF Carbon Black	48
SRF Carbon Black	24
Zinc Oxide	5
Petrolatum	3
Wax	4
Sulfur	1
TDEDC[a]	0.5
MBT[c]	0.5
ZDMDC[b]	1.5

[a]TDEDC = Tellurium diethyldithiocarbamate.
[b]ZDMDC = Zinc dimethyl dithiocarbamate.
[c]MBT = Mercaptobenzothiazole.

The ozone resistance of butyl rubber coupled with the moisture resistance of its essentially saturated hydrocarbon structure finds utility as a high quality electrical insulation. A cable insulation formulation for use up to 50 KV employs the lowest unsaturated butyl and the p-quinone dioxome (GMF) cure system:

Enjay Butyl 035	100	Masterbatch mixed at 300°F
Zinc Oxide	5	
Calcined Clay	100	
Pb_3O_4	5	
130°F m.p. Wax	5	
Low Density Polyethylene	5	
SRF Carbon Black	10	
MBTS[a]	4	Added as a separate cooler mix
P-quinone dioxime	1.5	

[a] MBTS = Mercaptobenzothiazyl disulfide.

Vibration Damping. The viscoelastic properties of butyl rubber are a reflection of the molecular structure of the polyisobutylene chain. This molecular chain with two methyl side groups on every other chain carbon atom possesses greater delayed elastic response to deformation. This damping and absorption of shock has found wide application in automotive suspension bumpers.[16] An elastomer with higher damping characteristics also restricts vibrational force transmission in the region of resonant frequencies. Transmissability is the ratio of output force to input force under impressed oscillatory motion.

In the region of resonance where the impressed frequency of vibration is equal to the natural frequency of the system, the more highly damped butyl

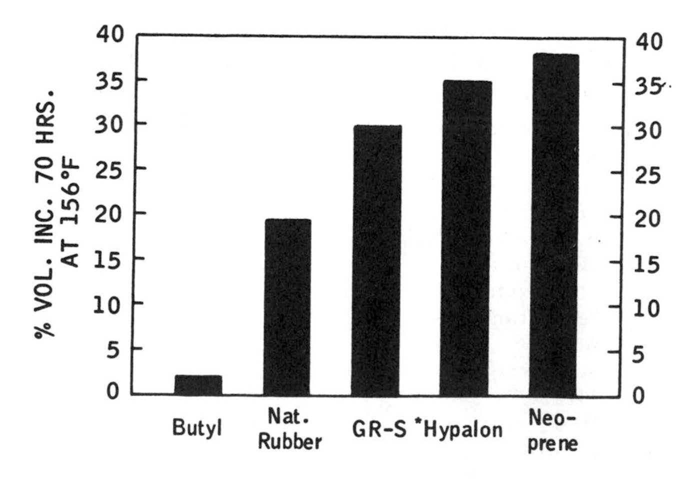

Fig. 10.5 Swelling of butyl and other elastomeric vulcanizates in phosphate ester fluids.
*Registered trademark of the Du Pont Company for its synthetic rubber

compositions more effectively control vibrational forces.[17] These frequencies are in the region of vehicular body shake, and as a result, butyl compositions are employed in the fabrication of automotive body mounts. In theory, more highly damped systems will less effectively isolate vibration at very high frequencies. However, the damping coefficients of viscoelastic polymers decrease with increasing frequencies in actual practice. This partially alleviates higher frequency deficiencies of butyl systems. In addition at higher frequencies, dynamic stiffness becomes a controlling factor governing transmissibility, and this dynamic behavior can be greatly influenced by compounding variations, as well as by the size and the shape of the molded part. For this reason it is difficult to provide a typical butyl body mount compound, but the one listed below can be considered representative of a 50 Shore A hardness vulcanizate:

Enjay Butyl 268	100
HAF Carbon Black	45
MT Carbon Black	15
Paraffinic Oil	20
Zinc Oxide	5
CdDEDC[a]	2
MBTS	0.5
Sulfur	1

[a]Cadmium diethyl dithiocarbamate.

Higher damping behavior of an elastomer can be associated with higher coefficients of friction between a rubber and a surface with a measurable degree of roughness or undulations.[18] This property of butyl has potential for improving the coefficient of friction of tire tread materials against a variety of road surfaces.[19]

Chemical and Moisture Resistance. The essentially saturated hydrocarbon nature of butyl obviously imparts moisture resistance to compounded articles. This is utilized in applications such as electrical insulation and rubber sheeting for external use. This same hydrocarbon nature provides useful solubility characteristics that can be applied to a variety of protective and sealant applications. These useful solubility characteristics are again based upon the hydrocarbon nature of the elastomer backbone which is expressed as a solubility parameter of 7.8.[20] This value is similar to the solubility parameters of aliphatic and some cyclic hydrocarbons, 7 to 8, but very different from the solubility parameters of more polar oxygenated solvents, ester type plasticizers, vegetable oils, and synthetic hydraulic fluids, 8.5 to 11.0. Thus while butyl vulcanizates will be highly swollen by hydrocarbon solvents and oils, they are only slightly affected by oxygenated solvents and other polar liquids. This behavior is utilized in elastomeric seals for hydraulic systems using synthetic fluids, as Fig.10.5 demonstrates.

The low degree of olefinic unsaturation in the saturated hydrocarbon backbone also imparts mineral acid resistance to butyl rubber compositions. After 13 weeks immersion in 70% sulfuric acid, a butyl compound experiences little loss in tensile strength or elongation while a natural rubber or SBR will be highly graded.

Vulcanization

Regular butyl rubber is commercially vulcanized by three basic methods. These are accelerated sulfur vulcanization, crosslinking with dioxime, and dinitroso-related compounds and the resin cure. The three methods will be briefly reviewed, but more detail can be found in the volume *Vulcanization of Elastomers.*[21]

Accelerated Sulfur Vulcanization. In common with more highly unsaturated rubbers, butyl may be crosslinked with sulfur, and activated by zinc oxide and organic accelerators. In contrast to the higher unsaturated varieties, however, adequate states of vulcanization can only be obtained with the very active thiuram and dithiocarbamate accelerators. Other less active accelerators such as thiazole derivatives may be used as modifying agents to improve processing scorch safety. Most curative formulations include the following ranges of ingredients:

	Parts by wt
Butyl Elastomer	100
Zinc Oxide	5
Sulfur	0.5–2.0
Thiuram or dithiocarbamate accelerator	1–3
Modifying thiazole accelerator	0.5–1

Thiurams with the structure $(R_2NC(=S)\text{—}S)_2$, and dithiocarbamates, $(R_2NC(=S)\text{—}S)_xM$, where M is a metallic element, provide the primary accelerating activity which promotes the most efficient use of sulfur. Thiazoles, generally mercaptobenzothiazoles,

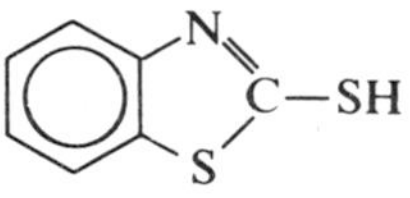

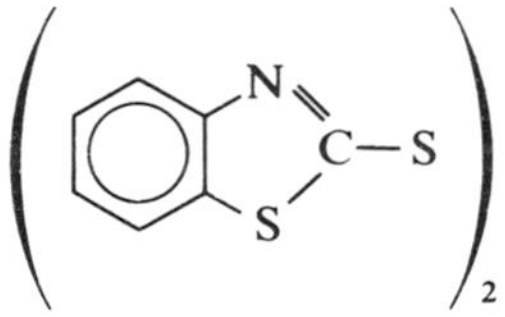

(benzothiazole ring)C—SH   ((benzothiazole ring)C—S)$_2$

reduce scorch during processing. High levels of dithiocarbamates and low levels of sulfur favor the formation of more stable monosulfidic crosslinks.[22] The use of sulfur donors in place of elemental sulfur also promotes simpler monosulfidic bonds.

Vulcanization temperatures may range from 275°F to 375°F, with temperature coefficients of vulcanization of 1.4 per 10°F for carbon black containing compounds.[23] This means that the vulcanization time, to a given state of cure, is multiplied by 1.4 for every 10°F *decrease* in temperature. Conversely, for every 10°F *rise* in temperature the vulcanization time is divided by 1.4.

The Dioxime Cure. The crosslinking of butyl with p-quinone dioxime or p-quinone dioxime dibenzoate proceeds through an oxidation step that forms the active crosslinking agent, p-dinitroso-benzene:

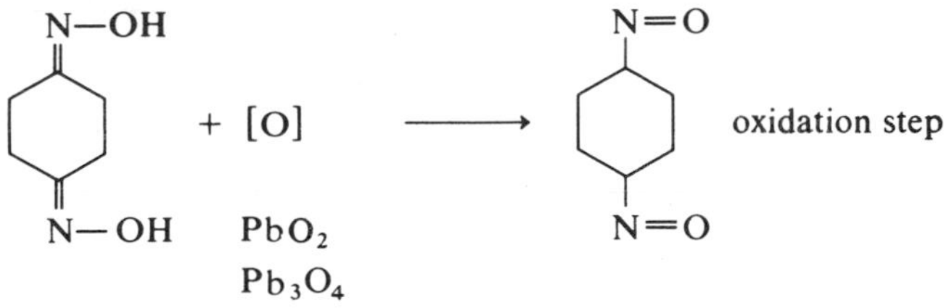

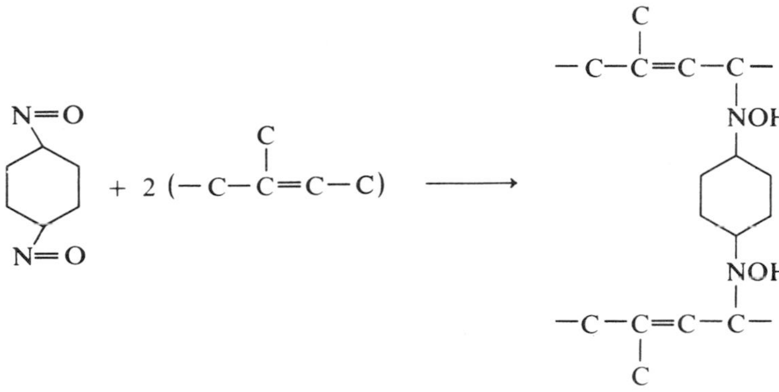

N—OH (p-quinone dioxime) + [O] (PbO_2, Pb_3O_4) ⟶ N=O (p-dinitroso-benzene) oxidation step

p-dinitroso-benzene + 2 (—C—C(C)=C—C) ⟶ —C—C(C)=C—C(NOH—C₆H₄—NOH)—C—, crosslinked to —C—C(C)=C—C—

The use of PbO_2 as the oxidizing agent results in very rapid vulcanizations, which can produce room temperature cures for cement applications. In dry rubber processing, the dioxime cure is largely used in butyl rubber electrical insulation formulations as outlined in a preceding section, to provide a maximum of ozone resistance and moisture impermeability to the vulcanizate.

The Resin Cure. The crosslinking of butyl rubber (and other elastomers containing olefinic unsaturation) is dependent upon reactivity of the phenol-methylol groups of reactive phenol-formaldehyde resins:

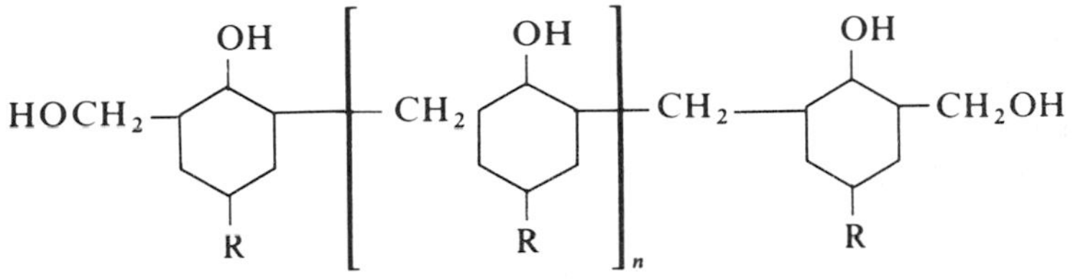

$$HOCH_2-C_6H_2(OH)(R)-\left[CH_2-C_6H_2(OH)(R)\right]_n-CH_2-C_6H_2(OH)(R)-CH_2OH$$

Two mechanisms involving the isoprenoid unit have been postulated. One involves reaction with allylic hydrogen through a methylene quinone intermediate,[24] and the other an actual bridging of the double bond.[25] The low levels of unsaturation of butyl require resin cure activation by halogen-containing materials such as $SnCl_2$

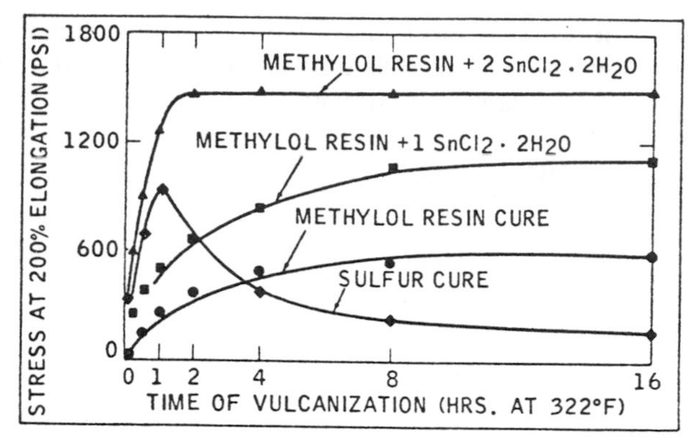

Fig. 10.6 The relative rates of cure and reversion of resin- and sulfur-cured butyl compounds.

or halogen-containing elastomers such as neoprene.[14] A series of curves in Fig. 10.6 from reference 14 illustrates the activating effect of stannous chloride and the stability of the resultant crosslink, to reversion upon prolonged heating. This feature of the resin cure is utilized in the fabrication of tire curing bladders as previously mentioned.

A more reactive resin cure system requiring no external activator is obtained if some of the hydroxyl groups of the methylol group are replaced by bromine. Such resins are produced commercially as SP 1055*; the ordinary reactive resin commonly used is Ambersol ST 137.** Examples of formulations were discussed for tire curing bladders in the section on "Thermal Stability."

CHLOROBUTYL RUBBER

The introduction of a small amount of chlorine (1.2 wt %) in the butyl polymer has two basic purposes:

(1) To amplify the reactivity of the functional units of the butyl molecule without changing their number. This implies that chlorobutyl has essentially the same structure as butyl and hence identical dynamic properties. A change from an elastic to a more plastic behavior would require a much larger amount of chlorine atoms in the polymer. The 1.2 wt % Cl_2 level in chlorobutyl is not sufficient to cause interaction of the polar groups between chains or on the same chain nor to change the flexibility of the butyl molecule.

(2) To increase the possibilities for vulcanization. In addition to the various cure systems acting via the double bonds (sulfur-accelerators, sulfur donor, quinoid, methylol resins) a variety of new cure systems effective through the allylic chloride can be used in chlorobutyl.

*Schenectady Varnish Co.
**Rohm and Haas.

This has two very important consequences:

(1) Reversion-resistant vulcanizates can easily be obtained with chlorobutyl, since crosslinks involving labile polysulfide bonds can easily be avoided.

(2) The possibility of covulcanization with highly unsaturated elastomers is enhanced. The vulcanization of butyl, due to its very small amount of isoprenoid units, is statistically unfavored in the presence of highly unsaturated polymers. This is not the case with chlorobutyl because it is vulcanizable through a completely different cure system than the ones used for crosslinking highly unsaturated elastomers.

The first CHLOROBUTYL became commercially available in 1960. The two major grades are specified as in Table 10.4.

TABLE 10.4.

	CHLOROBUTYL, ENJAY BUTYL[a]			
	HT 10–66		*HT 10–68*	
Grades	*Min.*	*Max.*	*Min.*	*Max.*
Mooney Viscosity				
ML(1 + 8 min) at 100°C (212°F)	51	60	–	–
ML(1 + 3 min) at 126.6°C (260°F)	–	–	50	60
Chlorine, wt %	1.1	1.3	1.1	1.3

[a]Currently CHLOROBUTYL is marketed in two molecular weight grades and these are designated ENJAY BUTYL 10-66 and 10-68, the latter being the higher molecular weight. HT stands for High Temperature, implying high temperature resistance of these vulcanizates.

CHLOROBUTYL Structure

CHLOROBUTYL is prepared by introducing a continuous stream of chlorine gas in a hexane solution of butyl. For each chlorine molecule reacted, one molecule of hydrogen chloride is evolved and one atom of chlorine appears in the polymer. Both HT 10-66 and HT 10-68 contain 1.1–1.3 weight percent chlorine which is slightly higher than 1 chlorine per isoprenoid unit.

The chlorination of butyl in both polar and nonpolar solvents to a level of 1 Cl per double bond is a very rapid electrophilic substitutive reaction on the isoprene moiety. Up to 2 atoms of chlorine can be substituted per isoprenoid unit, but the rate of reaction for the second atom of Cl is much slower, although following the same mechanism. It was found by McNeil[26] that the reaction rate remains constant up to the introduction of 1.3 chlorine atoms per double bond within his minimum kinetic time period of 10 minutes. Beyond this point the reaction is slower and excess Cl_2 is necessary to reach a level of 2 chlorine atoms per isoprenoid unit. The chlorination mechanism, involving a chloronium ion intermediate, results in the following structures:

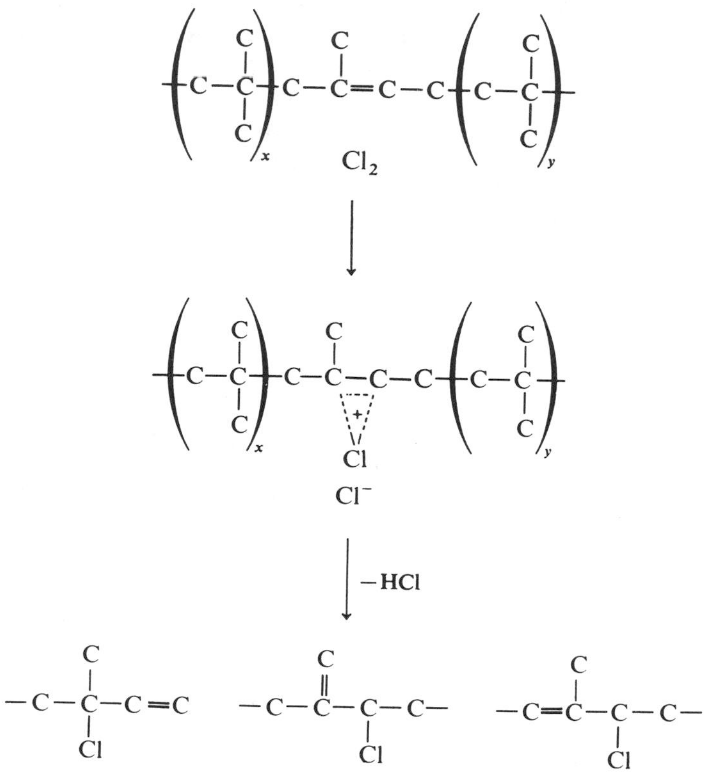

$$\left(\!-\mathrm{C}-\underset{\mathrm{C}}{\overset{\mathrm{C}}{\mathrm{C}}}-\right)_x-\mathrm{C}-\overset{\mathrm{C}}{\mathrm{C}}=\mathrm{C}-\mathrm{C}-\left(\mathrm{C}-\underset{\mathrm{C}}{\overset{\mathrm{C}}{\mathrm{C}}}-\right)_y$$

$$Cl_2$$

$$\downarrow$$

$$\left(\!-\mathrm{C}-\underset{\mathrm{C}}{\overset{\mathrm{C}}{\mathrm{C}}}-\right)_x-\mathrm{C}-\overset{\mathrm{C}}{\mathrm{C}}-\underset{\overset{+}{\mathrm{Cl}}}{\mathrm{C}}-\mathrm{C}-\left(\mathrm{C}-\underset{\mathrm{C}}{\overset{\mathrm{C}}{\mathrm{C}}}-\right)_y$$

$$Cl^-$$

$$\downarrow -HCl$$

$$-\mathrm{C}-\underset{\mathrm{Cl}}{\overset{\mathrm{C}}{\mathrm{C}}}-\mathrm{C}=\mathrm{C} \qquad -\mathrm{C}-\overset{\mathrm{C}}{\overset{\|}{\mathrm{C}}}-\underset{\mathrm{Cl}}{\mathrm{C}}-\mathrm{C}- \qquad -\mathrm{C}=\overset{\mathrm{C}}{\mathrm{C}}-\underset{\mathrm{Cl}}{\mathrm{C}}-\mathrm{C}-$$

I	II	III
tertiary allylic chloride	secondary allylic chloride	secondary allylic chloride

A kinetic study of the thermal decomposition of CHLOROBUTYL revealed that about 20% of the chlorine is present as tertiary allylic chloride showing a decomposition rate of about 45 times the rate of the less active (secondary) allylic chlorides.

On the other hand, NMR studies show that the methylenic and trisubstituted ethylenic unsaturation are about equal in concentration. This leads to the conclusion that structures I, II, and III are present in a ratio of approximately 1:2:2. The loss of molecular weight due to chain scission during the chlorination of butyl is small – from 3 to 9%. Below the 1 Cl level per C=C, chain scission is very small, but then increases significantly. Beyond 1 Cl/C=C, chlorination and scission rates are comparable.

Stabilization of CHLOROBUTYL

Allylic halogens are very reactive, particularly tertiary allylic halogens for two reasons. First, the double bond induces a higher electron density at the carbon atom attached to the halogen and thus weakens the carbon-halogen bond. Secondly, a more stable cation, as an intermediate, is formed once the halogen is removed.

The dissociation energies of the carbon-halogen bond decrease in the order C–Cl > C–Br > C–1. Therefore, one can expect a brominated butyl rubber to be less stable than CHLOROBUTYL, and this has been observed with laboratory prepared samples of the brominated product. The CHLOROBUTYL polymer can survive the very severe conditions of plant finishing operation only if a stabilizer is added. Short heat exposure times at temperatures between 175°C (347°F) and 200°C (392°F) are encountered in the plant. If not stabilized, CHLOROBUTYL would undergo dehydrochlorination. For this reason calcium stearate is added as a stabilizer. Its action can be explained by a mechanism similar to stabilization of PVC by means of certain metal soaps.[27,28]

Once the polymer is finished in the plant and cooled down to room temperature, no further changes occur in the polymer. Chlorobutyl can, therefore, be stored for an unlimited period without fear of chemical changes if one avoids extensive heat exposure. Normal masterbatch mixing, wherein stock temperatures may reach 270–320°F in ten minutes, does not inhibit vulcanization potentials.

Small amounts of a nonstaining inhibitor, Parabar 440 (2,6 ditert-butyl p-cresol), are added during manufacturing to inhibit oxidative breakdown. This inhibitor has FDA acceptance and, therefore, may be used in pharmaceutical applications (e.g., pharmaceutical stoppers).

Manufacture

Butyl is prepared by low temperature copolymerization of isobutylene and isoprene in methyl chloride. The cold slurry overflows from the reactor to a solution drum in which hot hexane replaces methyl chloride and solubilizes the butyl polymer. Remaining traces of methyl chloride and unreacted monomers are stripped from the polymer solution, which after further concentration is ready for chlorination.

Chlorination is done by addition of chlorine vapor to the polymer solution and thorough blending in a chlorine contactor. Since HCl is generated during the chlorination, the cement is neutralized with base followed by a separation of hexane and water phases. The cement is then introduced to the flash drum where it is contacted with steam and water. The addition of calcium stearate has a dual purpose. It aids in the formation of small solid elastomer particles and prevents the dehydrohalogenation during finishing. Most of the free water is removed on a vibrating screen and a rotating filter under vacuum. The polymer is then dried thoroughly by two stages of extrusion drying.

Vulcanization

The presence of both olefinic unsaturation and reactive chlorine in chlorinated butyl provides for a great variety of vulcanization techniques. While conventional sulfur-accelerator curing is possible and useful, in this section attention will be confined to two general vulcanization techniques not available to regular butyl.

INITIATION

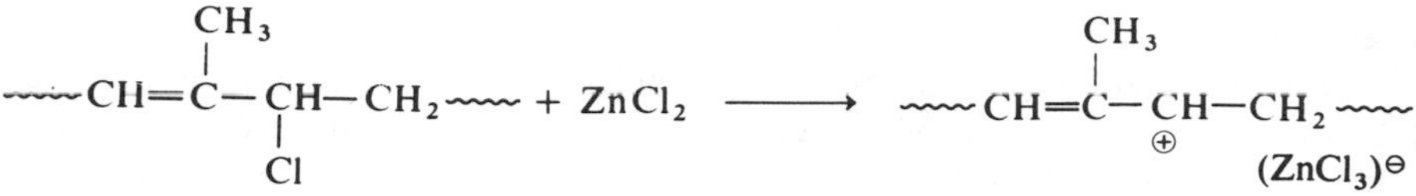

$$\sim CH{=}C(CH_3){-}CH(Cl){-}CH_2\sim \; + \; ZnCl_2 \longrightarrow \; \sim CH{=}C(CH_3){-}\underset{\oplus}{C}H{-}CH_2\sim \quad (ZnCl_3)^{\ominus}$$

PROPAGATION

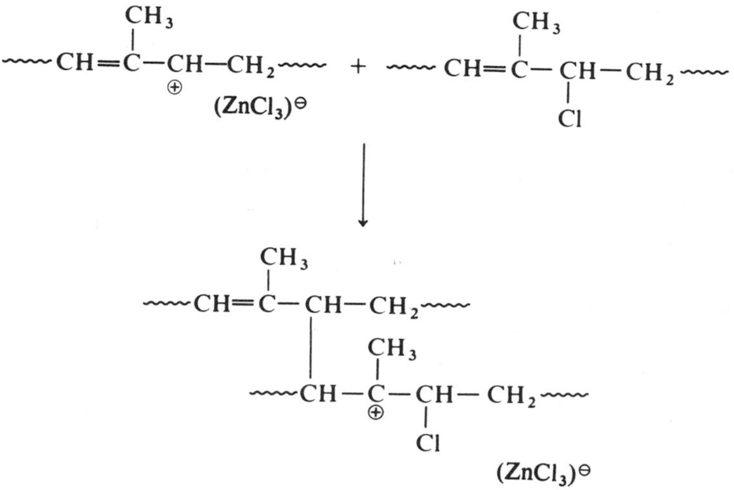

$$\sim CH{=}C(CH_3){-}\underset{\oplus}{C}H{-}CH_2\sim \; (ZnCl_3)^{\ominus} \; + \; \sim CH{=}C(CH_3){-}CH(Cl){-}CH_2\sim$$

$$\downarrow$$

$$\sim CH{=}C(CH_3){-}CH{-}CH_2\sim$$
$$\quad |$$
$$\sim CH{-}\underset{\oplus}{C}(CH_3){-}CH(Cl){-}CH_2\sim \quad (ZnCl_3)^{\ominus}$$

TERMINATION

$$\sim CH{=}C(CH_3){-}CH{-}CH_2\sim$$
$$\quad |$$
$$\sim CH{-}\underset{\oplus}{C}(CH_3){-}CH(Cl){-}CH_2\sim \quad (ZnCl_3)^{\ominus}$$

$$\nearrow$$

$$\sim CH{=}C(CH_3){-}CH{-}CH_2\sim$$
$$\quad |$$
$$\sim CH{-}C(CH_3)(Cl){-}CH(Cl){-}CH_2\sim$$
$$+ ZnCl_2$$

$$\searrow$$

$$\sim CH{=}C(CH_3){-}CH{-}CH_2\sim$$
$$\quad |$$
$$\sim C{=}C(CH_3){-}CH(Cl){-}CH_2\sim$$
$$+ HCl + ZnCl_2$$

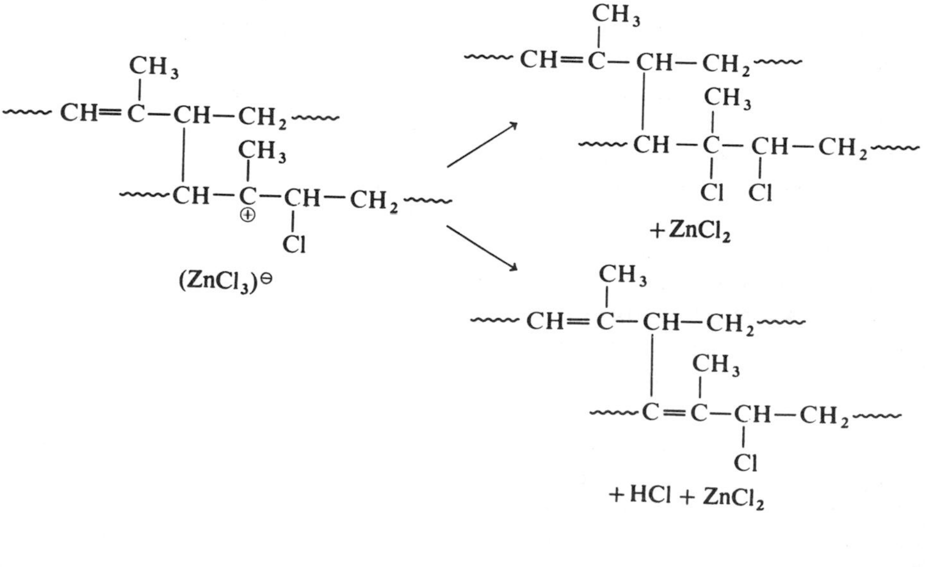

Zinc Oxide Cure and Modifications. Zinc oxide, preferably with some stearic acid, can function as the sole curing agent for CHLOROBUTYL. After vulcanization, most of the chlorine originally present in the polymer can be extracted as zinc chloride by methyl ethyl ketone. Further, the polymer can be vulcanized by the inclusion of zinc chloride instead of zinc oxide. A proposed mechanism by Baldwin[29] which gives rise to stable carbon-carbon crosslinks through a cationic polymerization route is given.

When zinc oxide is used as the vulcanizing reagent, the initiating amounts of zinc chloride necessary are likely formed as a result of thermal dissociation of some of the allylic chloride to yield hydrogen chloride. Subsequent reaction of the hydrogen chloride with zinc oxide would provide catalyst.

It is not likely that the propagation step proceeds very far, but for vulcanization purposes only one step is needed, particularly in view of the fact that both of the termination processes suggested result in the production of more catalyst.

The attainment of the full crosslinking potential of chlorobutyl by the ZnO system is relatively slow. This situation can be remedied by the inclusion of thiurams and thioureas into the curing recipe. It has been observed that chemical compounds with the grouping

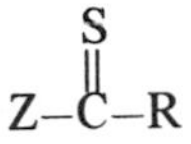

$$Z-\overset{\overset{\large S}{\|}}{C}-R$$

(where Z is some type of activating group) will accelerate the ZnO cure of CHLOROBUTYL. Examples are thiourea and tetramethyl thiuram disulfide.

The vulcanization of CHLOROBUTYL with this type of accelerator can proceed via a mechanism similar to the mechanism of polychloroprene vulcanization with substituted thioureas as proposed by Pariser.[30]

The increase in cure rate and modulus as compared to the straight ZnO cure is obtained without sacrifice in vulcanizate stability.

Vulcanization through Bis-alkylation. The other unique and valuable curing method for CHLOROBUTYL is that involving bis-alkylation reactions. This type of crosslinking reaction is perhaps best illustrated by crosslinking with primary diamines, a vulcanization reaction which proceeds rapidly to yield good vulcanizates. This crosslinking reaction is believed to occur by the mechanism shown below:

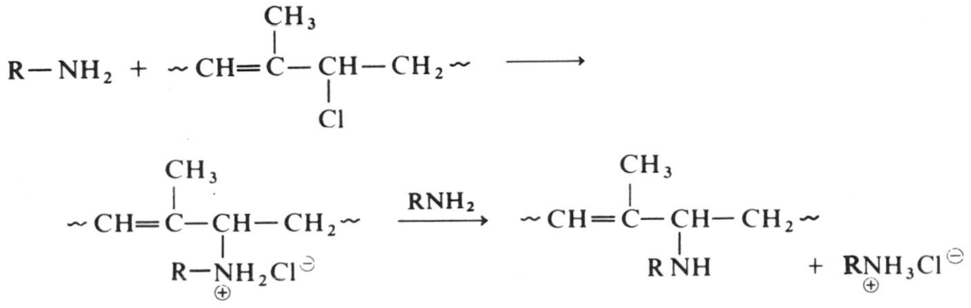

$$R-NH_2 + \sim CH=\overset{\overset{CH_3}{|}}{C}-\underset{\underset{Cl}{|}}{CH}-CH_2\sim \longrightarrow$$

$$\sim CH=\overset{\overset{CH_3}{|}}{C}-\underset{\underset{R-\underset{\oplus}{N}H_2Cl^{\ominus}}{|}}{CH}-CH_2\sim \xrightarrow{RNH_2} \sim CH=\overset{\overset{CH_3}{|}}{C}-\underset{\underset{RNH}{|}}{CH}-CH_2\sim \quad + \; R\underset{\oplus}{N}H_3Cl^{\ominus}$$

Obviously, in the presence of a diamine, reaction of both functional amino groups in the molecule with different polymer molecules results in crosslinking. Careful adjustment of diamine concentration is required for development of the highest crosslink density. Too little could not possibly provide the maximum crosslinks, whereas too much would allow for reaction with only one of the amino groups per diamino molecule. In the presence of a hydrogen chloride scavenger, maximum modulus is developed when the ratio of NH_2/Cl is very nearly unity.

Extensions of this bis-alkylation vulcanization technique are numerous and in general any molecule having two active hydrogens can, under the proper catalytic conditions, crosslink the polymer. Typical examples are vulcanization with dihydroxy aromatics such as resorcinol and with dimercaptans.

Resin Cure. Both CHLOROBUTYL and regular butyl are capable of vulcanization with heat-reactive phenolic resins which are usually characterized by 6–9 wt percent of methylol groups. Unlike conventional butyl no promotor or catalyst other than zinc oxide is needed for efficient vulcanization and a fast, tight cure is obtained with considerably less reagent.

Scorch Control. The modified zinc oxide cures are very fast and tend to be scorchy. As a general rule, acidic materials, such as channel blacks, activate the ZnO cure of CHLOROBUTYL while basic materials hamper or retard it. The retarding effects of some alkaline materials, such as magnesium oxide, can be used in a very practical way to provide processing safety. The effects of the addition of magnesium oxide to the TMTDS-zinc oxide cure system on both cure rate and processing safety in carbon black compounds are shown in Table 10.5 based on work of Ziarnik.[31]

Addition of 0.25 phr of MgO increases the margin against incipient vulcanization during processing at 126.5°C (260°F), judged by Mooney scorch measurements, from 5 to 15 minutes without greatly affecting tensile properties at the

TABLE 10.5. EFFECTS OF MAGNESIUM OXIDE ADDITION ON VULCANIZATION

Compound No.	*1*	*2*	*3*	*4*
TMTDS	1.0	1.0	1.0	0.5
Magnesium oxide	–	0.25	0.5	–
Mooney scorch measurements,[a] *time to 5 point rise at* 126.5°C:				
Minutes	5	15	30	8
Room temperature tensile properties,[a] *cured* 45 *minutes at* 153°C:				
Modulus at 300%, psi	2500	2370	1850	2360
Tensile strength, psi	2600	2720	2410	2580
Ultimate elongation, %	315	335	380	320

[a]Formulation (phr): Chlorobutyl–100, antioxidant 2246–1, HAF Black–50, stearic acid–1, ZnO–5, curatives as indicated.

vulcanization temperature of 153°C (307°F). However, if the concentration is increased to 0.5 phr, the cure rate is depressed significantly.

As a scorch retarder MgO is effective with all cure systems except the amine cure. In the latter case, it has the reverse effect since it prevents the hydrogen chloride, generated during crosslinking, from reacting with the curing agent. The choice of magnesium oxide type and concentration depends upon the type of compound used as well as the particular application involved.

Stability of CHLOROBUTYL Crosslinks

Since chlorobutyl has essentially the same structure as the butyl polymer all the properties inherent to the butyl backbone are found in chlorobutyl rubber. These properties include low gas and moisture permeability, high hysteresis, good resistance to ozone and oxygen, resistance to flex fatigue, and chemical resistance. Additionally, CHLOROBUTYL offers an appreciably higher level of heat resistance than regular butyl cured with conventional sulfur vulcanization systems.*

The ZnO/stearic acid accelerated CHLOROBUTYL compound displays a persistently increasing modulus or crosslink density with increasing vulcanization time. The problem of reversion does not exist in properly vulcanized CHLOROBUTYL. Berger[32] has compared the behavior of regular butyl and TMTDS-ZnO cured CHLOROBUTYL with respect to stress relaxation under conditions of fixed strain, which is a common method to follow the degradation of crosslinked networks. Gum vulcanizates were used and the recipes and curing times were adjusted to give approximately equal crosslink densities for both materials. The specimens were held in air at a fixed elongation of 50% and data were taken as a function of time at various temperatures. The curves shown in Fig. 10.7 represent the ratio of the stress (S) at any time, to that at 30 minutes (S_O) plotted as a function of time.

As the temperature increases, the relaxation of the butyl vulcanizates becomes more pronounced, as expected. The CHLOROBUTYL vulcanizate by contrast shows very little stress relaxation over the time period shown. Only the 100°C data are shown for the CHLOROBUTYL vulcanizate, since over the whole temperature range the behavior of this vulcanizate approximates that of sulfur-cured conventional butyl at the lowest test temperature. These data show basic stability of the crosslinks formed by this cure system, and the stability of the vulcanizate toward oxidative scission reactions.

The combination of TMTDS with ZnO for curing CHLOROBUTYL is effective in producing vulcanizates capable of withstanding temperatures up to 193°C (380°F). At that temperature most vulcanizates, other than CHLOROBUTYL, become excessively soft after a few hours of exposure, having lost all semblance of elastomeric behavior. Many CHLOROBUTYL cure systems, other than the ZnO-TMTDS cure described above, can be chosen for heat resistance such as straight zinc oxide, ZnO-ethylene thiourea (NA-22), and resin cure systems.

*Under many test conditions, resin-cured butyl provides the highest heat resistance obtainable with butyl-type elastomers but requires long cure cycles and large amounts of resin for optimum properties, which is not always practical.

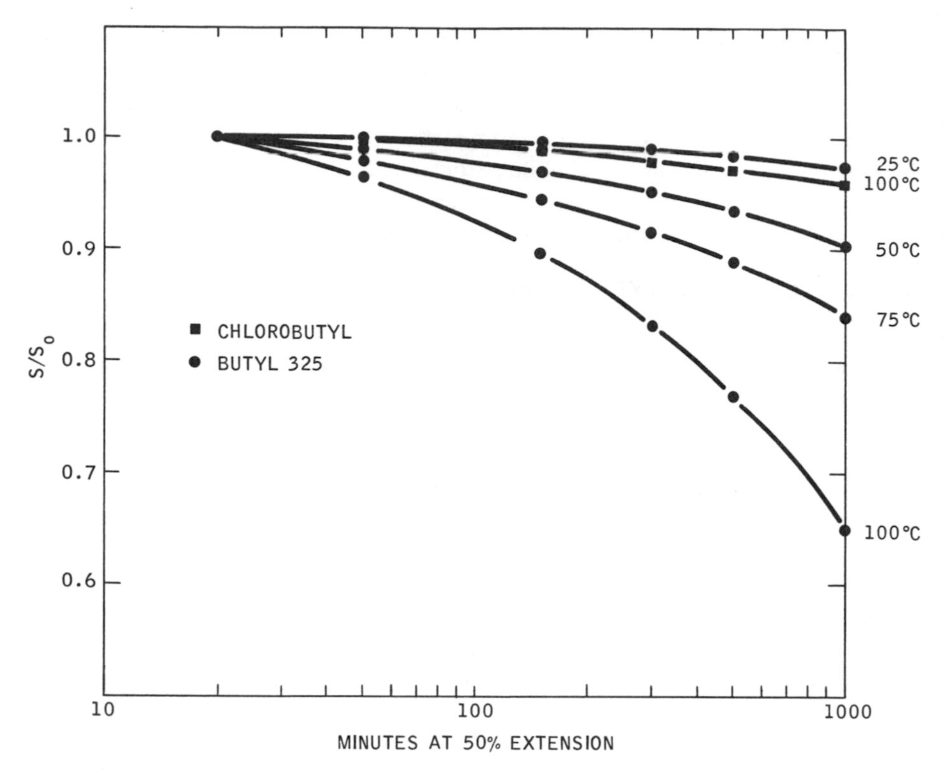

Fig. 10.7 Comparison of stress relaxation of gum vulcanizates: chlorobutyl and butyl.

Dithiocarbamates such as the lead and zinc derivatives have been successfully used in CHLOROBUTYL–zinc oxide systems.

Applications

Since its introduction in 1960, CHLOROBUTYL has proven to be highly useful in many commercial rubber products. These products generally take advantage of desirable characteristics generic to butyl polymers, such as the resistance to environmental attack and low permeability to gases. In addition, chlorobutyl offers the superimposed advantages of cure versatility, highly heat-stable crosslinks, and the ability to vulcanize in blends with highly unsaturated elastomers.

Innerliners for Tubeless Tires. The combination of low permeability, high heat resistance, excellent flex resistance, and ability to covulcanize with high unsaturation rubbers makes CHLOROBUTYL particularly attractive for use in innerliners for tubeless tires, and in Table 10.6 a typical chlorobutyl innerliner formulation is presented. 60–70 phr of Butyl HT 10-68 are necessary for optimum air barrier and heat resistance performance, while adequate building tack is provided by the inclusion of 20–30 phr of natural rubber. Cure systems based on Vultac No. 5 (alkyl-phenol disulfide) for black stocks and thioureas for mineral filled blends are widely used and strongly recommended to realize the best balance of properties.

TABLE 10.6. CHLOROBUTYL BLACK INNERLINER FOR PASSENGER TIRES

ENJAY BUTYL 10-68	65
Natural Rubber (No. 1 Smoked Sheet)	25
Whole tire reclaim	20
GPF Carbon Black	70
Paraffinic oil	12.5
Stearic Acid	1
Amberol ST-317X	4
Maglite K	0.5
Zinc Oxide	5.0
TMTMS[a]	0.25
Vultac 5 (Penwalt Co.)	0.7
MBTS	1.25
Permeability, Q × 10³ [b]	
Tested at 150°F	6.8
Adhesion to SBR-NR Carcass Stock Cured 45 Minutes at 287°F Pulled at a rate of 2 in./min	
Tested at 75°F, lb/in.	48
Tested at 250°F, lb/in.	24

[a]Tetramethylthiuram monosulfide.
[b]Q – Cubic feet of air (at 32°F and 29.92 in. Hg) diffusing through 0.001 in. of material under pressure differential of 1 psi/sq ft/day.

In a tubeless tire, the function of an innerliner is to provide an effective air barrier and minimize intracarcass pressure. Intracarcass pressure is a result of migration of air from the tire cavity into the cord area of the tire and the damming effect of the thick sidewall and tread exterior. The reduction in air (and oxygen) in the carcass serves to lessen oxidative degradation of the tire fabric and carcass rubbers. Intracarcass pressure and oxidation can weaken the body of the tire and effect tire failures through flex fatigue or loss of adhesion between components.

Tire performance data confirm the superiority of CHLOROBUTYL innerliners by increasing tire durability. In a severe accelerated wheel test, shown in the small table below, a 60/40 CHLOROBUTYL-natural rubber innerliner has increased mileage life by about 50% in a radial ply constructed tire.

TABLE 10.7. ASM WHEEL TEST (11.205″ DIAMETER WHEEL) 120% OF TRA LOAD, 28 PSI INITIAL INFLATION PRESSURE, 40 MILES PER HOUR

Innerliner	*Miles to Failure*	*Intracarcass Air Pressure*
SBR Control	3129	16.0 lb
CHLOROBUTYL Blend	4803	10.3 lb

(average of five tests of each type of tire)

The data in Table 10.7 clearly illustrate that the excellent heat and flex resistance of CHLOROBUTYL blends maintains the integrity and air barrier qualities of the liner, even under the most severe operating conditions.

Tire Sidewalls Components. Tire sidewall performance is critical from both an appearance and durability standpoint. Longer tire service life, particularly with the advent of belted bias and radial ply tires, and increasing amounts of atmospheric degradants are contributing to a higher service severity. Tire manufacturers find it more difficult and costly to obtain the desired performance requirements with antiozonants.

The substitution of CHLOROBUTYL either alone or in combination with ethylene-propylene terpolymer, for a portion of the high unsaturation polymers commonly in use, offers a simple, economical means to upgrade the weathering and flex resistance of tire sidewall components.

Heat Resistant Truck Inner Tubes. CHLOROBUTYL offers improved resistance to heat softening and growth while maintaining the desirable butyl property of excellent air retention in inner tubes. This is of particular importance in severe service conditions, such as high speed, heavy load trucks and buses where tire service temperatures exceed 280°F and sometimes reach as high as 300°F. After prolonged exposure butyl tubes will soften, whereas CHLOROBUTYL tubes will retain their original condition. A modified zinc oxide cure shown in the formulation below provides excellent thermal stability for a truck inner tube.

ENJAY BUTYL HT 10-68	100
GPF Carbon Black	70
Paraffinic oil	25
Stearic acid	1
Zinc oxide	5
TMTDS	0.2
Maglite D	0.1
(Cure 15′ at 350°F)	

Other Applications. CHLOROBUTYL is used in many rubber articles in addition to tires and tubes. In these applications, the cure versatility of CHLOROBUTYL and the stability of its vulcanizates are of particular importance. For example, CHLOROBUTYL can be cured with nontoxic cure systems, such as zinc oxide with stearic acid, for use in products which will contact food. Most cure systems provide fast, reversion-resistant cures with CHLOROBUTYL. This property facilitates the attainment of uniform cure state in thick articles. Due to its exceptional heat resistance and compression set properties, properly compounded CHLOROBUTYL will give good service at temperatures up to 150°C (300°F).

Examples of typical applications are: hose (steam, automotive), gaskets, conveyor belts, adhesives and sealants, tire curing bags, tank linings, truck cab mounts, aircraft engine mounts, rail pads, bridge bearing pads, pharmaceutical stoppers, and appliance parts.

REFERENCES

(1) J. P. Kennedy and E. Tornqvist, Ed., *Polymer Chemistry of Synthetic Elastomers*, Chap. 5 (by J. P. Kennedy), Interscience Publishers, New York, 1968.
(2) P. H. Plesch, Ed., *Cationic Polymerization and Related Complexes*, Chap. 4, Hefner, Cambridge, 1953.
(3) R. M. Thomas and W. J. Sparks, U.S. Patent 2,356,127, JASCO, Dec. 29 1937.
(4) P. J. Flory, *Principles of Polymer Chemistry*, Chap. VII (pages 273–314) Cornell University Press, 1953.
(5) S. G. Gallo, H. K. Wiese, and J. F. Nelson, *Ind. Eng. Chem.*, **40**, 1277 (1948).
(6) J. Rehner, Jr. and Priscilla Grey, *Ind. Eng. Chem.*, **17**, 367 (1945).
(7) J. Rehner, Jr., *Ind. Eng. Chem.*, **36**, 46 (1944).
(8) P. J. Flory, *J. Am. Chem. Soc.*, **65**, 372 (1943).
(9) R. L. Zapp and F. P. Baldwin, *Rubber Chem. Tech.*, **20**, 84 (1947).
(10) K. H. Altgelt and J. C. Moore, "Gel Permeation Chromatography," Chap. 34 of *Polymer Fractionation*, M. J. R. Cantow, Editor, Academic Press, New York, 1967.
(11) A. M. Gessler, *Rubber Age*, **74**, 59 (1953).
(12) A. M. Gessler and F. P. Ford, *Rubber Age*, **74**, 243 (1953) and A. M. Gessler and J. Rehner, Jr., *Rubber Age*, **77**, 875 (1955).
(13) G. J. Van Amerongen, *J. Appl. Phys.*, **17**, 972 (1946); *J. Pol. Sci.*, **5**, 307 (1950).
(14) P. O. Tawney, J. R. Little, and P. Viohl, *Rubber Age*, **83**, 101 (1958).
(15) D. J. Buckley and S. B. Robison, *J. Polym. Sci.*, **19**, 145 (1956).
(16) D. F. Kruse and R. C. Edwards, *SAE Transactions*, **77**, 1868–1895 (1968).
(17) R. C. Puydak and R. S. Auda, *SAE Transactions*, **76**, 817–840 (1967).
(18) D. Tabor, C. G. Giles, and B. E. Sabey, *Engineering*, **186**, 838 (1958).
(19) R. L. Zapp, *Revue Generale du Caoutchouc*, **40**, 265–274 (1963).
(20) C. J. Sheehan and A. J. Bisio, *Rubber Reviews, Rubber Chem. Tech.*, **39**, 149 (1966).
(21) G. Alliger and I. J. Sjothun, Editors, *Vulcanization of Elastomers*, Chap. 7, by W. C. Smith, Reinhold Publishing, N.Y. 1964.
(22) C. J. Jankowski, K. W. Powers, and R. L. Zapp, *Rubber Age*, **87**, 833 (August, 1960).
(23) J. G. Martin and R. F. Neu, *Rubber Age*, **86**, 826 (Feb., 1960).
(24) S. Vander Meer, *Rec. Trav. Chim.*, **65**, 149, 157 (1944).
(25) A. Greth, *Kunstoffe*, **31**, 345 (1941).
(26) I. C. McNeil, *Polymer*, **4**, 15 (1963).
(27) A. H. Frye and R. W. Horst, *J. Polym. Sci.*, **40**, 419 (1959).
(28) M. Onozuka and W. I. Bengough, *Polymer*, **6**, 625 (1956).
(29) F. P. Baldwin et al., *Rubber and Plastics Age*, **42**, 500 (1961).
(30) R. Pariser, *Kunstoffe*, **50**, 623 (1960).
(31) G. Ziarnik, *Rubber Chemistry and Technology*, **35**, 467, (1962).
(32) M. Berger, *J. Appl. Poly. Sci.*, **5**, 322 (1961).

11

SYNTHETIC POLYISOPRENE

GLENN R. HIMES
Shell Chemical Company
Torrance, California

INTRODUCTION

Most types of synthetic rubber represent man's attempt to reproduce or improve upon the physical behavior of natural rubber (NR). Only one, synthetic polyisoprene (IR), approximates the chemical composition of natural rubber. Its predominant structure is *cis*-1,4-polyisoprene, the same as that of natural rubber. It shares with NR the properties of good uncured tack, high pure gum tensile strength, high resilience, low hysteresis, and good hot tear strength. As might be expected, polyisoprene comes closest of all synthetics to being a complete replacement for natural rubber. It is a general purpose rubber which finds its largest applications in tires, mechanical goods, and footwear.

Although polyisoprene is chemically similar to natural rubber, minor differences exist which lead to certain advantages and disadvantages compared to NR. Polyisoprene is generally more economical to process, lighter in color, more uniform, and of higher purity. On the other hand, IR is lower in green (unvulcanized) strength, most noticeably in reworked stocks. In general, the synthetic polyisoprenes are lower in modulus and higher in elongation than the natural product. There are slight differences in vulcanizate properties among the individual types of isoprene rubber.

Thus far the synthetic polymer owes its market growth to its property advantages, ease of handling, and lower cost relative to the top grades of natural rubber. The development of a domestic isoprene rubber industry has, of course, provided protection against future shortages of natural rubber. The synthetic polymer's share of the total polyisoprene market is expected to grow markedly because of the fact that demand for high *cis* polyisoprene rubbers will exceed the rate of growth of the natural rubber supply.

HISTORY

An isoprene polymer was first reported synthesized in 1875. Through the years other polyisoprenes were made on a small scale but these polymers lacked the strength and versatility of natural polyisoprene. In molecular structure these polymers consisted of variable amounts of the various stereoisomers possible with a polyisoprene, whereas the natural product is 100% *cis* 1,4 isomer. After World War II, catalysts were developed which made possible the production of stereospecific polymers. Synthesis of the man-made equivalent of natural rubber was first reported by scientists of Goodrich-Gulf Chemicals, Inc., in 1954. They used a coordination catalyst of the type first developed by Ziegler and Natta. Using a catalyst based on finely dispersed lithium, Firestone Tire and Rubber Company reported preparation of a synthetic "natural" rubber in 1955. Both new polymers were solution-polymerized rubbers of greater than 90% *cis*-1,4-polyisoprene. Present commercial polyisoprene rubbers are produced with similar catalyst systems and have the same high level of *cis*-1,4 isomer.

COMMERCIAL POLYISOPRENES

Synthetic polyisoprene was introduced commercially in 1960 by Shell Chemical Company as "Shell Isoprene Rubber." Goodyear followed in 1962 with "Natsyn" and Ameripol, Inc. in 1968 with "Ameripol SN." At present these producers offer a total of 14 different types consisting of general purpose, easy-processing, high purity, and oil-extended grades (see Tables 11.1 and 11.2). At this writing no black masterbatches or oil-extended black masterbatches of polyisoprene have been commercialized. There is one commercial polyisoprene latex, produced by Shell Chemical Company. In all cases the rubber hydrocarbon is isoprene homopolymer of at least 91.5% *cis* configuration.

In addition to the domestic producers, polyisoprene is made in the Netherlands (by Shell Nederland Chemie) and in the Soviet Union.

Since 1964 *trans*-1,4 polyisoprene has been manufactured commercially by Polymer Corporation, Ltd. (Canada) as "TRANS-PIP." The *trans* isomer, unlike the *cis* isomer, is crystalline at room temperature and has a high hardness and high tensile strength without vulcanization. It is a substitute for natural balata and gutta percha and is used primarily in golf ball covers. This product is described

TABLE 11.1. DOMESTIC POLYISOPRENE PRODUCERS

Company	*Location*	*Annual Capacity, Thousands of Long Tons*
Ameripol, Inc.	Orange, Texas	50
Shell Chemical Company	Marietta, Ohio	37
Goodyear Tire & Rubber Company	Beaumont, Texas	65
Total		152

TABLE 11.2. POLYISOPRENE TYPES PRODUCED IN THE U.S.

(All types classed as nonstaining)

Trade Name	*Manufacturer*	*Nominal Mooney Viscosity, ML-1 + 4 at 212°F*	*Oil Content*		*Special Feature*
			Type	*Level, phr*	
IR-305	Shell Chemical	80[a]			General purpose
IR-307	Shell Chemical	85[a]			General purpose, Food & Drug grade
IR-309	Shell Chemical	65[a]			Easy processing, general purpose
IR-310	Shell Chemical	70[a]			Easy processing, general purpose, Food & Drug grade
IR-500	Shell Chemical	60[b]	Naphthenic	25	General purpose, oil-extended
IR-501	Shell Chemical	45[b]	Naphthenic	25	Easy processing, general purpose, oil-extended
IR-700	Shell Chemical	–			High solids latex
Natsyn 200	Goodyear Tire & Rubber	90[c]			General purpose
Natsyn 400	Goodyear Tire & Rubber	85[c]			General purpose
Natsyn 405	Goodyear Tire & Rubber	85[c]			Low gel
Natsyn 410	Goodyear Tire & Rubber	60[c]			Easy processing, general purpose
Natsyn 450	Goodyear Tire & Rubber	50[c]	Naphthenic	25	General purpose, oil-extended
Natsyn 2200	Goodyear Tire & Rubber	90[c]			General purpose
Ameripol SN 600	Ameripol, Inc.	82[c]			General purpose
Ameripol SN 601	Ameripol, Inc.	70[c]			Easy processing, general purpose

[a]Compound viscosity of 50 phr HAF stock.
[b]Compound viscosity of 40 phr HAF stock.
[c]Viscosity of raw polymer

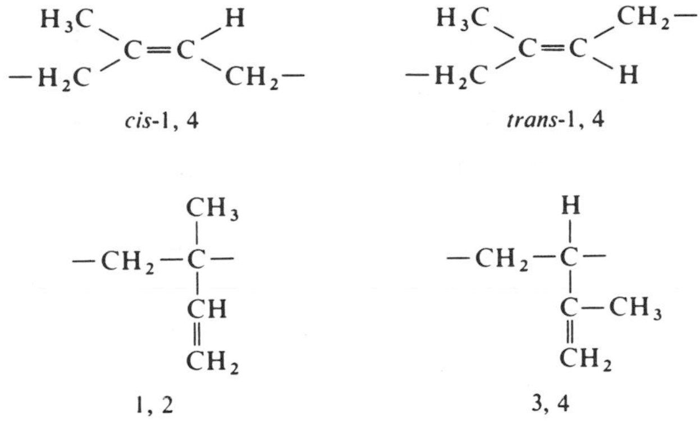

$$\begin{array}{ccc} H_3C & & H \\ & C{=}C & \\ -H_2C & & CH_2- \end{array} \qquad \begin{array}{ccc} H_3C & & CH_2- \\ & C{=}C & \\ -H_2C & & H \end{array}$$

cis-1, 4 *trans*-1, 4

$$\begin{array}{c} CH_3 \\ | \\ -CH_2-C- \\ | \\ CH \\ \| \\ CH_2 \end{array} \qquad \begin{array}{c} H \\ | \\ -CH_2-C- \\ | \\ C-CH_3 \\ \| \\ CH_2 \end{array}$$

1, 2 3, 4

Fig. 11.1 Stereoisomers of polyisoprene repeating unit.

further under "*Trans*-1,4-Polyisoprene" (page 300). Unless otherwise indicated the discussions in this chapter will refer to the *cis*-1,4-polyisoprene elastomers.

MOLECULAR STRUCTURE

The four isomers of the polyisoprene repeating unit which are of interest to the rubber technologist are *cis*-1,4; *trans*-1,4; 1,2; and 3,4 (shown in Fig. 11.1).

The relative proportions of these structures can be measured by infrared spectroscopy or nuclear magnetic resonance (NMR). The synthetic general purpose polyisoprenes consist of about 92 to 98% of the *cis*-1,4 structure with the remainder 3,4 and/or *trans*-1,4. Synthetic balata is 95-98% *trans*-1,4-polyisoprene with a small amount of *cis*-1,4. Hevea natural rubber is 100% *cis*-1,4, while natural balata and gutta percha are essentially 100% *trans*-1,4. The small differences in molecular structure between the natural and synthetic products lead to small but significant differences in physical properties. These differences will be discussed in later sections.

METHODS OF MANUFACTURE

Much of the information on commercial polyisoprene manufacture is proprietary in nature and unpublished. Therefore the following discussion is based in part on reports of laboratory and pilot plant scale studies and patents. Commercial practices would be expected to be similar in principle.

The main steps in the manufacture of polyisoprene are raw material preparation and purification, polymerization, catalyst deactivation and removal, solvent recovery, polymer drying, and baling (shown schematically in Fig. 11.2).

The preparation of raw materials is crucial to the successful production of isoprene rubber. Polar compounds (e.g., oxygen, water, certain oxygenated organic compounds) which would inactivate the catalysts must be removed from the

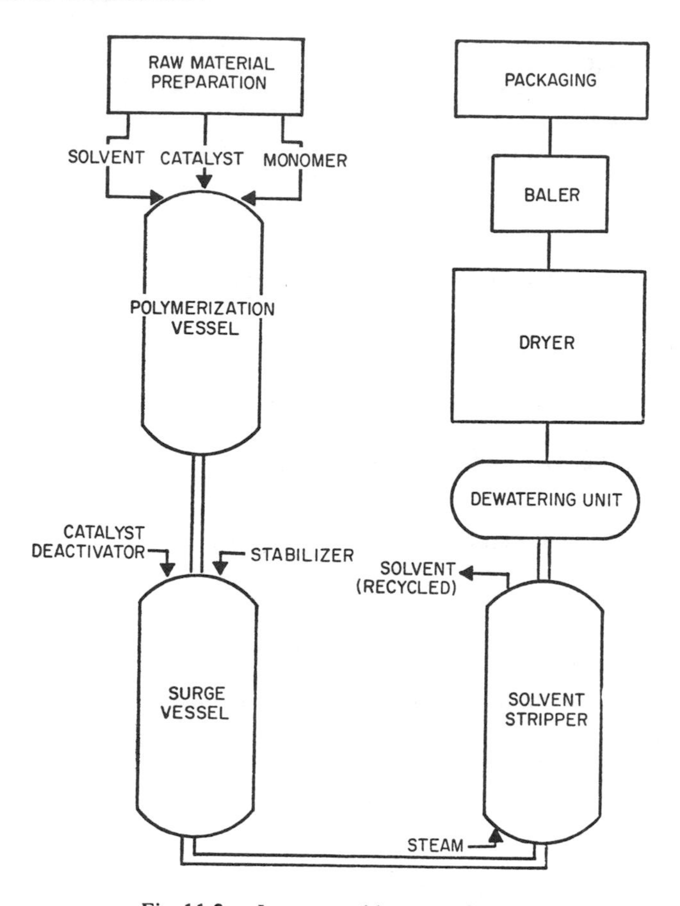

Fig. 11.2 Isoprene rubber manufacture.

isoprene monomer and the solvent used for reaction medium. Similarly, active hydrogen compounds and certain hydrocarbons (acetylenes, cyclopentadiene, cyclopentene) must be avoided. Laboratory purification procedures may include distillation from alkali metals and treatment with alumina.

Isoprene monomer used in commercial production of polyisoprene is a low boiling (93°F), colorless liquid usually manufactured from petroleum starting materials:

$$\begin{array}{cccc} H & & H & H \\ | & & | & | \\ C & = C - & C & = C \\ | & | & & | \\ H & CH_3 & & H \end{array}$$

Isoprene

In the U.S. isoprene is produced chiefly by the propylene dimer process and by the dehydrogenation of 2-methylbutenes and/or isopentane. The propylene dimer process consists of three basic steps: (1) dimerization of propylene to 2-methyl-1-pentene, (2) isomerization of the latter to 2-methyl-2-pentene, and (3) pyrolysis of the latter product to isoprene and methane. The dehydrogenation route to isoprene production is similar to the process for the production of butadiene. A refinery fraction containing isopentane and/or 2-methylbutenes is dehydrogenated in a Houdry Process reactor. The C_5 fraction is recovered from the reaction product and isoprene separated by extractive distillation. Development of the propylene dimer and dehydrogenation processes to produce isoprene was an essential step in the commercialization of synthetic polyisoprene, since these processes ensured a source of monomer at an ecomically attractive cost.

Isoprene can also be made from isobutylene and formaldehyde as starting materials (the Institut Francais du Pétrole process) and from acetone and acetylene (developed by Societa Nazionale Metanodotti of Italy).

The polymerization catalysts used in commercial production of polyisoprene are either of the coordination (Zeigler) or alkyl lithium types. The coordination catalysts are usually trialkyl aluminum/titanium tetrachloride mixtures at an Al/Ti mole ratio of about 1:1. The Al/Ti mole ratio is critical for achievement of high *cis*-1,4 content and high yields of solid polymer. Sometimes the trialkyl aluminum is complexed with an ether to yield a more active catalyst. Catalyst composition and method of preparation are carefully adjusted to give desired molecular weights with a minimum of low molecular weight diluents and no excessive gel. The coordination catalysts used in production of isoprene rubber are insoluble and function in the form of a finely divided suspension. They are capable of producing 96 to 98% *cis*-1,4 structure.

Alkyl lithium compounds are commercially important polymerization catalysts for isoprene because, in addition to high *cis* content, they give ready control of molecular weight distribution, a virtual absence of gel, and they leave no residues which are deleterious to rubber aging properties. The type of alkyl group has little effect on the *cis* content of the polyisoprene, but a significant effect on initiation rate. Butyl lithium is commonly used. Unlike the coordination catalysts, alkyl lithium systems are soluble and are characterized by the complete absence of a termination reaction, i.e., polymerization continues as long as monomer is present. Certain impurities, however, can react with the catalyst or the growing chain. Catalyst level is carefully controlled, since it has a strong influence on microstructure as well as molecular weight. Alkyl lithium-catalyzed polymers have *cis* contents of 92 to 95%.

Lithium metal will also cause the formation of high-*cis* isoprene polymers; however, the actual catalyst in such systems is an alkenyl lithium formed by reaction of the metal with isoprene. The initiation rate using lithium metal is much slower than with an alkyl lithium such as butyl lithium.

Alkali metals other than lithium will catalyze isoprene polymerization, as will organoalkali metal compounds other than alkyl lithiums. However, the resultant polymers do not have the desired high *cis* configuration. The unique properties of lithium which enable it to catalyze stereospecific polymerizations are thought to be its small ionic radius, its ability to function (in some reactions) as a polyvalent atom, and the stability of its steric configurations in many reactions.

The literature indicates that isoprene polymerizations are taken to high monomer conversion levels, i.e., 80-100%. Undesirable reactions tend to occur in coordination-catalyzed polymerizations above 80% conversion. There appears to be no upper limit in alkyl lithium polymerization.

Isoprene polymerization is carried out in an inert hydrocarbon solvent. Low molecular weight aliphatic solvents are favored because they lead to polymer solutions of lower viscosity and allow use of reflux cooling to control the reaction temperature. Based on available information, isoprene coordination-catalyzed polymerizations are normally carried out in the neighborhood of 120°F.

After polymerization the catalyst is deactivated and an antioxidant added to protect the polymer during subsequent operations and storage. In some cases the catalyst is at least partially removed with alcohol or alcohol/water solution. The method and degree of removal varies among different isoprene rubber types. Excess residues of coordination catalysts are detrimental to the aging stability of the polymer. The solvent remaining after polymerization can be stripped off by sparging with steam and is later purified and reused. The final slurry of polymer crumb in water is processed in equipment similar to that used in finishing styrene-butadiene rubber. The crumb is washed, dewatered, dried, baled, and packaged in suitable containers.

PROPERTIES OF THE RAW POLYMER

Commercial synthetic *cis*-1,4 polyisoprene is a light amber or water-white, essentially odorless elastomer. Except for oil-extended grades, isoprene rubber is almost 100% rubber hydrocarbon; it contains no fatty acids, resins, or proteins as are found in natural rubber (about 6%). Nonrubber constituents in IR consist chiefly of stabilizer, residual moisture, and trace quantities of inorganic materials. At present all commercial types contain a nonstaining, nondiscoloring stabilizer. The specific gravity of synthetic polyisoprene is 0.90-0.91, which is about equivalent to that of natural rubber (0.906-0.916).

The synthetic polyisoprenes are supplied at a relatively low viscosity compared to that of natural rubber. The general purpose grades of Natsyn and Ameripol SN polyisoprenes have Mooney viscosities of approximately 80 to 90 ML-4 at 212°F. Several easy processing types are available with lower viscosity (50-70 ML-4) (see Table 11.2). The viscosities of Shell Chemical Company raw polyisoprenes are reported in terms of intrinsic viscosity of a toluene solution (86°F). Intrinsic viscosities range from 6 to 12 dl/g for the non-oil-extended grades and 9-14 dl/g for the base polymer of oiled types. Mooney viscosity is considered an unreliable test with these polymers because of cleavage in the viscometer. Raw polymer with little or no mastication gives a fictitiously low Mooney value. If the polymer is masticated on a mill or in a Banbury mixer the apparent Mooney viscosity increases until the molecular weight is reduced to about 800,000, after which the Mooney value declines with further mastication. The Mooney test can be used on compounded stocks if a selected compounding procedure is rigorously followed; otherwise, nonreproducible Mooney values may be obtained. Typical values for HAF standard test compounds are given in Table 11.2.

In common with most polymers, isoprene rubbers are non-Newtonian in bulk viscosity behavior, i.e., their viscosities vary with shear rate or shear stress. However, alkyl lithium-catalyzed, linear polyisoprenes have been found to approach Newtonian behavior at low shear rates.

Because of the high degree of stereoregularity in synthetic isoprene rubbers, crystallinity is observed in the polymer under some conditions. However, the rate of crystallization in the synthetics is slower than in natural rubber. Crystallite formation is discussed further under "Processing" and "Vulcanizate Properties."

The molecular weight of polyisoprene can be characterized by any of the usual methods, i.e., osmotic pressure (gives a number-average, $\overline{M}_n$), light scattering (weight-average, $\overline{M}_w$), ultracentrifugation ($\overline{M}_w$), or intrinsic viscosity (viscosity-average, $\overline{M}_v$). The latter is the simplest and most popular technique. The molecular weight of polyisoprene can be calculated from its intrinsic viscosity using the Mark-Houwink equation, $[\eta] = K\overline{M}_v^{\alpha}$ where $[\eta]$ is the intrinsic viscosity and K and α are empirically-determined constants. For alkyl lithium-catalyzed polyisoprene in toluene at 86°F the equation takes the form,[1] $[\eta] = 2.00 \times 10^{-4} \overline{M}_v^{0.728}$, where $[\eta]$ is expressed in dl/g. This expression is very similar to the intrinsic viscosity-molecular weight relationship for natural rubber.

In general the average molecular weight of synthetic polyisoprene is substantially lower than that of natural rubber. Unmasticated alkyl lithium-catalyzed IR has a viscosity-average molecular weight of about 2,000,000, while that of the coordination-catalyzed polymer is about 1,000,000. A definite figure for the average molecular weight of Hevea is difficult to obtain, because of large molecular weight differences between trees and between clones, variable molecular weight changes during latex storage, and the presence of gel.

Molecular weight is readily controlled during polyisoprene manufacture and consequently the synthetic polymer is considerably more uniform in this regard than natural rubber. In some IR grades molecular weight has been intentionally lowered for better processing. In oil-extended grades the molecular weight has been increased to compensate for oil dilution.

Polyisoprenes, like almost all polymers, are polydisperse, that is, the molecules have a variety of sizes. The breadth of the distribution of molecular weights and the nature of individual fractions can have significant effects on processing and on the properties of the end-use article (discussed further in the sections on "Processing" and "Vulcanizate Properties"). A relative measure of the polydispersity of a polymer is the ratio of its weight-average molecular weight to number-average molecular weight, i.e., $\overline{M}_w/\overline{M}_n$. For a monodisperse polymer this ratio is 1.0. Alkyl lithium polyisoprenes have a fairly narrow distribution, with a $\overline{M}_w/\overline{M}_n$ ratio less than 2, while that of coordination-catalyzed types is somewhat wider, $\overline{M}_w/\overline{M}_n$ 2 to 3. $\overline{M}_w/\overline{M}_n$ of NR is greater than 3 and widely variable. Under mastication the different polyisoprene types tend to approach a common molecular weight distribution; i.e., the relatively broad distributions of NR and coordination-catalyzed IR become more narrow and the distribution of alkyl lithium-catalyzed IR becomes wider. This behavior results from the fact that the longer polymer chains have a greater probability of rupture during mastication.

Like natural rubber, coordination-catalyzed polyisoprene normally contains an appreciable quantity of gel, i.e., crosslinked polymer which is insoluble in common rubber solvents. The gel content of general purpose, Al/Ti-catalyzed isoprene rubber ranges from about 5% to 25%. Most of the gel in coordination-catalyzed IR breaks down with mastication, similarly to the "loose" gel* which constitutes the main portion of gel in NR. At present there is one grade of specially prepared Al/Ti-catalyzed IR, Natsyn 405, which is essentially gel free. The gel content of NR is widely variable and usually higher than that of the general purpose grades of coordination-catalyzed synthetic polyisoprene. Alkyl lithium-catalyzed polyisoprene contains no gel.

The glass transition temperature of synthetic *cis*-1,4 polyisoprene has been found to be equivalent to that of Hevea, about −98°F.

The chemistry of raw *cis*-1,4-polyisoprene is essentially the same as that of its natural counterpart. The IR molecule will undergo cyclization, maleic anhydride addition, halogenation, and hydrohalogenation in reactions which closely resemble the same reactions with natural rubber.

The reader is referred to reference sources on natural rubber for physical constants which have not yet been published for synthetic isoprene rubber.[2,3] Values for the coefficient of expansion and the thermal, optical, electrical, and mechanical properties of the synthetic polymer are expected to be similar to those of natural rubber.

PROCESSING

In general, synthetic polyisoprene behaves like natural rubber during mixing and subsequent processing steps. Since the two polymers are so similar chemically, similar processibility would be expected. Thus, both IR and NR show a relatively rapid decrease in viscosity with mastication (see Fig. 11.3) compared to most other elastomers. Both polymers undergo predominantly chain scission during mixing and subsequent operations, with a corresponding reduction in molecular weight. At high temperatures (> 175°F) in the presence of air, both polymers break down primarily by oxidative degradation. At room temperature the main mechanism of molecular weight reduction is mechanical chain scission. Neither IR nor NR change in microstructure during normal processing.

In spite of these similarities there are some differences in the way isoprene rubber and Hevea process. Most of these differences trace directly or indirectly to

*"Loose" gel may be roughly defined as that having a swelling index (weight ratio of solvent-swollen gel to dry gel) of greater than 20. Gel of markedly lower swelling index ("tight" gel) is deleterious to stress-strain and hysteresis properties.

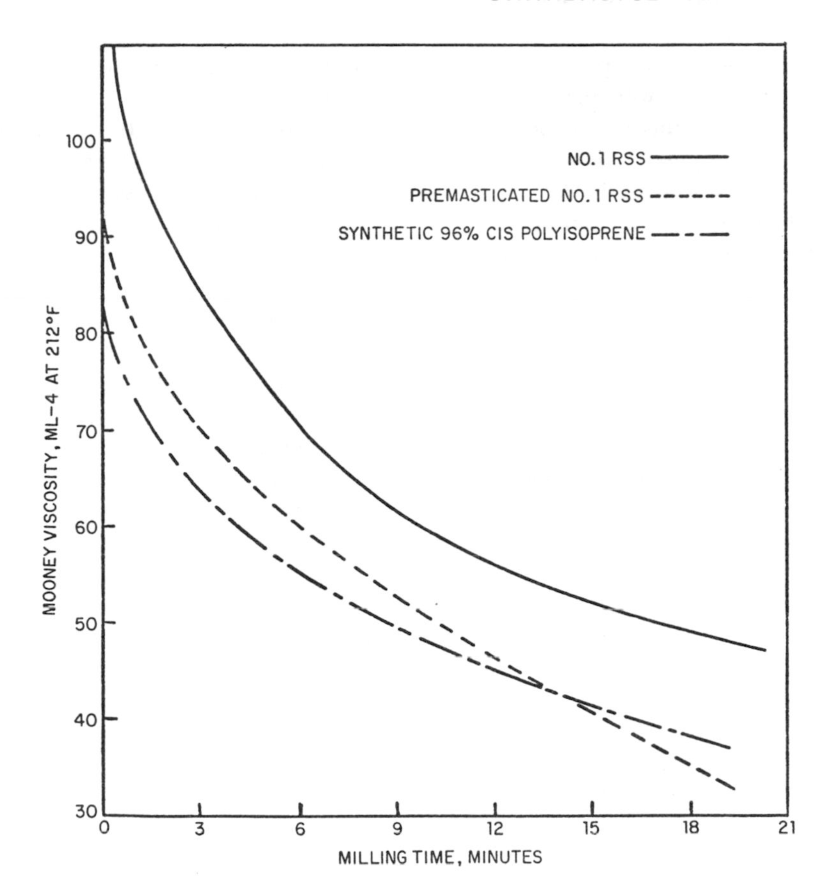

Fig. 11.3 Effect of milling on viscosity of synthetic and natural polyisoprenes. Mill roll temperature: 158° F. (Courtesy Goodyear Tire & Rubber Co.)

the lower viscosity, slightly lower *cis* content, and greater uniformity of the synthetic product. There are both advantages and disadvantages for isoprene rubber in processing characteristics.

The lower initial viscosity of synthetic polyisoprene obviates the necessity of a separate breakdown step which is normally used with natural rubber. It also eliminates the need for hot room storage and thawing out of rubber stored at low temperatures. As a rule IR processes like premasticated natural rubber. It thus requires less total mixing time. The time to incorporate carbon black or other fillers is about the same as with premasticated NR. Polyisoprene tends to become sticky and lose green strength if overmixed or subjected to prolonged milling. This is overcome by aiming for a compounded Mooney viscosity equal to or slightly higher than a corresponding NR compound, and by minimizing mastication periods.

Because of the lower viscosity and somewhat lower *cis* content of isoprene rubber (and therefore slower rate of stress-induced crystallization during mixing), the shear forces developed and heat generated during mixing are not as great as with natural rubber. Power consumption, Banbury dump temperature, and heat build-up in processing equipment are lower than with NR. Scorch safety of IR compounds is consequently greater than that of natural rubber compounds. These effects are more noticeable with the 92% *cis* synthetic polymer than with the 96% *cis* product. Manufacturers of the latter product recommend early addition of stearic acid in the Banbury to prevent stickiness and reduce power consumption.

A typical Banbury cycle for isoprene rubber in a 50 phr HAF carbon black stock is shown in Table 11.3. Because of the relatively soft nature of IR in mixing, a slightly larger than usual batch size (by about 10%) is frequently necessary to give good ram action. Maximum ram pressure and full coolant are used with IR stocks.

Ordinarily chemical softeners, or peptizers, are not necessary nor advised for use with isoprene rubber, since the polymer breaks down rapidly. However, they are sometimes employed to improve pigment wetting and reduce mixing time. Zinc-2,2′-dibenzamidodiphenyl disulfide* is particularly effective in softening polyisoprene.

TABLE 11.3. TYPICAL TWO-STAGE BANBURY CYCLE FOR A 50 PHR HAF BLACK STOCK

First Stage *Cumulative Time, Minutes*	*Operation*
0	Load isoprene rubber
1½	Add zinc oxide, stearic acid, antioxidant, and carbon black
3	Add softener
4	Scrape down ram
5	Dump
Second Stage	
0	Load masterbatch prepared in first stage and curatives
2-3	Dump at 220 to 230°F

A typical mill mixing procedure for synthetic polyisoprene is given in Table 11.4. Early addition of stearic acid prevents stickiness and back-rolling tendencies, and early antioxidant addition helps control oxidative chain scission. Accelerators are added early in the cycle with stocks that tend to be soft, to ensure thorough dispersion and uniform cure in the final product. With lightly-loaded 92% *cis* polymers, both sulfur and accelerator can be added early without danger of scorch.

*Pepton 65, American Cyanamid Co.

TABLE 11.4. TYPICAL MILL LOADING CYCLE

	Sequence of Operations
1.	Load isoprene rubber and band to smooth sheet
2.	Add stearic acid
3.	Add antioxidant, accelerator
4.	Add zinc oxide, fillers, and softener
5.	Add sulfur
6.	End roll and sheet off

When early addition of critical ingredients is impractical, a good dispersion is obtained by incorporating them via IR or NR masterbatches.

Recommended mill roll temperatures for polyisoprenes vary from about 130°F to 180°F, depending on the type of stock being processed and previous degree of mastication. The 92% *cis* polymers process well at 130-160°F, while the 96% *cis* types perform better at slightly higher temperatures, about 150-180°F. The easy processing grades of the respective IR types are milled at temperatures toward the low end of these ranges. Mill temperature is adjusted as necessary to control bagging and reduce stickiness or back-rolling.

Isoprene rubber is frequently used in a blend with natural rubber or other elastomers. In mixing polymers the normal practice of blending at similar viscosities is followed. In the case of IR/NR blends, isoprene rubber is preferably added to natural rubber, rather than the reverse. If it is necessary to add natural rubber to the synthetic polymer, the NR is first broken down to a viscosity slightly below that of the polyisoprene.

Isoprene rubber exhibits less nerve, i.e., elastic recovery, during processing than natural rubber. This behavior leads to faster banding times on the mill, lower die swell in extrusion, and lower shrinkage in calendered stocks. The lower nerve of IR is illustrated in Table 11.5 and in Fig. 11.4, where the mill shrinkages of various polymers are plotted. The mill shrinkage of polyisoprene approaches that of

TABLE 11.5. MILLING AND EXTRUSION DIE SWELL PROPERTIES

	General Purpose Polyisoprene[a]	*Easy Processing Polyisoprene*[b]	*Oil-Extended Polyisoprene*[c]	*RSS No. 1*
Raw Mooney Viscosity, ML-4 at 212°F	81	53	50	96
Time to Form Smooth Band (158°F lab mill), sec	60	15	15	180
Extrusion Die Swell[d] (3/8 inch round die, 225°F), %	61	28	43	71

[a]Natsyn 400, Goodyear Tire & Rubber Company.
[b]Natsyn 410, Goodyear.
[c]Natsyn 450, Goodyear
[d]Die swell evaluated using test compound containing 25 phr HAF black.

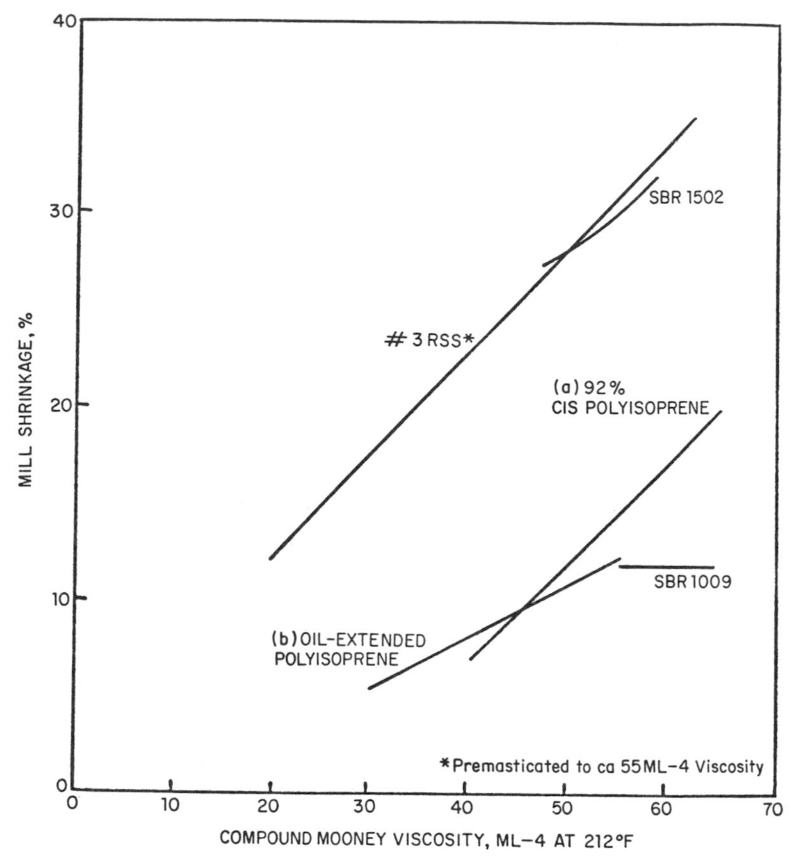

Fig. 11.4 Mill shrinkages of various polymers in a typical calendered shoe soling formulation. Mill shrinkages and Mooney viscosities were determined with samples cut directly from the mill under carefully controlled conditions at successive intervals of milling time. (a) Shell Isoprene Rubber 309, Shell Chemical Co. (b) Shell Isoprene Rubber 500, Shell Chemical Co. (Courtesy Shell Chemical Co.)

crosslinked SBR (Type 1009). The low nerve of IR also contributes to good injection molding properties since its residual stresses are less, after injection, than with NR.

Synthetic polyisoprene is superior to natural rubber and most synthetic elastomers in mold flow and reproduction of mold detail. This property is of particular importance in injection molding applications. IR also shows a faster injection time and lower temperature rise on injection molding than other general purpose rubbers. Compared to NR, the low, uniform viscosity and the relatively long scorch time of IR aid in maintaining consistent molding conditions and short cycle times.

There is a significant difference between synthetic polyisoprene and natural rubber in uncured strength, or "green" strength. Green strength is important in tire-building and other applications in which the unvulcanized stock must withstand considerable stress. The lower green strength of the synthetic polymer is attributed to its reduced capacity to undergo stress-induced crystallization. Although 100% IR tires have been made, the usual practice is to substitute IR initially at a small percentage and increase its level incrementally, keeping enough NR in the compound to maintain adequate green strength. Green strength is partially dependent on the compound viscosity and consequently is conserved by minimizing cycle times and avoiding overmastication.

IR is a tacky polymer compared to most synthetics, a factor which favors its use in laminated structures and adhesives. Its apparent tack is slightly less than that of NR; however, this behavior may be due to a lower green strength in interfacial layers rather than a true deficiency in tack.

The uniformity and ease of handling of synthetic polyisoprene are major advantages over natural rubber. No cross-blending of different lots is necessary as is often done with natural rubber to achieve uniform viscosities or cure rates. In-plant handling costs are lower since less inspection, bale cutting, and inside transportation are necessary. The competitive edge of the synthetic in these areas has recently spurred efforts among natural rubber producers to present NR in a more uniform and convenient form.

COMPOUNDING

The compounding technology of isoprene rubber is very much like that of natural rubber. Compounding ingredients such as sulfur, accelerators, antioxidants, fillers, activators, and plasticizers generally produce the same effects in synthetic polyisoprene as they do in the natural product. The differences in compounding practices which do exist are usually due to the lower content of nonrubber constituents in the synthetic polymer, the lower viscosity of isoprene rubber, the slight differences in microstructure between IR and NR, and the higher purity and greater uniformity of the synthetic product. The naturally-occurring nonrubber materials in Hevea include agents which act as accelerators of vulcanization (proteins and protein breakdown products), activators of acceleration (fatty acids), and stabilizers against oxidation (sterols and other compounds). The total of nonrubber ingredients in NR varies considerably but is generally around 6-7%. Since these materials are absent in synthetic polyisoprene, differences in cure rate and final properties often result if a 100% substitution of IR for NR is made without recipe adjustment. For example, a level of 2-3 phr of stearic acid is usually required for proper cure activation in isoprene rubber, whereas 1-2 phr is sufficient in natural rubber. Similarly, most types of commercial isoprene rubber require about 10% more accelerator to achieve the same rate of cure as exhibited by NR. Differences between IR and NR in cure characteristics are affected by type of curatives and type of loading. Differences are smaller if amine accelerators comprise part of the acceleration system. Furnace blacks have an activating effect on cure and tend to minimize cure rate differences. Acidic materials such as clays retard cure and lead to larger differences in rate and state of cure between IR and NR. In addition, cure rates differ among different grades of IR (owing, in some instances, to the presence of deactivated catalyst residues of different types) and proper acceleration should be determined for the individual grade by test before use. Cure rate within an individual IR grade is uniform.

In pure gum or lightly loaded stocks, especially with 92% *cis* IR types, full development of properties is aided by ensuring complete dispersion of accelerator and sulfur. Use of low melting or liquid accelerators helps achieve a molecular dispersion of these ingredients in soft stocks which do not build up much heat in processing. Early addition or masterbatching of curatives also helps, as discussed under "Processing." Adequate dispersion of all types of curatives is generally achieved in IR stocks containing filler, since the filler increases shear forces within the mixing batch. Inclusion of some amine-containing activator or accelerator in isoprene rubber cure systems, especially in gum stocks, promotes flat cures and good heat aging. Their participation in vulcanization may resemble that of protein breakdown products in natural rubber.

Sulfur requirement for synthetic polyisoprene is about the same as that of natural rubber and therefore higher than that for SBR and BR. In filled stocks 2 to 3 phr are normally used, with slightly more in highly loaded compounds. In gum stocks 1.5-2.0 phr are employed. Sulfur levels toward the low side of these ranges will give higher tensile strengths, higher tear strengths, and better aging resistance. Higher sulfur contents yield higher modulus, better hysteresis properties, and higher resilience. In soft isoprene rubber stocks when thorough sulfur dispersion is difficult, a low sulfur level gives more uniform properties. In the final analysis, as with any rubber, the amount of sulfur used depends on the balance of properties desired in the vulcanizate.

Isoprene rubber can also be cured with sulfur-bearing organic compounds such as tetramethylthiuram disulfide and dipentamethylenethiuram tetrasulfide. These agents give vulcanizates with superior aging properties since crosslinks consist primarily of thermally stable mono- and disulfides. The tensile strength of polyisoprene stocks cured in this manner is usually lower than that of natural rubber unless a small amount of elemental sulfur, e.g., 0.5 phr, is added

Isoprene rubber is occasionally cured with peroxides. Because of its high purity, synthetic polyisoprene can produce peroxide-cured gum stocks with glasslike transparency. Peroxide curing generally gives inferior physical properties to sulfur curing and the vulcanizates are often malodorous.

Because of the absence of naturally-occurring activators and accelerators in IR and the inherent lower heat generation during processing, isoprene rubber is less scorchy than natural rubber. This leads to less use of softeners (e.g., oil, resins) to reduce batch temperatures and less frequent necessity for retarders. If such materials are added, the same types that are used in natural rubber are employed. In comparison with SBR, IR tends toward greater scorchiness and higher levels of retarders may be required in substituting IR for SBR.

Isoprene rubbers predominantly exhibit a slightly lower modulus and hardness than natural rubber in equivalent formulations. This is due primarily to the slower rate of crystallization of the synthetic polymer under stress. In some applications the lower values are desirable. If, however, it is desired to match more nearly the properties of Hevea, a higher loading (about 10%) of reinforcing filler is used. In black stocks a change to a finer particle size, higher structure black produces the same effect. Omission or reduction of softener also serves to increase modulus and hardness. In some applications the lower hardness of IR can be turned to economic advantage, since a higher loading of inexpensive filler is possible for a given hardness.

All commercial polyisoprenes contain a stabilizer which protects the polymer from degradation during manufacturing and warehousing. This agent corresponds to the natural antioxidants which stabilize Hevea during storage and preliminary compounding. As with other elastomers, it is necessary to add an antioxidant to isoprene rubber during compounding. Since the mechanism of degradation of natural and synthetic polyisoprenes is the same, the same type and level of antioxidants used for Hevea are appropriate for IR. Secondary aryl amines are most effective, but are staining and discoloring. Hindered phenols are best for light-colored stocks.

If the finished rubber article is expected to be used in a stressed condition, it may be desirable to add an antiozonant during compounding to retard cracking. Methods of protection are the same for IR as for NR. A combination of wax and a chemical antiozonant, such as a p-phenylenediamine, is best for static ozone aging. The wax blooms to the surface and forms a protective film. Waxes are not effective in dynamic exposure to ozone because the wax film is destroyed by flexing. Aging resistance of isoprene rubber is discussed further under "Aging Properties".

The light color and high purity of synthetic polyisoprene have led to its use in many gum stock applications and light-colored filled stocks. Lighter-colored gum stock articles are possible with isoprene rubber than with even the best grades of pale crepe natural rubber. In white and colored stocks isoprene rubber requires less titanium dioxide or other pigments to achieve the desired shade. This frequently provides a significant saving in compound raw materials cost.

Although synthetic polyisoprene duplicates natural rubber in many properties and has been found to be acceptable as 100% replacement for natural rubber in many applications, it is very often used in blends with natural rubber at various ratios. The reasons for blending IR and NR are basically the same as for using any blend of polymers, i.e., to obtain a combination of properties not possible with a single polymer. Isoprene rubber manufacturers sometimes recommend starting with an IR/NR blend, for example in tires, to familiarize factory personnel with the slightly different handling characteristics of IR. At low IR levels (say, one-third or less of total polyisoprene content) usually no correction of cure rate is necessary. Use of some natural rubber in IR compounds will yield a higher modulus and hardness, if these are desired. Isoprene rubber will contribute faster processing, better mold flow, and frequently lower cost to the blend.

Synthetic polyisoprene is also blended with other general purpose elastomers, such as SBR and *cis*-1,4-polybutadiene. IR is added to SBR compounds to raise tear strength, improve hysteresis properties, increase resilience, and raise tensile strength. SBR or BR is added to isoprene rubber stocks to increase resistance to reversion on overcure or to reduce the tendency for oversoftening during repeated or prolonged processing. Polyisoprene can be used to improve the tensile and tear strengths of BR compounds.

Blends of isoprene rubber and ethylene-propylene-diene monomer (EPDM) rubbers have been found to exhibit outstanding ozone resistance (with 30 phr or more of EPDM) combined with good tack and cured adhesion not available in the EPDM polymer alone. The EPDM thus constitutes a highly effective nonstaining, nondiscoloring antiozonant for polyisoprene. Conventional sulfur accelerator cures, similar to those used in 100% IR, are used in 70/30 IR/EPDM blends. Proper covulcanization and wide selection of curatives are favored by employing recently introduced fast-curing EPDM rubbers (see Chap. 9). Thorough mixing of the two polymers is important in achieving the maximum possible ozone resistance.

Alkyl lithium-catalyzed isoprene rubber has been found to have the somewhat surprising ability to improve the ozone resistance of polychloroprene (neoprene) and chlorobutyl rubbers. Although the latter two polymers have exceptionally good ozone resistance compared to general purpose unsaturated elastomers, it is improved even further by blending 20 to 40 parts of alkyl lithium-catalyzed isoprene rubber with 80 to 60 parts polychloroprene, or 20 parts IR with 80 parts chlorobutyl. The improvement in ozone resistance imparted to the blend is believed to be due to the relatively low modulus of the polyisoprene. Ozone cracking occurs in areas under stress; for a given strain, compounds containing the IR are under a lower stress. Isoprene rubber also improves the processibility and scorch resistance of polychloroprene stocks, and reduces the cost of both polychloroprene and chlorobutyl-based compounds.

Compounding practices for injection molded isoprene rubber stocks are basically the same as for compression molded stocks. However, because of the high temperature exposure and limited cycle times in injection molding, cure systems must be chosen with somewhat more attention to scorch safety. Preferred accelerators are the delayed action type, such as N-cyclohexyl-2-benzothiazole sulfenamide or N,N-diethylthiocarbamyl sulfide, accompanied by an ultra accelerator such as a thiuram or dithiocarbamate. This combination provides good process safety and the desired fast, flat curing characteristics. Fillers which provide the best flowing characteristics for injection molding are the less reinforcing types, such as MT black or hard clay. FEF black and hydrated sodium aluminosilicate impart moderate flow properties. HAF black and finely divided silica give relatively inferior flow but can be used in some applications.

The superior flow properties which synthetic polyisoprene exhibits in injection molding are sometimes utilized to improve the performance of less processible elastomers. Blending of a small proportion of IR (e.g., 25%) with SBR, for example, gives significantly decreased cycle times and higher production rates. Blends of IR with various other rubber types also provide combinations of injection molding qualities and physical properties which are not available in any single elastomer.

VULCANIZATE PROPERTIES

Vulcanized *cis*-1,4-polyisoprene is widely used for its high resilience, high gum tensile strength, low heat build-up, and general good balance of physical properties. It presents roughly the same profile of advantages and deficiencies as vulcanized natural rubber.

Representative isoprene rubber formulations and their properties are compared with those of natural rubber in Tables 11.6 through 11.9. In general the modulus,

TABLE 11.6. PROPERTIES OF TIRE TREAD COMPOUNDS

Formulations: Polymer Type	*92% Cis Polyisoprene*[a]	*92% Cis Polyisoprene, Oil-Extended*[b]	*96% Cis Polyisoprene*[c]	*No. 1 RSS*
		Parts by Weight		
Polymer	100	100	100	100
HAF Carbon Black	50	50	50	50
Zinc Oxide	3	5	3	3
Stearic Acid	3	3	2	2
Antioxidant	1	1	1	1
Pine Tar	–	–	3	3
Sulfur	2.25	2.00	2.00	2.00
N-Cyclohexyl-2-benzothiazole Sulfenamide	0.4	0.6	0.9	0.9
Trimene Base[d]	0.3	–	–	–
Treated Urea	–	–	0.3	–
Total Parts	159.95	161.6	162.2	161.9
Properties:				
Mooney Viscosity (ML-4 at 212°F)	–	–	70.0	73.5
Mooney Scorch (MS at 250°F), min to 5 pt. rise	–	–	19.9	19.5
Optimum Cure	20′/293°F	20′/293°F	25′/287°F	25′/287°F
Tensile Strength, psi	3500	2900	3550	3950
Elongation, %	630	650	435	480
300% Modulus, psi	1310	1100	2330	2450
Hardness, Shore A	60	54	67	67
Tear Strength, pli, Die C: 73°F	330	420	360	590
212°F	250	280	310	470
Tensile Strength at 212°F, psi	2280	–	2800	3050
Yerzley Resilience, %	66	72	–	73
Goodyear Healey Rebound, %	–	–	61	60
Compression Set (22 hr at 158°F), %	–	–	17	22
Heat Build-up, Goodrich Flexometer, ΔT, °F	26	32	41	38

[a]Shell Isoprene Rubber 309, Shell Chemical Company.
[b]Shell Isoprene Rubber 500, Shell Chemical Company; contains 20% naphthenic oil.
[c]Natsyn 400, Goodyear Tire and Rubber Company.
[d]Reaction product of formaldehyde, ammonia, and ethyl chloride; Chemical Division, Uniroyal.

TABLE 11.7. PROPERTIES OF PURE GUM STOCKS

Formulations: Polymer Type	*92% Cis Polyisoprene*[a]	*96% Cis Polyisoprene*[b]	*No.1 RSS*
	Parts by Weight		
Polymer	100	100	100
Zinc Oxide	3	3	3
Stearic Acid	3	2	2
Antioxidant	1	0.5	0.5
Sulfur	1.5	1.5	1.5
N-cyclohexyl-2-benzothiazole Sulfenamide	0.3	–	–
Ridacto[c]	0.3	–	–
Hepteen Base[d]	0.1	–	–
MBTS[e]	–	0.4	0.4
DOTG[f]	–	0.6	0.4
Total Parts	109.2	108.0	107.8
Properties:			
Mooney Scorch (MS at 250°F), min to 5 pt. rise	–	19.8	15.6
Optimum Cure	15′/292°F	20′/284°F	20′/284°F
Tensile Strength, psi	3700	4300	4100
Elongation, %	910	805	740
300% Modulus, psi	160	225	250
Hardness, Shore A	35	37	38

[a]Shell Isoprene Rubber 309, Shell Chemical Company.
[b]Natsyn 400, Goodyear Tire and Rubber Company.
[c]Amine reaction product, Spencer Products Company, Inc.
[d]Heptaldehyde, aniline reaction product, Chemical Division, Uniroyal.
[e]Benzothiazyl disulfide.
[f]Di-ortho-tolyl guanidine.

hardness, and tear properties of the synthetic polyisoprenes are slightly lower, and elongation slightly higher, than observed in natural rubber. The tensile strength of IR is within about 10% that of NR; it may be lower or higher depending on compounding and processing conditions and structural purity.

The relative differences between IR and NR in vulcanizate properties are modified by level and type of filler loading. The differences in modulus, elongation, and hardness, for example, are relatively small in stocks filled with a highly reinforcing black, but somewhat larger in clay-loaded stocks (compare data in Tables 11.6 and 11.9).

Like other elastomers, the vulcanizate properties of IR are greatly influenced by molecular weight, molecular weight distribution, type and degree of dispersion of compounding ingredients and state of cure. In addition, some properties, particularly tensile strength, tear strength, and modulus, are affected by the

TABLE 11.8. PROPERTIES OF TRUCK TIRE CARCASS STOCKS

Formulations:	*Parts by Weight*	
Synthetic Polyisoprene (96% *cis* content)[a]	50	–
No. 1 RSS	50	100
FEF Carbon Black	35	35
Processing Oil	8	8
Antioxidant	1	1
Zinc Oxide	8	8
Stearic Acid	1	1
2-Benzothiazyl-N,N-diethyldithiocarbamyl Sulfide	0.7	0.7
Sulfur	2.5	2.5
Total Parts	156.2	156.2
Properties:		
Mooney Scorch (MS at 280°F), min to 5 pt. Rise	11	11
Optimum Cure	20'/293°F	20'/293°F
Tensile Strength, psi	3500	3550
Elongation, %	595	555
300% Modulus, psi	1100	1200
Hardness, Shore A	53	54
Tear Strength, pli, Die C: 73°F	220	205
212°F	150	180
Tensile Strength at 212°F, psi	1110	950
Elongation at 212°F, psi	330	280
Heat Build-up, Goodrich Flexometer, ΔT, °F	19	13

[a]Ameripol SN 600, Ameripol, Inc.

stereoregularity of the polyisoprene molecule. Because of its high percentage of *cis*-1,4 configuration, IR crystallizes on stretching. The rate of crystallization decreases somewhat with decreasing *cis* content. IR is superior to SBR and other amorphous elastomers in gum tensile and tear strength because of its self-reinforcing character.

The influence of stereoregularity on physical properties varies with state of cure and molecular weight distribution. At low crosslink densities, for example, the narrower molecular weight distribution of alkyl lithium-catalyzed polyisoprene (92% *cis*) leads to superior stress-strain and hysteresis properties. Carbon black and other reinforcing fillers tend to reduce the dependency of tensile strength and modulus on *cis* content. Differences owing to *cis* content are minimized by ensuring a thorough dispersion of cure system ingredients. At high crosslink densities the higher *cis* content synthetic and natural polyisoprenes have a higher tensile strength and modulus and are less dependent upon compounding practices for development of optimum properties.

Vulcanized synthetic polyisoprene has outstanding dynamic properties which in

TABLE 11.9. PROPERTIES OF NONBLACK STOCKS

Formulation:		*Parts by Weight*
Polymer		100
Zinc Oxide		5
Stearic Acid		2
Antioxidant		0.5
Hard Clay		75
Sulfur		3
Accelerator		Variable
	Synthetic Polyisoprene[a]	**No. 1 RSS**
N-Cyclohexyl-2-benzothiazole Sulfenamide	1.0	1.0
Treated Urea	0.5	–
Properties: (Cured 15 min at 293°F)		
Tensile Strength, psi	3650	3900
Elongation, %	670	570
300% Modulus, psi	525	1320
Hardness, Shore A	57	61

[a]Natsyn 2200, Goodyear Tire and Rubber Company.

some respects are superior to those of NR. The heat build-up of the synthetic polyisoprenes, especially the alkyl lithium-catalyzed types, is usually lower than that of Hevea. Heat build-up is more sensitive to network structure and molecular weight than structural purity. The very low level of branching and the high initial molecular weight of alkyl lithium-catalyzed IR are responsible for its lower heat build-up compared to that of 96% *cis* polyisoprene or NR. The narrower molecular weight distribution (and fewer elastically ineffective chain ends) of this polymer may also be a factor in its low heat build-up. At high crosslink densities (and high modulus) the heat build-up properties of all polyisoprenes decrease and eventually become equal (see Fig. 11.5); however, because of the loss of other properties (e.g., tensile strength) at high crosslink densities, practical use of this approach is limited.

The resilience of synthetic polyisoprene is approximately equivalent to that of NR. It is higher than that of SBR and slightly lower than that of *cis*-1,4-polybutadiene.

Isoprene rubber has been found suitable for use at high percentages as a replacement for natural rubber in the tread of truck, aircraft, and off-the-road tires. It is essentially equivalent to natural rubber in abrasion resistance, groove cracking, rib tearing, cold flex properties, and weathering resistance. The wear resistance of the 92% *cis* polyisoprene is maintained near that of natural rubber by a minor

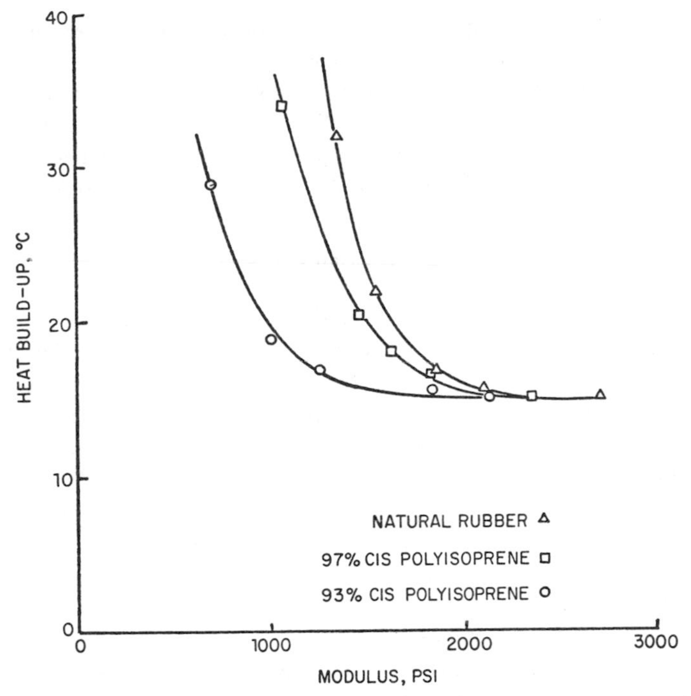

Fig. 11.5 Heat build-up vs. modulus at 300% elongation for 50 phr HAF black stocks. Determined with a Goodrich flexometer by ASTM D623-62. (From M. Bruzzone, G. Corradini, A. DeChirico, G. Giuliani, and G. Modini, "SRS 4," No. 3, December, 1969, Rubber and Technical Press, Ltd.)

increase in black level to compensate for its lower hardness. IR is not used appreciably in passenger tire treads because, like NR, its abrasion resistance is lower than that of SBR and polybutadiene. The use of IR and NR in truck treads derives chiefly from the low heat build-up properties required in large tires.

As would be expected, oil extension of synthetic polyisoprene reduces some vulcanizate properties moderately. The approximate magnitude of this reduction is illustrated in Table 11.6 by the data for oil-extended 92% *cis* polyisoprene versus that for a nonextended polymer. The reduction is most pronounced in tensile strength, modulus, and hardness. There is little or no reduction in dynamic properties or tear strength. The loss of properties is greater in gum stocks and compounds with nonreinforcing fillers than in highly reinforced stocks.

The properties of injection molded isoprene rubber vulcanizates are slightly different from those of compression molded vulcanizates. Tensile strength and elongation are usually increased while modulus and hardness are decreased by injection molding. This behavior is thought to be due to a lower state of cure achieved in injection molded specimens. The high temperatures used in this process (360°F or higher) favor the chain scission reactions which accompany vulcanization, resulting in a lower crosslink density.

Because of its great purity, synthetic polyisoprene exhibits exceptionally low water absorption and outstanding electrical properties. In one comparison, a pure gum vulcanizate of isoprene rubber absorbed less than 0.1%w water after 32 days immersion, whereas a pale crepe natural rubber gum vulcanizate absorbed 2%w. The IR stock consequently showed a substantially lower dielectric constant and dissipation factor, and a higher volume resistivity, than the NR stock.

AGING PROPERTIES

Synthetic polyisoprene contains reactive double bonds which react with oxygen and ozone in the atmosphere and thus allow gradual deterioration of useful physical properties. Other general purpose elastomers such as natural rubber, SBR, and polybutadiene share this behavior. The mechanisms of reaction of isoprene rubber with oxygen and ozone are believed to be the same as those of natural rubber. In oxidative degradation of raw synthetic polyisoprene chain scission predominates and therefore bulk viscosity decreases. SBR and BR primarily crosslink during oxidation, leading eventually to gel formation or resinification.

Oxidation is begun by formation of peroxides from oxygen and hydrocarbon free radicals. The latter can be generated by heat, light, or mechanical chain scission. The peroxides decompose and start a free radical chain reaction which results in rupture of polymer chains. Oxidative degradation is accelerated by heat, ultraviolet light, and the presence of certain metals such as iron, manganese, cobalt, and copper. It is retarded by chemical additives which decompose peroxides to nonreactive products, interrupt the chain reaction, deactivate metals, or absorb ultraviolet radiation. Combinations of different types of oxidation inhibitors sometimes function more effectively than an equal concentration of a single inhibitor. Most common antioxidants function by interrupting the free radical chain reaction. Antioxidants suitable for use in natural rubber perform similarly in isoprene rubber.

Addition of compounding ingredients and vulcanization can modify the oxidation process. Carbon black and dithiocarbamate accelerators exhibit moderate antioxidant behavior. Sulfur is deleterious to aging resistance, especially at high concentrations.

The changes in vulcanizate properties of synthetic polyisoprenes depend on aging conditions, state of cure of the compound, and efficiency of the antioxidant used. Under some aging conditions further crosslinking of vulcanizates occurs and modulus/hardness properties increase. Depending on state of cure, tensile strength may also increase and elongation decrease. If scission reactions predominate, modulus and hardness decrease. The retention of properties of properly compounded isoprene rubber after aging is equal to that of natural rubber. This indicates that deleterious catalyst residues have been removed or deactivated in commercial polyisoprenes.

The reaction of synthetic polyisoprene with atmospheric ozone severs the polymer chain and causes cracking if the rubber article is under stress. Ozonolysis is not a free radical reaction and antioxidants normally do not function as antiozonants. The most effective antiozonants are N,N′-disubstituted p-phenylene-diamines. These agents react directly with ozone and with ozonized rubber and form a surface film which apparently retards further attack. As a result, the critical energy required for cracking to occur is increased and/or the rate of crack growth is reduced. Waxes incorporated into rubber compounds protect against ozone attack under static conditions.

APPLICATIONS

Currently the tire market accounts for 60% of U.S. polyisoprene consumption. The largest share is used in the carcass and tread of truck, off-the-road, and aircraft tires, where the low hysteresis properties of IR are of advantage. Polyisoprene is employed at a lesser percentage in passenger tire carcasses.

Mechanical goods and footwear comprise most of the remainder of isoprene rubber end uses. Specific items fabricated with polyisoprene include automotive bushings and motor mounts, flat belting, gaskets, O-rings, sheet rubber, flooring, and rubber bands. Other important applications are cut thread, sporting goods, adhesives, rubber-covered rolls and battery separators. Because of its high purity and high gum tensile strength, polyisoprene is extensively used in articles employed in the medical field and in items required to be in contact with food. It is particularly suited for baby bottle nipples, milk tubing, hospital sheeting, and pharmaceutical sundries. Its high purity also makes IR attractive as a starting material for preparation of various chemical derivatives.

Oil-extended polyisoprene is used in many of the same areas as the nonextended types where maximum physical properties are not required. Largest applications consist of tire carcasses, sidewalls, shoe soles and heels, mechanical goods, and sponge.

The end use pattern of synthetic *cis*-1,4-polyisoprene not surprisingly follows that of natural rubber. IR can be expected to continue its penetration of the market for high tensile strength, low hysteresis polymers, as compounders gain experience with its use.

POLYISOPRENE LATEX

A general purpose synthetic polyisoprene latex was introduced in 1962 by Shell Chemical Company. This product, Shell Isoprene Latex 700, is in the same class of latex as Hevea, being a *cis*-polyisoprene latex of high polymer molecular weight, large particle size, broad particle size distribution, high solids content and low nonpolymer solids content (see Table 11.10). As in NR latex, the high stereoregularity of the polymer in polyisoprene latex imparts to it a high wet gel strength and high vulcanizate tensile strength. The synthetic latex has a range of applications similar to natural rubber latex.

Synthetic polyisoprene latex is different from natural rubber latex in being

TABLE 11.10. COMPARATIVE PROPERTIES OF POLYISOPRENE LATICES

	Polyisoprene Latex[a]	*Centrifuged Hevea Latex*
Total Solids, wt%	65.5	64.5
pH	10.5	10.5
Brookfield Viscosity, cp (No. 3 spindle at 20 rpm)	200	500
Average Particle Size, microns	0.75	0.6
Surface Tension, dynes/cm	41.0	46.0
Specific Gravity	0.94	0.95
Mechanical Stability, sec	> 1500	900
Odor	Mild	Ammoniacal
Color	White	White

[a]Shell Isoprene Latex 700, Shell Chemical Company.

gel-free and in the simplicity of its colloidal system. The polymer particles in polyisoprene latex are held in suspension with a low level of anionic stabilizer. Natural rubber latex contains a complex mixture of proteinaceous and fatty materials which stabilize the particles, and have antioxidant and cure-activating effects. These differences make it necessary to prepare and compound the synthetic latex somewhat differently from natural rubber latex. However, properly compounded synthetic polyisoprene latex exhibits vulcanizate properties quite similar to those of natural rubber latex.

As in the case of solid synthetic polyisoprene, the synthetic latex is more uniform from lot to lot in processing characteristics and physical properties than the natural latex. Good mechanical stability and light color are other advantages. No de-ammoniation step is required in processing the synthetic latex, since it does not contain ammonia.

Prior to compounding polyisoprene latex, it is necessary to add additional stabilizer to maintain the suspension. Fatty acid soaps, alkyl aryl sulfonates, and most other anionic stabilizers are suitable for this purpose. A small amount of potassium hydroxide may also be added to enhance mechanical stability. Ammonium caseinate, which is commonly added to natural latex before compounding, is not required with the synthetic latex.

Since polyisoprene latex as manufactured does not contain an antioxidant, the amount of antioxidant added during compounding is slightly higher than normal. Substituted phenolic antioxidants perform satisfactorily in polyisoprene latex. A wax is added if ozone resistance is important.

In formulating a cure system for polyisoprene latex a fatty acid, added in the form of a soap during compounding, is essential for activation and a secondary accelerator is recommended to enhance the relatively slow cure rate. An accelerator containing an amino group in combination with a thiazole, sulfenamide, or dithiocarbamate provides an efficient system.

Zinc oxide is required for proper vulcanization of polyisoprene latex. In foam about 3.0 phr is recommended; in dipped goods, 0.25-0.50 phr is required. Zinc carbonate can be substituted if transparency is desired in dipped films.

Sulfur level requirement is similar to that of Hevea latex. Low levels provide maximum tear strength and aging resistance.

When compounded with fillers, such as clays, carbon black, and silicates, the normal loss in physical properties attendant with filler loading occurs. Fillers are usually added in slurry form to preserve compound stability.

In foam rubber applications polyisoprene latex is an excellent blending agent. In blends with SBR foam, it produces good mold stripping characteristics, high resilience, and good resistance to cigarette burning (required in mattress and other cushion applications). In blends with Hevea latex, it reduces compression-deflection properties and thus imparts the soft feel desired in some applications.

TABLE 11.11. POLYISOPRENE LATEX IN DIPPED FILMS

Formulations:	*General Purpose*	*High Modulus*
	Parts by Weight	
Polyisoprene Latex[a]	100.0	100.0
Anionic Stabilizer[b]	0.5	0.5
Dibutyl Ammonium Oleate	0.75	0.75
Antioxidant[c]	2.0	2.0
Zinc Oxide	0.5	0.5
Sulfur	1.5	1.5
Zinc Dibenzyldithiocarbamate	0.5	–
Zinc Dimethyldithiocarbamate	–	0.75
Hepteen Base[d]	0.25	0.25
Properties:		
Cure Conditions	30′/212°F	30′/248°F
Tensile Strength, psi	5100	5000
Elongation, %	1050	890
700% Modulus, psi	550	1700
Aged Properties:		
(After aging in oxygen bomb, 300 psi at 158°F for 96 hours)		
Tensile Strength, psi	4560	4050
Elongation, %	990	780
700% Modulus, psi	800	2950

[a]Shell Isoprene Latex 700, Shell Chemical Company.
[b]Nekal BA-75, Antara Chemical Company.
[c]2,2′-Methylenebis(4-ethyl-6-tert-butylphenol), such as Plastanox 425, American Cyanamid Company.
[d]Accelerator, heptaldehyde-aniline condensation product, Chemical Division, Uniroyal.

Polyisoprene latex is an attractive material for dipped goods because of its high wet gel strength, high vulcanizate tensile strength and elongation, and high sensitivity to coagulants. Unlike other synthetic latices, polyisoprene latex has a large average particle size which gives it a deposition rate equivalent to that of Hevea latex. Representative compound formulas for dipped films are shown in Table 11.11.

In addition to dipped goods and foam rubber, polyisoprene latex is used in adhesives, extruded thread, and carpet backing. In the adhesives field the initial tack or "quick grab" characteristics of polyisoprene latex are of significant advantage.

TRANS-1,4-POLYISOPRENE

Synthetic *trans*-1,4-polyisoprene is polymerized using a coordination catalyst based on trialkyl aluminum/titanium tetrachloride combined with either ferric or vanadium trichloride. This system yields a polymer with 97% *trans*-1,4 microstructure, with the remainder being the *cis*-1,4 isomer. Synthetic *trans*-1,4-polyisoprene exists at room temperature as a hard, crystalline material with high unvulcanized strength. There are two crystal forms in the unstressed polymer and a third in stressed specimens. X-ray diffraction indicates there is no difference in the crystal structures of natural and synthetic balatas. The synthetic polymer melts at about 140°F. The glass transition temperature of synthetic *trans*-1,4-polyisoprene is approximately −76°F.

Trans-1,4-polyisoprene can be processed in conventional rubber fabricating equipment above its melting temperature. It has a Mooney viscosity similar to that of natural balata, about 30 ML-4 at 212°F, and has essentially equivalent processing characteristics. It can be compression molded, injection molded, extruded, and calendered. It may be compounded with fillers and blended with other polymers. However, addition of large quantities of these materials reduces the ability of *trans*-polyisoprene to crystallize and therefore decreases properties which depend on crystallinity, such as tensile strength and hardness. The polymer is usually vulcanized with accelerated sulfur systems to improve strength at high temperatures and impart solvent resistance. Vulcanization tends to lower tensile strength and extensibility. Synthetic *trans*-1,4-polyisoprene is equivalent to its natural counterpart in raw and vulcanizate physical properties.

In contrast to the *cis* isomer, *trans*-1,4-polyisoprene is highly resistant to ozone attack. It is also resistant to attack by some concentrated acids, alkalis, oils, and fats. It reacts with concentrated nitric and sulfuric acids. Like *cis*-polyisoprene, the *trans* isomer is subject to oxidation from atmospheric oxygen and must be protected by addition of an antioxidant.

The largest application for *trans*-polyisoprene is in golf ball covers, in which it is a direct replacement for natural balata. Additional areas in which it may find important uses are adhesives, sealants, caulking compounds, and coatings.

OTHER POLYMERS OF ISOPRENE

The best known copolymer of isoprene is butyl rubber. Isoprene is present in butyl at 1.4 to 4.5% to provide unsaturation and facilitate vulcanization by sulfur. Butyl rubber is discussed in detail in Chap. 10.

Isoprene undergoes polymerization with styrene in emulsion systems in reactions which resemble those of butadiene and styrene. Although the isoprene-styrene copolymers excel the butadiene-styrene copolymers in hysteresis properties, they are inferior in abrasion resistance and low temperature properties and have never achieved commercial importance.

Recently the development of anionic polymerization techniques has made possible the preparation of isoprene-styrene polymers with unique properties. Block polymers of the styrene-diene-styrene type exhibit rubbery elasticity without chemical crosslinking. The polystyrene end blocks associate in domains which act as physical crosslinks below the glass-transition temperature of polystyrene. A series of commercial styrene-diene-styrene block polymers have been introduced by Shell Chemical Company under the KRATON® Thermoplastic Rubber trademark. These products have rubberlike properties without vulcanization and process like thermoplastics. One of them, KRATON 1107 Thermoplastic Rubber, is a styrene-isoprene-styrene block polymer which has outstanding properties for use in adhesives. Elastomeric block polymers are discussed further in Chap. 20.

An emulsion-polymerized copolymer of isoprene and acrylonitrile, Krynac 833, is produced commercially by Polymer Corporation, Ltd., of Canada. This product is an oil-resistant rubber having an isoprene/acrylonitrile ratio of approximately 66/34 and a Mooney viscosity (ML-4 at 212°F) of 70. A similar polymer is available in latex form. The latex, Polysar 763, is used for oil-resistant dipped goods and thread.

REFERENCES

(1) W. H. Beattie and C. Booth, *J. Appl. Poly. Sci.* **(A2)**, **4**, 663 (1966).
(2) L. A. Wood, *Rubber Chem. & Technol.*, **39**, 132 (1966).
(3) J. Brandrup and E. H. Immergut, eds., *Polymer Handbook*, Interscience: John Wiley, New York (1966), p. IV–58ff.

12

NITRILE AND POLYACRYLATE RUBBERS

JOHN P. MORRILL
Technical Service Manager, Elastomers
B. F. Goodrich Chemical Co.
Cleveland, Ohio

INTRODUCTION

The polymers of the nitrile and polyacrylate-types are part of a larger classification of products often referred to as "special-purpose rubbers." What are special-purpose rubbers? They usually are considered to fit one or more of the following requirements: (1) they must be vulcanizable elastomers; (2) they are used primarily for applications other than tires and tubes; (3) they perform some special function that natural and SBR* type rubbers will not perform.

The first rubber chemists had to work only one basic product – natural rubber. Their work appeared easier and yet, at the same time, more difficult than ours. It was easier because there was only so much one could do with natural rubber, and their compounds were limited to this one basic polymer. The compounders did not have to be familiar with all of the present rubbers that are now taken for granted, nor were they faced with the necessity of compounding to meet the requirements of the many specifications which are in general use today. Their work was more difficult because of the chemical and physical limitations of natural rubber. The obvious need for a serviceable product for certain applications for which natural rubber was not suited led to the search for new polymers.

NITRILE RUBBER

What special function is performed by nitrile rubber? It is of interest primarily because it exhibits a high degree of resistance to attack by oils both at normal and elevated temperatures.

History. Before proceeding to a general discussion of the characteristics of the special-purpose oil-resistant rubbers, it may be well to review briefly their history.

*Styrene-butadiene rubber (GR-S or SBR)

For many years after the commercialization of vulcanized rubber products, chemists realized the need for special-purpose oil-resistant rubbers. Research and development started as early as 1910, but it was not productive until 1920 when Patrick, while attempting to prepare ethylene glycol, found that the reaction product of ethylene dichloride and sodium polysulfide yielded a rubbery type product. It was introduced to the trade by Thiokol Corporation in 1929–1930. This was the first truly oil- and solvent-resistant polymer to be commercialized.

The first literature reference to nitrile rubbers is found in a French patent issued in 1931, covering the polymerization of butadiene and acrylonitrile. However, it was not manufactured or commercialized until 1935. I. G. Farbenindustrie, in Germany, was the first to make and sell a reaction product of butadiene and acrylonitrile. It was known as "Perbunan."

However, as early as 1936, considerable research and development was initiated by the various larger rubber companies in the United States, toward the production of this type polymer. In January, 1939, The B. F. Goodrich Company produced the first American-made nitrile rubber. Production was at a rate of 250 pounds per day. On July 31, 1940, The B. F. Goodrich Company joined with Phillips Petroleum Company to form the Hydrocarbon Chemical and Rubber Company to manufacture and sell nitrile rubber under the trade name, "Hycar." During the same year, Goodyear Tire and Rubber Company built a production plant to manufacture a similar type of product under the trade name, "Chemigum." Likewise, Firestone, at about the same time, introduced its nitrile-type rubber and named it "Butaprene." The Standard Oil Company, which had worked very closely with I. G. Farbenindustrie, made available to the American trade a nitrile polymer under the same trade name as that in Germany, "Perbunan."

Thus, within a very short time, there were four manufacturers of nitrile rubber polymers in the United States. The outstanding oil and solvent resistance of the nitrile rubbers had just started to create an expanding industrial market when the United States entered the war in December, 1941. The properties of nitrile rubber were essential for bullet-sealing tanks, fuel hose, and many other military applications. From that time to the end of the war in August 1945, all nitrile rubber production was placed on strict allocation for military uses only. During the war, nitrile rubbers were among the most critical of all elastomeric materials, and although produced at a rate of over 2,500,000 pounds per month, no material was permitted for commercial development during this period. At the end of the war, the military demands ceased, leaving the suppliers of nitrile rubber with a large postwar inventory and few commercial applications.

In November of 1945, the Phillips Petroleum Company sold its interest in the Hydrocarbon Chemical and Rubber Company to The B. F. Goodrich Company, which became the sole owner of the company and turned over the operations to B. F. Goodrich Chemical Company. In January of 1950, Standard Oil of New Jersey sold its manufacturing plant at Baton Rouge to the Naugatuck Division of United States Rubber Company. With that historical background on special-purpose oil-resistant rubber, we now take a closer look at the nitrile polymers.

What is a Nitrile Rubber? A nitrile rubber may be broadly defined as a copolymer of a diene and an unsaturated nitrile. This portion of the chapter will be limited to copolymers of butadiene and acrylonitrile, since the majority of nitrile rubbers which are available today are made by copolymerizing these two monomers.

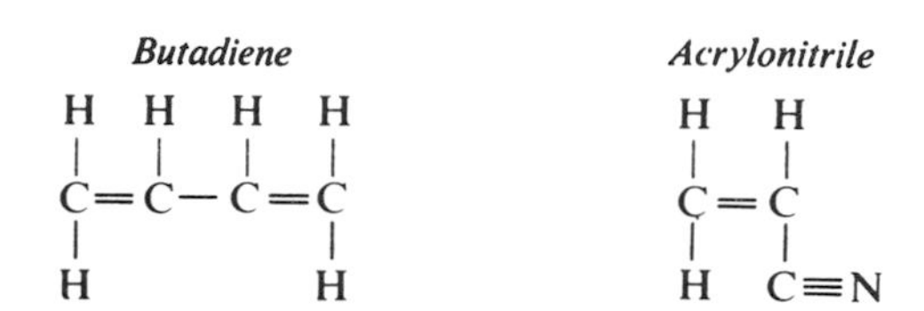

Acrylonitrile may be considered as a 2-carbon chain with a double bond in which a cyanide group, CN, has been substituted for one hydrogen atom. Thus acrylonitrile is a vinyl monomer, sometimes called "vinyl cyanide." Since the molecular weights of these two monomers are nearly equal, their proportions by weight are practically the same as their molar proportions. The statement that a nitrile rubber has a butadiene-acrylonitrile ratio of 2 to 1, indicates proportions either by weight or by number of molecules.

Oil resistance is the most important property of nitrile rubbers and is the reason for their extensive use even when they cost substantially more than natural rubber or general-purpose synthetic rubbers. Oil resistance refers to the ability of the vulcanized product to retain its original physical properties, such as modulus, tensile strength, abrasion resistance, and dimensions, while in contact with oils and fuels.

Suppliers

At the present time there are five major producers of nitrile rubbers in the United States and eight outside the United States. These suppliers and their corresponding trade names of their nitrile rubber are given below:

Supplier	Trade name
Copolymer Rubber and Chemical Corp.	Nysyn
Firestone Synthetic Rubber and Latex Co.	FR-N
B. F. Goodrich Chemical Co., A Division of The B. F. Goodrich Company	Hycar
Goodyear Tire & Rubber Co.	Chemigum
Uniroyal Chemical Division of Uniroyal, Inc.	Paracril
BP Chemicals Ltd. Plastics Dept. (U.K.)	Breon
N. V. Chemische Industrie (Netherlands) Aku-Goodrich	Hycar
Doverstrand Limited (U.K.)	Butakon
Farbenfabriken Bayer AG (Germany)	Perbunan-N
Japan Synthetic Rubber Co. Ltd. (Japan)	JSR
The Japanese Geon Co., Ltd. (Japan)	Nipol
Plastimer (France)	Butacril, Ugitex-N
Polymer Corporation Limited (Canada)	Krynac

Manufacture of Nitrile Rubbers. The basic steps involved in the manufacture of a dry rubber are polymerization, coagulation, washing, and drying. In case a latex is desired, the steps are polymerization, stabilization, and generally, concentration. In either case, the reactor or polymerizer is a jacketed water-cooled vessel capable of withstanding up to 150 psi pressure. The size may vary from 100 to 3700-gallon capacity. This vessel must be equipped with an efficient means of agitation, and in the case of cold polymerization, refrigeration must be utilized. Various inlets for introduction of the raw materials are provided for as well as means for withdrawing samples during the reaction. Various control instruments are necessary – in particular, temperature and pressure recorders.

Following the polymerization cycle, the charge is transferred to a blowdown tank, at which point the shortstop and antioxidant are added. The pH is also usually adjusted to the proper level at this stage. Procedure to this point is the same for both dry rubber and latex.

In the case of dry rubber production, the next step is the transfer of the latex to coagulating tanks, at which point the latex is coagulated into fine crumbs by the addition of various salts and acids. This slurry of rubber particles is then generally extracted with a caustic solution, washed, de-watered, and finally dried.

In the production of latex, it is usually necessary to add additional stabilizing agents in the blowdown tank prior to the stripping operations which remove the excess butadiene and acrylonitrile. The final step is concentration or removal of water to the desired final total solids.

The chemistry involved in emulsion polymerization is rather complicated and, for a more detailed discussion, the reader is referred to the discussion in Chap. 7.

Grades and Uses. Nitrile rubber is available in several grades of oil resistance based on the acrylonitrile content of the polymer ranging from about 50 to 18%. Generally, the grades are referred to as very high or ultra high, high acrylonitrile content, medium high, medium, medium low, and low. Various properties of nitrile polymers are directly related to the proportion of acrylonitrile in the rubber. The more important property trends which are influenced by acrylonitrile content are listed in Table 12.1. The many nitrile polymers available give the rubbber

TABLE 12.1. PROPERTY TRENDS INFLUENCED BY % VCN IN THE COPOLYMER

	% Acrylonitrile 18 ⟷ 50
Oil Resistance Improves	⟶
Fuel Resistance Increases	⟶
Tensile Strength Increases	⟶
Hardness Increases	⟶
Abrasion Resistance Improves	⟶
Gas Impermeability Increases	⟶
Heat Resistance Increases	⟶
Low Temp. Resistance Increases	⟵
Resilience Increases	⟵
Plasticizer Compatibility Increases	⟵

compounder a wide choice from which to select the optimum combination of properties desired for any given application.

The very high and high acrylonitrile polymers are used where the utmost in oil resistance is required such as oil well parts, fuel cell liners, fuel hose, and other applications requiring resistance to aromatic fuels, oils, and solvents. The medium grades are used in applications where the oil is of lower aromatic content or where greater swelling of the rubber is tolerable. The low and medium low acrylonitrile grades are used in cases where low temperature flexibility is of greater importance than oil resistance.

The nitrile rubbers are also available in various physical forms: sheet, slab, crumb, powder, and liquid.

The sheet or slab is by far the most widely used. The other forms are designed for more specific uses. The crumb is desirable for the preparation of cements by direct solution in a solvent or modification of vinyl, phenolic resins, or other plastics. The powder is ideal for preparing mixes with PVC and phenolic resins or in binder systems where finely divided polymer can be put to advantage. Both crosslinked and noncrosslinked versions of both the crumb and powder are available. The liquid form enables the compounder to soften a nitrile stock without sacrificing oil resistance. The liquid is also useful in providing good processibility and building tack to nitrile rubber stocks.

Oil Resistance of Nitrile Rubbers

Effect of Polymer Structure. The oil resistance of a nitrile rubber vulcanizate is determined by two primary factors: first, the proportion of acrylonitrile in a copolymer and, second, the chemical nature of the oil or fuel in question. Other factors which may have a bearing on the oil resistance of a vulcanizate are softeners, pigments, state of cure, degree of crosslinking, polymerization temperature, and homogeneity of the polymer. Volume change after immersion in test oils and fuels is used as the measure of oil resistance, because this measurement can be made accurately in the laboratory and has been proven by experience to be very useful in predicting the serviceability of commercial products.

There is a rather specific relationship between the chemical composition of a rubber and the chemical composition of the particular immersion fluid being considered. The closer these two compositions are to each other, the more effect there is of the fluid on the polymer. For instance, natural rubber and SBR, both hydrocarbon rubbers, are badly swollen and disintegrated by hydrocarbon fluids like gasoline and kerosene. The polar polymers are virtually unaffected by hydrocarbons. Conversely, natural rubber and SBR, two nonpolar polymers, are unaffected by strongly polar solvents such as acetone and methyl ethyl ketone, while the polar rubbers will swell very badly in these fluids.

Accordingly, it is extremely important that design engineers take into consideration the type of fluid that is going to be used in a particular service and specify the rubber which is the least affected by that fluid. In general, the nitrile types are considerably more resistant to a very much wider variety of fuels and lubricants than are any of the other rubbers. Aniline, a polar aromatic liquid, swells nitrile rubbers much more than any of the other rubbers except Thiokol, with which it appears to react chemically. This explains why the aniline point of an oil, which is a measure of the solubility of the oil in aniline, correlates with the swelling effect of an oil on nitrile rubber. An oil with a *lower* aniline point is *more polar* or aromatic and therefore swells nitrile rubber to a greater extent.

Test oils and fuels which are commonly used and their corresponding aniline points are given below:

	A.S.T.M. Oils			Fuels	
	No. 1	No. 2	No. 3	A	B
Aniline point, °C	124	93	70	45	0

Thus, A.S.T.M. Oil No. 3, having a low aniline point, will exhibit a greater swelling effect on nitrile rubber than will the high aniline point A.S.T.M. Oil No. 1. Immersions are usually run according to A.S.T.M. Procedure D471-57T, with the immersions in oil being for 70 hours at 100°C and 157°C and in the fuels for 70 hours at room temperature.

Compounding. Since oil resistance is the outstanding characteristic of nitrile rubbers, it is important that the fundamental factors affecting oil resistance be examined. These factors are discussed very briefly here and explored in more detail in the section on building an oil-resistant compound. The following very general statements can be made.

Carbon blacks show very little difference in their effect on the swelling of the vulcanizates. Softeners and plasticizers have a pronounced effect. Some softeners might be extracted by the immersion media and result in shrinkage. Other types of plasticizers may experience very little, if any, extraction, and the overall effect of immersion may be a slight swelling of the rubber part.

Nitrile rubber has poor compatibility with natural rubber, but it can be blended successfully in all proportions with butadiene-styrene rubber. This reduces the overall oil resistance of the part, but it is sometimes used as a technique to counteract shrinkage when a tight seal is desired for an oil with a high aniline point.

In general, state of cure has very little effect on oil resistance of the high acrylonitrile content polymers. The less oil-resistant polymers (low acrylonitrile content) are more sensitive to state of cure, and a badly undercured part will show a much higher degree of swell than one which has had an optimum cure.

Factors Inherent in the Raw Polymer. Until now, only those factors controlled by the compounder of a nitrile rubber have been considered. Some factors are controlled by the manufacturer of the nitrile rubber.

Two properties which have no appreciable effect on the oil resistance of a nitrile rubber are the degree of crosslinking, or "gel content," and the temperature of polymerization. Table 12.2 shows the volume changes in the three A.S.T.M. oils of a standard compound without softener, based on three different nitrile rubbers, each having the same 2:1 ratio of butadiene to acrylonitrile. Polymer "H" is a

TABLE 12.2. OIL SWELL OF NITRILE POLYMERS PRODUCED USING THREE DIFFERENT METHODS

Compound			*% Volume Change 70 Hours at 100°C A.S.T.M. Oil No.*		
Polymer	100.0				
Zinc Oxide	5.0		1	2	3
FEF Black	40.0				
TMTD	3.5	H	−2	+6	+15
Stearic Acid	1.0	C	−1	+6	+19
	149.5	S	−2	+6	+17

regular easy-processing type with some crosslinking in the raw polymer chain. Polymer "C" is polymerized at low temperature, using refrigeration equipment. Polymer "S" is a soluble type designed to have a minimum amount of crosslinking for convenience of processing and in the manufacture of cements. Their swelling characteristics are almost identical.

The resistance to fuels and oils is influenced by the composition of the polymer. The higher the acrylonitrile content, the lower the swell in oil. Low temperature flexibility presents the opposite picture. As acrylonitrile content increases, low-temperature flexibility becomes poorer. In other words, improved oil resistance is generally obtained at some sacrifice in low temperature flexibility. Table 12.1 depicts these trends as well as several others which are experienced in employing polymers having various ratios of acrylonitrile and butadiene. Various compounding techniques may be used to counteract, at least partially, these trends.

Compounding

On the basis of the information presented, how can a nitrile rubber compound be prepared to meet specification requirements? Basically, nitrile rubbers are compounded in much the same way as crude rubber or SBR. No radical changes in basic ideas or practices are necessary. However, there are a few important differences, as compared with natural rubber, which should be kept in mind. Failure to achieve uniform results may frequently be traced to one of these factors, which will be discussed in more detail as we build a nitrile rubber compound.

Polymer. What rubber should be used? If maximum oil resistance is specified, a high acrylonitrile polymer is indicated. If flexibility at very low temperatures is a requirement, it will be necessary to use a low-acrylonitrile polymer. A specification requiring no shrinkage of the rubber following oil immersion may necessitate the use of small quantities of SBR blended with the nitrile rubber.

The nitrile rubber is less plastic than natural rubber and will develop more heat when mixed under similar conditions. Preliminary breakdown of nitrile rubber is more important than that of natural rubber. The most efficient method requires breakdown on a cold tight mill for 5 to 10 minutes. If in a Banbury, it should run as cool as possible.

Activation. Following breakdown, zinc oxide, usually 5 parts, is added for proper activation.

Sulfur. Activation is followed by the addition of sulfur. The higher the acrylonitrile content of the polymer, the less sulfur is generally required for proper cure. One to two parts is usually sufficient. Solubility of sulfur in nitrile rubber is much lower than in natural rubber. Good sulfur dispersion is essential, and the sulfur is added at this point to allow maximum time for dispersing in the rubber. Sulfur masterbatch is often used.

Pigment. Nitrile rubber does not crystallize when stretched to its ultimate elongation and consequently is not self-reinforced. Proper choice of loading pigments is particularly important in nitrile rubber formulations because most of the reinforcement and tensile strength are achieved by these pigments. Channel and reinforcing furnace blacks are generally used to give optimum tensile strength, abrasion, and tear resistance. Thermal and semi-reinforcing furnace blacks permit higher loadings for equivalent hardness stocks, thereby obtaining cost savings in the compounds.

The light-colored pigments which give varying degrees of reinforcement to the vulcanizate include treated carbonates, clay, and silicates. They, of course, must be used when white or light-colored stocks are required. Inert pigments which provide little reinforcement include whitings, barytes, talc, diatomaceous earth, and slate flours. They are used to improve processing, and provide certain specific properties, as well as to reduce the cost of the compound.

Channel blacks and clay retard cure, and it is usually necessary to adjust the sulfur-accelerator combination when they are used. The first portions of the pigment are added slowly and well dispersed. Later portions may then be added more rapidly.

Softeners. The primary functions of softeners or plasticizers in Hycar formulations are to improve mixing and processing and to decrease the hardness and nerve of the compounds. In addition, the particular properties required of the finished part in service are correspondingly influenced by the choice of softener.

The high degree of oil resistance of nitrile rubber limits the selection to certain types of softeners. Ester types, aromatic oils, and polar-type derivatives are generally suggested. Two of the major factors affecting the compatibility of ester type softeners in nitrile rubber are the type of softener and the acrylonitrile content of the rubber. The degree of compatibility decreases as the acrylonitrile content increases. A softener which can be used at 30 PHR in a low-acrylonitrile content rubber may bleed if added in the same amount to a high-acrylonitrile rubber. Such a softener may be satisfactory, however, if used in combination with a second, more compatible plasticizer.

In compounding for good low-temperature properties, best results are usually obtained with a combination of two or three ester-type softeners. Certain types, however, are more strongly recommended than others. The adipates, sebacates, tributoxyethyl phosphate monomeric fatty acid ester (Synthetics L-1, Hercules Powder Co.), methyl acetyl ricinoleate, di(butoxy-ethoxyethyl) formal (TP-90B,

Thiokol Chemical Corp.), di(butoxyethoxyethyl) adipate (TP95, Thiokol Chemical Corp.), and triglycol ester of vegetable oil fatty acid (Plasticizer SC, Harwick Standard Chemical Co.) are specifically recommended for low-temperature properties. Thirty parts of softener with 100 parts of rubber polymer is the practical maximum for low-temperature service. Larger amounts have little additional effect and may result in bleeding.

Unfortunately, the ester-type softeners are generally extracted when the part is in contact with certain fuels and oils. At times, this extraction is not critical. However, it will be objectionable where shrinkage of the part cannot be tolerated and where contamination of the immersing medium is undesirable. It may be necessary to blend relatively nonextractable softeners such as the polymeric-polyesters or coumarone-indene resins with ester-type softeners in order to obtain the desired balance of low-temperature flexibility and nonextractability. Extraction or shrinkage may also be compensated by the addition of an oil-swelling polymer such as SBR. The amount should be controlled so that the swelling counteracts the shrinkage caused by the extraction of the softener. The over-all oil resistance of such blends is, of course, poorer than for a straight nitrile rubber.

When an application calls for the best possible heat resistance, the choice of softener is again of major importance. Coumarone-indene resins and polyester-type plasticizers are suggested. Other compounding ingredients, such as the accelerator and the vulcanizing agent, must also be selected carefully.

Nitrile rubbers are inherently low-tack rubbers. Thus where improvement in tack is required the following plasticizers are recommended: Coumarone-indene resins, rosins, modified phenolics, tetrahydronaphthalene, and low-molecular weight monomeric esters such as dibutyl phthalate and dibutyl sebacate.

Acceleration. A study of the physical properties, particularly compression set, is the best method of determining the state of cure and for choosing the type of acceleration. The use of tetramethylthiuram monosulfide with sulfur or tetramethylthiuram disulfide (no sulfur) generally produces vulcanizates with the lowest compression set characteristics. A tetramethylthiuram monosulfide-sulfur cure is an excellent general-purpose system. It produces vulcanizates which process safely and do not bloom.

Because of the absence of free sulfur, the tetramethylthiuram disulfide cure is ideal for applications which must withstand extended heat-aging. A very small amount of free sulfur (0.1 to 0.2 part added sulfur per 100 rubber polymer) will improve the state of cure without reducing the heat resistance excessively. Another widely used general-purpose cure system is 1.5 MBTS–1.5 sulfur. For improved aging, a reduction in sulfur and increase in MBTS to 3 MBTS–0.5 sulfur is recommended. For a faster cure, substitution of the MBTS with MBT or a 1.5 MBT–1.5 sulfur system is suggested while the system 1.5 MBT–1.5 Zn DMB–1.5 sulfur is still faster. A lower cost but still very satisfactory system is 0.4 TMTM–1.5 sulfur.

A truly nonsulfur cure of nitrile rubber can be accomplished by that of peroxides as the curing agent. Dicumyl peroxide (DiCup, Hercules Powder Co.), lead peroxide, and 2,5-bis (tert. butylperoxy)-2,5 dimethylhexane (Varox, R. T. Vanderbilt Co., Inc.) all produce satisfactory vulcanizates. Peroxide-cured nitrile rubber produces excellent aging, shiny, bloom-free, low compression set parts which, because of the complete lack of sulfur, do not stain silver nor discolor in the presence of lead. Peroxide has other advantages since it can cure blends of nitrile rubber and other polymers that would not ordinarily cure with the same system.

The most recent addition to nitrile cure systems, the cadmium/magnesium cure, provides a major improvement in heat resistance. With this system, nitrile polymers can be compounded to give excellent heat-resistance at 300°F. The basic composition of the cadmium/magnesium cure consists of five ingredients. The first, cadmium oxide, has a function similar to that of zinc oxide. Because of its higher atomic weight, cadmium has a greater activating power than zinc. For the best heat-resistance, cadmium oxide should be used at 5.0 phr. Sulfur is the second ingredient. Its concentration can be varied from 0 to 1.0 phr, depending on the desired elongation, scorch time, and heat resistance. The next ingredient is magnesium oxide. Maglite D (Merck & Co., Inc.) and Stan Mag 100 (Harwick Standard Chem. Co.) are the most efficient of the various grades. The level of magnesium oxide is fixed at 5.0 phr. The fourth, at 2.5 phr, is cadmium diethyldithiocarbamate (Cadmate – R. T. Vanderbilt). This ultra-accelerator is unique since it provides stable crosslinks for protection of unsaturated rubbers at elevated temperatures. Finally, MBTS acts as a delayed-action accelerator and is most useful at 1.0 phr.

For the greatest efficiency, this system should include an aromatic amine such as AgeRite Stalite (R. T. Vanderbilt). Ideally, the amine should be added to the latex after polymerization. This requirement can also be met by adding 2.5 phr of AgeRite Stalite S (R. T. Vanderbilt) during mixing, when a polymer not stabilized with Stalite is used.

The cadmium/magnesium cure system has many advantages for compounding nitrile polymers. The most important is heat resistance, but properties such as excellent compression-set, fast cure-rate, and shelf stability are achieved. The versatility of this cure system is unmatched for nonblack compounds, injection-molding compounds, and even SBR/Nitrile blends. While some sulfur is normally used, nonsulfur cures can be obtained by using high levels of cadmium diethyldithiocarbamate. Rubber parts which will be exposed to oil at 300°F can be made from compounds cured with the cadmium/magnesium system. Due to the tightness of this cure, initial elongations are generally lower than other systems. However, reducing the free sulfur by as little as 0.1 phr will increase the elongation substantially.

Low-temperature flexibility and resistance to oil or other immersion media are generally unaffected by the acceleration. When these are the determining factors, any of the commonly used accelerators are satisfactory. Since resistance to flexing failure is improved by slight undercure and seriously impaired by overcure, the choice of acceleration has some importance in applications where flex failure is encountered. For best flex life, a relatively slow acceleration and careful control of time and temperature during vulcanization are advisable.

Most nitrile rubber formulations exhibit a comparatively level cure plateau, and the amount of sulfur and accelerator is generally not critical. However, when ultra-accelerators or overacceleration are used to reduce cure time, the conventional precautions should be taken to avoid precure during mixing, processing, and bin storage.

Antioxidants & Stabilizers

Nitrile rubbers degrade by crosslinking, giving boardy and brittle stocks having poor flexibility and elongation. The uncompounded nitrile polymers are protected with antioxidants for many heat resistant applications. However, if service conditions are expected to be extremely severe, it is advisable to use one or two parts of a nonvolatile antioxidant such as AgeRite Resin D (R. T. Vanderbilt) for additional protection.

Nitrile rubber is quite susceptible to ozone attack, thus it is sometimes necessary to provide added protection in the preparation of the vulcanizate. This can be accomplished by the addition of 1 to 3 parts of an antiozonant such as Antozite 67 (R. T. Vanderbilt) and 1 to 3 parts of a wax to function as a migratory aid. Nitrile rubbers are also protected from ozone attack by combining them with PVC resin. Several polyblends of prefluxed nitrile rubber and PVC resins in various proportions with and without plasticizer modification are available.

Radiation attack on nitrile rubber causes a loss of tensile strength and increases in modulus. Although nitrile rubber is inherently fairly resistant to radiation attack, it can be further protected by the incorporation of an "antirad" during compounding. The most effective "antirad" of those studied is N-cyclohexyl-N-phenyl-p-phenylenediamine (Flexzone 6H, Uniroyal, Inc.).

Lubricants. Usually one part of stearic acid is added, both for activation and processing.

Recommended batch size for satisfactory mill mixing is as follows:

Length of roll (in.)	48	60	84
Batch weight (lb)	30–50	45–75	110–160

Much more time could be spent on building compounds and discussing the properties obtained; but if the general principles just set forth are kept in mind, no great difficulties will be encountered. In general, compounding principles which apply to SBR will also apply to nitrile rubbers. Each supplier of nitrile rubbers has available a great deal of specific compounding information which is readily available. Pigmentation studies, softener studies, acceleration studies, etc., have been made on all commercially available polymers. In addition, all suppliers have done a great deal of laboratory work on compounding nitrile rubbers to meet various automotive, military, A.M.S., and A.S.T.M. specifications and information is available on request.

Carboxy Nitrile Rubber. A unique product in the oil-resistant polymer field is the high-strength carboxylic-modified nitrile rubbers. This type of polymer contains in addition to acrylonitrile and butadiene, one or more acrylic-type acids as part of the comonomer system. The polymerization of these monomers produces a chain similar to a normal nitrile rubber except for the carboxyl groups which are distributed along the chain with a frequency of about 1 to every 100 to 200 carbon atoms.

This type polymer is unique in that it can be cured or vulcanized by reactions of the carboxylic group as well as normal sulfur-type vulcanization. One method of curing is to crosslink the chains by neutralizing the carboxylic groups with the oxide or salt of polyvalent metal. The Zn^{++}, Ca^{++}, Mg^{++}, and Pb^{++} ions are capable of effecting such a vulcanization reaction. Since the polymer chains also contain double bonds such as occur in butadiene-acrylonitrile rubbers, the normal types of sulfur vulcanization can also be employed. However, since most sulfur-curing systems also contain zinc oxide, the vulcanizate is likely to be that of a combination of the two curing systems. The metal oxide and sulfur-curing systems are applicable to both the dry rubber and latex form of the carboxylated nitrile rubbers.

As compared with a conventional nitrile rubber of equivalent oil resistance, the carboxy modification exhibits much higher tensile strength and modulus, lower elongation, higher hardness, much improved hot tear and tensile, better low-temperature brittleness, improved ozone resistance, and better retention of physical properties after hot-oil and air aging.

Latest Developments

(1) There are the improvements in processing being offered by many of the newer polymers on the market today. This is especially evident in the increasing array of very low Mooney nitrile rubbers now being marketed. These polymers offer an ease of processing directed at cooler running stock and ones that will be ideally suited for the increasing use of injection molding equipment which is just beginning to make an impact in this country. These new low Mooney products also have a minimal loss in inherent physical properties which had been a previously prohibitive quality of low Mooney polymers.

(2) Nitrile rubbers have recently been introduced in finely divided or powdered form. Powdered nitriles have been available for many years but these were only crosslinked varieties which detracted from physical properties. Now, however, noncrosslinked versions are available, having a particle size of 40 mesh, which opens a new chapter in compounding and end uses. Powder rubber can be easily blended with PVC powders to form polyblends of any ratio desired, and a compounder is not limited to the few ratios presently available. Now the possibility exists that nitrile rubber and modified PVC can be handled on plastic processing equipment where powdered PVC is now being used.

Finely divided powders can be combined with all compounding ingredients including curatives, fillers, and plasticizers to form a free flowing product. Such a

powdered compound, because of its finely divided state, makes up in an interna
mixer or mill in only a fraction of the time and at significantly lower temperature
than is now possible with slab and conventional mixing techniques. A great cos
saving is the end result. Polymers and processing techniques will be developed in th
future where such powdered compounds will be capable of complete mixing by
merely passing through an extruder, or ultimately by-passing any mixing step sav
that experienced when processing in injection molding equipment.

(3) A series of newer products which are also designed to provide process cos
savings to the rubber goods manufacturer are the preplasticized nitrile rubbers an
polyblends. By incorporating large volumes of plasticizer in the rubber at the poin
of manufacture, these products have been designed to save extensive mixing tim
especially when very low durometer products are being produced. Soft rolls is a
specific market where these types of products are now seeing service.

POLYACRYLATE RUBBER

When an unsaturated polymer is vulcanized, approximately 1 in 30 of the
unsaturated, double bonds is combined with sulfur in order to provide a sof
vulcanized rubber. If most of the double bonds are reacted, a rigid, hard rubber is
produced. The reaction of sulfur and rubber occurs normally with accelerators in 5
to 15 minutes at 300 to 320°F. When service in modern automobiles approache
this temperature, two things happen which cause unsaturated rubbers to fail:

(1) The temperature is above that normally safe for oils, making necessary oi
additives in order to protect the oil from degradation. These oil additives ofte
contain active sulfur which causes any rubber surface in contact with oil at hig
temperatures to be vulcanized to a hard rubber.

(2) Since it is possible to crosslink or vulcanize unsaturated rubbers wit
peroxides as well as with sulfur, the air which is in contact with the hot oil an
rubber causes similar vulcanization of the rubber surface, resulting in casehardening
Failure of these rubbers in service is almost unavoidable at temperatures abov
350°F, if there is oxygen present.

In order to get around this very serious trouble, the polyacrylate rubbers wer
developed. They can be cured with amines or sulfur systems, although th
mechanism is not fully understood. However, since there is no residual unsaturatio
in the polymer, it is quite resistant to sulfur or oxygen. Generally, the polyacrylat
rubbers are useful for temperatures 100 to 150°F higher than nitrile rubbers. Th
most outstanding feature of polyacrylate rubbers is their ability to perform i
sulfur-modified oils at temperatures in excess of 300°F.

History Unmodified polyacrylate rubbers were first developed in the earl
1940's by the B. F. Goodrich Company. The Department of Agriculture's Easter
Regional Research Laboratory and the University of Akron Government Labor
atories poineered the development of modified polyacrylate elastomers, which wer
identified as "Lactoprene EV" and "Lactoprene BN" types. The first commercia
products of both modified and unmodified types were manufactured and sold by B

F. Goodrich Chemical Company in 1948. They were identified as "Hycar PA" and Hycar "PA-21."

Current Polymers There are currently available polyacrylic polymers from three major suppliers in the United States and one outside this country.

These suppliers and their products are as follows:

B. F. Goodrich Chemical Co.	Hycar
American Cyanamid, Rubber Chemicals Dept.	Cyancryl
Thiokol Chemical Corp.	Thiacril
Polymer Corporation Limited (Canada)	Krynac

Method of Manufacture Polyacrylic polymers are either emulsion-polymerized to form a latex or polymerized in suspension, which is subsequently coagulated, washed, and dried in sheet form.

Characteristics of the Uncompounded Polymer Polyacrylic rubber is supplied as white sheets, and has a specific gravity of approximately 1.1. Even when stored for extended periods at elevated temperatures approaching 300°F, it does not deteriorate.

Typical Properties of the Vulcanizates

(1) Temperature resistance from –50° to +400°F (dependent on polymer choice).
(2) Very resistant to oxidation at normal and elevated temperatures.
(3) Excellent flex life.
(4) Very resistant to sunlight degradation.
(5) Excellent ozone resistance at normal and elevated temperatures.
(6) Good resistance to swell and deterioration in oils and aliphatic solvents, particularly sulfur-bearing oils at high temperatures.
(7) Resistance to permeability of many gases.
(8) Permanence of color in white or pastel shades.
(9) Physical properties in the following ranges:

Tensile	500-2400 psi
Elongation	100-400%
Hardness	40-90 Duro A

Most polyacrylic polymers are not generally recommended for use in water, steam, or water-soluble materials such as methanol or ethylene glycol. However, some of the newer versions do have improved water resistance. Polyacrylate rubbers decompose in alkaline media and swell in acid solutions. A special feature of polyacrylate elastomers is their resistance to lubricants of the extreme pressure type, particularly at high temperatures.

Applications Due to the excellent combination of heat and oil resistance, polyacrylate rubber compounds find applications in oil hose, automotive gaskets, search-light gaskets, and O-rings, especially for applications involving sulfur-containing oils and high-pressure lubricants, belting, tank linings, white or pastel-colored goods, cements, and latex coatings for cloth.

Polyacrylate polymers are now in general use in numerous automotive applications such as transmission seals and lip seals. In these applications relatively high temperatures are encountered, due to the desire for smaller, lighter weight, and more efficient transmission systems. Also, oil additives needed to extend the service temperature of automotive fluids are usually sulfur bearing, which causes degradation of unsaturated polymers. The excellent high-temperature resistance of polyacrylate rubber, combined with good physical properties, good compression set, and absence of unsaturation, makes it a natural for these applications.

Compounding Ingredients Polyacrylate rubbers do not contain any carbon-carbon double bonds like diene based rubbers. therefore crosslinking or vulcanization can not occur in the same manner as the chemically unsaturated "conventional" rubbers. However, this thermoplastic rubber is responsive to certain processes that convert it from thermoplastic to a thermosetting or "cured" state. Although polyacrylic polymers are saturated polymers, antioxidants are sometimes used in compounding certain types for certain applications.

Curatives The choice of curative is somewhat specific for the particular polyacrylic being compounded. Crosslinking is accomplished by introduction into the polymer at the point of manufacture of a small percent of reactive monomer. Certain curatives which are suited for crosslinking the reactive monomer used in a given polymer are not satisfactory for crosslinking all available products. Thus it is important to ascertain which cure systems are recommended by the supplier of the polymer before a compounder undertakes the addition of curatives.

Polyacrylic elastomers that were first developed and are still available today are cured with amines: trimine base, triethylenetetramine, tetraethylenepentamine, Diak No. 1 (E. I. duPont), and a system comprising Red Lead and NA-22 (E. I. duPont) are the most widely used. The first three, being volatile amines, do not offer the most efficient shelf stability or efficiency during processing. The last two are more generally used to overcome these deficiencies. Sulfur and sulfur-bearing materials act as retarders in this type of cure and also act as a form of age resistor. The presence of such materials aids the retention of tensile strength, elongation, and hardness on high-temperature oil and heat aging.

Newly introduced polyacrylic polymers can be cured with certain amines but are more versatile and are responsive to a broad range of curative systems. As mentioned previously, it is recommended that the supplier of a given product be contacted for his suggestions. The systems most commonly used include alkali metal stearate/sulfur or sulfur donor, Methyl Zimate (R. T. Vanderbilt), and ammonium adipate.

Reinforcing Agents. As in most synthetic rubbers, reinforcing agents such as carbon black and certain white inorganic pigments are necessary to bring out optimum physical properties. However, the selection of these agents is much more critical. Pigments which are acidic in nature must be carefully avoided as they interfere seriously with the curing systems, which are usually basic. Thus, neutral or basic ingredients are required.

The most effective black reinforcing agents are the N-550 and N-330 types. They provide a good balance of original physical and aged properties. Optimum loadings are obtained in the range of 35 to 80 parts per 100 polymer. Using these loadings, tensiles from 1500 to 2400 psi, elongations of 100 to 400% and hardness from 40 to 90 Duro A can be obtained.

Light-colored compounds made with polyacrylate rubber offer advantages for many applications because of their exceptionally good resistance to discoloration. Precipitated silicas (HiSil, PPG Industries), calcium silicate (Silene EF, PPG Industries), and chemically altered clay (Zeolex, J. M. Huber Corp.) or combinations are most commonly used. Although white compounds have, in general, lower tensile strength and heat resistance than black-loaded stocks, these properties are superior to those available with "conventional" rubbers. White polyacrylate rubber stocks are essentially unaffected by ultraviolet light, oxygen, and ozone.

Pastel shades are usually obtained through the use of organic pigments; Benzidene Yellow, Lithol Red, Lima Blue, and Maroon Toner (Harshaw Chemical Co.) are recommended because of their good heat resistance and lack of inhibition of cure. In general, the inorganic metallic oxide color pigments should not be used because of their detrimental effect on cure rate and heat aging.

Softeners. The use of softeners and plasticizers in polyacrylate compounds presents somewhat of a problem. By selecting the proper plasticizer and polymer, untempered vulcanizates can be made to pass −50°F on freeze tests. TP-90B (Thiokol Chem. Corp.) and thioethers and certain adipates (TP-95) are examples of this type of plasticizer. However, good low-temperature plasticizers volatilize during the tempering process, and their effect is easily lost. A nonvolatile plasticizer with good low-temperature properties has not as yet been discovered. Most other softeners are also fairly volatile and therefore have limited use. Again, care must be exercised to avoid acidic materials, and softeners should be limited to 15 phr.

Certain extenders such as the higher-melting point coumarone-indene resins find usage in that they do not adversely affect the freeze properties of a compound and do not volatilize at tempering or most aging temperatures. These materials are therefore still available for extraction in oils if needed to meet certain specifications and to aid processing.

Tackifiers. Many extenders also serve as tackifiers. Polyacrylate rubber compounds have little inherent tack, and tackifiers must be added as needed. The coumarone-indene resins and Koresin (GAF Corp.) are good tackifiers, particularly the Koresin. Plasticizer TP-90B (Thiokol Chem. Corp.) also seems to have some properties along these lines, more so than the other low-temperature plasticizers. Fifteen parts appear to be most effective.

Processing Aids. Such materials as stearic acid and TE-80 (Technical Processing Co.) are quite helpful in mill-mixing and general processing. Though stearic acid is acidic, its use in 1 phr quantities does not adversely affect the cure. Acrawax C (Glyco Chemicals, Inc.) is often helpful in aiding the release of articles from the molds and may be used up to 5 parts without adverse effect on cured properties.

Curing. Polyacrylate rubber compounds are cured in virtually the same manner as "standard" polymers. Cure temperatures from 290 to 338°F are recommended along with cure times of 4 to 45 minutes, depending upon the thickness of the article. The newer alkali metal stearate/sulfur cure system can be used when injection molding polyacrylic parts. Cure times of 60 to 120 seconds at 400 to 450°F are suggested. Alkali metal stearate/sulfur cured stocks show improved shelf stability. After removal from the mold, hot articles should not be quenched in water but should be air-cooled if at all possible. Polyacrylate rubber stocks containing amine curatives should be cured within 1 week after mixing.

In some cases, mold release difficulties are encountered. Sticking and mold discoloration are the most common problems. Generally, the use of a good silicone lubricant, such as DC-35 (Dow Corning Corp.) preferably added as a spray, will eliminate the problem. In more severe cases, the incorporation of 1 to 2 parts of Acrawax C into the recipe will help.

Tempering. Many applications call for excellent compression set properties. At present, polyacrylic compounds as they come from the mold do not have the maximum values obtainable. No curing agents or combinations have yet been found which will accomplish this. Heating the cured articles in an air oven for 3 hours at 350°F will bring the set down to an excellent level. This time and temperature may be varied to suit individual factory conditions. Tempering also improves the resilience.

Tempering has an effect on the other physical properties as well. Tensile strength is usually unaffected; elongation, however, is lowered and the hardness increased. Tempering also is detrimental to flex life and causes the partial volatilization of any good low-temperature plasticizers which may be present. But since compression set is generally the most important item to be considered, tempering is standard practice.

Processing

Mill Mixing. Employing the nonamine cure systems, polyacrylate rubber compounds require mill-mixing techniques similar to standard rubber stocks. The following mill mixing procedure is suggested: (1) polymer; (2) reinforcing agent, black or white; (3) stearic acid and sulfur or other similar materials (may be added with the reinforcing agent); (4) process aid or other softeners and tackifiers; (5) curative.

Depending on the amount of loading, the following batch weights should be used:

60″ Mill – 45 to 60 lb
84″ Mill – 100 to 120 lb

The polymer should be added to a cool mill opened sufficiently to band the polymer with a very small bank. Pigment may be added as soon as the polymer has banded, as no breakdown occurs or is necessary.

The dry reinforcing agents may be added fairly rapidly but should not be dumped onto the bank unceremoniously. This is only standard rubber practice.

Sulfur, stearic acid, or other dry pigments may be added with the main reinforcing agent. The band should not be cut with free pigment on the rolls. It may be necessary to open the mill slightly to keep the bank rolling. When all the pigment is in, the batch can be cut and worked with ease. After a few cuts each way, any softeners or resins present should be added. Again, cut and blend after all the plasticizer is in.

Finally, the curative should be added slowly to the rolling bank. If the batch is too cool, the stock becomes tough to cut and work.

The two most important rules to follow in mixing polyacrylate rubbers are: (1) Do not attempt to work the stock before pigment addition; (2) Maintain a stock temperature of 160 to 200°F.

A mill mix should take between 20 and 40 minutes, depending on the particular batch and the mill man running it.

Banbury Mixing. Polyacrylic is ideal for Banbury mixing. This method avoids the mill-sticking problem sometimes associated with liquid amine cures, and again, no breakdown is needed. A typical procedure is as follows:

#9 Banbury
Slow Speed
Cooling water half on

(1)	Polymer in	0 minutes
(2)	1/2 Black	1 1/2–2 1/2 min
(3)	1/2 Black, stearic acid, sulfur 1/2 Resin, or oil if present	3 1/2–4 1/2 min
(4)	Rest of oil, if any	6–7 1/2 min
(5)	Dump	9–10 min
	Dump temperature	250–300°F

Add curative on the dump (sheet off) mill or on second pass through Banbury. A one-step Banbury mix is obtainable with certain polymer and cure system combinations.

Stock Preparation. Standard polyacrylate stocks with little or no softener extrude sufficiently well for most stock preparation purposes.

Compounding for Specific Applications

Extrusion. Temperatures at 100–300°F in the barrel and 170°F in the die are standard. In order to obtain good smooth extrusions, more loading and lubrication are necessary than for molded goods due to the inherent nerve of the polymer. Loadings from 60–100 parts, plus some "inert" filler, are used along with 2–3 parts of lubricants previously mentioned and the normal 1 part of stearic acid. Combinations of black and nonblack loading pigments can be used where compression set is not of prime importance.

Calendering. Generally, those compounds which extrude well will calender well. Higher loadings of pigments and softeners are necessary in order to reduce the nerve. About 100–160°F temperatures are best. Mill warm-up is recommended, and 10–mil films are possible.

Friction and Roll Building Stocks. Tackifying agents are very important for both applications. Koresin has proved to be very satisfactory. The coumarone-indene resins are good, as is TP–90B when used at 15 parts:

Typical Friction Stock

Hycar 4021	100.0
Sulfur	1.0
Koresin	15.0
N–601 Black	20.0
S–300 Black	10.0
Stearic acid	1.0
Trimene base	2.5
TETA	2.0
Chlorowax 40	7.5

Physical Properties Obtained (45 min at 310°F untempered)

Tensile	1400 psi
Elongation	650%
Duro "A"	40

Molding. The most common application for polyacrylic polymers is in various automotive seals and gaskets. A typical general purpose seal compound follows which meets many of the specifications written for this type of part:

Hycar 4041	100.0
Acrawax C	2.0
N–550 Black	65.0
Sulfur	0.3
Flexichem PHS	3.0

Physical Properties Obtained (Cured 4 min @ 338°F and tempered 8 hours @ 347°F)

Tensile	2050 psi
Elongation	180%
Duro "A"	78

Rubber Solutions. Polyacrylic rubber possesses interesting possibilities for use in solutions in the paint industry and for adhesives. It is generally soluble in alcohols, ketones, aromatic hydrocarbons, and esters. The following specific solvents are recommended:

Acetone
Ethyl acetate
Toluene
Butyl acetate
Methyl alcohol
Benzene
Carbon tetrachloride
Methyl ethyl ketone
Perchlorethylene

LATEXES

Both the polyacrylate and nitrile polymers are available as latexes. Time will not permit a detailed discussion of these materials here, but the following will illustrate the many and varied applications in which nitrile and polyacrylate latexes are useful:

Textiles

(1) Textile backcoatings
(2) Nonskid backing on rugs
(3) Fiber binder for nonwoven fabrics
(4) Pigment binder in printing pastes
(5) In blends with urea or melamine-formaldehyde resins to improve the tear strength and abrasion resistance while maintaining a soft hand

Paper

(1) Binder for pigment coatings
(2) Coating for water- and greaseproofness
(3) Saturant in abrasive papers, masking tape, gasket papers
(4) Wet-end addition for high-strength papers

Leather Finishing

(1) Binder for pigments
(2) Increased abrasion resistance

Adhesives

(1) General adhesion work where bond is exposed to oil
(2) In blends with phenolic resins
(3) Laminating or bonding adhesives for fabrics
(4) Flocking adhesives

13

NEOPRENE AND HYPALON®

DON B. FORMAN*
Elastomers Chemicals Department
E. I. du Pont de Nemours
Wilmington, Delaware

Neoprene is the generic name for chloroprene polymers (2-chloro-1,3-butadiene manufactured since 1931 by E. I. du Pont de Nemours & Company. HYPALON® synthetic rubber, which is chlorosulfonated polyethylene and manufactured by the Du Pont Company, has been made since 1952. The two specialty synthetic rubbers are reviewed here separately.

NEOPRENE

After nearly 40 years of development and manufacture, the beginning of the 1970's finds polychloroprene made by at least seven manufacturers, namely: 1) E. I. du Pont de Nemours & Company (plants in USA, Northern Ireland); 2) Showa Neoprene K.K. (Japan, operated by the Du Pont Company and Showa Denka K.K.); 3 Petrotex (USA); 4) Bayer-Baypren (Germany); 5) Denki Kagaku–Denka (Japan) 6) Plastimere–Butaclor (France); and 7) Skyprene–Toya Soda (Japan). Each company uses emulsion polymerization; and makes a number of variations of the product. Varieties are marketed both as solid polymers and as latices. The solid types vary over a wide voscosity range. In some instances, either nonstaining or staining antioxidants may be included at the time of manufacture.

There is a general summary on the development of 2-chlorobutadiene polymers published in the *Encyclopedia of Polymer Science and Technology, Volume 3.* This treatise is one of the most comprehensive ever published. It includes considerable information on the theoretical aspects, both polymerization and polymer characteristics.

Commercial Polychloroprene

Lists of available elastomers vary almost day to day. In early 1972 the seven manufacturers of neoprene were making the polymers shown in Table 13-1. Some manufacturers also make the equivalent blend of some of these specific types.

The chronological development of the polychloroprenes is a classic story of how

*Present address: 1313 Chadwick Road, Wilmington, Delaware

TABLE 13-1. COMMERCIAL POLYCHLOROPRENES[a]

Du Pont Neoprene	*Showa Neoprene KK*	*Petrotex Neoprene*	*Bayer Baypren*	*Denki Kagaku Denka*	*Plastimere Butaclor*	*Toya Soda Skyprene*
AC		–	321, 331	TA-85, 95	MA-41H, K	
AD		–	320, 330	A-70, A-120	MA-40S, T	
AF		–	–	–	–	
AG		–	–	–	–	
AJ		–	–	–	–	
FC		–	–	–	–	
FB		–	–	–	–	
FM		–	–	–	–	
GN		S-1	–	–	–	
GNA		S-2	–	PM-40	–	
GRT		S-3	610	–	SC-10	R-10
GS		S-5	710	PM-40NS	SC-22	R-22
GT		S-4	–	PT-60	SC-11	R-11
HC		–	–	–	–	
KNR		–	–	–	–	
S		–	–	–	–	
W		M-1	210, 220	M-40, M-41	MC-30	B-30
W-M1		M-1.1	211	M-30	MC-31	B-31
WB		M-6	214	EM-40	ME-20	Y-20B
WD		–	130	–	MH-10	
WHV		M-2	230	M-120, 130	MH-30	Y-30
WHV-100		M-2.7	230	M-100	MH-31	Y-31
WK		–	124	ES-70	–	
WRT		M-3.2	110	S40V	MC-10	B-10
WX		M-3.8	112	S40	MC-20	B-20
TW		–	215	MT-40	–	
TW-100		–	235	–	–	
TRT			115			

[a]The polychloroprene made in Russia is called Nairit. Available sparse information indicates the following:
Nairit, probably equivalent to W.
Nairit HE, probably equivalent to GNA.
Nairit M, ester extended, no equivalent.
Nairit HN, an adhesive grade.
Nairit HT, an adhesive grade with considerable crystallinity.
Chlornairit, a chlorinated polychloroprene.

man has created variations in a synthetic rubber to satisfy the ever changing needs of the rubber industry, including greater reliability during processing and many severe property demands in end products.

The solid neoprenes are classified general purpose, adhesive, or specialty types. *General purpose* types are used in a variety of elastomeric applications – particularly molded and extruded goods, hose, belts, wire and cable, heels and soles, tires, coated fabrics and gaskets. The *adhesive types* are unusually adaptable to the manufacture of quick setting and high bond strength adhesives. *Specialty types* have unique properties such as exceptionally low viscosity, high oil resistance, extreme toughness, or balata-like characteristics. These properties make the specialty neoprenes useful in unusual applications: for example, vulcanizable plasticizers, crepe soles, prosthetic applications, high solids cements for protective coatings in tanks and turbines, etc.

General Purpose Neoprenes

General purpose types of neoprene can be processed with standard rubber machinery. The specific gravities range between 1.23 and 1.25 at 25/4°C. Color varies from amber for the *G* types to degrees of whiteness for the *T* and *W* types.

Neoprene GN is the oldest general purpose neoprene being produced today. Compounds prepared from Neoprene GN cure rapidly and develop highly resilient vulcanizates with good tear resistance. As uncompounded Neoprene GN polymer ages, its viscosity declines; at the same time the polymer becomes nervier, faster curing, and more prone to scorch. For these reasons, close attention must be paid to inventories of Neoprene GN so that the polymer will be processed within a reasonably short time after it is received, otherwise processing difficulties can be expected.

Neoprene GNA physical properties are similar to those of GN. It is used in black compounds. In applications where discoloration or staining of finishes is a problem, Neoprene GNA should not be used since it contains a staining type of antioxidant which is added to stabilize the raw polymer.

Neoprene GRT is more resistant to crystallization than either GN or GNA. It contains a nondiscoloring, nonstaining stabilizer. Neoprene GRT is especially well suited for friction and skim stocks, sheet goods, or other products dependent upon good tack retention.

Neoprene GT is very stable as a raw polymer. It approaches the stability of Neoprene W. During storage its viscosity, scorch tendencies, and cure rate remain relatively constant. Compound viscosity can be controlled to meet different processing requirements through the addition of modifiers during compounding. Vulcanizate properties resemble those of Neoprene GN. Resistance to crystallization is similar to that of Neoprene GRT.

Neoprene W has excellent polymer stability. This is characteristic of all W types. As with natural rubber, it can be milled over a wide temperature range without sticking to the rolls. Neoprene W contains no staining stabilizer. Specifications requiring better than average resistance to compression set or heat can be met with compositions based upon Neoprene W.

Neoprene WHV is similar to W in all respects except its Mooney viscosity, which is considerably higher. This makes WHV particularly adaptable for high-quality, low-cost compounds highly extended by fillers and softeners. WHV is blended with other neoprenes, particularly W, to obtain desirable viscosity for processing operations.

Neoprene WRT is one of the most crystallization resistant neoprenes. It has the same viscosity as standard Neoprene W, shows a little more nerve upon milling, and develops slightly lower tensile and tear strength. Neoprene WRT requires 25 to 50% more organic accelerator than does W for a comparable rate of cure.

Neoprene WD is a high viscosity form of WRT, bearing the same relationship to WRT that WHV does to W. Neoprene WD is equivalent to WRT in crystallization resistance, cure rate, and vulcanizate properties. It is particularly well suited for compounds containing large amounts of ester plasticizers for flexibility at extremely low temperatures when maximum crystallization resistance also is required. The high viscosity of WD allows such compounds to be processed with minimum sticking.

Neoprene WX has a crystallization resistance intermediate between that of W and WRT and about equal to that of GN and GNA. The range of processing and vulcanizate properties of WX are between those of W and WRT.

Neoprene WB compounds have outstandingly good extrusion characteristics. Although manufactured in the same viscosity range as W, Neoprene WB yields firmer, more collapse-resistant compounds, which extrude cooler, smoother, and faster than compounds made with any other neoprene. The crystallization resistance of WB is equivalent to that of WX. Vulcanizates of WB are equivalent to or better than those of W in resistance to heat, oil, ozone, and compression set; but they are lower in tensile strength, tear strength, and resistance to flex cut growth and abrasion. Because of its lower physical strength properties, WB is often used in blends (20 to 40%) with other neoprenes to improve their processing.

Neoprene WK is a highly crystallization resistant, medium high viscosity, general purpose chloroprene polymer having excellent processing characteristics. It is especially well suited for extruded and calendered products requiring compounds that are firm as well as low in nerve.

Neoprene TW is an easy processing polychloroprene which can be compounded to produce vulcanizates having fine physical properties. Neoprene TW greatly facilitates the manufacture of both calendered and extruded items.

Neoprene TRT is a crystallization-resistant polychloroprene. Its crystallization resistance is greater than that of either Neoprene GRT or WRT. Its processing and vulcanizate property characteristics are both similar to those of Neoprene TW.

Adhesive Type Neoprenes

Neoprene AC, AD, and AF are used in the general manufacture of adhesives. Neoprene AC and AD have a high degree of crystallinity. Neoprene AF has a low degree of crystallinity. The specific gravities of these polychloroprenes lie between 1.23 and 1.25 at 25/4°C. Neither Neoprene AC, AD, nor AF contain a staining type of stabilizer.

Toluene is a satisfactory solvent for dissolving AC or AD in either chip or milled form. With AF, some polar solvent should be used with toluene to increase polymer solution stability.

Neoprene AC is a cream colored, rapidly crystallizing sol polymer. The viscosities of Neoprene AC and its solutions change very little during storage. Low viscosity grades readily go into solution without milling. Clear solutions stored in metal containers slowly discolor.

Neoprene AD is cream to pale amber in color. It is a readily crystallizing sol polymer. Solubility characteristics are similar to those of Neoprene AC. Color stability and solution stability are considerably better than Neoprene AC, even when stored in metal containers.

Neoprene AF contains small amounts of carboxyl functionality and it is essentially a noncrystallizing polymer, designed for one-part adhesives. Films from AF adhesives rapidly develop high room or elevated temperature bond strengths. Adhesive solutions made with Neoprene AF are highly resistant to phasing.

There are two neoprenes designed for use in specific adhesive applications.

Neoprene AG is a chloroprene polymer of high gel content, which exhibits thixotropy when dissolved or dispersed in solvents. It is especially suited for high solids, high viscosity mastics, which require easy extrudability and resistance to slump. Low viscosity adhesives prepared from Neoprene AG spray more easily than those based on other neoprene polymers, and offer the performance properties required for many contact adhesive applications. Generally, Neoprene AG can be compounded in the same fashion as Neoprene AC and AD.

Neoprene AJ is a crystallization-resistant, chemically peptizable, carboxylated polychloroprene. It is used for pressure sensitive and other one-way adhesives requiring high peel strength, high creep resistance, flame resistance, and aging resistance.

Special Purpose Neoprenes

Each of the special purpose neoprenes has an outstanding feature which distinguishes it from the other polychloroprenes and makes it particularly suitable for specific applications.

Neoprene FB and FC (specific gravity 1.23 ± .02 at 25/4°C) are amber colored, soft, partially crystalline solids at room temperature. Their viscosities drop rapidly as temperature is increased. Both polymers become pourable at 130°F. They are used in adhesives and caulks and as nonvolatile, vulcanizable plasticizers for other neoprenes. Caulking compounds are readily mixed in open, light-duty mixers of the sigma blade type. Filler dispersions in Neoprene FB or FC are improved by grinding the mix on paint mills. FB and FC dissolve readily in aromatic hydrocarbons, chlorinated hydrocarbons, or low molecular weight ketones and esters.

Neoprene HC is a sol polymer having a specific gravity of 1.25 ± .02 at 25/4°C. It is slightly pink in color. It is very crystalline and somewhat balata-like. At room temperature the raw polymer is hard and crystalline. It becomes soft and plastic as the temperature is elevated; for example, the Mooney viscosity of a 10 pass mill sample is 25 ± 5 (ML 1 ± 2 1/2 at 212°F). It is used in golf ball covers, heat sealable adhesives, sealants, and prosthetics.

Neoprene KNR has a specific gravity of 1.23 ± .02 at 25/4°C. Its viscosity is reduced readily by mill mastication with or without added chemical peptizer. It becomes very soft and sticky. Neoprene KNR is used to make high solids (25 to

85%) solutions to be used as protective coatings. Physical strength properties are somewhat lower than those of Neoprene GN but its other properties are similar.

Neoprene S is a tough, highly elastic, noncrystallizing, amber, gel polymer having a very high molecular weight. It is used as crepe soling. This requires minimum compounding and the soles are unvulcanized. It is insoluble in solvents normally used for other neoprene polymers. It cannot be milled to a smooth sheet. It contains a nondiscoloring stabilizer.

Crystallization

Crystallinity is an inherent property of polychloroprene rubbers.[1,2] It varies in degree as some of the neoprenes crystallize more readily and to a greater extent than others. An understanding of this phenomenon is essential to the understanding of polychloroprene technology.

Neoprene, as well as natural rubber, butyl, and other linear elastomers, becomes highly oriented upon stretching and shows X-ray diffraction patterns indicating varying degrees of crystallinity. The crystalline structure of neoprene is indicated by a sharp peak in diffraction intensity at a Bragg angle of 19°30′. As crystallization develops, a small decrease in volume occurs and stressed specimens tend to relax and elongate in the direction of the stress. Crystallization does not take place at high temperatures because these forces are overcome by vigorous molecular motion. Crystallization rate slows down at low temperatures because thermal stiffening inhibits movement and the incidence of favorable alignment is reduced. Anything that aids mobility of molecules at low temperature actually encourages crystallization at low temperatures. Crystallization is a completely reversible phenomenon; warming a crystallized specimen to a temperature above that at which the crystallites were formed destroys them. When neoprene is decrystallized, it returns to its original softness and flexibility and recovers from any relaxed condition which may have developed as a result of crystallization under stress.

Crystallization, which is time dependent, should not be confused with thermal stiffening and embrittlement effects which occur at very low temperatures and are a function of temperature only. Neoprene, as with all elastomers, becomes progressively stiffer as it is cooled. This type of stiffening is evident just as soon as thermal equilibrium is established. An unplasticized neoprene compound, for example, will have an *ASTM D-746* brittleness temperature of about −40°F. This temperature is the same for either crystallized or uncrystallized vulcanizates.

It would be well to show a comparison of the relative crystallization rates of all different neoprenes. Dry neoprenes are produced with widely different crystallization rates so that the user can avoid crystallization if it would be a liability in his product, or he can exploit it if it would be an advantage. Uncompounded polymer crystallizes faster than its vulcanizate. The relative crystallization rates of the neoprenes are presented in Table 13-2. This table is based upon room temperature observations which are important to adhesives; however, it should be noted that maximum crystallization will occur at 14°F.

TABLE 13.2. ROOM TEMPERATURE CRYSTALLIZATION RATES (MEASUREMENTS MADE ON FRESHLY MILLED POLYMER)

Relative Time	*Neoprene*
Minutes	HC
Hours	AD, AC, CG, FC
Up to 3 days	W, TW
Up to 7 days	FB, GN, GNA, WX
One to two weeks	AF
More than two weeks	GRT, WRT, WD, AG, WB, WK, S, TRT

Compounding Neoprene

Neoprene products require certain engineering properties usually associated with strength or working environment. Raw neoprene is converted to these products by mixing selected ingredients into the neoprene and curing (i.e., vulcanizing) the resulting compound. Minimum requirements for a practical compound include the following type of ingredients:

chloroprene polymer
processing aid
antioxidant*
metallic oxide*
curing agent and/or accelerator*
filler or reinforcing agent
physical softener (plasticizer)

Optional ingredients may include:

antiozonants
retarders
extenders
resins
other elastomers
blowing agents

Metallic Oxides. Magnesium oxide and zinc oxide are used extensively and together in the compounding of neoprene. Two of the lead oxides, specifically litharge and red lead, are used for products requiring maximum resistance to swell and deterioration by water. Other oxides have been tested, but they are not practical. The quality of magnesium oxide is critical and selected types must be used. Any good grade of "rubber makers" high purity zinc oxide or lead oxide is satisfactory.

The grade of magnesia used in neoprene compounds plays an important role, particularly in influencing the degree of processing safety, cure rate, and vulcanizate quality. A considerable amount of research and development effort has been

*Essential ingredients for all compounds

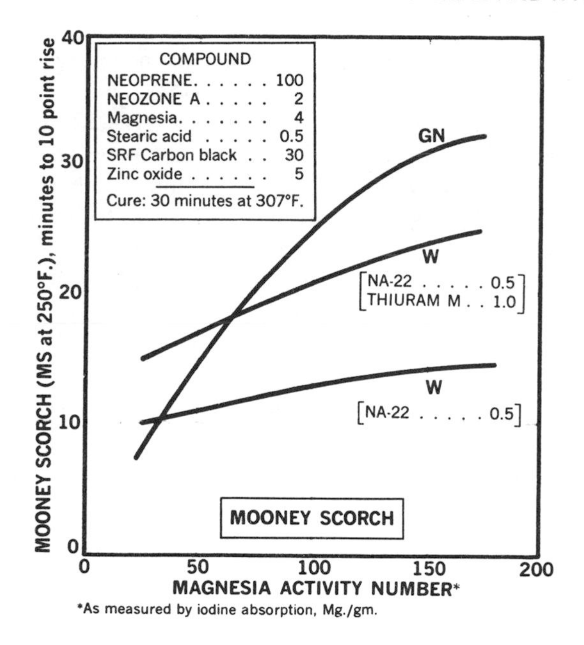

Fig. 13.1 Effect of Magnesia activity on processing safety. Neozone® A–phenyl-a-naphthylamine.

expended by magnesia producers to develop grades especially well suited for use in neoprene. "Neoprene grades" of magnesia have two characteristics in common. First, they are precipitated (not ground) and calcined after precipitation. Second, they are very active, having a high ratio of surface to volume.

For magnesia to be fully effective, it should disperse rapidly and uniformly in the neoprene. The chemical and physical properties of magnesia, such as bulk density, particle size, and purity (MgO content), cannot be correlated with performance in neoprene. On the other hand, magnesia activity, as measured by iodine absorption, appears to be a valid criterion – in general, the higher the magnesia activity, the greater the processing safety and the better the vulcanizate properties obtained in neoprene compounds.

Compounds based on the W types of neoprene are much less sensitive to magnesia activity than are those based on the G types (see Fig. 13.1).

Exposure to atmospheric moisture and carbon dioxide, even briefly, can cause a considerable loss in activity. Therefore, it is highly important to protect magnesia in moistureproof, air-tight containers until it is added to the compound.

Accelerators and Curing Agents. Amines, phenols, sulfenamides, thiazoles, thiurams, thioureas, and sulfur are the common accelerators and/or curing agents for polychloroprene. Since metallic oxides also function as accelerators and/or curing agents, the application of accelerators and curing agents can be better understood after a brief review of the mechanism of cure.*

*See *The Neoprenes* or *Encyclopedia of Polymer Science and Technology*, Volume 3.

Zinc oxide and magnesium oxide are used in most neoprene formulations. These two oxides are sufficient to cure Neoprene GN, GNA, and GRT in the absence of any organic accelerators. An accelerator, such as NA-22 (ethylene thiourea), accelerates the vulcanization process when very fast cures are required. The cure rate of the W-types with metallic oxides alone is impractically slow; accordingly, an organic accelerator such as NA-22 is always included in formulations of Neoprene W. Organic acceleration is included in formulation of Neoprene TW.

Curing mechanism for organic curing agents other than thioureas is reviewed by Kovacic.[3] Sulfur vulcanization is probably similar to those of natural rubber, polyisoprene, and synthetic rubber involving butadiene. The use of sulfur in neoprene is described in *The Neoprenes.*[1]

Probably all "rubber chemicals" and many more have been evaluated as accelerators and/or curing agents for polychloroprene. As with natural rubber, there are many correct selections depending upon the need. Accordingly, only a guide would be helpful and Table 13.3 is designed as such. It shows what type of acceleration could be used as a starting point with the key types of neoprene.

In compounding the adhesive or specialty types there are other considerations. In adhesives and coatings it is sometimes desirable to use two-part systems so that very active accelerators do not cause "set-up" in cans, etc. The accelerators selected

TABLE 13.3. ACCELERATION GUIDE FOR NEOPRENE COMPOUNDS (QUANTITIES ARE PARTS PER HUNDRED RUBBER)

		Neoprene	
Requirement	*G Type*	*W Type*	*T Type*
Delayed action or slow cure; bulky items	MBTS – 1.0	(THIONEX® –1.0 (DOTG –1.0 (Sulfur –1.0	(CONAC S –0.75 (NA-101 –1.25
Intermediate cure rate – conventional or standard extruded and molded items	None	(NA-22 –1.0 (THIURAM M –0.5	(CONAC S –0.75 (TA-11 –2.00 (Na-101) 1.00
Fast cures; wire and cable, continuous curing, or injection molding	NA-22 – 1.0 or PERMALUX® –0.5	NA-22 –1.0	(NA-22 –1.5 (MBTS –0.5

MBTS – Benzothiazyl disulfide
THIONEX® – Tetramethyl thiuram monosulfide
DOTG – Diorthotolyl guanidine
NA-22 – Ethylene thiourea
CONAC S – N-cyclohexyl-2-benzothiazylsulfenamide
TA-11 – A modified amine (Du Pont)
PERMALUX® – Diorthotolyl guanidine salt of dicatechol borate
NA-101 – Tetramethyl thiourea

must be compatible with the solvents. New and current information on accelerator and compounding systems can always be obtained from raw material suppliers. The recommendations will meet the broad basic requirements:

(a) Scorch resistance
(b) Bin storage stability
(c) Good cure rate
(d) Good resistance to heat aging
(e) Very low compression sets
(f) Generally good performance as an accelerator for a mineral filled compound as well as carbon black.

Antioxidants and Antiozonants. All neoprene compounds must contain an antioxidant added at the time of compounding. Only a small number of polychloroprenes have an antioxidant included at the time of manufacture but this is insufficient for practical usage and full protection.

Antiozonants are added to neoprene compounds as needed. Basically neoprene has very good resistance to ozone attack, but it is not as good as the superior ozone resistance of elastomers having little or no unsaturation in their molecular backbones. AKROFLEX® AZ, a diaryl-p-phenylene diamine, provides protection from ozone and has no adverse effects on the stability of mixed, uncured stock.

The data in Table 13.4 vividly show the necessity of having an antioxidant in neoprene compounds.

TABLE 13.4. AGING OF NEOPRENE W GUM STOCK (SPECIMEN CURED 20 MINUTES AT 307°F)

Stress-Strain Properties	*No Antioxidant*	*1 phr NEOZONE® A*
Original		
Modulus at 400% Elongation, psi	440	390
Tensile Strength, psi	3025	2900
Elongation at Break, %	980	990
Hardness, Durometer A	44	44
After 7 days in Oxygen Bomb		
Modulus at 400% Elongation, psi	Melted	790
Tensile Strength, psi	in	3375
Elongation at Break, %	3	960
Hardness, Durometer A	days	52
After 28 days in Oxygen Bomb		
Modulus at 400% Elongation, psi	No	970
Tensile Strength, psi	Data	1850
Elongation at Break, %		830
Hardness, Durometer A		49

Selecting the antioxidant or antiozonant requires some special consideration. The following is a useful check list:

(a) Relative effectiveness on either a weight or cost basis.
(b) Does it affect processing safety, cure rate or bin storage?
(c) Does it cause staining?
(d) Does it cause discoloration?
(e) How effective is it at high temperatures (i.e., will it enhance heat resistance)?
(f) Is it soluble in neoprene to the extent used? If insoluble, bloom will occur as well as possible migration.

Some preferential antioxidants and antiozonants with recommended quantities are listed in Table 13.5.

Fillers and Softeners. Any one of the wide choice of available fillers for the rubber industry can be used in neoprene compounding. Some fillers have specific

TABLE 13.5. ANTIOXIDANTS AND ANTIOZONANTS

	General Aging & Weathering	
NEOZONE® A or D		2 phr
AKROFLEX® DAZ		3 phr
	Nonstaining	
Antioxidant 2246		2 phr
	Heat Resistance	
Octamine	4 phr)	5 phr total
Aranox	1 phr)	
	Ozone Resistance	
AKROFLEX® DAZ		3 phr
NEOZONE® A	2 phr)	5 phr (superior)
AKROFLEX® AZ	3 phr)	
	Flex Resistance	
NEOZONE® A	2 to 5 phr)	3–7 phr total
AKROFLEX® CD	1 to 2 phr)	

(Use when flex resistance is extremely important)

NEOZONE® A – Phenyl-a-naphthylamine
NEOZONE® D – Phenyl-b-naphthylamine
Antioxidant 2246 – 2,2′-Methylene-bis-) 4-methyl-6-tert-butylphenol)
Octamine – Octylated diphenylamine (Uniroyal)
Aranox – N-phenyl-N′-(p-toluene sulfonyl)-p-phenylene diamine
AKROFLEX® AZ – Undisclosed composition (Du Pont)
AKROFLEX® CD – Mixture: 65% phenyl-b-naphthylamine and 35% N,N′-diphenyl-p-phenylene diamine
AKROFLEX® DAZ – Mixture: 50% Neozone D and 50% Akroflex AZ

effects on certain properties and these effects are described in considerable detail in *The Neoprenes,*[1] *The Economies in Compounding and Processing Neoprene,*[4] and *Profile of Cabot Carbon Blacks in Neoprene.*[5] These three references are recommended as aids in specification compounding. Each utilizes the principle of contour compounding. The reasons for using fillers are summarized:

Carbon Blacks

(1) Improve abrasion resistance
(2) Improve tear resistance
(3) Increase modulus
(4) Increase hardness
(5) Improve ozone resistance
(6) Improve weather resistance
(7) Reduce swell in oils
(8) Improve processibility.

Mineral Fillers

(1) Improve heat resistance (whitings)
(2) Improve flame resistance
(3) Increase hardness
(4) Increase modulus
(5) Reduce gas permeability (mica)
(6) Improve tear resistance (silicas and silicates)
(7) Improve properties at high temperatures
(8) Improve electrical properties (resistivity)
(9) Improve resistance to swell in water (silica)

Plasticizers (or softeners) are almost always included in a compound. Petroleum derived oils are the most frequently used softeners. The utility of the oil is controlled by adjusting the paraffinic, aromatic, and naphthenic components of the oil. Esters, coal-tar products, resinous materials, animal oils, and vegetable oils are used for specific purposes.

All plasticizers will lower vulcanizate hardness levels and all tend to improve processing, reduce swell of vulcanized products when immersed in oils, and improve service at low temperatures. Excepting esters, most softeners reduce compounding costs.

Base Compound. Obviously it is impossible to designate a base compound meeting all requirements. However, for general purpose neoprenes, a starting formula could be:

Neoprene	100
Antioxidant	2
Magnesium oxide	1 to 4
Zinc oxide	5
Accelerator and/or curing agent*	0 to 3

*Can be one material or a combination (see Table 13.3).

Processing and Curing Neoprene

As with all elastomers and synthetic rubbers there are "rules" on the processing of polychloroprene. In all operations it is most important to avoid precure or scorching as a result of too much heat history. This means short mixing cycles at the lowest possible temperatures. Accordingly, mixing cycles call for processing aids, stabilizers, antioxidants, fillers with softeners, and finally, zinc oxide with accelerators and/or curing agents.

The neoprenes can be mixed in internal mixers or on mills. The compounds can be calendered or extruded as required to produce thousands of end products. The G types tend to be softer and tackier. They make good friction stocks. The W types are "dry and nervy." Their viscosity during storage remains relatively constant. Their calendered and extruded surfaces may be a little rough. The T types are low in nerve and easy to process, but may require a tackifier in some operations.

The general purpose polychloroprenes can be cured by any of the standard techniques used in the rubber industry. This includes molding, continuous procedures, ovens, autoclaves, and even extended room temperature exposures.

Detailed guidelines and "trouble shooting" procedures for both processing and curing are given in *The Neoprenes*.[1]

Applications

Application and end products of polychloroprene are probably wider than any other type of specialty synthetic rubber. Some of the more important uses are adhesives, transportation industry, wire and cable, construction industry, hose, and belting.

Adhesives. There are hundreds of different kinds of neoprene-based adhesives available for use by the fabricator of finished goods. Manufacturers of shoes, aircraft, automobiles, furniture, building products, and industrial components rate neoprene adhesives as the most versatile of material for joining. The reason is threefold.

(1) Neoprene-based adhesives are available in fluid or dry film form. The fluid types are classed as solvent, latex, or as 100% solids compositions. A *solvent adhesive* consists of the neoprene polymer and suitable compounding ingredients dissolved in an organic solvent, or combination of solvents (toluene, ethyl acetate, naphtha, methyl ethyl ketone). A *latex adhesive* is composed of particles of neoprene and compounding ingredients dispersed in water. Both solvent and latex types are normally supplied in ready-to-use "one-part" adhesive systems. (Included here are contact-bond adhesives, the "workhorses" of industry.) When specified "two-part" adhesive systems can also be produced to meet specific end-use requirements. Solvent and latex neoprene adhesives achieve their bonds through evaporation of the fluid and subsequent *crystallization and curing* of the elastomeric residue.

(2) Solid adhesives are based on fluid neoprene polymers and contain neither solvent nor water. Adhesives of this type are normally used in specialty applications where fluidity is required, yet where volatile loss or shrinkage cannot be tolerated.

They possess little "green" strength and usually require application of either heat or catalytic agent to develop an adhesive bond. *Dry-film adhesives* based on neoprene are 100% solids materials normally supplied in the form of a tape or sheeting. This type of adhesive is compounded so that it softens when heated. Cooling then resolidifies the material and forms the adhesive bond.

(3) Neoprene adhesives are made in a range of consistencies, i.e., from very thin liquids through heavy-bodied "putties" to dry films. These adhesives can be controlled over a wide range of properties: for example, tack, "open tack time," storage stability, bond-development, flexibility and strength, heat and cold resistance, and all of the basic characteristics of polychloroprene polymers.

Transportation. In the *automotive* field neoprene is used to make window gaskets, V-belts, blown sponge door gaskets, shock absorber covers, wire jackets, molded seals, motor mounts, adhesives, and many other items. In *aviation* it is used in mountings, wire and cable, gaskets, deicers, seals, etc. In *railroading* it is used in track mountings, car body mountings, air brake hose, flexible car connectors, etc.

All of these uses require physical strength, compression set, elasticity, fluid resistance, water resistance, and aging and weather resistance.

Wire and Cable. Jackets for electrical conductors (low and high voltages) are one of the oldest uses. The abrasion, oil, and weather resistance make neoprene very desirable in this field.

Construction. This is a relatively new application area. Outstanding items are highway joint seals, bridge mounts, pipe gaskets, and high rise window wall seals and roof coatings.

Hose. Neoprene hose has been made almost continuously since 1932. Polychloroprene is used in cover, cushion, and tube stock. All types of hose are involved, including industrial and automotive, garden, oil suction, dredging, fire, gasoline curb pump, oil delivery, and air hose.

Belts. Polychloroprene's heat and flex resistance make it excellent for this application, including V-belts, transmission belts, conveyor belts, and escalator hand rails.

NEOPRENE LATEX

Variation in polymerization techniques results in different dry neoprenes or various neoprene latices. Theory, compounding, and application of these latices are described in detail by J. C. Carl.[6]

Description of General Purpose Latices. The characteristics of several general purpose latices follow.

Latex 400 products are characterized by high modulus, high tensile strength, high tear strength, and good electrical properties. Its films offer excellent resistance to heat, weather, and ozone. Applications for Latex 400 include bonded fibers, coatings, adhesives, and treated paper.

Latex 571 products are characterized by high tensile strength and very good oil resistance. This latex is the preferred neoprene latex for dipped goods and coatings.

Latices 750 and *650* formulations produce films having fine stress-strain properties as well as good wet gel extensibility. *Plastimere Latex* is similar to Latex 650. The chloroprene polymer in these latices resists crystallization. Formulations made from Latices 750 and 650 are used extensively in the manufacture of dipped goods and adhesives.

Latices 842A, 601A and Denka LM-50, LM-60 are counterparts. Compared with Latex 571 they are more resistant to crystallization and have a faster curing rate, but tensile strengths are lower. These latices are used in a variety of applications, e.g., bonded fibers, paper products, and dipped coatings.

There are several special purpose neoprene latices. Comments on unique properties of the several types are presented.

Latex 450 is a sol copolymer of chloroprene and acrylonitrile. Is products are crystallization resistant and very oil resistant. It is used for treating cellulose in paper and as a wet end additive* for binding asbestos in sheeting.

Latex 572 and Denka LA-50 contain a neoprene copolymer which readily crystallizes. It is used in adhesives.

Latex 950 and Denka LK-50 are the only cationic neoprene latices. They are used in elasticized concrete and in selective fiber treatments.

Latex 460 contains a low molecular weight, crystallization resistant sol polymer. It imparts high strength and flexibility when used as a saturant for paper.

Latices 735, 736, and 635 are based on a sol type polymer. Films from these latices have low moduli and high elongation. *Latex 735* is principally used as a wet end binder for asbestos cellulose and other fibers. *Latex 736* is Latex 735 with additional stabilizers to facilitate its use as a wet end additive. *Latex 635* is used in adhesive formulations requiring higher solids content than is obtainable with Latex 735.

Baypren Latex SK is rated as intermediate between Latices 572 and 635.

Baypren Latex MKB features fine heat resistance, high tear strength, and rapid development of bond strength.

Baypren Latex 4R is a special latex having a very low pH 6. It requires unusual compounding techniques.

Baypren B Latex is used to improve Bitumen and to aid in mine fire fighting.

Baypren T Latex is particularly recommended for dip goods.

Applications for Neoprene Latices. *Gloves, meteorological balloons,* and *automotive* parts are foremost amongst products made by *dipping processes.* The weatherability, strength, flame, and oil resistance of articles made from neoprene latex formulations are very desirable for these applications. Impregnation and/or coating of cloth and paper is another large field of use for neoprene latex formulations; typical products include *shoe insole board, tape backings, belting, protective clothing,* and *canvas gloves.* Neoprene latices also are widely used in *binders* for *curled hair* and *abrasives.*

*Wet end additives are added to the paper pulp slurry at the beater stage and should not be confused with coatings added to a wet or dry sheet.

Compounding and Curing Neoprene Latices. Neoprene latex formulations always include a metallic oxide (zinc oxide) to help vulcanize the polymer, to improve aging, and to act as an acid acceptor. At least five parts zinc oxide and two parts antioxidant per 100 parts neoprene are necessary in neoprene latex formulations.

Accelerators (e.g., thiocarbanilide, thiuram disulfides, and dithiocarbamates) are used to increase cure rate. Combinations of tetraethyl thiuram disulfide and sodium dibutyl dithiocarbamate are often used, especially when low modulus and high tensile strength are required. The mechanism of cure is similar to that described for dry neoprene.

Small amounts (3 to 10 parts) of light petroleum process oils often are used in neoprene latex compounds as plasticizers to improve the "hand" of cured films. Conventional fillers are used in quantities ranging from 10 to 50 parts. Thickeners (methyl-cellulose and polyacrylate types) are added as required.

Processing Neoprene Latices. Anionic neoprene latices are processed similarly to natural rubber latex. Compounding ingredients not too readily soluble in water are added to neoprene latices in the form of fine particle size, stable, aqueous emulsions or dispersions to avoid the destabilization of the colloidal system of the latex. Latex compounds are usually held at a pH above 10.5 (control with 2 to 5% NaOH), and soft (or distilled) water must be used to ensure compounded latex stability. Neoprene latex articles are cured at temperatures ranging from 120 to 140°C.

HYPALON®

HYPALON® synthetic rubber, which is chlorosulfonated polyethylene, is characterized by ozone resistance, light stability, heat resistance, weatherability, resistance to deterioration by corrosive chemicals, and good oil resistance. Present available types contain from 25 to 43% chlorine and from 1.0 to 1.4% sulfur. HYPALON® has the following basic structure:

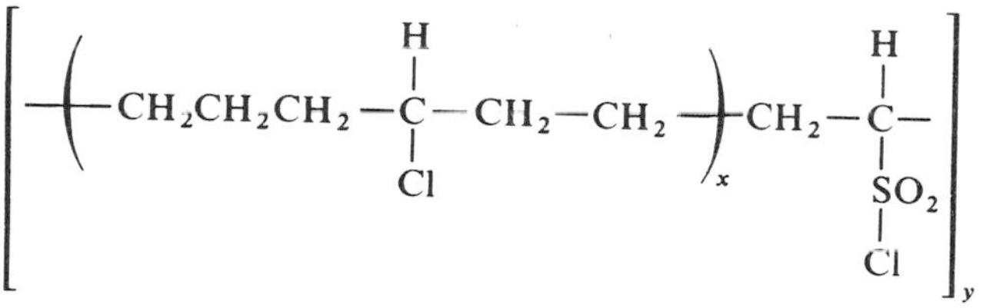

$$\left[\left(CH_2CH_2CH_2-\underset{Cl}{\overset{H}{C}}-CH_2-CH_2 \right)_x CH_2-\underset{\underset{Cl}{SO_2}}{\overset{H}{C}}- \right]_y$$

The ratio y/x varies with the specific type of HYPALON®.

Types of HYPALON®

HYPALON® 20 and *30* are used primarily for solution coating applications because they are more soluble in organic solvents than the other types. Solutions with high solids content can be formulated with relatively low viscosities. Coatings of HYPALON® 20 generally are used over flexible substrates while those of HYPALON® 30 are used over rigid substrates. Films from solution compositions of HYPALON® 30 are harder, drier to the touch, have glossier surfaces, and are more resistant to soiling than those of HYPALON® 20. HYPALON® 20 is often added to HYPALON® 30 to improve the latter's low temperature flexibility.

HYPALON® 40 is the general purpose type with a good balance of processing and vulcanizate properties. It is made in varying viscosity grades, making it easy to meet different requirements for processing as well as providing more latitude in the amount and type of filler and plasticizer that can be used. Low viscosity polymer is used for processing in high hardness vulcanizates or in compounds containing low levels of plasticizer. High viscosity polymer is useful in highly extended or low durometer formulations offering improved processibility and extensibility.

HYPALON® 45 is more thermoplastic than HYPALON® 40. It has a lower viscosity at processing temperatures but gives harder vulcanizates. It is used to obtain high durometer hardness vulcanizates with moderate loading. HYPALON® 45 can also be compounded to give remarkably good stress-strain properties in uncured stocks. These compounds may be used in applications such as cove base, magnetic door closures, roofing film, and ditch liners.

HYPALON® 48 also is more thermoplastic than HYPALON® 40, having a higher viscosity at low temperatures while being softer at curing temperatures. The chemical, oil, and solvent resistance of vulcanizates of HYPALON® 48 is superior to that of the other types; it is thus suitable for use in fluid handling hose, tank linings, and rolls. Uncured compounds based on HYPALON® 48 have physical properties similar to those obtained with HYPALON® 45.

Raw Material Considerations

The various types of HYPALON® have excellent storage stability. Their esthetic values are high, for example: color stable to light and no odor. Physically they maintain their chip form at cool storage temperatures, but due to characteristic thermoplastic qualities, HYPALON® may mass if the storage temperature is too high.

Uncured HYPALON® synthetic rubber is more thermoplastic than other commonly used elastomers. It is generally tougher at room temperatures, but softens more rapidly as temperatures are increased. Viscosity-temperature relationships for HYPALON® are shown in Fig. 13.2.

The low viscosity of HYPALON® 40 at elevated temperatures permits extrusions that are low in die swell, smooth at fast extrusion rates, and sharply defined, especially at thin edges. This softness makes HYPALON® practical for use in building operations, giving good ply adhesion and knitting without the use of tackifiers. However, because of the softness, immediate cooling of extrusions of HYPALON® is necessary to prevent distortion. The high viscosity of HYPALON® 40 at low temperatures makes it necessary to warm stocks before extruding or calendering for more uniform results. Variations in stock temperature cause variable extrusion rates and calendered sheet with uneven gauge and a rough surface.

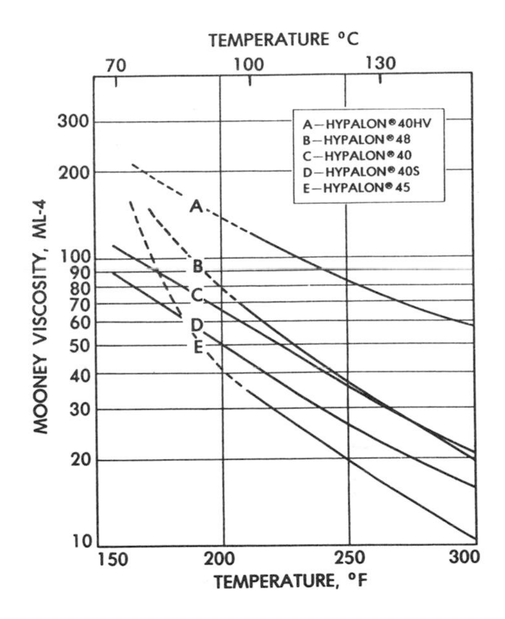

Fig. 13.2 Viscosity-temperature relationship.

The processing of either very soft or very stiff compounds can be improved by using one of the other HYPALON® 40 type polymers. The high viscosity of HYPALON® 40HV at elevated temperatures helps eliminate trapped air and blistering during the processing and curing of highly extended or low durometer stocks. It also helps minimize distortion of these compounds during extruding and open steam curing. The low viscosity of HYPALON® 40S permits easier processing of stock containing highly reinforcing fillers and small amounts of plasticizers. Note that HYPALON® 45 and 48 are more temperature-sensitive than any of the HYPALON® 40 types. The properties of HYPALON® are summarized in Table 13.6.

Compounding HYPALON®

Two basic types of reaction are important in the curing of HYPALON®. The first involves the interaction of the sulfonyl chloride junction, traces of water or pentaerythritol, and a divalent metallic oxide to form sulfonate salt bridges. The second involves the reaction of organic rubber accelerators with active alkyl chloride structures in the presence of metallic oxides to form covalent, probably sulfur, crosslinks. Stevenson[7] and Maynard and Johnson[8] have described these reactions in detail.

Three practical curing systems for HYPALON® are shown in Table 13.7. These and their modifications to meet specific requirements are applicable to all types of HYPALON®.

The metal oxide chosen is particularly important to vulcanizate properties.

TABLE 13.6. **DESCRIPTION OF TYPES OF HYPALON®**

			HYPALON®		
	20	*30*	*40*	*45*	*48*
Description					
Chlorine Content, %	29	43	35	25	43
Sulfur Content, %	1.4	1.1	1.0	1.0	1.0
Physical Form	Chips	Chips	Chips	Chips	Chips
Color	White	White	White	White	White
Odor	None	None	None	None	None
Specific Gravity	1.12	1.28	1.18	1.11	1.27
Mooney Viscosity	30	30	45, 55 & 115	40	55–85
Storage Stability	Excellent	Excellent	Excellent	Excellent	Excellent
Processing Characteristics					
Extruding	Fair	Fair	Excellent	Good	Good
Molding	Good	Fair	Excellent	Excellent	Good
Calendering	Fair	Fair	Excellent	Excellent	Good
Solution Properties					
Solubility	Good	Excellent	Very Limited	Very Limited	Fair–Good
Viscosity	Intermediate	Low	—	—	High

Vulcanizate Properties					
Hardness, Durometer A	45–95	60–95	40–95	65–98	60–95
Tensile Strength, psi					
Carbon Black Stocks	Up to 3000	Up to 3500	Up to 4000	Up to 4000	Up to 4000
Gum Stocks	Up to 1200	Up to 2500	Up to 4000	Up to 3500	Up to 3500
Color Stability	Excellent	Excellent	Excellent	Excellent	Excellent
Low Temperature Properties	Good	Poor	Good	Good	Poor
Tear Strength	Fair	Fair	Good	Good	Good
Resistance to Abrasion	Very Good	Very Good	Excellent	Excellent	Excellent
Chemicals	Excellent	Good	Excellent	Good	Excellent
Compression Set	Fair	Poor	Good	Good	Fair–Good
Flame	Fair	Very Good	Good	Fair	Very Good
Heat Aging	Very Good	Fair	Very Good	Very Good	Good
Ozone	Excellent	Excellent	Excellent	Excellent	Excellent
Petroleum Oils	Fair	Excellent	Good	Poor	Excellent
Weathering	Excellent	Excellent	Excellent	Excellent	Excellent
Resilience	Good	Poor	Good	Good	Poor
Handling Precautions	No unusual precautions required				

TABLE 13.7.

	Litharge	*Litharge/ Magnesia*	*Magnesia/ Polyol*
HYPALON®	100	100	100
Litharge	25	20	–
MBTS	0.5	0.5	–
TETRONE® A	2.0	0.75	2.0
Magnesia	–	10	4
Pentaerythritol	–	–	3
Recommended Use	General Purpose	Heat Resistance	Non-black

Litharge or organic lead bases are used to provide resistance to water, chemicals, and compression set. The litharge should be predispersed to ensure a good dispersion and reproducible results. Several commercial dispersions satisfy this requirement. If powdered litharge is used, good results can be obtained by dispersing it in HYPALON®. Litharge-based systems are not suitable for white or colored products since the lead sulfide formed discolors the vulcanizate.

Magnesia-based systems are used when colorability is desired. They are low cost, have good processing safety, and give vulcanizates with better elongation than obtained with litharge systems. These systems are suggested for general use particularly when water, chemical, and compression set resistance are not critical properties. High activity magnesias are preferred and should be kept dry. Exposure to atmospheric moisture can cause a considerable loss in activity. A combination of magnesia and litharge is used when the maximum resistance to heat degradation is desired.

Except where HYPALON® is used as part of a blend with other elastomers, zinc oxide is undesirable because degradation reactions catalyzed by the formation of zinc chloride lead to poor weathering and heat aging.

Epoxy resins are substituted for litharge or are combined with magnesia to give water resistant vulcanizates suitable for colored applications.

A few significant compounding variations and their effects follow:

(a) Increasing the level of litharge in any system improves water resistance with some loss in processing safety. Conversely, reducing the level of litharge improves processing safety, but decreases resistance to compression set and water. Electrical insulation resistance and dielectric strength, both before and after heat aging, are improved by increasing the litharge level.

(b) A significant improvement in heat resistance, over that obtained with the general purpose systems, can be obtained by adding 3 phr NBC; however, water and chemical resistance is poorer and there is some loss in processing safety.

(c) An increase in the level of either the magnesia or the pentareythritol causes an increase in the state of cure (as indicated by higher modulus) and better compression set and oil resistance. Processing safety, however, is poorer and tensile strengths and elongations are lower. For applications requiring a CV cure, the level of magnesia should be raised to at least 8 phr. When the amount of TETRONE® A is reduced, processing safety is improved but the state of cure is lowered.

(d) An acceleration system composed of 5 phr magnesia, 3 phr pentaerythritol, 2 phr THIURAM M, 1 phr sulfur, and 5 phr of an epoxy minimizes lead press discoloration. It is fast curing with good processing safety but gives lower modulus and hardness as compared with the general purpose systems.

Antioxidants are not required as a compounding ingredient for HYPALON® since the polymer is chemically saturated and contains no double bonds.

Conventional fillers and plasticizers can be used to obtain specific physical properties in vulcanizates. In colored compounds care must be taken to omit ingredients characterized by discoloring or staining tendencies.

Processing

Uncompounded and uncured HYPALON® is a thermoplastic elastomer. It is not necessary to break it down during milling or mixing operations. It can be processed on all standard rubber processing equipment. All operations should be run at the lowest possible temperatures to avoid excessive softening and to keep the polymer below temperatures at which small quantities of moisture might cause some precure. HYPALON® stocks can be mixed as quickly as ingredients can be incorporated.

Calendering operations can be conducted within the range of normal temperature variations used in general rubber processing.

Compounds of HYPALON® containing reinforcing type fillers produce the best extrudates. Lubricants aid the passage of stocks through tubers.

Curing

Conventional curing and molding techniques apply to composition of HYPALON®. HYPALON® 40 is preferred in molded goods manufacture because its stocks are firm. Compression, injection, and transfer molding techniques are all used successfully. Parts of HYPALON® cure rapidly in open steam vulcanizers because of the catalytic effect of the water.

The effect of the curing temperature on rate of cure can be seen by comparing the oscillating disc rheometer curves shown in Fig. 13.3 for litharge-cured black/oil filled compounds. From these data, it can be seen that a 15 minute cure at 298°F (50 lb steam) is equivalent to 9 minutes at 307°F (60 lb steam) or 5 minutes at 344°F (110 lb steam) or 2.5 minutes at 379°F (180 lb steam).

A similar comparison is shown in Fig. 13.4 for a magnesia/pentaerythritol cured clay/oil filled compound. For this compound, a 25 minute cure at 298°F gives the same state of cure, as indicated by the torque values, as 17 minutes at 307°F or 7 minutes at 344°F or 4 minutes at 379°F.

The time of cure at any one temperature has a significant effect on the vulcanizates resistance to compression set. For example, the compression set resistance of the litharge-based compound can be improved by about 25% by

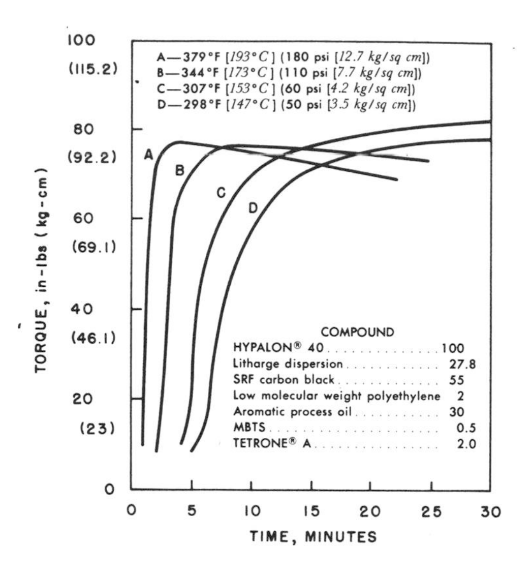

Fig. 13.3 Effect of cure temperature on rate of cure (oscillating disc rheometer) (Carbon black/litharge compound).

increasing the cure time at 307°F from 15 to 30 minutes (see Fig. 13.5). This same change in cure time will result in a similar improvement in the compression set resistance of the clay/oil filled compound. This effect can be seen in Fig. 13.6.

In addition to the preceding conventional curing methods, HYPALON® also can

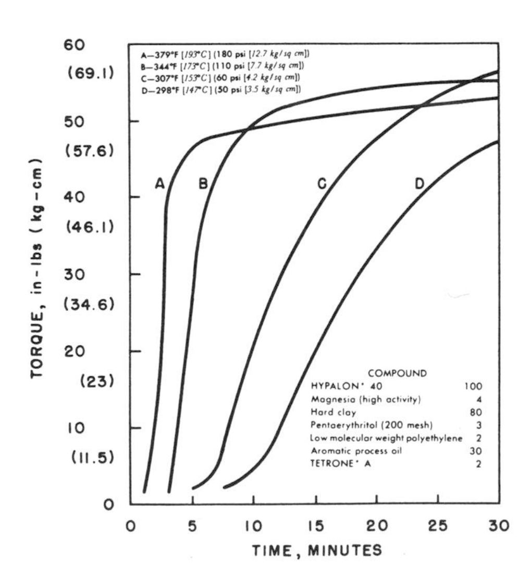

Fig. 13.4 Effect of cure temperature on rate of cure (oscillating disc rheometer) (hard clay/magnesia compound).

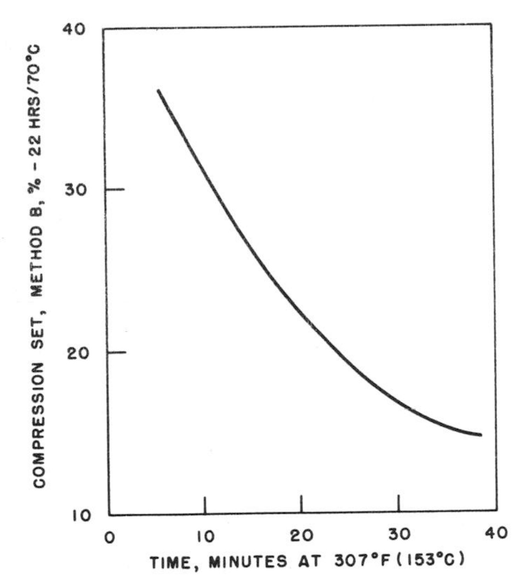

Fig. 13.5 Effect of cure time on compression set resistance (carbon black/litharge compound).

be cured by the following techniques:

(a) Compositions of HYPALON® *can be cured by exposure to moisture.*[9] Cure time is a function of moisture concentration and temperature, ranging from weeks at 122°F (50°C) and 50% relative humidity to a few hours in boiling water. It is postulated that water accelerates cure by hydrolyzing the sulfonyl chloride groups in the HYPALON® molecule, making them reactive with the metallic oxide curing agent.

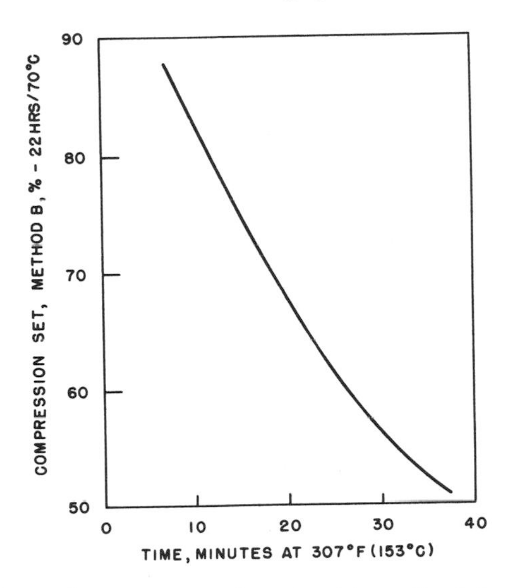

Fig. 13.6 Effect of cure time on compression set resistance (hard clay/magnesia compound).

(b) HYPALON® can also be cured by immersion in a concentrated *ammonia* solution or by exposure to an ammonia atmosphere.[10]

(c) Compounds of HYPALON® can be cured by *electron irradiation.* Cure cycles are short – similar to those used for CV and LCM curing with the added advantage of excellent processing safety because the rubber accelerator is removed from the compound. Electron accelerators are commercially available in the range of 0.3 to 3.0 MeV, and might prove economical for curing large volumes of products such as wire coverings, hose, and sheet goods.

(d) The LCM (*Liquid Curing Medium*) process is a practical method for continuously curing extruded products based on HYPALON®. It is possible to cure many cross-sections (various shapes and thicknesses) in 90 seconds at temperatures of 400 to 425°F. This curing system eliminates cutting extrusions to length for pan curing and the need for loading and unloading of a steam vulcanizer.

Extruded products cured by the LCM process require special compounding. Since it is necessary to eliminate entrapped air during processing, firm compounds are required. These can be obtained by using the high viscosity HYPALON® 40HV. In addition to using a high viscosity polymer, it is necessary to include 5 phr of calcium oxide in the compound to remove moisture present in the compounding ingredients and water evolved by the vulcanization reaction. Volatile ingredients must be avoided.

Curing temperatures should be kept below 425°F since at higher temperatures the formation of gases due to thermal degradation will cause sponging and blistering.

Applications

The unique combination of properties (e.g., colorability, weather resistance, resistance to corrosive materials, flame resistance, toughness, and environmental durability) makes HYPALON® adaptable in many domestic and industrial uses. Some of the more outstanding are reviewed.

Recent *automotive* uses have included headliner coatings, spark plug boots, primary and ignition wire, tarpaulins for trailers, and hose (as colored jacket or as liner in power steering hose).

HYPALON® is an ideal material for hot conveyor *belting*; for example, a belt of HYPALON® designed to handle hot salt as it comes from a kiln at 275 to 324°F has shown no signs of deterioration after 15 months of service. It is used to provide colored belt covers for domestic purposes and hand rails.

Coated fabrics are a major product. In addition to tarpaulins, they are used in making rain wear, boat covers, radomes, and inflatable structures.

Solutions of HYPALON® are used in *coatings* because the resulting films can be made in a wide color range and their weatherability is excellent. HYPALON® 30 is designed especially for use in coatings and is readily soluble in solvents having low to medium hydrogen bond strengths and solubility parameters within a range of 8.0

to 10.0. HYPALON® 20 is preferable in roof coatings when low temperature exposure is anticipated.

Solutions can be made from mill mixed compositions or directly from individual ingredients. Ball mills and rubber mills are used to ensure satisfactory dispersions of solid components. The shelf life of a solution of HYPALON® is directly dependent upon the base formulation of the compound. It is sometimes necessary to use two-part systems to avoid premature "set-up."

The color stability and long life of HYPALON® have proven its merit in such *construction materials* as curtain wall gaskets and roofing. Surfacing of fluid coatings based on HYPALON® offers a practical means of protecting steeply pitched, gracefully contoured roofs that highlight contemporary architecture. HYPALON® can be applied cold to the roof by brush, spray, or roller. It air cures to form a lightweight, elastic membrane over almost any substrate (plywood, cement asbestos board, metals, portland cement concrete).

HYPALON® is a *sheathing* for many types of wire and cable. HYPALON® is used for automotive ignition and primary wire, control cable, mine trailing cable, service drop wire, stationary power cable, RHW–RHH building wire, 90°C and 105°C appliance wire, and motor lead wire. HYPALON® also is used in cable connectors, sockets, line hose, insulating hoods and blankets, and other electrical accessories.

Insulating blankets made of HYPALON® wrapped around the tops of telephone poles during installation eliminate danger to workmen when the poles come in contact with live wires. Natural rubber blankets, sometimes used in this service, become badly cracked and cut from exposure to ozone and corona. HYPALON® can be used for line hose, where it effectively insulates sections of power cable during maintenance and repair operations.

Discharge *hose* made of HYPALON® has safely delivered 60° and 66° Be sulfuric acid at 30 psi pressure for over two years. Conventional rubber hoses char and contaminate the acid after two months of service. Hose made of HYPALON® has been successful in plating shops using chromic acid. It is finding increased usage in automotive and industrial hose where oil resistance is needed.

Fluid containment and control is an essential part of pollution abatement. *Liners* are needed as fluid tight barriers to contain thousands to millions of gallons of fluid. These liners are made of uncured HYPALON® synthetic rubber film. HYPALON® is excellent in this service as it resists oil, water, and most wastes. HYPALON® can be employed to *line tanks* used in manufacture of certain steel products.

HYPALON® is used for *industrial roll* coverings and continues to be used in white sidewall tires. In both instances HYPALON® is used for colorability, ozone resistance, and weatherability.

REFERENCES

(1) R. M. Murray and D. C. Thompson, *The Neoprenes*, 1963.
(2) *First and Second Order Transitions in Neoprene*, Du Pont Report BL-373, October, 1961.
(3) P. Kovacic, *Ind. Eng. Chem.*, **47**, 1090 (1955).
(4) Economies in Compounding and Processing Neoprene, Du Pont Neoprene Report, March, 1967.
(5) Cabot Technical Report KC-126.
(6) J. C. Carl, *Neoprene Latex*, 1962.
(7) G. Alliger and I. J. Sjothun, *Vulcanization of Elastomers*, Reinhold Publ., 1964, Chap. 8, p. 273–279.
(8) J. T. Maynard and P. R. Johnson, *Rub. Chem. & Tech.*, **36**, 963 (1963).
(9) C. E. McCormack and A. H. Fernandes, *Moisture Curing of Du Pont HYPALON®*, Du Pont Report, August, 1966.
(10) P. R. Johnson, I. C. Kogan, and F. J. Rizzo, *Crosslinking Chlorosulfonated Polyethylene with Ammonia*, Du Pont HYPALON® Report, October, 1964.

14

POLYSULFIDE RUBBERS

J. R. PANEK
Manager, Technical Communications
Thiokol Chemical Corporation
Trenton, N.J.

EARLY HISTORY

The first introduction of a polysulfide rubber for commercial application was made in 1930 by the Thiokol Chemical Corporation located at Yardville, New Jersey. This product was originally designated as THIOKOL ® Type A, and was based on the reaction product of ethylene dichloride and sodium tetrasulfide.

The condensation reaction of ethylene dichloride and sodium tetrasulfide is typical of a large number of reactions involving various organic halides and alkaline polysulfide solutions, as shown by Patrick.[1] The results of these studies brought about the introduction of a number of polysulfide rubbers by the Thiokol Chemical Corporation, to include the following standard types: "A," "B," "D," "F," "DX," "FA,"® "N," "ST," "PR-1," "WD-2 Crude," and "S-102." The Thiokol Chemical Corporation also marketed THIOKOL Type RD during the years 1942 – 1947. This rubber was not a polysulfide type, but rather an acrylonitrile based synthetic rubber. Many of these polymers were transitional and were replaced by polymers with either improved physical or chemical properties. In some cases, difficulties in production brought about the discontinuation of several products.

Type A, the original ethylene tetrasulfide polymer, introduced in 1930 for commercial applications, found immediate acceptance in a number of applications where its excellent solvent resistance was needed. Its high sulfur content of 84% made it resistant to most solvents, and because of this reason "A" is still being used today in a few applications where excellent solvent resistance is needed. With "A" there were many undesirable properties, such as tear gasing during milling and molding, poor low and high temperature resistance, and poor processing.

® Registered Trademark – Thiokol Chemical Corporation
TM – Trademark – Thiokol Chemical Corporation

Considerable formulation studies have been made and handling was improved by introducing from 5 to 20% of natural or nitrile rubber. There was an obvious loss in solvent resistance, but the products were still satisfactory in this respect. A second polymer Type "B" was introduced to overcome some of the undesirable features of "A." An excellent review covering Types A and B, as well as comparing these polymers with other polymers introduced up to 1940, is given by Wood.[2] Type B was introduced early in 1934 and was the condensation product based on dichlorodiethyl ether and sodium tetrasulfide. This polymer had a sulfur content of 64%, which resulted in a lowering of the solvent resistance properties. However, the low temperature flexibility was improved to −35°F and compounds based on "B" gave considerably less gasing and odor during milling and molding. The polymer gave poor physical properties and was more expensive to manufacture. Furthermore, production yields were poor due to the formation of thioxane in an undesirable side reaction during polymerization. The removal of thioxane during washing also caused odor problems. "B" found special application because of its unique resistance to mustard gas. Type B compounds were also readily dispersable in trichloroethane and such compounds were employed in a number of cements used for coating and impregnation fabrics. This polymer was discontinued and was eventually replaced.

Continued research in the polysulfide polymeric compositions resulted in an improved polymer based on the use of dichlorodiethyl formal and ethylene dichloride. This polymer was designated as "FA" and was introduced in 1939. The polymer was made as a disulfide. The substitution of dichlorodiethyl formal for dichlorodiethyl ether in the original reaction eliminated the formation of the undesirable thioxane. The polymer also had an improved low temperature flexibility of −50°F and was easier processing. The sulfur content was approximately 47%. It was expected that "FA" would replace all other types except "A." While its properties are not ideal, "FA" has found wide usage in hose, printing rolls, paper impregnated gaskets, molded goods, and flexible putties because of its excellent solvent resistance properties.

Type ST was introduced in 1943 and was the first radial departure from "A" through "FA." Whereas all the other types exhibited very poor compression set resistance, "ST" was modified to give compression set resistance. The polymer was based on dichlorodiethyl formal, using 2% trichloropropane, and was reacted with a blend of polysulfide solutions, which resulted in an average number of 2.0 sulfur atoms per repeating unit. The crude was further prepared in a millable plasticity and the polymer units had mercaptan terminals which necessitated an entirely different cure mechanism than formerly used. Another principal advantage regarding its use was that compounds prepared from this polymer could be dropped hot in a press, which resulted in considerable time savings. "ST," because of its different polymer structure and method of manufacture, is more expensive than "FA." It gives improved low temperature flexibility to −60°F. Because of its lower sulfur content of approximately 37%, "ST" compounds give poorer solvent resistance than "FA." "ST" has been somewhat restricted in its use because of poor

TABLE 14.1. HALIDES USED IN PREPARING POLYSULFIDE RUBBERS

Polysulfide Rubber Types	*Ethylene Dichloride*	*Dichloro-Diethylether*	*Dichloro-diethyl-formal*	*Cross-linking Agent*	*Sulfur Rank*[a]	*Sp. G.*	*Sulfur Content*
"A"	x				4	1.60	84%
"B"		x			4	1.51	64%
"FA"	x		x		2	1.34	47%
"ST"			x	x	2	1.27	37%

[a]Sulfur rank is the average number of sulfur molecules in the average repeating polymeric unit.

bin stability of both crude and mixed compounds. Where the bin stability of "FA" crude and mixed compounds is in the order of years, the stability of "ST" crude is in the order of months; and the stability of mixed compounds in the order of weeks, unless refrigeration is used. The chemical components used in manufacturing the various polysulfide rubbers are given in Table 14.1 along with other pertinent data. "A" has the highest sulfur content of the various commercial polysulfide rubbers sold by Thiokol Chemical Corporation.

The solvent resistance of several typical formulations based on the different polysulfide rubbers developed over the years is given in Table 14.2. These formulations are given in Table 14.3. The formulation based on Type A contains 20 parts of nitrile rubber, and is only slightly affected by the solvents listed. If the nitrile rubber were omitted, the solvent swell in benzene would be reduced to 2% after one month at room temperature, while all other values would be reduced to 0%. The sulfur content plays a major role in the solvent resistance. The higher sulfur content polymers, such as Types "A" and "B," gave better solvent resistance than the remaining types.

TABLE 14.2. SOLVENT RESISTANCE OF VARIOUS POLYSULFIDE RUBBER COMPOSITIONS (Values are expressed in % volume swell after one month minimum at 75°F)

Polysulfide Crude Formula	*Type A 1620*	*Type B 1707*	*Type FA 3000*	*Type ST 3600*
Solvent				
#2 Fuel Oil	5	0	9%	10%
Kerosene	2	0	9%	3%
Benzene	30%	54%	91%	114%
CCl_4	20	26%	39%	48%
Duco thinner	8	11%	64%	23%
Turpentine	2	6%	9%	5%
80/20 gasoline/ benzene	6	14%	12%	12%
50/50 gasoline/ benzene	13%	30%	26%	30%

TABLE 14.3. FORMULATIONS FOR THE VARIOUS POLYSULFIDE RUBBER COMPOSITIONS

Polymer Formula	*Type A 1620*	*Type B 1707*	*Type FA 3000*	*Type ST 3600*
Polymer	100	100	100	100
BTDS[a]	0.50	–	0.3	–
TMTDS[a]	–	0.25	–	–
DPG[c]	0.15	–	0.10	–
Zinc Oxide	10	10	10	–
Stearic Acid	0.5	0.5	0.5	–
SRF[d]	30	50	60	60
ZnO_2	–	–	–	4
Nitrile Rubber	20	5	–	–
Lime	–	–	–	1

[a]BTDS – benzothiazyldisulfide.
[b]TMTDS – tetramethylthiuramdisulfide.
[c]DPG – diphenylguanidine.
[d]SRF – semi-reinforcing furnace carbon black.

The effect of sulfur content also plays a major role in the low and high temperature resistance of the various polymers. Type A is very poor for low temperature, and has a brittle-point of 0°F. Because of the high sulfur content, "A" also after-hardens or crystallizes at room temperature with time, which makes the polymer more difficult to handle. "A" is not recommended for use above 120°F because of its greater susceptibility to cold flow. As the sulfur content decreases, the upper limit on temperature is increased. "ST" is recommended for continuous use at temperatures up to 212°F and for intermittent use at temperatures up to 300°F.

A comparison of the physical properties of the polysulfide polymers is given in Table 14.4. The table shows the poorer physical properties of types A and B, as compared to the other polymers. This was partially due to the high sulfur content, which gave the higher hardness, and to the hydrocarbon portion of the polymer.

EARLY APPLICATIONS FOR POLYSULFIDE RUBBER COMPOSITIONS

The introduction of THIOKOL Type A in 1930 created quite an impact on the rubber world since it was the first synthetic rubber made available for industrial application. Because of its excellent solvent resistance, it was immediately evaluated for use in all applications requiring any degree of solvent and chemical resistance. Type A was used and suggested for use in a multitude of applications in spite of problems encountered in processing as well as the physical limitations of the polymers. "A" was used until either a better handling polysulfide polymer with improved physical properties was introduced or until other synthetic rubbers made their appearance. In each case a reevaluation of the economic aspects had to be

TABLE 14.4. PHYSICAL PROPERTIES OF VARIOUS POLYSULFIDE RUBBER COMPOSITIONS

Polysulfide Polymers Formula	*Type A 1620*	*Type B 1707*	*Type FA 3000*	*Type ST 3600*
Tensile, psi[a]	790	560	1250	1250
100%, Modulus, psi	350	260	450	450
300%, Modulus, psi	740	500	800	1200
Elongation, %	370	370	400	310
Hardness, Shore A	78	78	71	70
Drop cold	yes	yes	yes	yes
Drop hot	no	no	no	yes
Compression set resistance[b]	no	no	no	yes
Gas formation on milling	yes	slight	no	no
Cure	50'/287°F	40'/287°F	40'/298°F	30'/287°F
Low Temp.[c] Flexibility limit	0°F	–40°F	–50°F	–60°F

[a] ASTM-D412-64 – Tension testing of vulcanized rubber.
[b] ASTM-D395-61 – Compression set of vulcanized Rubber, method B 22 hours at 158°F.
[c] ASTM-D1053-65 – Measuring low temperature stiffening.

made, along with a comparison of the advantages and disadvantages of the polymers being compared before a decision was reached.

In many cases, one polysulfide polymer was replaced by another that had better processibility and improved physicals. In others, the polymers were replaced by other synthetic rubbers to include the neoprenes, the acrylonitrile polymers, the butadine-styrene polymers, the polyisobutylene polymers, etc., because of improved physical properties where ultimate solvent resistance was not needed. For these reasons, many of the early applications involving polysulfide rubbers were short lived.

The oil and paint industries found immediate use for "A" for various hose applications such as paint spray hose, oil suction and discharge hose, gasoline hose, hose nozzles, etc. As the other polysulfide rubbers such as Types "B," "D," "F," "FA," "ST," etc., became available, comparisons were made with these polymers as well as with each synthetic rubber that was introduced. With the introduction of "A," considerable work was done on combinations of polysulfide rubbers with natural rubber to reduce cost and improve processibility and other properties.

One very early unique application for "A" was as a coating for asbestos fabric for use in the Wiggins floating roof[4] for minimizing volatilization of hydrocarbons from vast gasoline storage tanks. Because of the very low water permeability of "A" and other polysulfide polymers, they were used in various cable cover compounds both underwater and underground. "A" was found to prevent sulfur crystallization when used in small percentages and considerable use of such compositions has been made in sulfur cements. These compositions were also used as pipe joint sealing compounds and as a mortar in acid pickling tanks.

Several very natural applications for polysulfide rubber compounds were found where they were used in airplane fuel tanks as sealants,[5] as linings, concrete fuel storage tank linings, tank car linings, self-sealing aircraft tanks, and de-icers on wings. One very successful method was devised to line underground concrete tanks with a THIOKOL FA compound. A description covering the manufacture and installation of these linings is given by Crosby.[6] The thermoplastic nature of polysulfide polymers was used to best advantage in the use of these compositions as flame spray powders[7] and for preparing cavitation erosion resistant coatings on ships. On June 28, 1939, the *Dixie Clipper*, a Pan American Airways plane, made the first regular passenger ship crossing of the Atlantic Ocean. This airplane used polysulfide rubber in a number of strategic areas.[8]

In 1941, the Chrysler model used polysulfide based compositions in twenty-two different locations covering gaskets, hose, and hose assemblies. In 1944, bullet-proof gasoline tanks for aircraft were made using neoprene exterior coating for toughness, a gum rubber inner layer that had a very rapid swell in gasoline, and a polysulfide inner coating for gasoline resistance. Polysulfide rubbers went into the production of Mareng Cells, which were collapsible gasoline storage tanks. They were also used to line the interior of boxcars to convert this equipment for the conveyance of various fuels. Prestressed concrete underground gasoline storgae tanks were first lined with a Type FA compound in order to preserve octane ratings and prevent leakage.

PLASTICIZATION AND CURING

The three polysulfide rubbers presently sold can be divided into two distinct classes for the purpose of discussing plasticization and vulcanization.

Class I. The first class is comprised of the following polysulfide rubbers: Types "A" and "FA." These polymers are prepared as a tough composition that requires chemical plasticization for proper processing. These polymers are all characterized by the fact that they must be cooled in the mold before release and are all essentially cured through the use of zinc oxide. Because of their high molecular weight, they have very few terminals, which have been established by Fettes[3] to be hydroxyl groups. The methods of plasticization and vulcanization, along with the polymeric structure, have all contributed to the poor compression set resistance of these polymers.

Class I polymers will redistribute with disulfide linkages in accelerators such as MBTS (benzothiazyl disulfide) to chemically plasticize the polymers as shown in the following equation:

$$\underset{\text{Polymer}}{(\text{-SS-R}'\text{-S-S-R}''\text{-SS})} + \underset{\substack{\text{Rubber}\\\text{Accelerator}}}{\text{R}'''\text{SSR}'''} \longrightarrow \underset{\text{Softened Polymer}}{(\text{-SS-R}'\text{-SS R}''')} + \underset{\text{Softened Polymer}}{(\text{R}'''\text{ SSR}''\text{-SS-})}$$

This would also be true of TMTDS (tetramethylthiuram disulfide), TETDS (tetraethylthiuram disulfide), and other disulfide rubber accelerators. Other rubber accelerators, such as MBT, also have some activity.

The use of thiazole type accelerators as chemical plasticizers had an additional feature in that such plasticized compounds could be vulcanized with zinc oxide. The use of the thiuram accelerators as plasticizers gave compounds that would not vulcanize with heat with zinc oxide, so that the plasticization was permanent, except in the case of Types "A" and "B" which could be vulcanized. Even with these polymers, there was little advantage derived by using the thiuram accelerators which were more expensive, and the use of MBTS with DPG became fairly consistent. Where permanent putties were desired, both thiazoles and thiuram accelerators were used in higher quantities in the absence of zinc oxide. Such compounds remained putty-like after years of exposure.

The data shown in Table 14.5 illustrate the effectiveness of MBTS as a softener for a formulated "FA" compound. Williams plasticity values above 140 signify compounds that are tough and difficult to process. The range 120 to 140 is the range for calendering and tubing while 130 is considered to be optimum. Values below 120 signify compounds that are soft. Such compounds, if extruded in a tuber while hot, would collapse. Values in the range of 60 to 70 are desirable for permanent putties. The use of 0.3 grams of MBTS and 0.1 grams of DPG has been

TABLE 14.5. THE EFFECT OF VARIABLE MBTS[a] ON THE PLASTICITY OF THIOKOL TYPE FA

Weight of MBTS	*Williams Plasticity*[b]	
.15 g	188	
.20 g	164	
.25 g	145	
.30 g	129	ideal stock
.35 g	115	
.40 g	105	
.50 g	88	
.60 g	70	putty range

Base Formula

THIOKOL Type FA	100 g
Zinc oxide	10
SRF[c]	60
Stearic Acid	0.5
MBTS[a]	as shown above
DPG[d]	0.10

[a]MBTS – Benzothiazyldisulfide.
[b]ASTM Designation D 926-56 – Plasticity of parallel plate plastometer.
[c]SRF – Semi-reinforcing furnace carbon black.
[d]DPG – Diphenylguanidine.

standardized in the "base formula" shown in Table 14.5 for control purposes. Production lots of "FA" are tested for Williams plasticity in this formula and compounds giving values falling within the range of 120 to 140 are designated "M" and are segregated for formulations in calendering, tubing, and molding applications. Designations "S" and "H" for Soft and Hard are segregated for use in putty formulations and other applications.

The vulcanization of this group of polysulfide polymers is obtained by means of two simultaneous reactions. One involves the removal of the benzothiazole segments from the terminal sites and reforming the polymer, while the second has to do with further reacting the few hydroxyl terminals to give a cured composition with rubber-like properties. The removal of the benzothiazyl terminals where MBTS is used as a chemical plasticizer is accomplished by the use of zinc oxide as shown in the following equation:

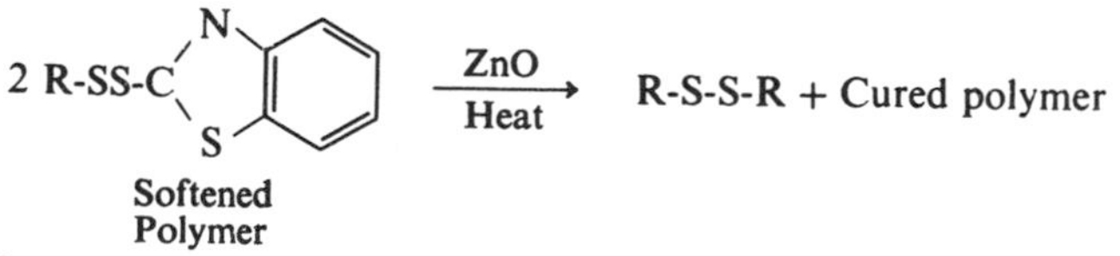

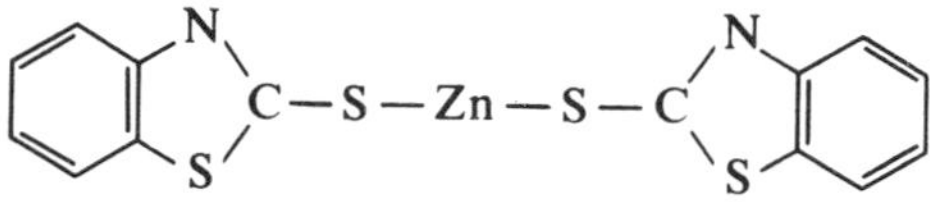

2 R-SS-C(benzothiazolyl) $\xrightarrow[\text{Heat}]{\text{ZnO}}$ R-S-S-R + Cured polymer

Softened Polymer

(benzothiazolyl)C—S—Zn—S—C(benzothiazolyl)

This reaction is the predominating one in the vulcanization process. The further polymerization of the polymer by reacting the hydroxyl terminals is also accomplished by zinc oxide.

The presence of the zinc salt of MBTS in the cured polymer is believed to be one of the reasons for the somewhat thermoplastic nature of this class of polymers. The fact that the zinc salts of several rubber accelerators show some activity as chemical plasticizers during processing point to possible reversible reactions under controlled conditions. The amount of zinc oxide generally recommended is somewhat more than that required to obtain an optimum cure. However, experiments have shown that this excess is desirable since it acts as a mildly alkaline buffer and improves the heat stability properties of cured compositions. Other metallic oxides have been found equally satisfactory as substitutes for zinc oxide. Among these are lead oxide, lead dioxide, cadmium oxide, zinc hydroxide, zinc borate, and zinc stearate. For various reasons, zinc oxide has been recommended in all standard formulations. Each of the various curing agents has an optimum time and temperature for most efficient vulcanization, and for optimum physical properties. The recommended cure time and temperature for the standard polymers and formulae are shown in Table 14.3.

Class II. The second class is presently comprised of Type "ST." This class of high molecular polysulfide polymers has mercaptan terminals. The "ST" does not require any chemical plasticization since it is made in the proper plasticity range and is vulcanized with oxidizing agents giving the cured polymer which can be dropped hot and has compression set resistance.

In manufacture, the "ST" has been depolymerized to give lower molecular weight polymers. The general reactions are as follows:

$$\text{R-S-S-R} + \text{NaSH} \longrightarrow \text{R-SSNa} + \text{R SH}$$
$$\text{R-S-S-Na} + \text{NaHSO}_3 \longrightarrow \text{R SH} + \text{Na}_2\text{S}_2\text{O}_3$$

Since the splitting of disulfide groups occurs along the chain, all of the polymer segments become dimercaptans. As the batches are manufactured, the Mooney viscosity is determined, and blends are made to fall within the range of 25 to 35. With these polymers, there is a small but gradual increase in the molecular weight with time, which becomes evident by an increase in the Mooney viscosity. This is due to the ease with which the mercaptan terminals can be oxidized. In case

TABLE 14.6. CURING AGENTS FOR MERCAPTAN–TERMINATED POLYSULFIDE RUBBERS

Class	*Compounds*	
Metallic Oxides	ZnO	
	PbO	BaO
	FeO	Cr_2O_3
	CaO	CuO
	Sb_2O_3	CoO
		Pb_3O_4
	As_2O_3	MgO
Metallic Peroxides	ZnO_2	SeO_2
	PbO_2	TeO_2
	CaO_2	BaO_2
	MgO_2	MnO_2
Inorganic Oxidizing Agents	$ZnCrO_4$	$(NH_4)_2Cr_2O_7$
	$PbCrO_4$	$NaCrO_4$
	$KCrO_4$	
Organic Oxidizing Agents	urea peroxide	
	tertiary butyl hydroperoxide	
	tertiary butyl perbenzoate	
	dinitrobenzene	
	trinitrobenzene	
	benzoyl peroxide	
	cumene hydroperoxide	
	stearoyl peroxide	
	lauroyl peroxide	
	methyl ethyl ketone peroxide	
Quinoid Compounds	p-quinone dioxime	quinone
	dimethylglyoxime	other dioximes
Sulfur-Containing Compounds	benzothiazyldisulfide	
	dinitrobenzene salt of mercaptobenzothiazole	mercaptobenzothiazole

softening is necessary, small amounts of liquid polymers LP-2 or LP-3 can be added during processing.

The vulcanization of Type "ST" is accomplished by oxidation of the mercaptan terminals as illustrated:

$$X\ \text{HSR-SSR-SH} + [\text{O}] \longrightarrow (\text{-S-R-SS-R-S})_x + H_2O$$

Because of the relatively few mercaptan terminals present on these polymer crudes, a number of oxidizing agents have been found that will vulcanize these polymers satisfactorily, but will not completely cure the lower molecular weight mercaptan terminated polysulfide liquid polymers. General classes of compounds which cure these polymers are shown in Table 14.6.

The use of many curatives is not completely satisfactory. Many mixed compounds are very unstable and cure before the compounds can be properly processed. Except for this one feature, lead dioxide would be an excellent curing agent. In other cases, many compounds do not exhibit any compression set resistance upon cure.

Relatively few systems have been found which give satisfactory performance with respect to compression set resistance, bin stability, and heat resistance. One of these systems is based on the use of 1.5 parts of p-quinone dioxime and 0.5 parts of zinc oxide. A second system is based on 6 parts of zinc peroxide. A third system which has better heat resistance and compression set resistance, but is poorer in bin

TABLE 14.7. THE EFFECT OF VARIABLE CARBON BLACK ON THE PHYSICAL PROPERTIES OF THIOKOL TYPE FA

Carbon Loading	*Tensile Strength*	*Elongation, %*	*Hardness, Shore A*[d]
0 grams	155 psi	450	40
10	300	550	45
20	600	700	49
40	1000	600	58
60-most practical	1200	380	68
80	1200	230	78
100	100	210	82

All Compounds cured 40 minutes at 298°F

Base Formula

Thiokol Type FA	100 grams
Zinc oxide	10
Stearic Acid	0.5
MBTS[a]	0.3
DPG[b]	0.1
SRF carbon black[c]	variable as shown above

[a]MBTS – benzothiazyldisulfide.
[b]DPG – diphenylguanidine.
[c]SRF – a semi-reinforcing furnace carbon black.
[d]Hardness – as measured by ASTM Designation D-676-49T, "Indentation – of Rubber by Means of a Durometer."

stability, is based on 4 parts of zinc peroxide and 1 part of lime to 100 parts of "ST."

REINFORCEMENT

All the polysulfide rubbers exhibit very poor properties when vulcanized without the use of reinforcing fillers. A good illustration of the effect of increasing carbon black loading on the physical properties of "FA" is shown in Table 14.7. As indicated, the tensile properties are very poor until at least 30 parts of carbon black are used. Most polysulfide rubbers are formulated with 60 parts of carbon black, since this loading of carbon black gives a lower cost and displays optimum physical properties.

As with other rubbers, both natural and synthetic, the particle size of the carbon blacks has a great effect on the physical and chemical properties. Table 14.8 shows

TABLE 14.8. THE EFFECT OF VARIOUS REINFORCING FILLERS ON THE PHYSICAL PROPERTIES OF THIOKOL TYPE ST

Parts Loading	*Filler Description*	*Tensile psi*	*300% Modulus*	*Elongation %*	*Hardness*
60 parts	Med. thermal black	800	300	600	50
60 parts	Fine thermal black	940	630	520	55
60 parts	Semi-reinforcing furnace black	1275	1130	390	62
60 parts	Fine furnace black	1675	1390	380	67
60 parts	Easy processing channel black	1685	1485	340	70
60 parts	Med. processing channel black	1850	1555	350	70
60 parts	Hard processing channel black	2010	1560	380	70
65 parts	Hydrated alumina	1070	395	650	62
120 parts	Brown iron oxide	1050	340	650	57
70 parts	Calcium carbonate	155	155	340	45
70 parts	Clay	70	70	640	18
145 parts	Chromium oxide	505	380	500	51
110 parts	Titanium dioxide	725	300	710	52

Cures on the above formulations were 30 minutes at 287°F

Base Formula

Thiokol Type ST	100
Stearic Acid	3
GMF[a]	1.5
Zinc oxide	0.5
Reinforcing pigment (equivalent to 60 parts carbon by volume)	

[a]GMF – p-quinonedioxime.

the effect of various carbon blacks, as well as nonblack fillers, on the physical properties of an "ST" compound. All fillers have been varied so that they are equivalent to 60 parts volume loading of carbon black.

The types of carbon blacks that can be used with the various polysulfide crudes are limited. The very soft and large particle size carbon blacks such as an MT carbon black or an FT carbon black do not give optimum physical properties. They are used, however, where high volume loading is necessary in order to lower costs. The semi-reinforcing carbon blacks are most commonly used. The finer sized carbon blacks such as EPC or HPC are seldom used because the mill mixed compounds become very hard, dry, and difficult to process.

SOLVENT RESISTANCE

The major importance of the polysulfide polymers is their excellent solvent resistance. These polymers have been evaluated against a number of solvents and chemical solutions. Representative solvents are listed in Table 14.9 for three standard formulations shown in Tables 14.3 and 14.4 and are compared in their respective volume swell values. The three standard formulations are very satisfactory in most solvents, but are not recommended for use against strong oxidizing acids in any concentration. The compounds will withstand dilute hydrochloric acid and sulfuric acid. Compounds 3000 FA and 3600 ST are not generally recommended for use with benzene or cresylic acid unless the high volume swell can be tolerated. Otherwise both 3000 FA and 3600 ST are satisfactory for use against the remaining solvents in Table 14.9.

The chief advantage of blending polysulfide rubbers with other synthetic rubbers is that the resulting compounds give improved processing. With the exception of SBR, there is little serious affect on the physical properties of the polysulfide rubbers where the modifying synthetic rubber is kept at 20% or less. There is no particular advantage obtained in going to higher percentages of the modifier since the solvent resistance properties approach the values for the modifier. Solvent resistance data are shown for "FA" modified with 20 parts of various synthetic rubbers in Table 14.10. As shown in the table, other synthetic rubbers which have poor resistance in certain classes of solvents carry this property into the blend. Blends of "FA" with nitrile rubber show poorer resistance to ketones and ester, but are not greatly affected in the chlorinated solvents, whereas the reverse is true with Neoprene W. Knowing the performance required would greatly help in selecting the modifier. Blends with butyl rubber were sluggish in cure, but otherwise were good in solvent resistance. Blends with Hypalon gave shorter elongations, which is undesirable.

MISCELLANEOUS PROPERTIES

Ozone Resistance. Tests on relative ozone resistance were conducted on several polysulfide compounds in the Mast Development apparatus. The ozone

TABLE 14.9. COMPARISON OF SOLVENT RESISTANCE OF THREE STANDARD POLYSULFIDE RUBBER COMPOUNDS (% VOLUME SWELL AFTER ONE MONTH AT 77°F)[a]

Solvent	*1620 AH*	*3000 FA*	*3600 ST*
Benzene	30%	96%	114%
Toluene	24	55	79
Xylene	17	31	39
Carbon Tetrachloride	20	36	48
SR-6[b]	6	10	10
SR-10[c]	−2	1	1
Glacial acetic acid	10	21	18
Cresylic acid	27	83	123
Butyl acetate	17	17	35
Dibutyl phthalate	6	7	8
Linseed oil	0	0	0
Ethyl alcohol	0	2	5
Ethyl glycol	0	1	3
Glycerol	0	2	1
Methyl ethyl ketone (MEK)	34	28	49
Methyl isobutyl ketone (MIK)	24	13	25
10% HC1	2	2	2
100% HC1	D	D	D
20% NaOH	1	2	2
10%HNO_3	D	D	D
10% H_2SO_4	2	2	2
50% H_2SO_4	D	D	D
Water	3	5	5

D=Decomposed
[a]ASTM Designation D-471-64T "Change in Properties of Elastomeric Vulcanizates Resulting from Immersion in Liquids."
[b]40% Aromatic blended reference fuel – 60% isooctane, 20% xylene, 15% toluene, 5% benzene.
[c]100% isooctane.

concentration was 50 pphm using an air change of 6 cfm at 100°F. Observations were made on a triangular test specimen following procedure ASTM D-1171–61. The specimens were observed on the exposed ridge for cracking not visible at 1X magnification, but visible at 2X magnification. A comparison is shown in Table 14.11 with GR-S 1502, a butadiene styrene copolymer, natural rubber, and several butyl rubber formulae.

Oxidation Resistance. One study is shown in Table 14.12 where formulations based on "ST" are aged in an oxygen bomb and then compared in physical properties. It is interesting to note that certain cure mechanisms are better than others in imparting oxygen resistance. In this study, the lead dioxide cure is the most resistant. The GMF cure is not very satisfactory on exposure to oxygen.

Vapor Permeability. Polysulfide compounds have low vapor permeability to many solvents. Mueller shows considerable data which compare the specific permeability of a standard 3600 ST composition in methyl alcohol, carbon

TABLE 14.10. THE SOLVENT RESISTANCE OF 3000 FA MODIFIED WITH VARIOUS SYNTHETIC ELASTOMERS (% VOLUME SWELL AFTER ONE MONTH AT 77°F)

Solvent	*Control 3000 FA*	*20 Parts SBR Rubber*	*20 Parts Nitrile Rubber*	*20 Parts Neoprene W*	*20 Parts Hypalon S-2*	*20 Parts Butyl 215*
Benzene	96%	203%	107%	119%	99%	121%
Toluene	55	135	78	93	65	89
Xylene	31	100	51	78	25	75
SR-6	10	40	15	26	25	27
SR-10	1	8	2	5	1	5
Water	5	2	13	10	4	1
CCl_4	36	150	38	73	50	96
MEK	28	40	119	39	27	28
MIK	13	30	69	33	24	15
Butyl Acetate	17	28	33	25	18	18

TABLE 14.11. OZONE RESISTANCE OF POLYSULFIDE RUBBERS

Formula	*Hours to 2X Visible Cracking*[a]
Thiokol 1620 AH	312 hours
Thiokol 3000 FA	312 hours
Butyl 035	312 hours
Butyl 325	76 hours
Thiokol 3600 ST	16 hours
GR-S 1502	2 hours
Natural Rubber	2 hours

[a]Cracking not visible at IX magnification, but visible at 2X magnification. Tests run in triangular test strip or molded and tested under ASTM Designation. The above are typical formulations using carbon black with no protective natioxidants.

TABLE 14.12. OXYGEN BOMB AGING OF POLYSULFIDE RUBBER COMPOUNDS

	1	*2*	*3*
THIOKOL Type ST	100	100	100
SRF[a]	60	60	60
Stearic Acid	1.0	1.0	1.0
Zinc Oxide	0.5	–	–
GMF[b]	1.5	1	–
Zinc Chromate[d]	–	10	–
Lead Dioxide[c]	–	–	3
Original Physical Properties on Sheets Cured 30 Minutes at 287°F			
Tensile strength, psi	1400	1300	1575
200% Modulus, psi	975	1025	1225
Elongation, %	290	270	270
Hardness, Shore A	68	68	72
Physical Properties on Specimens Aged for 10 Days at 70°C			
Tensile strength, psi	25	1075	1325
200% Modulus, psi		725	900
Elongation, %	350	310	350
Hardness, Shore A	28	70	75

[a]SRF – a semi-reinforcing furnace carbon black.
[b]GMF – paraquinone dioxime.
[c]Lead Dioxide – technical grade.
[d]Zinc Chromate – commercial paint grade.

TABLE 14.13. SPECIFIC PERMEABILITY[a] OF 3600 ST IN VARIOUS SOLVENTS AT VARIOUS TEMPERATURES[c]

	Methyl Alcohol	*Carbon Tetrachloride*	*Ethyl Acetate*	*SR-6*[b]	*Benzene*	*Diisobutylene*
75°F	.005	.042	.150	.019	.540	0
100°F	.010	.065	.190	.029	.690	0
125°F	.025	.100	.270	.039	.920	0
150°F	.055	.160	.380	.056	1.30	0
180°F	.140	.210	.510	.080	1.80	.001

[a]Specific permeability – $\frac{(oz)(in.)}{(24\ hr)(ft^2)}$
[b]SR-6 is 60% diisobutylene, 20% toluene, 15% xylene, and 5% benzene by volume.
[c]Data obtained from Thesis for Master of Science Degree by William J. Mueller, Ohio State University, 1952.

tetrachloride, ethyl acetate, SR-6, benzene, and diisobutylene over a temperature range of 75 to 180°F. The 3000 FA is also compared to SB-R, neoprene, a low nitrile rubber, and a high nitrile rubber, in these solvents. Data in Table 14.12 compare specific permeability of 3600 ST in the various solvents. Table 14.13 compares 3600 ST with a low nitrile synthetic rubber, a high nitrile synthetic rubber, SB-R, and a chloroprene compound in specific permeability of SR-6 at various temperatures.

MAJOR APPLICATIONS

"FA" Rollers. Today "FA" finds one of its major uses in the manufacture of rollers. These rollers are employed for lacquering cans, roller and grain coating of paint on metal, and for the applications of quick drying inks for printing. The reason for the acceptance of "FA" in this field is its overall excellent resistance to the many active solvents involved, permitting long dimensional stability of the

TABLE 14.14. COMPARISON OF 3600 ST WITH STANDARD COMPOSITIONS[c] OF VARIOUS SYNTHETICS IN SPECIFIC PERMEABILITY[a] IN SE-6[b]

	3600 ST	*SBR*	*Low Nitrile*	*High Nitrile*	*Chloroprene*
75°F	.019	2.00	0.40	0.17	0.53
100°F	.029	2.30	0.52	0.19	0.62
125°F	.039	3.00	0.75	0.22	0.75
150°F	.056	3.80	1.10	0.30	1.10
180°F	.080	4.00	1.60	0.37	1.60

[a]Specific permeability – $\frac{(oz)(in.)}{(24\ hr)(ft^2)}$
[b]SR-6 is 60% diisobutylene, 20% toluene, 15% xylene, and 5% benzene by volume.
[c]Data obtained from Thesis for Master of Science Degree by William J. Mueller, Ohio State University, 1952.

TABLE 14.15. THIOKOL FA RUBBER COMPOUNDS FOR ROLLERS

Masterbatch

Thiokol FA	100
MBTS	0.35
DPG	0.10
Neophax A	20

Hardness, Shore A:	*30*	*35*	*40*	*50*	*60*
Masterbatch, pbw	120	120	120	120	120
Neoprene W	20	20	20	20	20
E. L. Magnesium Oxide	0.8	0.8	0.8	0.8	0.8
NA-22	0.1	0.1	0.1	0.1	0.1
Zinc Oxide	10	10	10	10	10
Stearic Acid	0.5	0.5	0.5	0.5	0.5
SRF Black	35	40	45	50	65
P-25 Cumar	10	10	10	10	10
Kenflex N	35	30	20	–	–

rollers. Roller compounds are usually mill mixed (sometimes Banbury), calendered, built ply-by-ply around a cemented metal core, wrapped tightly in nylon or cotton tape, open steam cured, and ground to the diameter desired. Five roller compounds for obtaining a Shore A hardness range of 30 to 60 are shown in Table 14.15.

As indicated, a masterbatch is employed blending the "FA" with MBTS, DPG and a factice (Neophax A in this case). The DPG activates the chemical softening effect of MBTS. The factice aids in processing ease and helps hold high levels of plasticizer, when used. It is recommended that the softer masterbatch be blended completely into the tougher neoprene before adding the dry ingredients. The oil goes in last. The Neoprene W and the Cumar provide the necessary tack for building operations. The hardness is controlled by the proper ratio of carbon black to oil (Kenflex N), or by the proper amount of black alone for the 50 and 60 hardness grades.

Table 14.16 provides the physical properties for the roller compounds as shown in the preceding table. The tensile strength and modulus go up with increasing hardness due to the increasing amount of carbon black and decreasing amount of oil, whereas the elongations tend to decrease.

TABLE 14.16. PROPERTIES OF ROLLER COMPOUNDS

Compound Hardness, Shore A	*30*	*35*	*40*	*50*	*60*
200% Modulus, psi	150	180	310	610	790
Tensile, psi	450	470	630	960	965
Elongation, %	530	430	410	290	230

TABLE 14.17. TYPICAL HOSE LINER FORMULA

		Neoprene Masterbatch	
Thiokol FA	100.0	Neoprene W	100.0
Zinc Oxide	10.0	ELC Magnesia	4.0
SRF Black	60.0	SRF Black	55.0
Stearic Acid	0.5	Stearic Acid	0.5
MBTS	0.4	Zinc Oxide	5.0
DPG	0.1		164.5
NA-22	0.1		
Neoprene Masterbatch	29.0		

	Original Physicals (Mold Cure: 40 min at 298°F)	**Physicals After Immersion in SR-6 for 48 hr at R.T.**
Tensile, psi	1380	900
Elongation, %	300	230
Shore A Hardness	74	–

Volume Swell after Immersion for 30 Days at 80°F

SR-6	20%	MEK	36%
SR-10	4	MIK	28
Ethyl Acetate	24	CCl_4	60
Acetone	18	Water	4

"FA" Hose Liner. Another major application is in hose liner. Again, the active solvents used in fluids are the reason for "FA" use. Hoses lined with "FA" are widely used for paints, paint thinners, lacquers, and aromatic hydrocarbons. A typical formulation is shown in Table 14.17.

The above formula produces an excellent liner for this end-use requirement. The Neoprene W role is to help the processing, particularly during the extrusion and braiding operations. However, rather recently with more precise control of compound viscosity, it has been found that neoprene can be entirely eliminated. Of course, in the case where the neoprene is not used, the NA-22 is also deleted from the formula in the left column. This neoprene removal permits better resistance to the solvents than is shown in the preceding table.

Type ST Gas Meter Diaphragms. An important application for Type ST is in the gas meter diaphragms. The diaphragms require excellent low temperature flexibility and resistance to drip oils, aromatics, and aliphatic hydrocarbons which come from the gas. The "ST" compound is calender-coated onto fabric at the thickness of 0.10 to 0.20 on each side, and then molded into the desired shape. Two of these are used in practically all of the gas meters in our homes today.

Type A Rubber/Sulfur Cements. The primary use for Type A is as a flexibilizer for sulfur. From 2 to 5 parts of Thiokol Type A, dissolved in molten sulfur, prevents straight sulfur from crystallizing. Sulfur, used as a mortar for acid pickling tanks, water sewers, and oil pipes, is brittle and can crack easily due to impact or thermal expansion and contraction. Type A prevents this breaking or cracking and still permits the molten sulfur to be poured into place. Probably the effect of Type A is to keep cooled sulfur in the amorphous state as opposed to its normal crystalline state.

REFERENCES

(1) J. C. Patrick, U.S. Pat. 1,854,423 (April 19, 1932); British Pat. 359,000 (September 19, 1929).
(2) L. A. Wood, *India Rubber World,* **102**, 33 (1940); *Chemical Abstracts,* **34**, 6481 (1940).
(3) E. M. Fettes, J. S. Jorczak, and J. R. Panek, *Ind. Eng. Chem.,* **46**, 1539 (1954).
(4) J. C. Patrick, British Pat. 360,890 (May 2, 1930); *Chemical Abstracts,* **27**, P1114 (1933).
(5) R. B. Gray, and H. W. Shealey (to Glen L. Martin Company) U.S. Pat. 2, 140,670 (December 20, 1938).
(6) J. W. Crosby, *Rubber Age,* **53**, 240 (1943).
(7) E. A. Bukzin, *Rubber Age,* **67**, 681 (1950); *Chemical Abstracts,* **45**, 1372 (1956).
(8) *Thiokol Facts,* Volume 1, No. 18, Thiokol Chemical Corporation (1939).

15

SILICONE RUBBER

W. J. BOBEAR
General Electric Company
Silicone Products Department
Waterford, New York

INTRODUCTION

Silicone rubber has attained an annual domestic sales volume of over 17.5 million pounds[1] due to a combination of properties that are quite unique relative to organic elastomers. These properties are, of course, dependent upon the unusual molecular structure of the polymer, which consists of long chains of alternating silicon and oxygen atoms encased by organic groups. These chains have an organic-inorganic nature; and, compared to organic rubber polymer chains, they have a large molar volume and very low intermolecular attractive forces. The molecules are unusually flexible and mobile, and can coil and uncoil very freely over a relatively wide temperature range.

It is, therefore, not surprising that the most outstanding property of the silicone elastomers is a very broad service temperature range that far exceeds that of any other commercially available rubber. The silicones can be compounded to perform for extended periods at −150°F to +600°F under static conditions, and at −100°F to +600°F under dynamic conditions.

Young's modulus of an extreme low temperature silicone rubber shows very little change down to −100°F, and eventually reaches 10,000 psi at −150°F. The tensile strength, measured at room temperature, is less than that of most organic rubbers; however, it is superior when measured at 400°F. Also, at 400°F, the silicone rubber has an estimated useful life of 2 to 5 years, while most organics will fail within a few days. It is important to note that the silicone rubber will maintain its elastomeric properties almost indefinitely at moderately elevated temperatures. Life has been estimated at 5 to 10 years at 300°F, and 10 to 20 years at 250°F.

Silicone rubber performs unusually well when used as a gasket or O-ring in sealing applications. Over the entire temperature range of −120°F to +500°F, no available elastomer can match its low compression set.

Many types of wire and cable are insulated with silicone rubber, mainly because its excellent electrical properties are maintained at elevated temperatures. Even when the insulation is exposed to a direct flame, it burns to a nonconducting ash which continues to function as insulation in a suitably designed cable.

The ozone and corona resistance of silicone rubber is outstanding, approaching that of mica. These properties are important in many electrical applications, and in exposure to outdoor weathering.

Many samples of elastomeric silicones have been exposed to outdoor weathering for 15 years with no significant loss of physical properties. This demonstrates unique resistance to temperature extremes, sunlight, water, and ozone and other gases. The rubber will not support fungus growth if properly cured, and it has good resistance to the low concentrations of acids, bases, and salts normally found in surface water. It has been estimated that a silicone elastomer will last in excess of 30 years under weathering conditions that would cause the best organic rubbers to fail within a few years.

Silicone rubber is odorless, tasteless, and nontoxic. When properly fabricated, it does not stain, corrode, or in any way deteriorate materials with which it comes in contact. Consequently, it has found application in gas masks, food and medical grade tubing, and even in surgical implants in the human body.

APPLICATIONS

Due primarily to its exceptional mechanical and electrical performance under extreme temperature conditions, silicone rubber is widely used in hundreds of commercial and military applications. In many uses, its inertness, nontoxicity, ease of processing, and resistance to ozone and weathering are also of critical importance.

The following applications, listed by industry, illustrate the amazing versatility of this unique specialty elastomer.

A. *Aerospace*
   - Aerodynamic balance and control surface seals
   - Airframe opening seals (doors, hatches, etc.)
   - Hot air ducts (for de-icing, cabin heating, etc.)
   - Dust shields and limit switch boots
   - Air and oxygen pressure regulator diaphragms
   - Jet engine starter hose
   - O-rings, seals, and gaskets for lubricating and hydraulic systems
   - Airframe and spacecraft body sealants
   - AN connector inserts
   - Aircraft and missile wire insulation
   - Missile external and base heating insulation

B. *Automotive Industry*
   - Spark plug boots
   - Ignition cable jacket
   - Transmission seals
   - Sealants
   - Hose

C. *Appliances*
Oven door and washer-dryer gaskets
Seals, gaskets, and insulation in steam irons, frying pans, coffee makers, etc.

D. *Electrical Industry*
Capacitor bushings
Rubber coated glass sleeving
Rubber tubing
Electrical potting, impregnation, and encapsulation
Unsupported and cloth supported electrical insulating tapes
Television corona shields
Apparatus lead wire
Appliance and fixture wire
Defroster harness wire
Electronic hook-up wire
Nuclear power cable

E. *Construction Industry*
Construction sealants for expansion joints and glazing applications
Weather coatings (wall, roof and deck)

F. *Miscellaneous Applications*
Rubber rolls
Sponge
Flexible mold fabrication
Prosthetic devices
Pharmaceutical stoppers
Medical tubing

THE SILICONE RUBBER POLYMER

The commercial synthesis of polydimethylsiloxane usually involves the direct process[2–4] for manufacture of the intermediate dimethyldichlorosilane from methyl chloride and elemental silicon.

1. $SiO_2 + 2C \rightarrow Si + 2CO$
2. $CH_3OH + HCl \rightarrow CH_3Cl + H_2O$
3. Direct Process
   $2\,CH_3Cl + Si \xrightarrow[\text{catalyst}]{\text{heat}} (CH_3)_2SiCl_2$

   Though it is possible to direct this reaction to produce large amounts of the dimethyldichlorosilane, a mixture of silanes is actually formed. The desired product must be distilled from the reaction mixture.
4. $(CH_3)_2SiCl_2 + 2H_2O \rightarrow (CH_3)_2Si(OH)_2 + 2HCl$
   The dimethylsilanediol is not stable, and it continues to condense, with the evolution of water, to form a mixture of linear and cyclic polydimethylsiloxanes of relatively low molecular weight. The cyclic siloxanes are separated, and then polymerized to high molecular weight by heat in the presence of acidic or basic catalysts.

The resulting polymer, used in the commercially available silicone rubber stocks, is a very high viscosity fluid or gum, which is composed mainly of linear polydimethylsiloxane chains. The commercial gums usually contain between 3000

$$(CH_3)_3SiO\left[\begin{array}{c} CH_3 \\ | \\ Si-O \\ | \\ CH_3 \end{array}\right]_{n = 3000 \text{ to } 10{,}000} Si(CH_3)_3$$

and 10,000 dimethyl siloxy units in the average chain. These polymers have essentially the most probable molecular weight distribution, although this can vary with the purity of the monomers.

If less than one half of one percent of the methyl groups are replaced by vinyl groups ($-CH=CH_2$), the resulting polymer makes more efficient use of peroxide vulcanization agents, requires less peroxide for cure, and forms a vulcanizate that is more resistant to the rearrangements that cause reversion and high compression set. As a consequence, nearly all commercial gums and compounds now contain the vinyl modified polymers.

Dimethyl silicone rubber tends to become stiff below −60°F. However, the low temperature flexibility may be improved by substitution of phenyl ($-C_6H_5$) or ethyl ($-CH_2CH_3$) groups for some of the methyl groups attached to the silicon atoms in the polymer chain. Replacement of only 5 to 10% of the methyl groups by phenyl groups will lower the crystallization temperature and extend the useful service temperature range to below −130°F.

The dimethyl rubbers swell more in aliphatic and aromatic hydrocarbons than they do in acetone and diesters. As shown in Fig. 15.1, this performance may be

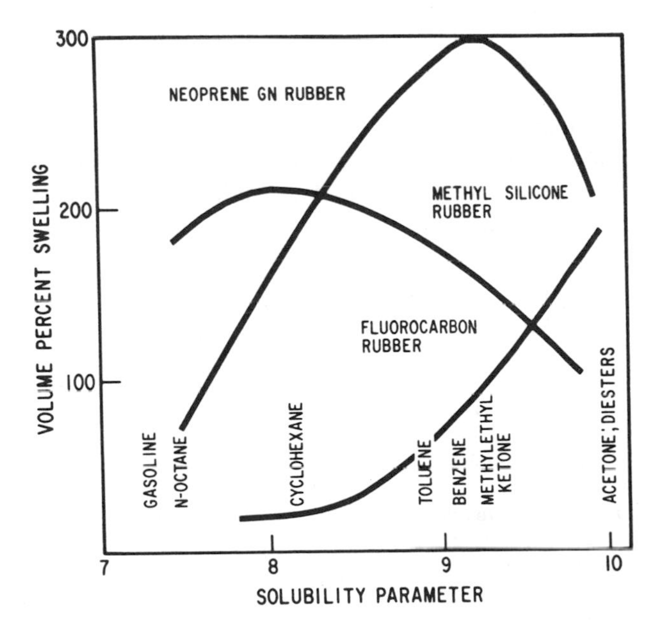

Fig. 15.1 Variation of volume swell with solubility parameter.

reversed by replacement of one methyl group on each silicon atom by a more polar group. Commercially available polymers of this type contain the trifluoropropyl group ($-CH_2CH_2CF_3$). These polymers have a brittle point of −80 to −90°F.

VULCANIZATION

Silicone rubber compounds are normally heat cured in the presence of one or more of the organic peroxides shown in Table 15.1. Some other peroxides are used to a limited extent; however, the first four in the table are the most important ones.

In the case of the cure of a dimethyl polymer with a diaroyl peroxide, the following cure mechanism has been fairly well established:[4,5]

$$1.\quad ROOR \longrightarrow 2RO^{\bullet}$$

$$2.\quad -\underset{\underset{CH_3}{|}}{\overset{\overset{CH_3}{|}}{Si}}-O- \; + RO^{\bullet} \longrightarrow -\underset{\underset{CH_3}{|}}{\overset{\overset{\dot{C}H_2}{|}}{Si}}-O- \; + ROH$$

$$3.\quad \begin{array}{c} CH_3 \\ | \\ -Si-O- \\ | \\ CH_2 \\ : \\ CH_2 \\ | \\ -Si-O- \\ | \\ CH_3 \end{array} \longrightarrow \begin{array}{c} CH_3 \\ | \\ -Si-O- \\ | \\ CH_2 \\ | \\ CH_2 \\ | \\ -Si-O- \\ | \\ CH_3 \end{array}$$

It is apparent that these reactions could produce no more than 1 mole of chemical crosslinks per mole of peroxide. The hydrogen abstraction reaction has been estimated at 50% efficient, and the ethylene bridge formation, at 40% efficient. And, indeed, independent work[6] has shown that the actual crosslink yield

TABLE 15.1. ORGANIC PEROXIDES USED FOR SILICONE RUBBER VULCANIZATION

Peroxide	*Temperature, °F, for half life = 1 minute*
Bis(2,4-dichlorobenzoyl) peroxide	234
Di-benzoyl peroxide	271
Di-cumyl peroxide	340
2,5-dimethyl-2,5-bis(t-butyl peroxy) Hexane	354
Di-tertiary butyl peroxide	379

in an unfilled polymer is 0.1 to 0.3 moles of chemical crosslinks per mole of diaroyl peroxide. This corresponds to a chemical crosslink density of 0.6 to 1 x 10^{-5} moles of crosslinks per gram of polymer. It is not practical to increase the crosslink concentration much further by raising the peroxide level.

The cure mechanism for the methyl vinyl siloxy containing co-polymers is unknown. However, one of the proposed mechanisms[4,7,8] seems to account for available experimental observations:

1. $$\text{ROOR} \xrightarrow{\text{heat}} 2\text{RO}^{\bullet}$$

2. $$\begin{array}{c} CH_3 \\ | \\ -Si-O- \\ | \\ CH{=}CH_2 \end{array} + RO^{\bullet} \longrightarrow \begin{array}{c} CH_3 \\ | \\ -Si-O- \\ | \\ {}^{\bullet}CH \\ | \\ CH_2OR \end{array} \qquad \begin{array}{c} CH_3 \\ | \\ -Si-O- \\ | \\ CH_3 \end{array} \longrightarrow \begin{array}{c} CH_3 \\ | \\ -Si-O- \\ | \\ CH_2 \\ | \\ CH_2 + RO^{\bullet} \\ | \\ CH_2 \\ | \\ -Si-O- \\ | \\ CH_3 \end{array}$$

3. Additional cure steps followed by termination.

This mechanism predicts more than 1 mole of crosslinks per mole of peroxide, and not more than 1 mole of crosslinks per mole of vinyl groups.

In the case of unfilled methyl vinyl siloxy containing copolymers, actual crosslink yields have also been measured.[6] Data on a polymer with Vi/Si = 0.0026 are shown in Table 15.2.[6, 9] Efficiencies are high, and these results are consistent with the "trimethylene bridge" cure mechanism.

2,5-dimethyl-2,5-bis (t-butyl peroxy) hexane, di cumyl peroxide, and di-t-butyl peroxide are usually considered to be "vinyl specific" in that they will give good cures only with vinyl containing polymer. The two diaroyl peroxides, benzoyl and bis (2,4-dichlorobenzoyl), will cure both vinyl and nonvinyl containing gums. This difference is illustrated in Fig. 15.2, again using the polymer with Vi/Si = 0.0026.[6] Di-tertiary butyl peroxide seems to be truly "vinyl specific," in that the chemical crosslink density is constant in the range 0.2 to 6% peroxide. However, the crosslink density is dependent upon peroxide concentration in the case of the bis (2,4-dichlorobenzoyl) peroxide.

Figure 15.3, drawn from data in Reference 6, gives more insight into the effect of vinyl on vulcanizate properties. Using 1.95 x 10^{-5} moles of bis (2,4-dichlorobenzoyl) peroxide/g polymer, one obtains 0.6 x 10^{-5} moles chemical crosslinks/g polymer. This can be increased to 0.74 x 10^{-5} moles crosslinks per g polymer by doubling the peroxide concentration, or to 2 x 10^{-5} moles crosslinks per g polymer by adding 0.2 mole % methyl vinyl siloxy units to the polymer. It seems, therefore, that with a vinyl containing polymer, it is possible to reach a much higher level of primary valence crosslinking than with a nonvinyl containing polymer. In addition,

TABLE 15.2. CROSSLINKING EFFICIENCY
Polydimethylomethylvinylsiloxane (Vi/Si = 0.0026)

Peroxide	*% Peroxide (Optimum Level)*	*(Moles Chemical Crosslinks per Gram Polymer)* $\times 10^5$	*Moles Chemical Crosslinks per Mole Peroxide*	*Moles Chemical Crosslinks per Mole Vinyl*
2,5-dimethyl-2,5-bis(t-butyl peroxy) Hexane	0.315	3.2	3.2	0.9
Di-cumyl peroxide	0.315	3.3	3.0	0.9
Di-t-butyl peroxide	0.21	3.3	2.4	0.9

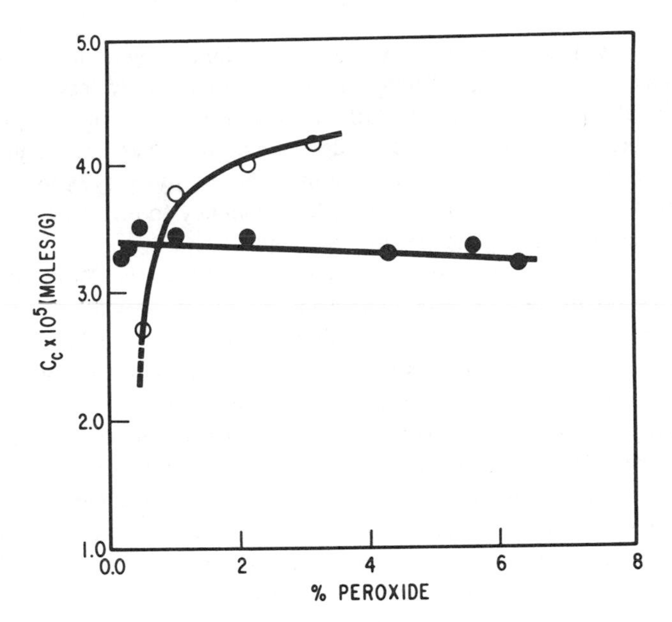

Fig. 15.2 Variation of chemical crosslink density with peroxide concentration: ○, bis(2,4-dichlorobenzoyl) peroxide; •, di-t-butyl peroxide.

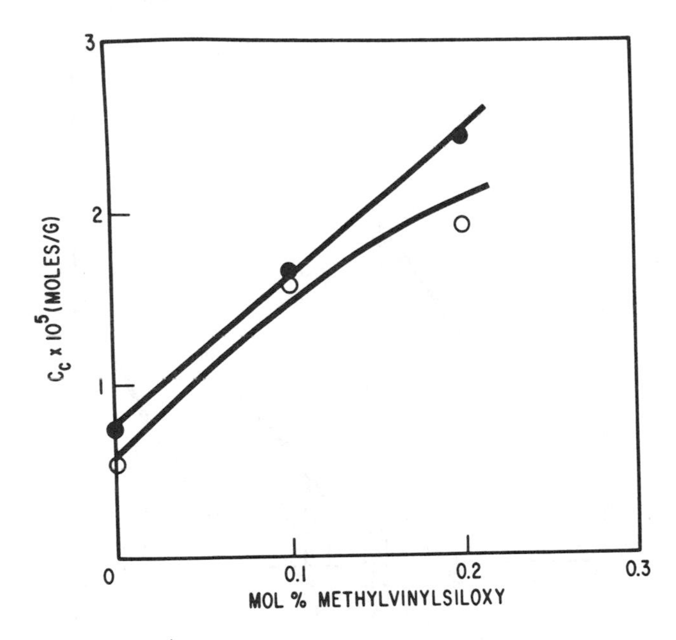

Fig. 15.3 Variation of chemical crosslink density with vinyl level and concentration of bis(2,4-dichlorobenzoyl) peroxide: ○, 1.95×10^{-5} moles peroxide/g gum; •, 3.90×10^{-5} moles peroxide/g gum.

this can be done with significantly lower concentrations of peroxide. Since there are more bonds to be broken, a "tight" network should show less tendency to revert than a barely formed one, particularly if the concentration of acidic peroxide decomposition products, which can degrade the polymer, has been significantly reduced. And, indeed, the compounds containing the methyl vinyl copolymers do show improved vulcanization characteristics, less tendency to revert in cure or post bake, and lower compression set at elevated temperatures.

Vulcanization rate is conveniently studied by means of the Monsanto Rheometer. This instrument, described elsewhere in this book, provides a continuous measurement of complex dynamic shear modulus while a rubber is being cured in a mold under heat and pressure. This is accomplished by measurement of the torque on a conical disk rotor that is embedded in the rubber, and is being sinusoidally oscillated through a small arc. The torque is a linear function of crosslink density as determined by swelling measurements, though the proportionality constant varies with the stock. The torque readings can, therefore, be considered as relative crosslink densities. From a rheograph, it is possible to get an idea of how the compound will flow in the mold before cure starts (minimum viscosity and scorch time), cure rate (and consequently time to various degrees of cure), and a measure of the final crosslink density. If a series of these curves are run at different temperatures with the various peroxides, the data may be used to estimate hot mold residence times required to effect a given degree of cure at a given temperature with a given peroxide. Figures 15.4, 15.5, and 15.6 may be used to make

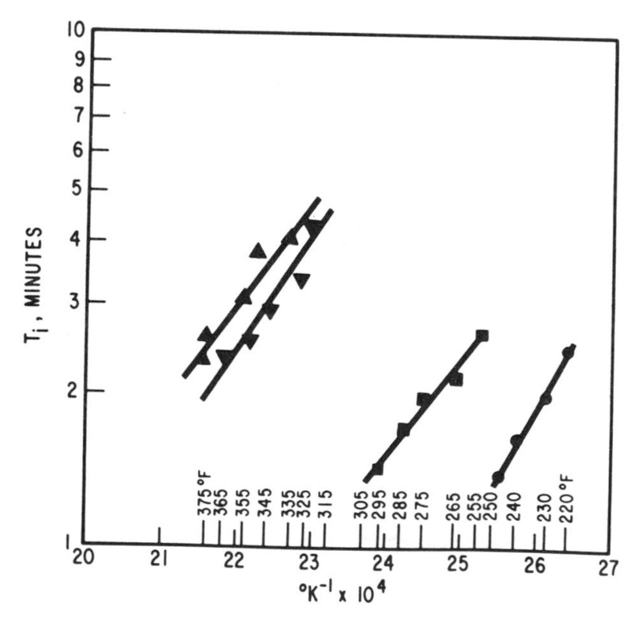

Fig. 15.4 Induction times for cure with various peroxides: ●, 0.65 phr bis(2,4-dichlorobenzoyl peroxide; ■, 0.35 phr benzoyl peroxide; ▲, 0.6 phr 2,5-dimethyl-2,5-bis(t-butyl peroxy) hexane; ▼, 0.8 phr dicumyl peroxide.

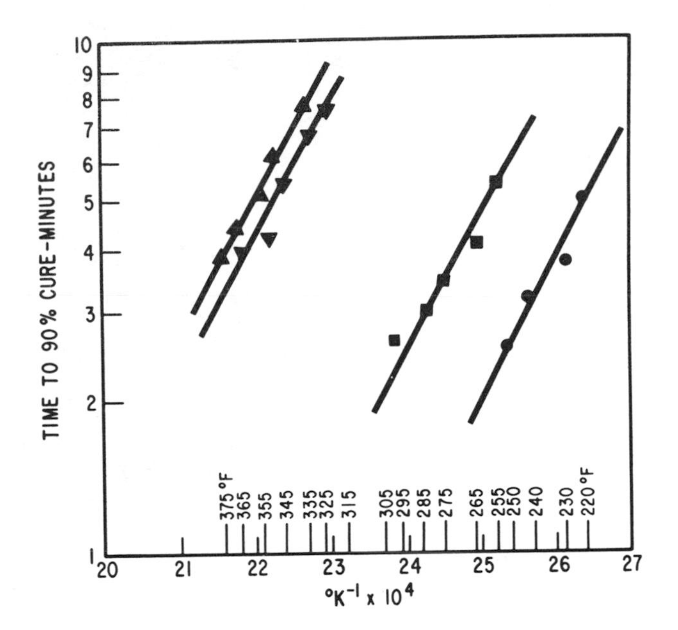

Fig. 15.5 Time to 90% cure with various peroxides: •, 0.65 phr bis(2,4-dichlorobenzoyl) peroxide; ▪, 0.35 phr benzoyl peroxide; ▴, 0.6 phr 2,5-dimethyl-2,5-bis(t-butyl peroxy) hexane; ▾, 0.8 phr dicumyl peroxide.

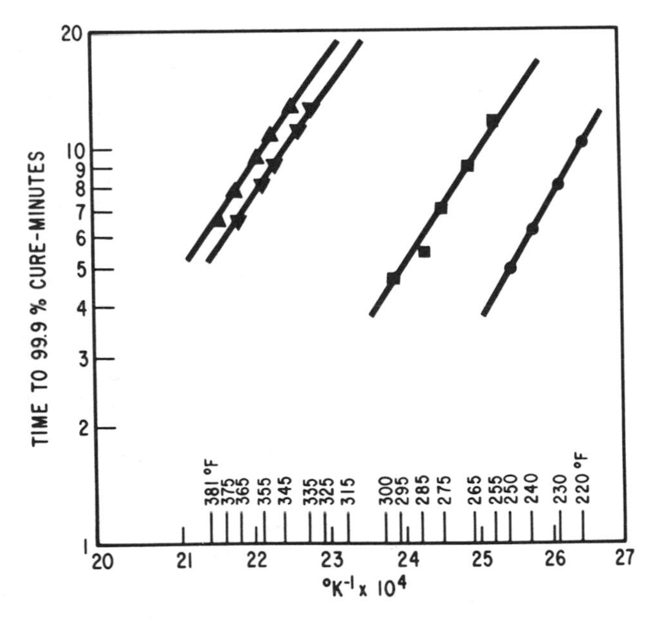

Fig. 15.6 Time to full cure with various peroxides: •, 0.65 phr bis(2,4-dichlorobenzoyl) peroxide; ▪, 0.35 phr benzoyl peroxide; ▴, 0.6 phr 2,5-dimethyl-2,5-bis(t-butyl peroxy) hexane; ▾, 0.8 phr dicumyl peroxide.

TABLE 15.3. SILICONE GUMS

ASTM D-1418 Designation	Description	Specific Gravity	Shrinkage[a]	Commercial Designation G.E.[b]	D.C.[c]	U.C.[d]
Si	General Purpose	0.98	High	SE-76	Silastic 400	W-95
Si	General Purpose	0.98	Low	SE-30	–	–
V Si	General Purpose, Low Compression Set	0.98	High	–	–	W-96
V Si	General Purpose, Low Compression Set	0.98	Low	SE-33	Silastic 430	W-98
PV Si	Extreme Low Temperature	0.98	High	–	Silastic 440	–
PV Si	Extreme Low Temperature	0.98	Low	SE-54	–	–
FV Si	Solvent Resistant	1.30	–	–	Silastic LS-420	–

[a]High shrinkage ≅ 5–6%; low shrinkage ≅ 3%.
[b]General Electric Company, Silicone Products Department, Waterford, N. Y. 12118.
[c]Dow Corning Corporation, Midland, Michigan 48640. Silastic is a registered trademark of Dow Corning Corp.
[d]Union Carbide Corporation, Chemicals and Plastics Division, 270 Park Ave., New York, N. Y. 10017.

such estimates for the vulcanization of a high tear strength methyl vinyl compound. For example, with bis (2,4-dichlorobenzoyl) peroxide at 0.65 phr, and at a temperature of 230°F, the induction period is 2 minutes. The hot mold residence time is 4 minutes for 90% cure, and 8 minutes for full cure. A technically satisfactory cure (in terms of durometer, tensile, and elongation) is obtained by introduction of 90% of the crosslinks; however, a full cure is required to obtain the lowest compression set.

In order to obtain full cure in 8 minutes, the mold temperature should be 230°F for bis (2,4-dichlorobenzoyl) peroxide, 270°F for benzoyl peroxide, 255°F foı dicumyl peroxide, and 365°F for 2,5-dimethyl-2,5-bis (t-butyl peroxy) hexane.

COMPOUNDING INGREDIENTS

A typical silicone rubber formulation contains a silicone polymer, reinforcing and (or) extending fillers, process aids or softeners to plasticize and retard crepe-aging, special additives (e.g., heat aging additives and blowing agents for sponge), color pigments, and one or more peroxide curing agents.

Silicone Gums. Pure silicone rubber polymers, differing from one another in polymer type and molecular weight, are available from the basic suppliers. These gums are listed and described in Table 15.3.

Silicone Reinforced Gums. Although pure polymer may be used, it is generally easier and more economical for the rubber fabricator to compound from silicone rubber reinforced gums. These are mixtures of pure gum, process aids, and highly reinforcing silica that have been specially processed. In some cases they contain additives for special effects, such as improved heat aging or bonding properties. The characteristics of some of the most commonly used reinforced gums are described in Table 15.4. The manufacturers will supply other reinforced gums for special purposes.

Reinforcing Fillers. The fume process silicas (Table 15.5) reinforce silicone polymer to a greater extent than any other filler. Due to the high purity of the silica, the rubbers containing it have excellent electrical insulating properties, especially under wet conditions.

Silica aerogels impart moderately high reinforcement. Rubber containing these silicas has relatively high water absorption due to the small amounts of water, alcohol, sodium sulfate, and free acid on the filler surface. As a result, these compounds are generally inferior to fume silica compounds in wet electrical properties, compression set, and reversion resistance.

Carbon black gives moderate reinforcement. However, the blacks inhibit cure with the aroyl peroxide vulcanizing agents; and this limits their use mainly to the production of electrically conductive or semiconductive rubber. The high structure blacks, such as Shawinigan Black, are suitable for this purpose.

Semi-reinforcing or Extending Fillers. The extenders (Table 15.6) are important for use in compounds containing reinforcing fillers in order to achieve an optimum balance of physical properties, cost, and processibility.

TABLE 15.4. SILICONE REINFORCED GUMS

ASTM D-1418 Designation	*Description*	*Specific Gravity*	*Commercial Designation*		
			G.E.	*D.C.*	*U.C.*
V Si	General Purpose	1.10	SE-404	Silastic 432	KW-1300
V Si	General Purpose (makes stiffer and drier compounds than SE-404)	1.10	SE-406	–	
V Si	General Purpose (low compression set, long freshened life, 600°F capability)	1.08	SE-421	Silastic 433	KW-1320
V Si	General Purpose (accepts high loadings of extending filler)	1.09	SE-463	Silastic 437	
V Si	General Purpose (requires no post bake)	1.09	SE-465	Silastic 740	
PV Si	Extreme Low Temperature, High Strength	1.12–1.20	SE-505	Silastic 446	
PV Si	Extreme Low Temperature, High Strength (improved physicals and processing)	1.13	SE-517	–	
FV Si	Solvent Resistant	1.38	–	Silastic LS-422	

TABLE 15.5. REINFORCING FILLERS FOR SILICONE RUBBER

Filler	Type	Particle Size, Mean Diameter (millimicrons)	Surface Area (Square Meters per Gram)	Specific Gravity	Supplier	Reinforcement Produced in Pure Silicone Gum	
						Tensile Strength Range, psi	Elongation Range, %
CAB-O-SIL HS-5	Fumed Silica	10	300–350	2.20	Cabot Corporation 125 High Street Boston, Mass.	600–1800	200–800
CAB-O-SIL MS-7	Fumed Silica	15	175–200	2.20	Cabot Corporation 125 High Street Boston, Mass.	600–1200	200–600
SANTOCEL CS and FRC	Silica Aerogel	30	110–150	2.20	Monsanto Chemical Co. Inorganic Chemicals Div. St. Louis, Mo.	600–900	200–350
SHAWINIGAN Black	Acetylene Black	45	75–85	1.85	Shawinigan Products Corp. Empire State Bldg. New York, N. Y.	600–900	200–350

NOTE: The following are trade marks of the companies shown: CAB-O-SIL – Cabot Corporation;
SANTOCEL – Monsanto Chemical Co.

TABLE 15.6. SEMI-REINFORCING OR EXTENDING FILLERS FOR SILICONE RUBBER

Filler	Type	Particle Size, Mean Diameter (microns)	Surface Area (Square Meters per Gram)	Specific Gravity	Supplier	Reinforcement Produced in Pure Silicone Gum: Tensile Strength Range, psi	Reinforcement Produced in Pure Silicone Gum: Elongation Range, %
BLANC ROUGE	– –	1–5	<5	2.65	Illinois Mineral Co. Distributed by Berkshire Chem. Inc. 630 Third Ave. New York 17, N. Y.	100–400	200–300
CELITE SUPER FLOSS	Flux Calcined Diatomaceous Silica	1–5	<5	2.30	Johns-Mansville 22 E. 40th St. New York 16, N. Y.	400–800	75–200
CELITE 350	Calcined Diatomaceous Silica	1–5	<5	2.15	Johns-Manville	400–800	75–200
Dicalite PS	Calcined Diatomaceous Silica	1–5	<5	2.25	Dicalite Division Great Lakes Carbon Corp. 18 East 48th St. New York 17, N. Y.	400–800	75–200
Dicalite White	Flux Calcined Diatomaceous Silica	1–5	<5	2.33	Dicalite Division Great Lakes Carbon Corp.	400–800	75–200
Iron Oxide RO-3097	Iron Oxide	1		4.80	Chas. Pfizer & Co., Inc. 235 E. 42nd St. New York, N. Y.	200–500	100–300
Iron Oxide RY-2196	Iron Oxide	<1		4.95	Chas. Pfizer & Co., Inc.	200–500	100–300

Isco 1240 Silica	Ground Silica	5–10		2.65	Innis Speiden & Co., Inc. 420 Lexington Ave. New York 17, N. Y.	100–400	200–300
MIN-U-SIL							
5μ	Ground Silica	5	<5	2.65	Pennsylvania Pulverizing Co.	100–400	200–300
10μ		10	<5	2.65	7 Gateway Center		
15μ		15	<5	2.65	Pittsburgh, Pa.		
NEO NOVACITE	Ground Silica	1–10		2.65	Malvern Minerals Co. Box 1246 Hot Springs, Ark.	100–400	200–300
SUPERPAX	Zirconium Silicate			4.50	Titanium Alloy Mfg. Div. National Lead Co. 111 Broadway New York 6, N. Y.	400–600	100–300
THERMOMIST		10–20	<5	2.60	Indian Mountain Minerals Hot Springs, Ark.	100–400	200–300
TITANOX RA	Titanium Dioxide	.3		4.2	Titanium Pigment Corp. 111 Broadway New York 6, N. Y.	200–500	300–400
Whitetex Clay	Calcined Kaolin	1–5	<5	2.55	Southern Clays Inc. 33 Rector St. New York 6, N. Y.	400–800	75–200
WITCARB R	Precipitated Calcium Carbonate	.03–.05	32	2.65	Witco Chemical Co. 260 Madison Ave. New York 16, N. Y.	400–600	100–300
ALBACAR 5970	Precipitated Calcium Carbonate	1–4	8	2.71	Chas. Pfizer & Co., Inc. 235 E. 42nd St. New York, N. Y.	400–600	100–300
Zinc Oxide XX-78	Zinc Oxide	.3	3.0	5.6	New Jersey Zinc Co. 2045 City Line Rd. Bethlehem, Pa.	200–500	100–300

NOTE: The following are trademarks of the companies shown:

CELITE	Johns-Manville
MIN-U-SIL	Pennsylvania Pulverizing Co.
NEO NOVACITE	Malvern Minerals Co.
SUPERPAX	National Lead Co.
THERMOMIST	Indian Mountain Minerals
TITANOX	Titanium Pigment Corp.
WITCARB	Witco Chemical Co.
ALBACAR	Chas. Pfizer & Co., Inc.

TABLE 15.7. COLOR PIGMENTS FOR SILICONE RUBBER

REDS	Red (RY-2196)[a]
	Red (RO-3097)[a]
	Red (F-5893)[g]
	Maroon (F-5891)[g]
	Dark Red (F-5892)[g]
GREENS	Chromium Oxide Green (X-1134)[b]
	Chromium Oxide Green (G - 6099)[a]
	Yellow Green (F-5688)[g]
	Blue Green (F-5687)[g]
	Turquoise (F-5686)[g]
BLUES	Cobalt Aluminum Blue (F-6279)[g]
	Medium Blue (F-5274)[g]
	Dark Blue (F-6279)[g]
	Violet Blue (F-5273)[g]
ORANGES	Mapico Tan #20[c]
	Orange Red (F-5894)[g]
	Orange (F-5895)[g]
	Light Orange (F-5896)[g]
WHITES	Titanium Dioxide (TITANOX RA)[d]
	Titanium Dioxide (TITANOX ALO)[d]
YELLOWS	Cadmolith Yellow[b,e]
	Cadmium Yellow (F-5897)[g]
	Lemon Yellow (F-5512)[g]
BUFFS	Dark Buff (F-6115)[g]
	Buff (F-2967)[g]
BLACK	P-33 Carbon Black[f]
	Black Iron Oxide (Drakenfeld 10395)[h]
BROWNS	Light Yellow Brown (F-6109)[g]
	Medium Brown (F-6111)[g]
	Red Brown (F-6112)[g]

[a]Chas. Pfizer & Co., Inc., 235 E. 42nd St., New York, N.Y.
[b]Hercules, Inc., Imperial Color Division, Glens Falls, N. Y.
[c]Cities Service Corp., Columbian Division, Drawer 4, Cranberry, N.J. 08512
[d]Titanium Pigments Corp., 111 Broadway, New York 6, N. Y.
[e]Chemical & Pigment Div. of Glidden Co., 1396 Union Commerce Bldg., Cleveland 14, Ohio
[f]R. T. Vanderbilt Co., Inc., 230 Park Ave., New York 17, N. Y.
[g]Ferro Corp., 4150 East 56th St., Cleveland, Ohio
[h]Standard Bronze Works, Inc., Drakenfeld Div., 159 W. 25th St., New York, N. Y.

Ground silica and calcined kaolin do not provide significant reinforcement. As a consequence, they can be added to a reinforced gum or compound in relatively large quantities in order to reduce pound-volume cost. These extenders are satisfactory in either mechanical or electrical grade rubber.

The reinforcement obtained with calcined diatomaceous silica is greater, though quite modest, than that obtained with any other extender. Therefore, as an extender, it is not as useful as ground silica. However, it is used in electrical stocks, low compression set stocks, and in general mechanical stocks to reduce tack and modify handling properties.

Calcium carbonate and zirconium silicate are special purpose extenders, used mainly in pastes that are coated on fabrics from solvent dispersion.

Zinc oxide is used as a colorant, and as a plasticizer. It imparts tack and adhesive properties to a compound.

Additives. Organic colors, and many inorganic colors, have adverse effects on the heat aging of silicone rubber. The inorganic pigments listed in Table 15.7 have been found suitable for use. Usually 0.5 to 2 parts per 100 parts of compound are sufficient for tinting purposes. It is often desirable to masterbatch color pigments in order to get good dispersion and close color matches.

Red iron oxide is used as a color pigment, and as a heat aging additive. Two to four parts per hundred parts of gum will give improved heat stability at 600°F.

Process aids are used with highly reinforcing silica fillers. These have a softening or plasticizing effect, and they retard the "crepe aging" or "structuring" or "pseudocure" of the raw compound that occurs due to the high reactivity of the reinforcing filler with the silicone polymer.

The characteristics of the two commonly used blowing agents for silicone rubber sponge are shown in Table 15.8. Upon decomposition, UNICEL ND releases nitrogen and formaldehyde. The residue is hexamine and inert filler. Nitrogen and a residue of dimethyl terephthalate are formed by the decomposition of NITROSAN. The two blowing agents are often used in combination to control cell structure. Both should be added in masterbatch form, although UNICEL ND disperses more readily than NITROSAN.

Curing Agents In commercial practice, it has been found that none of the six commonly used peroxides is a universal curing agent. The three aroyl peroxides (Table 15.9) may be considered general purpose in that they will cure both nonvinyl and vinyl containing polymers. However, no one of them is suitable in all types of fabrication procedures. The three dialkyl type peroxides (Table 15.10) are "vinyl specific" since they will give good cures only with vinyl containing polymers.

Bis (2,4-dichlorobenzoyl) peroxide normally requires a curing temperature of about 220–250°F. Its decomposition products (2,4-dichlorobenzoic acid and 2,4-dichlorobenzene) volatilize relatively slowly at commercial curing temperatures; and, consequently, compounds containing it may be cured without external pressure, provided that air removal and forming have been done (e.g., by extrusion or calendering) prior to the application of heat. In fact, one of the main uses of this peroxide involves hot air vulcanization of extrusions in a few seconds at

TABLE 15.8. BLOWING AGENTS FOR SILICONE RUBBER SPONGE

Commercial Designation	*Composition*	*Decomposition Temperature*	*Some Characteristics*	*Supplier*
UNICEL	42% N,N'-dinitroso-pentamethylene tetramine 58% inert filler	300–500°F	Can use t-butyl perbenzoate and (or) benzoyl peroxide Blow at >300°F Large cell sponge	E. I. Dupont de Nemours Co. Elastomer Chemicals Dept. 329 W. State St. Trenton, N. J. 08618
NITROSAN	70% N,N'-dimethyl-N,N'-dinitroso-terephthalamide 30% Mineral Oil	220°F	Can use bis(2,4-dichloro-benzoyl) peroxide and t-butyl perbenzoate. Blow at >240°F Fine cell sponge	E. I. Dupont de Nemours Co. Explosive Dept. Chemical Products Sales Div. Wilmington, Delaware 19898

NOTE: UNICEL and NITROSAN are trademarks of E. I. Dupont de Nemours Co.

TABLE 15.9. PEROXIDE CURING AGENTS FOR SILICONE RUBBER: GENERAL PURPOSE

Peroxide	Commercial Designation	Form	Assay (%)	Cure Temperature (°F)	Decomposition Products	Uses
Bis(2,4-dichloro-benzoyl) peroxide	CADOX TS-50[a] LUPERCO CST[b]	Paste Paste	50 50	220–250	"Nonvolatile" Acidic	*Hot Air Vulcanization* Continuous Steam Vulcanization Autoclave Thick Section Molding Low Compression Set
Benzoyl Peroxide	CADOX BSG-50[a] LUPERCO AST[b] CADOX 99[a] (200 mesh)	Paste Paste Powder	50 50 99	240–270	Volatile Acidic	*Continuous Steam Vulcanization* *Autoclave* *Tower Coating* Low Compression Set Thin Section Molding
Tertiary Butyl Perbenzoate	t-Butyl Perbenzoate[b]	Liquid	95	290–310	Volatile Acidic	Usually with other peroxides; *sponge*; generally for high temperature activation

[a]McKesson & Robbins, 42–15 Crescent St., Long Island City, N. Y. 11101. CADOX is a trademark of Chemetron Corp.
[b]Lucidol Division, Pennwalt, Corp., 1740 Military Road, Buffalo, N. Y. LUPERCO is a trademark of Lucidol Division, Pennwalt Corp.

TABLE 15.10. PEROXIDE CURING AGENTS FOR SILICONE RUBBER: VINYL SPECIFIC

Peroxide	Commercial Designation	Form	Assay (%)	Cure Temperature (°F)	Decomposition Products	Uses
Dicumyl Peroxide	DICUP R[a]	Solid	95	300–320	Fairly Volatile	Thick Section Molding
	DICUP 40C[a]	Powder	40			Low Compression Set
						Carbon Black
2,5-Dimethyl-2,5-Bis(t-butyl peroxy) Hexane	VAROX[b]	Powder	50	330–350	Volatile	*Thick Section Molding*
	LUPERSOL 101[c]	Liquid	95			*Low Compression Set*
	LUPERCO 101XL[c]	Powder	50			*Carbon Black*
Di-tertiary Butyl Peroxide	Di-t-butyl Peroxide[c,d]	Liquid	97	340–360	Volatile	*Thick Section Molding*
						Low Compression Set
	CW-2015[e]	Powder	20			Carbon Black

[a]Hercules, Inc., 972 Market St., Wilmington, Delaware. DICUP is a trademark of Hercules, Inc.
[b]R. T. Vanderbilt Co., 230 Park Avenue, New York, N. Y. VAROX is a trademark of R. T. Vanderbilt Co.
[c]Lucidol Division, Pennwalt Corp., 1740 Military Road, Buffalo, N. Y. LUPERSOL is a trademark of Lucidol Division, Pennwalt, Corp.
[d]Shell Chemical Corp., 380 Madison Ave., New York, N. Y.
[e]Harwick Standard Chemicals, 60 S. Sieberling St., Akron, Ohio.

temperatures of 600–800°F. Although this peroxide may be used for molding, it has some undesirable features. Since it starts crosslinking at an appreciable rate as low as 200°F, thin sections may start to gel before flow and air removal are complete. In addition, a thick section must be carefully programmed through a post vulcanization oven bake cycle in order to remove acidic decomposition products without degrading the interior of the part. The peroxide may be used for steam cures in autoclaves and in continuous steam vulcanizers. However, benzoyl peroxide is more desirable, particularly in the latter case, because of its higher cure temperature (240–270°F). With benzoyl peroxide, there is less tendency for the compound to scorch in the dies when extruding at high speed into a continuous steam vulcanizer that is operating at 100–200 lb of steam.

Due to the volatility of its decomposition products (benzoic acid and benzene), external pressure is required to prevent porosity when curing with benzoyl peroxide. The only exception is in the very thin sections involved when tower coating fabrics from solvent dispersions of silicone rubber. In this case, benzoyl peroxide is nearly always used because it has long shelf life in the dip tanks, and because it is not volatilized from the rubber during the solvent removal operation prior to cure.

Low compression set rubber may be produced with either of the diaroyl peroxides, provided that the polymer contains vinyl groups, and that the acidic decomposition products are removed by an oven post bake.

Unlike the aroyl peroxides, the so-called "vinyl specific" peroxides (Table 15.10) may be employed to vulcanize stocks containing carbon black. If dicumyl peroxide is used these stocks can be hot air vulcanized.

All three "vinyl specific" peroxides are good for thick section molding, with dicumyl being less preferable due to a slight tendency to "air inhibit" in the same manner as benzoyl peroxide. In addition, dicumyl peroxide has somewhat less volatile decomposition products (acetophenone and α,α-dimethylbenzyl alcohol) than do the other two. This means that external pressure during cure is less important, but still required; and that longer oven post bakes are needed for thick sections than in the case of the other two peroxides. Just as with the diaroyl peroxides, optimum physical properties require relatively close control of dicumyl peroxide concentration.

Due to their nonacidic decomposition products, the "vinyl specific" peroxides require less oven post bake after vulcanization; and, in addition, close-stepped program post bakes are not necessary. Lower compression sets are obtained with these peroxides than with the general purpose curing agents.

During vulcanization with di-t-butyl peroxide, external pressure is especially important due to the extreme volatility of the peroxide and its decomposition products (acetone and methane). The state of cure is determined primarily by the vinyl content of the polymer, and not by peroxide concentration; air inhibition is absent; prevulcanization or scorch is never a problem; and the innocuous decomposition products can be removed by short, high-temperature oven post bakes. This peroxide also produces a rubber with the best overall balance of

TABLE 15.11. SOME COMMERCIALLY AVAILABLE SILICONE RUBBER COMPOUNDS

ASTM D-1418 Designation	*Description*	*Durometer*	*Commercial Designation*[a] *G.E.*	*D.C.*	*U.C.*
V Si	General Purpose, Low	40	SE-4401	Silastic 241	K-1034
	Compression Set	40	SE-4404		K-1044R
V Si	General Purpose, Low	50	SE-4511	Silastic 745	K-1365
	Compression Set, No	60	SE-4611	Silastic 746	K-1366
	Post Bake Required	70	SE-4711	Silastic 747	K-1367
		80	SE-4811	Silastic 748	K-1368
V Si	General Purpose, High Tear, Resilient	50	SE-456	Silastic 55	
V Si	Low Compression Set	60	SE-3613	Silastic 2096	
		70	SE-3713	Silastic 2097	K-1037
		80	SE-3813	Silastic 2098	
V Si	Low Compression Set	75	SE-3701		
	High Temperature	70	SE-3715		
PV Si	Extreme Low Temperature	25	SE-5211	Silastic 6508	
		25	SE-525		
		50	SE-551	Silastic 6526	
PV Si	Extreme Low Temperature,	40	SE-5401		
	Low Compression Set	50	SE-540/SE-5601	Silastic 651	
		60	SE-5601		
		70	SE-5701	Silastic 675	
PV Si	Extreme Low Temperature,	50	SE-555	Silastic 916	
	High Strength	50	SE-557	Silastic 955	
		60	SE-565	Silastic 960	
FV Si	Solvent Resistant	60		Silastic LS 53	
		60		Silastic LS 63	
Si	Cloth Coating–		SE-100	Silastic 132	K-1014
	Electrical		SE-1170	Silastic 9119	
Si	Cloth Coating– Mechanical		SE-701	Silastic 6535	
PV Si	Flame Retardant	50	SE-5549	Silastic 2351	
V Si	Wire and Cable		SE-9008	Silastic 1601	
			SE-9011	Silastic 1602	
			SE-9016	Silastic 2083	
			SE-9028	Silastic 2287	
			SE-9035		
			SE-9044		
			SE-9058		
PV Si	Wire and Cable		SE-9025		
			SE-9090	Silastic 1603	

[a]Stocks from different manufacturers are not necessarily equivalent.

properties. Its main deficiency lies in its extreme volatility. A stock must be molded with external pressure shortly after the peroxide has been added.

Although it does not produce rubber with quite as good a balance of properties, 2,5-dimethyl-2,5-bis(t-butyl peroxy) hexane is quite similar to di-t-butyl peroxide in its performance. It has the advantage of lower vapor pressure at room temperature. In fact, it can be added to a compound 1 to 2 months before vulcanization. However, external pressure is still required during cure due to the volatility of the decomposition products.

COMPOUNDING

The basic suppliers have developed a very extensive body of compounding information in work that has resulted in compounds, gums, and reinforced gums with which the rubber fabricator can meet well over a hundred industrial and military specifications for silicone rubber. Some of the commercial available materials are listed in Tables 15.3, 15.4, and 15.11. Much of the compounding data is also available in the form of handbooks and data sheets.[10–13] Though this knowledge cannot be thoroughly covered within the present space limitations, it is possible, by means of selected formulations and property profiles, to illustrate certain fundamental principles of silicone rubber compounding, and to give an overall view of the properties of silicone rubber.

Formulation

1. Low compression set requires a vinyl containing polymer, preferably without phenyl or trifluoropropyl groups. Fume silica is best for reinforcement, and ground silicas and diatomaceous earth are satisfactory extenders. The "vinyl specific" peroxides are preferred over the aroyl peroxides. However, the latter may be used provided the rubber is properly post-baked after cure. A representative formulation, and the physical properties obtained with it, are shown below and in Table 15.12:

G.E.	SE-421	100
Cabot	CAB-O-SIL MS-7	5
P.P.C.	5μ MIN-U-SIL	50
C.C.	CADOX TS-50	1

2. Upon the addition of red iron oxide, most compounds will have a useful life at 600°F. This life can be extended to one or two weeks, provided that the filler loading is sufficiently low. To provide for cases where a color other than red is required, the manufacturers have built 600°F capability into some of the reinforced gums:

G.E.	SE-421	100	
G.E.	SE-406		100
Cabot	CAB-O-SIL	8	8
P.P.C.	10μ MIN-U-SIL		80
	Red Iron Oxide		2
C.C.	CADOX TS-50	1	1.8

(See Table 15.12 for physical properties)

TABLE 15.12. PHYSICAL PROPERTIES OF SELECTED FORMULATIONS

Property	*Low Compression Set*	*Extreme High Temperature*		*Extreme Low Temperature*		*High Strength Extreme Low Temperature*	*Solvent Resistance*	*Wire and Cable Insulation*		
Hardness, Shore A	60	46	68	54	50	63	50	60	68	72
Tensile Strength (psi)	925	1000	800	800	900	1500	1000	900	1400	900
Elongation, %	200	430	200	300	400	700	220	300	500	220
Tear Strength, Die B, pi	70	95	80	90	95	200	75			
% Compression Set (22 hr/350°F)	14	14	30	26	20	42	50			
Brittle Pt (°F)	−85	−85	−85	−150	−150	−150	−90			
Oil Resistance, % Volume Change										
7 days @ 160°F — Skydrol 500A		+10			+25		+30			
70 hr @ 300°F — ASTM #1	+7	+6	+5		+10	+10	+1			
70 hr @ 300°F — ASTM #3	+40				+90	+90	+5			
Heat Aging (48 hr/600°F)										
Hardness, Shore A		56	78							
Tensile Strength (psi)		400	520							
Elongation, %		194	100							
Volume Resistivity, ohm-cm									5×10^{15}	1×10^{15}
Electric Strength, vpm									650	650
Dielectric Constant, 60 cps									3.15	3.4
Power Factor, 60 cps									0.001	0.005

3. Extreme low temperature performance requires a phenyl containing gum. For good vulcanization characteristics and compression set, it is essential that it also contain vinyl groups:

G.E.	SE-54	100	
G.E.	SE-505		100
Cabot	CAB-O-SIL	35	
P.P.C.	5μ MIN-U-SIL		25
G.E.	81499 Process Aid	6	
W.T.	LUPERCO AST	1.2	
C.C.	CADOX TS-50		1.8

4. Though high tear VSi compounds have recently been developed (Table 15.11), a high strength compound normally requires a PVSi gum and a high surface fume silica:

G.E.	SE-505	100
Cabot	CAB-O-SIL HS-5	11
C.C.	CADOX TS-50	1.4

5. A VSi polymer is best for resistance to acetone and diester hydraulic fluids, but a FVSi gum provides the better performance in the widest variety of fluids, particularly aromatic solvents:

D.C.	Silastic	LS-422	100
	Cabot	CAB-O-SIL MS-7	7
		Red Iron Oxide	0.8
	W.T.	LUPERCO CST	1.6

6. The use of silicone rubber as wire and cable insulation is based on retention of its excellent electrical insulating qualities at elevated temperature, unusual resistance to ozone and corona, good voltage endurance, flame retardance, ability to burn to a nonconducting ash that continues to insulate, and excellent processibility. These properties are best obtained by compounding with fume silica and ground silica:

G.E.	SE-30	100		
G.E.	SE-404		100	100
Cabot	CAB-O-SIL MS-7	55	25	7.5
P.P.C.	5μ MIN-U-SIL			85
G.E.	81499 Process Aid	10	3.5	
W.T.	LUPERCO AST	3.2		
C.C.	CADOX TS-50		1.5	2

Mixing

Silicone rubber may be compounded in conventional equipment. Doughmixers and the two-roll rubber mill are preferred. A Banbury mixer works well with dry, highly loaded stocks, but is undesirable with many compounds that are too tacky to dump easily to the sheet-off mill.

For silicone rubber compounding, a two-roll rubber mill should have the following features.

1. Adjustable roll clearance through a movable fast roll.
2. Fast roll peripheral speed of 75 to 150 feet per minute.
3. Roll speed ratio of 1.2:1 to 1.4:1.
4. Fast roll positioned toward operator.
5. Nylon scraper blade on fast roll, and nylon plows on both rolls.
6. Water cooled rolls.
7. Bearings designed to keep grease and oil out of the compound and mill pan.
8. Exhaust system to remove fine silica dust over the mill.

In mixing, the gum is added to the two roll mill first, and is allowed to band. Process aids should go in next if they are to be used. Blend thoroughly. Then add the fillers evenly across the bank. As filler is added adjust roll separation to maintain optimum bank. In the case of reinforcing silicas, it is essential to avoid adding too much filler at any portion of the bank. This will form hard, overloaded particles that will not disperse. An extending filler may be added much faster, providing that the band is not allowed to dry out and break.

Any material that falls to the mill pan must be scraped out with a squeegee and returned to the batch frequently during compounding. Addition of this material upon completion of the batch may cause a dispersion failure.

Additives may be incorporated after or during the filler addition, frequently in the form of a masterbatch.

Curing agent goes in last. It is helpful to add colors with the curing agent. Cross-blending to a homogeneous color then gives assurance of good dispersion of the vulcanizer. Depending upon the curing agent used, stock temperature may have to be kept as low as 100°F in order to avoid partial peroxide decomposition or scorch.

Cleanliness is essential. Contamination by very small amounts of sulfur and (or) organic compounds (such as curing agents, accelerators, antioxidants, etc.) can interfere with cure and with heat stability.

Though most compounds can be fabricated immediately after mixing, a 24 hour minimum bin aging is recommended. For low compression set, a bin aging time of 1 to 2 weeks may be desirable.

Fabricating

Freshening is usually the first fabrication step. This is a remilling operation, essential to processibility in the case of those compounds that "crepe harden" or "structure" during bin aging. Before loading the mill, the roll clearance should be set at about 1/4″ for a low green strength compound. Usually it will band on the slow roll first (crumbling and lacing can be, but usually are not, encountered). As milling continues, the nip should be gradually tightened to a final roll clearance of 1/16″ to 1/8″. In the case of a hard compound, the original roll setting should be 1/8″ to 1/4″ (just wide enough to pass the compound a few times). Next the clearance should be set at 1/16″ for a short time. Then the rolls should be gradually opened to about 1/4″, followed by gradual tightening to the final clearance of 1/16″ to 1/8″. This setting, combined with the proper amount of stock on the mill (slight bank), minimizes air entrapment during further milling and blending. After a smooth sheet has formed on the fast roll, compounding ingredients may be added if desired. Milling is continued until the stock reaches the desired consistency. Under-freshened stock will flow poorly upon molding, and will form parts with a rough surface upon extrusion. Over-freshened compound loses green strength, and becomes sticky and hard to handle. Stock must not be allowed to warm up above 100–130°F after curing agent has been added.

At this point, the stock is freshened and ready for use in one of the following fabrication procedures.

Molding. Compression molding is the most widely used method for molding silicone rubber parts. However, transfer and injection molding are also employed. Compression and transfer molding are conducted at pressures of 800 to 1500 psi, and temperatures of 220 to 370°F. For a given cure time, the cure temperature is influenced mainly by the choice of peroxide curing agent, but also by the requirements for prevulcanization mold flow time as dictated by mold design and the press closing time. Mold cycles and choice of peroxide were discussed in previous sections.

Injection molding involves pressures of 10,000 to 12,000 psi, and temperatures of 370 to 410°F. Injection times are approximately 5 to 10 seconds. Molding time is in the range of 30 to 90 seconds. This method, relative to compression molding, gives less flash, better properties, and greater uniformity. It is particularly suited for high volume production of complex parts.

Steel molds are recommended. In the event that a high finish is desired, and in the case of transfer and injection molding, chrome plating is also recommended. Gates for both transfer and injection molding should be about 25% larger in area than those used with organic rubbers. Assistance in mold design is available from the basic silicone rubber suppliers.

Silicone mold release agents are unsuitable for use with silicone rubber. 0.5% water solutions of household detergents are recommended for spray or brush application to the mold.

Extrusion. Although short barrel extruders may be used, long barrel machines are preferred for silicone rubber extrusion. A single flight screw with diminishing pitch gives the best results. The compression ratio should be in the range 2:1 to 4:1. Feeding is one of the main problems with the low green strength silicone rubber stocks. However, it is easily solved by using roller feed from a "hat" (coiled strips of compound) to a screw with a deep flight.

A 40 to 150 mesh screen is recommended to ensure a clean extrusion. These screens are supported by a coarse backing screen set in a breaker plate. This assembly increases back pressure in the extruder to ensure air removal, and to assist dimensional control.

A spider flange is used to hold a universal die holder and a mandrel when extruding tubing. Centering is thus possible. Both flange and mandrel should be drilled so that low pressure air may be used to keep the tubing round, and to introduce talc inside the tube if desired.

In order to form wire and cable insulation, the extruder must be fitted with a cross head.

Hardened steel coin or plate dies are recommended, particularly when extruding compounds containing high loadings of ground quartz. Silicone rubber flows very freely in extrusion, and the die design should allow uniform flow through all parts of the die. There should be no dead spots on the die approach because trapped rubber may start to cure. Allowance must be made for die swell, and for the shrinkage that will accompany vulcanization and oven post bake. In so far as possible, dies should be designed to permit easy adjustments.

Silicone rubber extrusion is ordinarily done as close to room temperature as possible in order to avoid scorching. Temperature should not be allowed to rise above the range 100 to 130°F. Immediately following extrusion, the rubber should be vulcanized. Otherwise, the parts may deform or collapse entirely.

Using bis(2,4-dichlorobenzoyl) peroxide, silicone rubber may be extruded onto an endless belt that passes through a heated tunnel, typically six inches square and ten feet long. In the tunnel, hot air vulcanization of the rubber takes place in a few seconds at 600–800°F. HAV is the most common method of curing extrusions.

Both hot air and continuous steam vulcanization are widely used to cure wire and cable insulated with silicone rubber by extrusion. Continuous steam vulcanization (CV) gives good cures in a few seconds at steam pressures of 100 to 250 psi. Extrusion speeds up to 1200 feet per minute are possible. Benzoyl peroxide is ordinarily used. Bis(2,4-dichlorobenzoyl) peroxide may be used but compounds containing it tend to scorch in the extruder head due to proximity of the vulcanizer.

Steam autoclave vulcanization is often used for cure of short runs. It is economically feasible only for small batches. Extrusions are usually loosely wound on drums, or laid in pans of talc.

Calendering. Standard three or four roll calenders are used to fabricate long, thin unsupported or fabric supported sheets of silicone rubber. A variable speed main drive is recommended. Silicone rubber should be processed at slower speeds than organic rubber. 0.2 to 2 feet per minute is best for start up, and running speed should be in the range of 5 to 10 feet per minute.

The speed ratio of the center roll to the top roll should be 1.1:1 to 1.4:1. The bottom roll should run at the same speed as the center roll.

Silicone rubber has a tendency to transfer not only to the faster roll but also to the cooler roll. Therefore, it is desirable to have provisions for heating and cooling the rolls. Even though calendering is usually done at room temperature, it is often helpful to warm the center roll, or to warm the top roll and cool the bottom roll. However, the temperature must be kept well below 130°F in order to prevent scorching.

Silicone rubber is often calendered onto unprimed or primed fabrics such as glass, dacron, nylon, cotton, etc. A typical calendering set-up is shown in Fig. 15.7.

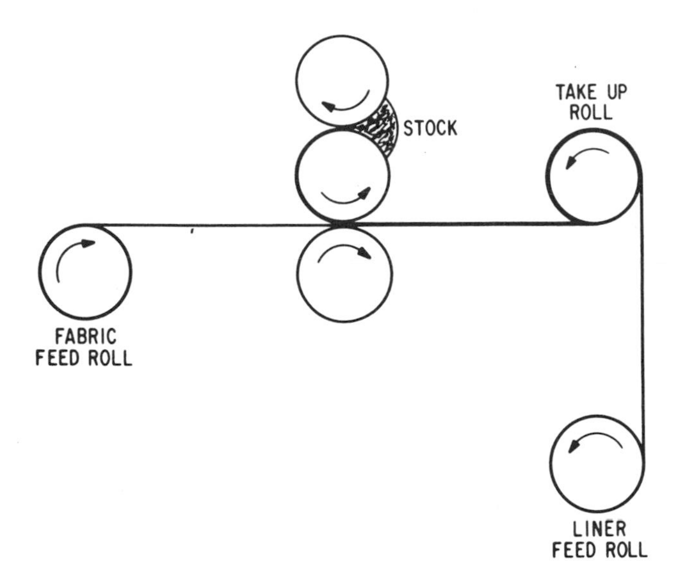

Fig. 15.7 Schematic diagram of set-up for calendering supported sheet.

The liner is used to prevent "blocking" in the roll. Cellulose acetate butyrate, holland cloth, polyethylene, and release treated paper are commonly used.

For unsupported sheet, the fabric feed is eliminated when using high green strength compounds. If the compound has low green strength, it is calendered onto the liner by feeding liner from the fabric feed roll. In either event, the liner is stripped off after cure, or just before use if the sheet is to be used "green."

Calendered sheet can be vulcanized in roll form in a steam autoclave provided that the roll is first pressure taped (for example with wet cotton or nylon). The sheet can also be cured or semicured by eliminating the liner feed, and by passing the supported or unsupported sheet over a hot drum placed between the calender and the take up roll.

Dispersion Coating of Fabric. This technique of fabric coating permits thinner coatings and provides more thorough penetration of the fabric than does calendering. A wash coat from a 5% to 15% silicone rubber dispersion will improve the strength and flex life of glass cloth, and provide a good "anchor coat" for calendering. Excellent high temperature electrical insulating materials, diaphragms, and gaskets can be made from glass cloth with thicker coatings. Silicone rubber can also be dispersion coated on organic fabrics, and then used in many applications such as aircraft seals, radome covers, and general purpose control diaphragms.

The basic manufacturers will supply silicone rubber dispersions or soft, readily dispersed pastes that have been especially designed for cloth coating. However, any silicone rubber compound can be dispersed in solvent.

Xylene and toluene are the recommended solvents, except for the fluorosilicones where the solvent must be a ketone such as methyl ethyl or methyl isobutyl ketone. Chlorinated solvents and solvents containing antioxidants, rust inhibitors, and similar additives should be avoided since they interact with the peroxide vulcanizing agent.

The coating pastes are readily dispersed with a propeller type mixer. Compounds containing reinforcing fillers should be freshened, sheeted off thin, cut into small pieces, and soaked overnight in just enough solvent to cover the compound. The mixture should then be stirred with the propeller mixer until uniform. The remaining solvent is then added, with mixing, in small portions.

The dispersion should be filtered through an 80 to 150 mesh screen (depending upon consistency) before use.

Benzoyl peroxide is normally used for curing because it has a high enough decomposition temperature and a low enough vapor pressure to permit the use of heat to remove solvent after coating. It is best to add crystalline benzoyl peroxide to the dispersion in the form of a 5% solution in toluene or xylene. It is essential to prevent overheating during or after peroxide addition.

Because of its strength, high temperature resistance and low moisture pick up, glass cloth is usually used in coating silicone rubber. Greige cloth gives maximum strength and is the most frequently used grade. A 210 finish glass cloth is employed for maximum electricals and maximum adhesion of the coating. Some loss of strength accompanies the partial removal of starch-oil size from greige goods to form the 210 finish.

Nylon, dacron, and other organic fabrics are used where high temperature resistance is not required.

Primers are not ordinarily applied to the fabrics. However, they are used in cases where maximum adhesion of the rubber coating is desired. Various primers are available from the basic suppliers; and, in some cases, a 5% to 15% dispersion of silicone rubber will act as a primer. The prime coat must be dried and cured before coating the fabric by either calendering or dispersion coating.

Dip and flow coating is used for priming and for applying thin coatings (for example, coating a 4 mil electrical grade glass cloth to 10 mils overall thickness). Thickness is controlled by coating speed and by dispersion solids concentration (ordinarily 5% to 25%).

Ordinarily, the uncoated cloth enters the dip tank at a 45° angle, passes under an idler roll at the bottom of the tank, and then up into a vertical oven or coating tower. Excess dispersion flows off the cloth between the tank and the tower. The coating tower is heated by hot air, and, ideally, is divided into three zones. Solvent is removed in the first zone at 150 to 175°F. Vulcanization takes place in the second zone at 300 to 400°F in the case of glass cloth, and at 250 to 300°F in the case of organic fabrics. When coating glass cloth, the third zone is maintained at 480 to 600°F. This removes final traces of volatiles, including peroxide decomposition products. Maximum bond to the cloth and optimum electrical properties are developed in the third zone.

Two zone towers are also used. In this case, the temperature setting for the first zone is critical. Both solvent removal and vulcanization must take place in the same zone without blistering the coating.

Time in the tower varies, but is usually in the range 10 to 30 minutes.

Dip and knife coating is similar to dip and flow. However, thicker dispersions (35% solids is typical) are used, and a knife or rod is placed between the dip tank and the tower. The coated fabric is pulled past the knife (a second knife can be placed on the opposite side of the fabric), which wipes off excess dispersion. Thickness is controlled by dispersion solids concentration and knife position relative to the fabrics. Heavier coatings can be obtained by this method than by dip and flow.

In reverse roll coating, the knife is replaced by a roll which rotates in opposition to the direction of cloth movement. This often improves penetration.

Ducts and Hose by Mandrel Wrapping. Uncured and semicured, unsupported and fabric-supported silicone rubber sheets and tape may be fabricated into ducts and hose by mandrel wrapping. Hollow aluminum mandrels are recommended. Release of the finished part is often a problem, particularly with complex shapes. The mandrel may be sprayed with a 0.5% to 5% solution of household detergent in water, and allowed to dry thoroughly before wrapping. Light dusting with talc or mica will also give good release. With complex parts, a collapsible mandrel, or one made with a low melting alloy, is often used.

Fabric reinforced tape and sheet should be cut on a 45° bias rather than straight cut. This will give the duct maximum flexibility. Sheet should be straight wrapped to the desired thickness, while tape should be overlapped in a spiral wrap. It is easier to get a tight, uniform, wrinkle-free, and air-free construction if the mandrel is turned on a lathe for wrapping. Smooth liner for hose may be made by butt wrapping uncured tape or sheet on the mandrel, or by using extruded and cured tubing.

After the hose has been built, it should be pressure taped with wet cotton or wet nylon. This is necessary to prevent sponging during cure, and to provide maximum ply adhesion.

Curing is usually accomplished by running steam into the hollow mandrel, or by placing the assembly in a steam autoclave. Curing time and temperature depend upon the peroxide used, and the heat capacity of the mandrel and hose assembly. Thin walled ducts on thin wall mandrels can often be cured in an air-circulating oven Following cure, the wrapped mandrel should be cooled to less than 200°F. The pressure tape should be removed, and the mandrel stripped, while the assembly is still warm. Release agent should be washed off before the part is oven post baked.

Bonding. Silicone rubber can be bonded to many surfaces, including titanium, vanadium, iron, nickel, copper, zinc, aluminum, various steels, ceramics, glass and organic fabrics, many plastics, and to itself.

In all cases, it is essential to clean thoroughly the surface to be bonded. Metal surfaces containing loose scale, oxides, and embedded dirt should be sand blasted. Metal surfaces should be degreased (e.g., with trichloroethylene) to remove all traces of moisture, grease, and oil. Some surfaces, such as plastics and vulcanized rubbers, should be roughened with abrasive. All surfaces should be cleaned with acetone and thoroughly dried just prior to application of primer. Primers are applied in dilute solution to all surfaces, except silicone rubber, in cases where a bond to unvulcanized silicone rubber or adhesive is desired. The only exception is when using the "self-bonding" silicone rubbers, which will bond to many metal surfaces without the use of primers.

Most primers for silicone rubber hydrolyze by contact with moisture in the air after the solvent has evaporated. Solvent removal and hydrolysis will take place in about 30 minutes. This can be hastened by heating at 300°F for 5 or 10 minutes. Suitable primers, as well as adhesives and tie gums, are available from the basic suppliers.

Peroxide pastes should not be used as curing agents in cases where maximum bond strength is desired because the oils migrate to the bond interface and interfere with bonding. Dicumyl peroxide, VAROX, and powdered benzoyl peroxide are recommended.

a. *Bonding Unvulcanized Silicone Rubber.* Curing agent should be added to the freshened compound. After preforming, the stock should be carefully laid on the freshly primed surface. Vulcanization must be done under heat and pressure at temperatures appropriate to the peroxide used. Approximately 15–30 minutes at 330–350°F is usually sufficient when using dicumyl peroxide or VAROX, and 15–30 minutes at 260–280°F, when using powdered benzoyl peroxide. Oven post bake is usually required to develop optimum bond strength.

The self-bonding compounds may be bonded to many metals in the same manner, except that no primer is required. Unlike bonding to primed surfaces, this technique is insensitive to part geometry and to degree of compound flow in the mold while bonding.

b. *Bonding Vulcanized Silicone Rubber.* Cured silicone rubber may be bonded to a primed surface by use of a 10–40 mil interlayer of heat curing silicone rubber adhesive. Vulcanization is conducted under heat and pressure. The pressure must not be so great as to squeeze out the adhesive. This bonding can also be accomplished under pressure at room temperature by use of a room-temperature vulcanizing silicone rubber adhesive. Both techniques can be used for splicing, or bonding cured silicone rubber to itself.

The self-bonding compounds are excellent for bonding cured silicone rubber to itself, and are recommended for splicing. These bonds will usually be stronger than those obtained with adhesives.

Sponging. Though all silicone rubber compounds can be sponged, specially designed compounds and reinforced gums are usually required to meet industrial and military specifications.

UNICEL ND and NITROSAN, either separately or in combination, are usually used as blowing agents. To ensure uniform dispersion, these should be added in the form of a 50% masterbatch in the sponge compound to be used. The usual quantity of blowing agent is in the range 2 to 6%.

A mixture of bis(2,4-dichlorobenzoyl) peroxide and t-butyl perbenzoate is the most commonly used curing agent. Typical formulations are shown in Table 15.13.

TABLE 15.13.

		Press Blown Sponge	*Free Blown Sponge*
	Compound	100	100
Dupont	UNICEL ND	3	
Dupont	Nitrosan	1	3.5
C.C.	CADOX TS-50	0.4	1.1
	t-butyl perbenzoate	1.5	1

a. Free Blowing. Free blown sponge is made by unconfined blowing of extrusions and calendered sheets in a hot air vulcanizing unit, an oven, or a steam autoclave. This technique usually requires a higher level of bis(2,4-dichlorobenzoyl) peroxide than does press blowing.

Deforming of parts is a problem. In the case of extrusion into an HAV unit, kinking can be prevented by slight tension on the extrusion as it leaves the unit. Warping of sheets can be prevented by turning them over occasionally in the oven, or by embedding fabric in the top and bottom surfaces. This latter technique will prevent the stock from expanding lengthwise and sidewise, and will permit vertical blowing only.

Typical cure cycles are 1.5 minutes at 650°F for HAV, 10 minutes at 400°F for an oven, and 10 minutes at 300°F for a steam autoclave.

b. Press Blowing. Confined blowing in a mold can produce parts with higher density, thicker skin, and closer dimensional tolerance than in the case of free blowing.

Stock is sheeted off a mill or calender at a thickness inversely proportional to the degree of expansion required. The preform, sized to allow 0.5 to 1 inch clearance on each side, is then placed in a chase or frame. The frame is sized to allow about 1 inch more expansion in length and width than required in the trimmed, finished part. The top and bottom of the chase are then covered with a fabric (dusted with talc or mica). The frame is then placed in a press, and pressure and heat are applied.

Blowing and vulcanization can be completed in the press at 300 to 400°F with UNICEL ND, or at 240 to 300°F with NITROSAN. Cure time depends upon thickness and temperature for a given curing agent. If the preform is sheeted nearly to the thickness of the mold it may be hot pressed just long enough to partially vulcanize it and form a skin; then it can be free blown in a hot air circulating oven.

A typical cycle for a compound containing 1 part of NITROSAN and 3 parts of UNICEL ND is 3 minutes at 300°F in the press, and 10 minutes at 400°F in the oven.

Oven Curing. With the exception of semicured tapes and sheets, tower-cured coated fabrics, and the no post-cure compounds, oven post baking is the final step in the fabrication of silicone rubber parts by the procedures that have been outlined.

Oven curing removes volatile materials that might make the inside of the parts soft and porous when heated in service. These include low molecular weight polymer and peroxide decomposition products resulting from the initial vulcanization. The oven post bake not only stabilizes the properties for high temperature service, but it also maximizes reversion resistance, compression set, electrical properties, chemical resistance, and the bond to cloth and other substrates.

Ovens should be indirectly heated, either electrically or by gas burners. Horizontal, forced air circulation with baffling to prevent dead air pockets should be provided. Air flow rate should be 2 cubic feet of fresh air per minute per pound of rubber charged, with venting to outside the building. This is necessary to keep the oven atmosphere outside the explosive limits. Temperature control should be ±10°F to a maximum of 480°F. A 600°F maximum oven would be useful, though not ordinarily necessary. The oven should be equipped with a temperature limit switch, and a safety switch should be provided to turn off the heat if the blower stops.

The rubber charge should be on expanded metal trays to provide maximum contact with the circulating air. The parts must not be in contact with one another. For some applications, no post bake is necessary. However, it is generally desirable to oven bake at least 1 hour at 300°F. This removes odor, flavor, and prevents "blooming" of peroxide decomposition products to the rubber surface.

Usually longer cures to at least 50°F above the service temperature are required. Sections thicker than 0.075 inches should be program cured, especially if the aroyl peroxides were used for initial vulcanization. 2,4-dichlorobenzoic acid and benzoic acid will cause internal reversion and sponging if the part containing them is not step cured. There are no set curing schedules that will cover every situation that a fabricator may encounter. In each case, the optimum curing schedule must be worked out. The compound, curing agent, vulcanization conditions, size and shape of the part, construction of the part, and the oven are among the variables that will influence the curing cycle. Table 15.14 is offered as a guide for determining the proper oven post bake.

TABLE 15.14. SUGGESTED OVEN POST BAKE SCHEDULES

	Hours at					
Cross Section	*250°F*	*300°F*	*350°F*	*400°F*	*450°F*	*480°F*
0.075"–0.125"	–	1	–	–	–	–[a]
0.125"–0.250"	–	1	1	3	–	–[a]
0.250"–0.500"	–	3	4	16	–	–[a]
0.500"–0.625"	4	3	4	16	–	–[a]
2 inches	16	8	16	24	8	–[a]

[a]Continue oven cure until desired physical profile is reached.

ROOM TEMPERATURE VULCANIZING (RTV) RUBBER

The first references to these liquid silicone rubber compounds appeared in the patent literature over 15 years ago. During the past 10 years, their technology and commercial use have developed to the point where they now represent a very substantial portion of the overall silicone rubber market.

Although simple in principle, these products are actually quite complex to manufacture. For this reason, the compounding and packaging technology are still mainly proprietary to the basic suppliers, and the RTV materials are marketed as "ready to use" products.

All the liquid rubber compounds are based on low molecular weight silicone polymer with reactive end groups. Just as in the case of the high molecular weight

$$XO\left[\begin{matrix} CH_3 \\ | \\ Si-O \\ | \\ CH_3 \end{matrix}\right]_n X \qquad n = 200 \text{ to } 1000$$

heat cured rubber polymers, some of the methyl groups may be replaced by phenyl groups for improved low temperature flexibility, or by trifluoropropyl groups for improved resistance to jet fuels. End Group reactivity depends upon the cure system, and X is typically hydrogen, or dimethylvinylsilyl, or alkyl (one to four carbon atoms). The polymer molecular weight depends upon the filler and plasticizers used, and upon the compound viscosity desired.

Suitable compounding ingredients include the usual reinforcing and extending fillers, color pigments and heat aging additives, as well as thickeners, plasticizers, and other additives peculiar to the RTV system.

A survey of the patent literature reveals well over 300 patents relating to the room temperature vulcanization of silicone rubber. This is usually achieved by the addition of low molecular weight polyfunctional silicone or silicate curing agents that are reactive with the polymer endgroups at room temperature. A vulcanization catalyst is often required.

Most of the curing systems for the RTV rubbers can be assigned to one of three broad general classifications.

Condensation Cure–Moisture Independent. In these systems, the reactive polymer endgroup is usually silanol (≡Si–OH). The curing agent must have a functionality greater than or equal to 3. This crosslinker may be a silanol containing silicone, in which case an organic base can be used as a condensation catalyst:[14]

$$\equiv Si-OH + HO-Si \equiv \xrightarrow[\text{base}]{\text{organic}} \equiv Si-O-Si \equiv + H_2O$$

An alkoxy containing crosslinker (e.g., ethyl o-silicate) requires a Sn soap catalyst:[15]

$$\equiv Si-OH + CH_3CH_2O-Si\equiv \xrightarrow[soap]{Sn} \equiv Si-O-Si\equiv + CH_3CH_2OH$$

If polyfunctional aminoxy silicon compounds are used as crosslinkers, a catalyst is not always required:[16,17]

$$\equiv Si-OH + R_2NO-Si\equiv \longrightarrow \equiv Si-O-Si\equiv R_2NOH$$

RTV products containing these, and similar, curing systems will all cure in deep sections, independent of atmospheric moisture. The compounds are known as "two-package RTV's," since the curing agent and (or) catalyst must be added just prior to use.

Applications include molds for plastic parts, coatings, adhesives, therapeutic gels, and encapsulants. Typical physical property ranges of two-package RTV's are given below:

Hardness, Shore A	15–70
Tensile Strength, psi	200–900
Elongation, %	100–800
Tear Strength, pi (Die B)	15–125

Condensation Cure–Moisture Dependent. Some of these RTV compounds are made by adding a polyfunctional silicon-containing curing agent to a compound containing a silanol terminated polymer. Others are made by compounding a polymer that is end-stopped with the curing agent. In most cases, a condensation catalyst is also added.

Condensation cure will take place when the compound is exposed to atmospheric moisture. Vulcanization will occur first at the surface, and then progress downward with the diffusion of moisture into the rubber. In the case of a polymer end-stopped with curing agent, the reactions may be generalized as follows (Y = one of the reactive groups on the curing agent that terminates the polymer chain):

$$\equiv Si-Y + H_2O \longrightarrow \equiv Si-OH + HY$$

$$\equiv Si-OH + Y-Si\equiv \longrightarrow \equiv Si-O-Si\equiv + HY$$

According to the patent literature, a wide variety of polyfunctional silicon containing curing agents can be utilized in the design of curing systems of this type. The useful, reactive functional groups include acyloxy,[18] alkoxy,[19] amino,[20] ketoximo,[21] aldoximo,[22] and amido.[23] Most of these require condensation catalysts.

The moisture dependent RTV compounds are known as "one-package RTV's," since the curing agent and catalyst are incorporated in the base compound at the time of manufacture. The main applications are in areas where deep section cure is not required. These products make excellent adhesive sealants, and can be used to form films from solvent dispersion. Excellent physical properties are obtainable in this system:

Hardness, Shore A	15–50
Tensile Strength, psi	200–900
Elongation, %	150–900
Tear Strength, pi (Die B)	25–150

Addition Cure. This curing system involves the metal ion-catalyzed addition of a polyfunctional silicon hydride crosslinker to a dimethyl-vinylsiloxy terminated silicone polymer:[24]

$$\equiv Si{-}CH = CH_2 + H{-}Si\equiv \longrightarrow \equiv Si - CH_2CH_2 - Si\equiv$$

Cure takes place without the formation of any volatile by products, and without dependence upon air or atmospheric moisture. The products are ideal for deep section cures in a confined space. They have excellent resistance to compression set, and to reversion, even when subjected to high pressure steam. The following physical property ranges are typical:

Hardness, Shore A	30–60
Tensile Strength, psi	600–800
Elongation, %	100–400
Tear Strength, pi (Die B)	50–120

The addition cure products are normally sold as two-package RTV's; however, a one-package system is possible by use of an inhibitor. In this case, the inhibitor must be volatilized, or deactivated, by heat in order to "trigger" the cure. Applications include flexible molds, dip coating, and the potting and encapsulation of electrical and electronic components.

The cure rate and physical properties of all the RTV compounds can be varied over a rather wide range by suitable choice of polymer type and molecular weight, filler, curing agent, and type and concentration of catalyst.

REFERENCES

(1) Staff report, *Rubber World*, **161**, 66 (1970).
(2) E. G. Rochow, *An Introduction to the Chemistry of the Silicones*, 2nd Ed., Wiley, New York, 1951.
(3) R. N. Meals and F. M. Lewis, Silicones, Reinhold, New York, 1959.

(4) F. M. Lewis, in *High Polymer Series, Vol. 23, Polymer Chemistry of Synthetic Elastomers*, Part 2, J. P. Kennedy and E. G. M. Tornquist eds., Wiley, New York, 1969, Chap. 8, Part B.
(5) S. W. Kantor, 130th Meeting, *Am. Chem. Soc.*, Sept., 1956.
(6) W. J. Bobear, *Rubber Chem. Technol.*, **40**, 1560 (1967).
(7) G. Alliger and I. J. Sjothun, Eds., *Vulcanization of Elastomers*, Reinhold, New York, 1964, p. 370.
(8) M. L. Dunham, D. L. Bailey, and R. Y. Mixer, *Ind. Eng. Chem.*, **49**, 1373 (1957).
(9) W. J. Bobear, unpublished data.
(10) *Silicone Rubber Handbook*, General Electric Co., Waterford, N. Y.
(11) "Compounding with Silastic® brand Silicone Rubber Gums and Bases," Dow Corning Corp., Midland, Mich.
(12) "Manufacturing Products with Silastic® brand Silicone Rubber," Dow Corning Corp., Midland, Mich.
(13) *Silicone Rubber Notebook*, Union Carbide Corp., New York, N.Y..
(14) U.S. Pat. No. 3,205,283.
(15) U. S. Pat. Nos. 2,843,555 and 3,127,363.
(16) Fr. Pat. Nos. 1,359,240 and 1,423,790.
(17) U. S. Pat. Nos. 3,341,486, 3,441,583, and 3,484,471.
(18) Brit. Pat. No. 835,790; Belg. Pat. Nos. 569,320 and 577,012; U.S. Pat. Nos. 3,035,016 and 3,133,891.
(19) U. S. Pat. Nos. 3,334,067 and 3,294,739.
(20) U. S. Pat. Nos. 3,032,528 and 3,291,772.
(21) Belg. Pat. Nos. 614,394 and 637,096; U. S. Pat. Nos. 3,184,427 and 3,189,576.
(22) Fr. Pat. No. 1,432,799.
(23) Belg. Pat. No. 659,254; Neth. Pat. No. 6,501,494.
(24) U. S. Pat. Nos. 2,823,218; 3,020,260; 3,159,601; 3,220,972; 3,425,967 and 3,436,366.

16

FLUOROCARBON RUBBERS

D. A. STIVERS
3M Company
St. Paul, Minnesota

INTRODUCTION

In the short space of time since the appearance of the first edition of *Introduction to Rubber Technology* a group of new rubbers based upon fluorine-containing monomers has become commercially important. The handling characteristics and physical properties of these relatively new elastomers, of which an estimated 2.2 million pounds were processed in 1969,[1] are herein presented. Because of their relative newness and the highly specialized nature of their application, their uniqueness with respect to other special purpose elastomers and the fundamentals of compounding are outlined so as to be of practical aid to the beginning rubber technologist. For the rubber technologist familiar with hydrocarbon rubbers, and in particular with special purpose rubbers, the advantage of the combination of physical properties which has established the fluorocarbon rubbers in today's market place will be recognized. For the compounder concerned with detailed formulating of fluorocarbon rubbers, a number of practical considerations have been included so that the development of specific physical properties and the handling of certain factory problems relating to the fluoroelastomers may be better accomplished either through the reading of this material or in reference to the resource material upon which it is based.

Because they have been, and are presently, the most expensive of the synthetic rubbers – ranging from \$10.00 to \$15.00 per pound at specific gravity as high as 1.85 – rubber technologists have tended to regard the fluoroelastomer as the rubber of last resort. However, their applications are growing, and the resistance of the market place to their high price is lessening because of their superior performance. At the same time, rubber technologists have learned that fluorocarbon rubbers are not all-purpose materials and that they, indeed, perform quite poorly under certain conditions and in certain environments where cheaper rubbers are necessarily selected because of their optimum performance. Thus, for a certain application, a particular specialty elastomer is selected because it possesses a combination of one or two or possibly three inherently superior physical properties. The compounder proceeds to utilize these properties by sustaining them through the development of adequate supporting properties and then fits the compound into a commercial processing setup. The principles of compounding outlined in this chapter are presented to assist in that purpose.

WHAT IS A FLUOROCARBON RUBBER?

The fluorocarbon rubbers are generally referred to as the fluorinated counterpart of the nonfluorinated source material: for example, fluorinated silicones and fluorinated acrylates. Table 16.1 identifies the fluorocarbon rubbers that have been produced commercially in the United States. By trade name, they are five in number; by classification, they are four.

Table 16.1 indicates that in addition to having established trade names, these fluorocarbon rubbers have also been designated by ASTM D-1418 with the following nomenclature: CFM-Polychlorotrifluoro-ethylene; FPM-Vinylidene Fluoride and Hexafluoropropylene Copolymer; FVSi-Silicone Elastomers having both methyl or vinyl and fluorine substituent groups on the polymer chain. They are further described as to physical properties of their cured vulcanizates in ASTM

TABLE 16.1. COMMERCIAL FLUOROCARBON RUBBERS

Approximate Price	*Trade Name*	*ASTM Designation*	*Supplier*	*Principal Repeating Units*
\$15.00/lb	"KEL-F" Brand Elastomer	CFM	3M Company	CH_2CF_2/CF_2CFCl
\$60.00/lb	Fluororubber 1F4		3M Company	$-CH_2-CH-$ with pendant $C{=}O$, O, $CH_2CF_2CF_2CF_3$ on CH
\$12.00/lb	Fluorosilicone LS-53	FVSi	Dow Corning	$-Si-O-$ with pendant CH_3 and $CH_2CH_2CF_3$ on Si
\$10.00/lb	"VITON" Brand Elastomer	FPM	DuPont	CH_2CF_2/CF_2CF with pendant CF_3 on CF
\$10.00/lb	"FLUOREL" Brand Elastomer	FPM	3M Company	CH_2CF_2/CF_2CF with pendant CF_3 on CF

TABLE 16.2. PRIMARY PHYSICAL PROPERTIES OF CURED FLUOROCARBON RUBBERS – TEMPERATURES OF EQUIVALENT RESISTANCE

Type	*Fluid Resistance*	*Heat Resistance (7 days)*	*Compression Resistance (3 days)*	*Low Temperature Flexibility (5 hours)*
"KEL-F" Elastomer	Acids, Aromatics, Functional	435°F	275°F	−30°F
Fluororubber 1F4	Mixed Solvents, Functional	325	250	−0
Fluorosilicone LS-53	Aliphatics, Functional	475	325	−80
"VITON" Elastomer	All-Round, except Ketones, Esters	525	400	−30
"FLUOREL" Elastomer	All-Round, except Ketones, Esters	525	400	−30

D-2000: HK-Fluorinated Elastomers, namely, "VITON,"* "FLUOREL"** Brand Elastomer, etc., including "KEL-F"** Brand Elastomer, and FK-Fluorinated Silicones.

Table 16.2 shows the primary physical properties of the cured fluorocarbon rubbers and indicates the temperatures of equivalent resistance to heat and cold and compression set. Their basic fluid resistance is also briefly indicated. Each product shown can be considered to have retained approximately the same percentage of their original properties, or in another way, the same degree of resistance, when tested under the conditions shown for the periods of time designated. Table 16.2 may be used to identify the specific strengths of the fluorocarbon type rubbers or it may be related to one's experience with the hydrocarbon types. For example, if a typical nitrile rubber compound were heated for a period of seven days at 275°F, it would retain about the same percent of its tensile and elongation and hardness as for a "FLUOREL" compound heated for the same period of time at 525°F. Likewise, if the nitrile compound were subjected to a 25% deflection at 250°F for three days, the percent set would be about the same as that obtained on a general purpose "FLUOREL" compound compressed for the same period of time at 400°F. With respect to low temperature properties, they would have about the same degree of flexibility or possess equivalent resistance to embrittlement if the nitrile compound were at −40°F and the "FLUOREL" compound at −30°F. These are broad generalizations, but nevertheless useful in categorizing elastomers for particular application areas.

* Registered Trademark of E. I. DuPont de Nemours, Wilmington, Del.
**Registered Trademark of 3M Company, Saint Paul, Minnesota.

TABLE 16.3. CHARACTERISTICS OF TYPICAL FLUOROCARBON ELASTOMER RAW GUMS

ASTM Designation	*Trade Name*	*Specific Gravity*	*Mooney Viscosity, Large Rotor, 4 Min. Readings/212°F*
CFM	"KEL-F" Elastomer 3700, 5500	1.85	200 +
FVSi	Fluorosilicone LS-53U	1.41	–
FVSi	Fluorosilicone LS-2332U	1.44	–
FPM	"VITON" AHV	1.82	170
FPM	"FLUOREL" 2140	1.85	110
FPM	"VITON" A	1.82	75
FPM	"VITON" B-50	1.86	50
FPM	"FLUOREL" 2146	1.84	37
FPM	"VITON" C-10	1.82	10
	"VITON" LM	1.72	Brookfield Viscosity (100°C) 2000 cp.

Table 16.3 lists the characteristics of typical fluorocarbon elastomer raw gums. The Mooney viscosities are also shown; a wide range is indicated. The table also indicates a predominance of the availability of the FPM type of the FPM type of fluoroelastomer. Because this type of fluoroelastomer is more widely applied and because its compounding and processing characteristics have been more thoroughly studied at the laboratory and plant levels, greatest emphasis will be placed upon its technology. Unless otherwise specially noted, the compounding to alter processing and physical properties will be confined to this class of fluoroelastomer.

RAW GUM FLUOROCARBON RUBBERS

With the exception of the fluorosilicones, the fluorocarbon rubbers can be produced by emulsion polymerization. They range widely in molecular weight. For example, the "VITON" fluoroelastomers A and AHV run from 100,000 to 200,000, providing a practical processing range of viscosity. Lower molecular weight grades have been produced for special processing requirements, including "VITON" LM, a waxlike material, having a molecular weight of less than 5,000. "FLUOREL" Elastomer is available in similar viscosities at an intermediate range of molecular weight. The "KEL-F" elastomers are extremely tough raw gums, from the standpoint of processing, and are greater than 200,000 molecular weight. On the other hand, "3M" Brand Fluororubber 1F4 Elastomer is a soft raw gum having a Mooney viscosity of approximately 15, although its molecular weight is approximately 1,000,000. The FPM (vinylidene fluoride based elastomers) may be prepared as described by Kennedy and Tornqvist.[2]

Processing of Fluorocarbon Rubbers

Throughout the various steps of converting a particular class of elastomer into a final cured compound which possesses the desired physical properties, it is also of great importance to recognize and maintain the proper balance of processing characteristics. Most commercial applications require not only that the finished product perform, but also that the compound from which it is made process within certain narrowly confined limits and even on occasion in somewhat automated systems. This has proved to be the case in the expansion of the usage of the fluorocarbon rubbers. In any given class of special purpose elastomers, there will be a greater number of commercial products made available in order to fit this class of elastomer into some particular processing requirement than will have been developed for the purpose of gaining an advantageous physical property in the finished part. This is also true of the fluorocarbon rubbers, as indicated by the varieties shown in Table 16.3. Except for the "KEL-F" elastomers, which remain extremely tough, they have been tailored considerably to fit a variety of modern processes used on a production scale in the fabrication of rubber parts. All of the fluorocarbon elastomer raw gums have been designed for mill mixing. Those in the Mooney viscosity range 30 to 120 are preferred. The "KEL-F" elastomers handle with difficulty, almost too tough to cut with the mill knife, except in thin gauges, while the fluorosilicone rubbers are soft and nonsticky and work with ease. Most of the FPM types process without sticking on a cool mill, and compounding ingredients are easily incorporated. No more than the usual precautions regarding good dispersion and methods of addition of ingredients respecting the scorch characteristics of the formula are required for successful mixing of these compounds. There is one widely used curing agent, the carbamate of hexamethylene diamine marketed as Diak #1, which sometimes requires remilling of the fully compounded stock on a tight mill in order to obtain the desired dispersion; this is especially true of Banbury mixed stock. Most formulations contain carbon black and metallic oxides, and faster mixing usually results when these raw materials are thoroughly blended prior to their addition on the mill.

Calendering. Most FPM elastomer compounds can be calendered into gum stock and frictioned or coated on the fabric, provided roll temperatures are maintained between 200 and 250°F. Gauges as thin as .003 inches are attainable under these conditions. In the temperature range from 110 to 200°F, problems of sticking are likely to be encountered; however, some formulations sheet out smoothly below these temperatures. In equivalent formulations, the higher molecular weight grades tend to stick less during calendering.

Extruding. Extrusion of the "FLUOREL"-"VITON" type of fluorocarbon rubber is widely practiced for the purpose of both producing finished goods and preforming uncured compounds for subsequent molding. The best practice for achieving surface smoothness, low-nerve, dimensionally stable extrusions is to use the right viscosity raw gum. This is because many end-use applications limit the choice of filler or the amount of physical plasticizer or chemical softener that would improve extrusion. Most compounds run satisfactorily in typical rubber extruders at temperature settings of 150 to 175°F on the die and barrel. The extrusion die should be designed having a relatively long land, i.e., the surface of the die which is parallel to the direction of flow. A screw compression ratio of approximately 1.25 to 1 and a length to diameter ratio of 10 to 1, or greater, are also preferred. Smooth extrusions may also be obtained at lower temperature settings by installing a breaker-plate. Because of a tendency towards high frictional heat buildup, it is preferable to operate with compounds which show a Mooney scorch time of 20 minutes or better to a 10 point rise when measured at 250°F. Safer compounds are especially required where re-extrusion is desired.

Molding. Most compounds can be molded either by compression, transfer, or injection molding, as with other elastomers. The press cure of 5 to 30 minutes at 300 to 350°F is mainly a forming operation in which the cure is completed only to the point where the molded piece has the dimensional stability required for removal from the mold and insertion in an oven for post curing at 400 to 500°F for 16 to 48 hours. Twenty-four hours at 400°F has found wide acceptance. Compression set resistance can be improved by using the higher temperature; however, the elimination of the post cure will not usually affect fluid resistance. The practice of step curing for periods of several hours at 50°F intervals from 200 to 400°F is often necessary to prevent blowing when the thickness of the part is greater than 0.25 inches. Sponging of thick sections prepared by laminating can also be reduced by washing the individual plys with a ketone solvent prior to laminating. This procedure is especially useful in roll building. For extremely thick sections, calcium oxide is more effective than magnesium oxide or zinc oxide in preventing porosity. In addition to the normal advantages of transfer molding, some success has been reported in the design and operation of these molds so as to quickly allow the opening of the mold immediately following the completed transfer of compound from the transfer pot and permit removal of the flash pad before continuing with the cure. Some compounds may in this way be remilled and worked away in small percentages in subsequent mill batches, providing an important cost saving, especially in instances where the amount of material in the pad is several times the amount of material contained in the finished part. Injection molding offers perhaps the greatest potential in low cost processing of fluorocarbon elastomers, especially for thin cross-section seals, where it is possible to obtain rapid cures.

Molding operations usually require the application of mold lubricant in order to provide satisfactory release at the press. Water-based lubricants containing Aquarex L, silicone oil, or polyethylene emulsion work well. The addition of about one part of zinc stearate to the compound often makes a significant improvement in mold release characteristics without harming physical properties. The fluorocarbon rubbers do not inherently stick to mold surfaces more than other elastomers; however, the volatiles formed during curing have a corrosive effect on ordinary carbon steel molds. This action causes a pitting effect which leads to sticking and hastens buildup on mold surfaces. As molds continue in use, frequent cleaning becomes necessary. By chrome plating carbon-steel molds or by making the entire mold out of more corrosive resistant stainless steels, the maintenance of molds and quality of parts can be improved. The lower molecular weight fluorocarbon rubbers tend to have poorer release characteristics than those having higher molecular weight.[3] This is especially true with press temperatures of 350°F and higher. Thus, among the FPM types, "VITON" AHV has sometimes shown superior mold release characteristics.

TABLE 16.4. FLUID RESISTANCE OF FLUOROCARBON RUBBERS COMPARED WITH HYDROCARBON RUBBERS (% VOLUME SWELL – 3 DAY IMMERSION)

Organic	*Temp. °F.*	*"Thiokol"*	*Neoprene*	*Nitrile*	*Acrylate*	*Silicone*	*"Hypalon"*	*Fluoro Silicone LS53*	*"Viton B"*	*"Fluorel"*
Acetone	77	25	40	130	250	15	18	181	300	375
Benzene	77	100	290	100	350	175	275	27	12	22
70/30 Isooctane/Toluene	77	10	75	22	40	200	110	22	2	3
Ethyl Acetate	77	25	60	105	250	175	60	140	280	375
Methylene Chloride	77	det.	150	300	300	150	250	150	25	30
Inorganic										
Hydrochloric Acid (conc.)	77	det.	4	11	det.	5	5	8	1	2
Nitric Acid (conc.)	77	det.	det.	det.	det.	–10	0	4	4	4
Sulfuric Acid (conc.)	77	det.	det.	det.	det.	det.	10	det.	4	3
Sodium Hydroxide (50%)	77	2	2	0	det.	–1	0	1	0	0
Steam	300	–	–	–	–	+2	–	–2	–	6
Functional										
ASTM Oil #3	300	–2	–	10	16	49	–	4	3	3
MIL-L-7808 Lubricant	400	–	–	–	–	31	–	13	10	12
OS-45 Hydraulic Fluid	400	–	–	–	–	80	–	13	3	3

Shrinkage. The maintenance of constant dimensions, especially in O-ring seals, has been somewhat of a problem in the industry due to the variations in compound shrinkage. In fact, it has become of such concern that recommendations have been made for the establishment of slightly wider tolerances with this class of elastomer. The causes for variation in dimensions are usually attributed to the following: the relatively low quantities of nonreinforcing fillers, which are used for the development of maximum resistance to compression set; borderline press cures, which enhance the variation caused by inherently high thermal expansion during oven post cure; the effect of the escapement of volatiles which are generated during the oven post cure, which vary in type and amount depending upon curing agent and acid acceptor used, as well as the method of post cure (that is, whether step cured); and, the high degree of impermeability of the elastomer to these volatiles.

Steam Cures. The "press cure" of rolls and finished extrusions has been performed by this method, which is also satisfactory for preparing for subsequent post curing in an oven. Some compounds harden excessively upon exposure to open steam cures, and formulations to avoid this will be discussed later.

Physical Properties of Cured Fluorocarbon Rubbers

As indicated by the physical properties shown in Table 16.2, the application of fluorocarbon rubbers has centered around four basic physical properties: (1) chemical resistance, (2) heat resistance, (3) resistance to compression set, and (4) low temperature flexibility. These are the main physical properties considered in the development of compounds for O-rings, seals, gaskets, etc., which are currently the major uses for this class of elastomer. These properties are also, however, fundamental to the application of most specialty elastomers for many other applications, and are therefore a suitable basis for use in considering the various aspects of their compounding.

Since the desire for obtaining a basic fluid resistance is perhaps the main reason for choosing a fluorocarbon rubber, Table 16.4 has been prepared to show the comparative fluid resistance of fluorocarbon rubbers with hydrocarbon rubbers.[4] Although the fluorocarbon rubbers are shown to have the best all around chemical resistance, certain solvents such as ketones and acetates attack them severely. The volume swell characteristics of the various elastomers shown in Table 16.4 are a reminder that the selection of the elastomer is an important first step because of the difficulty in modifying the basic fluid resistance of a given elastomer. The weathering characteristics of the fluorocarbon rubbers are outstanding due to their stability in sunlight and ozone.

Table 16.5 shows a group of general purpose formulations based on several "FLUOREL" elastomers. From this table, formulas may be selected that process satisfactorily under various conditions and provide resistance to heat and compression set typical of this class of elastomer. Table 16.6 lists additional physical properties of one of these general purpose compounds.[5] Table 16.7 shows the heat resistance of "VITON" fluoroelastomers at temperatures from 450 to 650°F. Compound B-7 has been reported to retain a tensile strength of 425 psi,

TABLE 16.5. TYPICAL PHYSICAL PROPERTIES OF SOME CURED "FLUOREL" COMPOUNDS

	A-5	*B-5*	*C-5*	*D-5*	*E-5*	*F-5*
"FLUOREL" 2140	100	–	–	100	–	–
"FLUOREL" 2141	–	100	–	–	100	–
"FLUOREL" 2146	–	–	100	–	–	100
Thermax	15	15	15	15	15	–
Magnesia	20	20	20	20	–	–
Calcium Oxide	–	–	–	–	20	–
Zinc Oxide	–	–	–	–	–	10
Dyphos	–	–	–	–	–	10
HMDA-C or Diak #1	1	1	1.4	–	1.2	0.9
Diak #3	–	–	–	3	–	–
Copper Inhibitor #50	–	–	–	–	–	–
Mooney Scorch, MS @ 250° F						
Minutes to 10 point rise	9	25	23	25+	15	25+
Minimum reading	55	57	18	44	50	13
Press Cure: Minutes/° F	30/300	30/300	20/320	30/300	30/300	20/320
Oven Post Cure: Hr/° F	24/400	24/400	24/400	24/400	24/400	24/400
Original Physical Properties						
Tensile Strength, psi	2400	1900	2430	2640	1750	1280
Elongation, %	310	330	290	305	195	420
Hardness, Shore A	68	67	67	66	73	57
100% Modulus, psi	410	360	435	460	875	210
Oven Aged (Hours/° F)	16/600	16/600	16/575	16/600	16/600	16/575
Tensile Strength, psi	1340	955	1930	1500	1460	1030
Elongation, %	265	270	240	160	170	280
Hardness, Shore A	70	70	72	79	80	58
Oven Aged (Days/° F)	15/500	2/600	2/600	2/600	2/600	2/600
Tensile Strength, psi	1450	1090	Brittle	Brittle	1160	Brittle
Elongation, %	150	50	–	–	105	–
Hardness, Shore A	82	85	98	–	87	–
Compression Set, ASTM Method B						
Time, Hr/Temp., °F	22/450	24/400	22/350	22/450	22/450	22/350
Per Cent	52	43	26	47	38	30

elongation of 280%, and Shore A hardness of 78 following two years' aging at 400°F. Table 16.8 lists some additional, important physical properties of "VITON" Compound B-7.[6] Since most applications involving fluorocarbon rubbers are at elevated temperatures, two effects upon their physical properties, other than simple resistance to embrittlement upon heat aging, should be noted. One effect is a loss in Shore A hardness of approximately 15 points between 250 and 500°F; therefore, in certain applications it is advisable to start with compounds of higher hardness. Another important consideration is the rapid loss of tensile strength as temperature increases, as shown in Table 16.8. Approximately a two-fold increase in tensile strength at 400°F may be obtained by the substitution of Cabolite P4 for medium thermal carbon black without affecting the original cured hardness. The high molecular weight copolymers also retain better physical properties when tested at

TABLE 16.6. ADDITIONAL PHYSICAL PROPERTIES OF "FLUOREL" COMPOUND A-5 IN TABLE 16.5

Tear Strength (ASTM D-624-54, B)lb/in.	180
Bashore resilience	5
Abrasion resistance	0.031
Taber, H-22, 1000 g at 1000 rev. wt loss g	

Low Temperature Properties

Gehman Stiffness, (ASTM D-1053-52T), °F	
T_2	+17
T_5	+7
T_{10}	+3
T_{100}	−11
Temperature Retraction Test, °F	
TR 10	−8
TR 30	−1
TR 50	+5
TR 70	+13
Brittle Point, (ASTM D746-52T)	
0.075" thickness, °F	−30
0.025" thickness, °F	−40

Electrical Properties

D.C. Resistivity	
at 50% Relative Humidity, ohm-cm 2×10^{13}	
at 90% Relative Humidity, ohm-cm 1.5×10^{13}	
Dielectric Strength (short time), volts/mil (20 mil specimen)	630
Dielectric Constant, 100 cps, 25°C	11.4
Dissipation Factor, 100 cps, 25°C	0.0125

Permeability Constants

Gas	*Test Pressure (psi)*	*Test Temp. (°F)*	*Permeability Constant (cc/mm-cm²-sec-cm Hg)*
Hydrogen	50	73	121×10^{-11}
Hydrogen	50	210	4500×10^{-11}
Nitrogen	50	73	25×10^{-11}
Nitrogen	50	210	1450×10^{-11}
Oxygen	50	73	28×10^{-11}
Oxygen	50	210	1920×10^{-11}
Freon 22		77	2×10^{-9}

elevated temperatures; for example, "VITON" AHV shows a tensile strength of 850 psi with elongation of 100% when tested at 300°F. The various "FLUOREL" and "VITON" raw gums may be used to develop compounds having generally equivalent cured physical properties, thus performing similarly under field service conditions. The physical properties that have been listed in Tables 16.5 through 16.8 are selected data which are meant to characterize the primary and supporting properties of this general type of fluororubber for their use in most applications.

TABLE 16.7. HEAT RESISTANCE OF VITON FLUOROELASTOMERS

	A-7	*B-7*
Viton A	100	
Viton B		100
Magnesia	15	15
M T Carbon Black	20	20
Diak #3	2	3
Press Cure: Min./°F	30/325	30/325
Oven Post Cure: Hr/°F	24/400	24/400
Physical Properties at 75°F		
Modulus at 100%, psi	350	550
Tensile Strength, psi	2175	2250
Elongation at break, %	470	410
Hardness, Durometer A	68	74
Oven Aged, Days/°F	100/450	100/450
Tensile Strength, psi	1000	625
Elongation at break, %	160	480
Hardness, Durometer A	87	75
Oven Aged, Days/°F	20/500	20/500
Tensile Strength, psi	1250	550
Elongation at break, %	100	400
Hardness, Durometer A	94	83
Oven Aged, Days/°F	2/600	2/600
Tensile Strength, psi	1050	500
Elongation at break, %	60	240
Hardness, Durometer A	91	83
Weight Loss, %	18	11
Oven Aged, Days/°F	1/650	1/650
Tensile Strength, psi	Brittle	575
Elongation at break, %		15
Hardness, Durometer A	99	91
Weight Loss, %	36	22

Permeability (Test sample post-cured @ 450°F)	Air	.50
Permeability X10^8, Tested at 75°F	Nitrogen	.33
	Helium	18.5

Radiation Resistance		
Dosage Megareps	10	100
Stress at 100% Elong., psi	400	–
Tensile Strength, psi	1675	1380
Elongation, %	240	80
Hardness, Shore A	69	76
Set at break, %	6	0

Thermal Conductivity

At 77°F	4.3×10^{-4} cal/cm x sec x °C
At 118°F	3.3×10^{-4} cal/cm x sec x °C

Ozone Resistance

At 150 ppm for 200 hours – no cracking by the bent loop method.

Weathering

No change in physical properties after 2 years exposure to outdoor industrial atmosphere.

Flammability

Will not support combustion.

Next to the consideration of fluid resistance, the resistance to compression set is the most important physical property utilized in the application of fluorocarbon rubbers. Although fluid resistance is little affected, provided a substantial press cure is given the molded part, the ability to recover following deflection at elevated temperatures is greatly affected by the time and temperature of oven post cure.

TABLE 16.8. , ADDITIONAL PHYSICAL PROPERTIES OF VITON COMPOUND B-7 FROM TABLE 16.7

Properties Measured at Elevated Temperatures

Tested at	75°F	300°F	500°F
Tensile Strength, psi	2450	500	300
Elongation, %	330	120	80
Hardness, Shore A	75	65	63
Tear Resistance (lb/inch)			
Winkelman (ASTM D624, Die B)	170	40	20
Graves (ASTM D624, Die C)	245		40
Trousers (ASTM D470)	18		1

Abrasion Resistance

Bureau of Standards (Method B with No. 1 Smoked Sheet Control)

at 75°F, 115 at 212°F, 54

Taber Abrasion (1000 gram weight, H-22 wheel)

Grams lost per 1000 revolutions, 0.143

Du Pont Abrasion

Percent of standard compound, 300

Flex Life

At 75°F, ASTM D813 (Demattia), 300 cycles/minute, 1/4" thick specimen

Time	*Crack Growth, %*
0	10
1	18
10	29
20	34
60	76

Kinetic Coefficient of Friction

Linear Velocity (ft/min)	*Kinetic Coefficient of Friction*
25	.67
50	.64
75	.55

Low Temperature Properties

Brittle Temperature, °F (ASTM D746) −49

Temperature Retraction, °F (ASTM D1329) – T_{10} − 17

Clash-Berg Stiffness Test, psi (ASTM D1043)

At 75°F	440
10	4,760
0	53,200

TABLE 16.9. EFFECT OF OVEN POST CURE TEMPERATURE ON COMPRESSION SET RESISTANCE

	A-9
"VITON" A	100
Magnesia	15
MT Carbon Black	20
Diak No. 1	1.5

Press cured 30 minutes at 300°F followed by oven cure 24 hours at 400°F

100% Modulus, psi	500
Tensile strength, psi	2200
Elongation at break, %	190
Hardness, durometer A	74

Pellets press cured 30 minutes at 300°F, post cured in oven as shown

Compression Set, % ASTM D-395, Method B
24 Hours at Temperature Shown

Compression Temp.	−20°F	32°F	158°F	250°F	350°F	400°F	450°F
Post Cured 24 Hours at							
350°F	100	43	17	15	38	–	–
400°F	100	33	17	13	17	41	95
500°F	100	53	15	10	9	20	60

Table 16.9 shows the effect of these conditions upon a "VITON" A compound when post cured for 24 hours over a temperature range of 350 to 500°F and tested over a temperature range from −20 to 450°F.[5] As with other elastomers, the duration of tests during compression affects the amount of set, although the fluorocarbon rubbers show relatively greater resistance with the extension of time. Table 16.10 shows the percent set taken by a "VITON" A compound when compressed for periods of time up to seven days and at temperatures up to 450°F.[7] Table 16.11 shows the physical properties of a "VITON" A compound compared with two lower molecular weight elastomers. In addition to possessing adequate stress-strain properties, the advantages of these low molecular weight materials in controlling scorch and in transfer molding procedures are indicated.[8]

Physical Properties of Fluorosilicone Rubbers. The fluorine-containing silicone rubbers have been developed into the second largest volume of the fluoroelastomer types. Table 16.12 shows the physical properties of two fluorosilicone rubbers, LS-53U and LS-2332U, which have been vulcanized with benzoyl peroxide. This type of silicone retains most of the useful qualities of the regular silicone rubbers and improves resistance to many fluids. Exceptions are ketones and phosphate esters; however, they may be blended with the regular dimethyl silicone types which have good resistance to these fluids at 300°F. They are most useful today where the best in low temperature properties is required in addition to fluid

TABLE 16.10. EFFECT OF TEST TEMPERATURE AND DURATION OF TEST ON COMPRESSION SET

	A-10
"VITON" A	100
Magnesia	15
MT Carbon Black	20
Diak No. 1	1.5

Cure: Press – 30 Minutes at 325°F, followed by:
Oven – 24 Hours at 400°F

Physical Properties	
100% Modulus, psi	500
Tensile strength, psi	2200
Elongation at break, %	190
Hardness, durometer A	74

Compression Set, % ASTM D-395, Method B
Compression Temperature

	−20°F	0°F	75°F	158°F	250°F	350°F	400°F	450°F
Compression Time								
1 Day	100	54	18	17	13	17	41	85
3 Days	100	62	20	–	15	30	80	95
7 Days	100	59	22	20	20	50	–	95

TABLE 16.11. COMPARISON OF VITON C-10 WITH "VITON"A AND "VITON" A-35

Compound	*A-11*	*B-11*	*C-11*
"VITON" A	100	–	–
"VITON" A-35	–	100	–
"VITON" C-10	–	–	100
Magnesia (low activity)	15	15	15
MT Carbon Black	20	20	20
Diak No. 1	1.5	1.5	1.5
Mooney Scorch at 250°F (MS)			
Minimum	34	25	19
Minutes to a 5 point rise	9	12	11
Minutes to a 10 point rise	11	14	14
Cure: 15 minutes at 325°F – 1,000 psi pressure – 1/16 inch orifice			
Transfer Molding			
Flow thru 1/8 in. sprue, inches	5/8	1–3/8	7–3/4
Flow thru 1/4 in. sprue, inches	1	2–1/2	9–1/4
Cure: Press – 30 minutes at 325°F Oven – Step + 24 hours at 400°F			
Physical Properties			
Original			
100% Modulus, psi	875	625	450
Tensile strength, psi	2775	2400	1900
Elongation, %	180	200	220
Hardness, durometer A	72	73	71

TABLE 16.12. PHYSICAL PROPERTIES OF FLUOROSILICONE RUBBERS

	A-12	*B-12*
	LS-53U[a]	*LS-2332U*[a]
Color	red	tan
Specific gravity	1.41	1.46
Molding Conditions		
Press Cure: minutes/°F	5/260	5/260
Post Cure: hr/°F	24/300	8/392
Hardness, Shore A Scale	55	55
Tensile Strength, psi	1000	1220
Elongation, %	170	470
Tear Strength, ASTM, D624-54 Die B, lb/in.	55	240
Compression Set, Method B 22 hr/300°F, %	22	20
Other Physical Properties Characteristic of the Silastic L. S. Grades		
Brittle Point, ASTM D746°F		−90
Stiffening Temperature, ASTM D795-58		
(Young's Modulus = 10,000 psi) °F		−75
Linear Shrinkage, %		3.0 to 3.4
Electrical Properties		
Electric Strength, v/mil		340 to 380
Dielectric Constant @ 10^2 cps		6.4 to 7.4
@ 10^6 cps		6.2
Dissipation Factor @10^2 cps		.01 to .07
@ 10^6 cps		.04
Volume Resistivity, ohm-cm		1×10^{12} to 10^{14}
Heat Aging (weeks/F)	20/300	0.5/392
Tensile Strength, psi	600	1100
Elongation, %	100	400
Hardness, Shore A	68	54

These properties are based on benzoyl peroxide cures.

[a] "SILASTIC" No.

resistance, although resistance to fuels – particularly those containing aromatics – is poorer than for the FPM type fluorocarbon rubbers. Improvements have been made in the general strength of the fluorosilicone rubbers, especially tear strength where 200 pounds per inch thickness can be obtained. Compared with the FPM fluoroelastomers, they have slightly higher tensile strength at elevated temperatures, though original tensiles are ordinarily one-half as much. Fluorosilicone compounds do not drop in hardness as rapidly with increased temperature.[9]

Physical Properties of "KEL-F" Brand Elastomers 5500 and 3700. Table 16.13 shows the physical properties of typical "KEL-F" elastomer compounds.[10] These high strength raw gum fluoroelastomers were hampered in their commercial development by their toughness in processing, and they were replaced by the easier

TABLE 16.13. PHYSICAL PROPERTIES OF "KEL-F" BRAND ELASTOMER 5500 AND 3700

	A-13	*B-13*
"KEL-F" Elastomer 5500	100	
"KEL-F" Elastomer 3700		100
Zinc Oxide	10	10
Dyphos	10	10
Benzoyl Peroxide	1.5	
HMDA-Carbamate		3
Press Cure: Min/°F	30/300	30/300
Oven Post Cure: Hr/°F	16/300	16/300
Original Properties		
Tensile Strength, psi	2430	2300
Elongation, %	410	300
Hardness, Shore A	60	60
300% Modulus, psi	400	2100
Oven Aged, 14 days/400°F		
Tensile Strength, psi	1380	2540
Elongation, %	630	490
Hardness, Shore A	69	70
Compression Set, Method B, %		
70 hr/250°F	50	25
70 hr/300°F	60	40
Chemical Resistance Time (days) at 77°F, Volume Change, %		
Red Fuming Nitric Acid (30)	24	32
Fuming Sulfuric Acid (30)	1	27
70/30 Isooctane-Toluene (30)	30	15
ASTM Oil #3 (30)	1	7
Steam @ 400°F (7)	22	
Steam @ 300°F (30)		20

processing "FLUOREL"–"VITON" types which also exhibited generally superior physical properties for most applications. However, the "KEL-F" elastomers continue to show outstanding performance in the presence of strong oxidizing acids, and under such conditions have demonstrated superior resistance to flex cracking under dynamic loads. They were also the first of the fluorocarbon rubbers to be approved for FDA applications. In addition, large volumes have been employed as binders in solid propellant systems.

Applications

Table 16.14 lists different kinds of applications of the fluorocarbon rubbers. Because of their effectiveness as sealing materials, a full size range of O-rings is commercially available. The actual applications for these are, of course, numerous since the application of the several outstanding physical properties can be made to

TABLE 16.14. APPLICATIONS OF FLUOROCARBON RUBBERS

1. Seals – Full size-range of O-rings; also flat gaskets, extrusions, and many special cross-sectional configurations for sealing purposes.
2. Valve and Pump Linings – Solid rubber inserts; rubber-to-metal bonded.
3. Hose – Rubber lined or rubber covered; tubing or fabric reinforced.
4. Rolls – 100% fluorocarbon rubber covered or laminated to other elastomers.
5. Fabric – Frictioned or coated; single ply or laminated; also applied from solution.
6. Nonflammable – For use in 100% oxygen at high pressure; solid rubber parts or protective overlay for flammable substrates (paper, wood, rubber, leather, plastics, fabric).
7. FDA – "KEL-F" Brand Elastomers and selected "FLUOREL" or "VITON" Elastomers may be compounded to meet the Food and Drug Administration requirements for processing edibles.
8. Standard Specifications – Society of Automotive Engineers (SAE); Aeronautical Material Specifications (AMS); American Society of Testing Materials (ASTM) – D 2000, D 1418; Military (Air Force, Navy); Industrial Specifications

the upgrading of the performance of the wide range of hydrocarbon types now performing to a lesser degree of satisfaction. The aircraft, petroleum, and chemical industries find many uses for these materials. Their resistance to hot aircraft engine lubricants is an outstanding example of how they have been employed to greatly increase service life while maintaining higher standards of safety. Fluorocarbon rubber rolls are used for processing paper, film, and fabric. Fuel and flame resistant compounds have been developed for protection of clothing and other flammable substrates. A number of compounds have now been approved for handling carbonated beverages and pharmaceutical items. In addition, the quality and performance of many compounds that have been fabricated into the items shown in Table 16.14 have been included under industrial and military specifications, as well as ASTM Standards, which may be referenced to specify special purpose requirements.

Compounding

Converting the thermoplastic, rubbery raw gums of the fluorocarbon elastomers to usable vulcanizates requires primary crosslinking agents, such as the amines, as well as acid acceptors, such as the metallic oxides of magnesium, calcium, lead, or zinc. The combined effects of these materials may also be regarded as one of "activation" since they promote the removal of hydrogen and halogen to produce the double bonds which are able to further participate in the crosslinking reactions that result in stable, useful vulcanizates. Most systems can also be sufficiently retarded; salicylic acid or hydroquinone are effective.

The curing of the "FLUOREL"-"VITON" type fluoroelastomers is reported to occur in three stages[11]: (1) Hydrogen fluoride is eliminated at relatively low

TABLE 16.15. CURING AGENTS FOR THE GENERAL PURPOSE FPM FLUOROCARBON RUBBERS

HMDA-C	(hexamethylenediamine carbamate) – usual range: 0.75 to 1.5 phr
"DIAK"	#1 (hexamethylenediamine carbamate) – equivalent to HMDA-C
"DIAK"	#2 (ethylenediamine carbamate) – usual range: 0.85 to 1.25 phr
"DIAK"	#3 (N,N'-dicinnamylidene 1,6 hexane-diamine) – usual range: 2-4 phr
"DIAK"	#4 (an amine salt) – usual range: 1.9 to 2.4
Copper	Inhibitor 50 (disalicylal propylene diamine) – usual range: (as secondary curing agent) 0.2 to 1.0 phr

temperature in the presence of basic materials to form regions of unsaturation; (2) during the press cure at 250°F, difunctional agents react through addition to double bonds or through substitution of an allylic fluoride atom to form chemical crosslinkages; (3) during the oven post cure at 400°F, conjugated double bonds are formed which undergo Diels-Alder condensation. Subsequent aromatization results in extremely stable aromatic crosslinkages. Oven cure is essential if long service life is required, particularly at temperatures in excess of 400°F. Table 16.15 shows the curing agents used with the general purpose FPM fluorocarbon rubbers.

These elastomers do not require "protective materials," that is, antioxidants as for natural rubber, to protect them from oxidative or thermal degradation during oven aging; however, the choice of metallic oxide used in promoting stable vulcanizates during oven post curing has a pronounced effect upon heat stability, and also resistance to specific chemicals. Table 16.16 shows some of the effects obtained when using various amines and oxides. Magnesium oxide is the most widely used because it provides the best balance of processibility and physical properties. The grade of magnesium oxide is somewhat critical and occasionally specific to a given elastomer. Table 16.16 suggests the variety of processing conditions and the specific types of physical properties that can be developed with two grades of "FLUOREL" elastomer, namely, "FLUOREL" 2140 and "FLUOREL" 2141. "VITON" A and AHV would respond similarly to "FLUOREL" 2140 in these curing systems.[12]

Table 16.17 shows effect of fillers on the physical properties of "VITON" A.[13] Fillers are added in amounts up to 20 parts to achieve smoothness in processing. Beyond this level, their chief function is to obtain high hardness or low cost. Medium thermal carbon blacks are used to achieve all these characteristics. Emtal 549 is a high modulus, low hardness talc pigment having good extrusion characteristics. Pyrax A and Mistron H-6055 also impart high modulus at low levels with minimum effect upon hardness. Increasing filler content is detrimental to heat

TABLE 16.16. SCORCH RATES OF "FLUOREL" 2140 AND "FLUOREL" 2141 ELASTOMERS USING VARIOUS AMINES AND OXIDES

	Basic Compound	Mooney Scorch Time
"FLUOREL" Elastomer	100	
Thermax	15	Small Rotor at 250°F
Metallic Oxide	20	Minutes to 10 point rise
Curing Agent	As Shown	above minimum

Metallic Oxide		**Maglite K**			**Calcium Oxide**		**Litharge**	**Zinc Oxide-Dyphos**	
Curing Agent	HMD-Amine	HMDA-C	Diak#3	HMDA-C	Diak#3	HMDA-C	Diak#3	HMDA-C	Diak#3
parts	1	1	3	1	3	1	3	1	3
"FLUOREL" Elastomer 2140	4	8	29	7	32	5	25	13	31
"FLUOREL" Elastomer 2141	13	24	16	15	12	9	15	26	19
Application		General Purpose			Heat Resistance Compression Set		Chemical Resistance	Steam Resistance	

TABLE 16.17. EFFECT OF FILLERS ON THE ORIGINAL PHYSICAL PROPERTIES OF "VITON" A ELASTOMER

"VITON" A	100
Zinc Oxide	10
Dibasic Lead Phosphite ("DYPHOS")	10
Filler	As Indicated
Diak No. 1	As Indicated

Press Cure: 30′/300°F Oven Cure: Step cure to 400°F and 24 hr/400°F
Original Physical Properties (Small dumbbells .025″ thick)

Filler	*Volume*	*Parts*	*Parts Diak #1*	*Modulus, psi 100% Elongation*	*Tensile Strength, psi*	*Elongation, %*	*Hardness, Shore A*	*Compression Set Method B 70 hr/ 121°C*
None	–	–	1	200	2200	400	58	19.4
MT Black	20	19.4	1	350	2625	350	65	16.9
	40	38.8	1	625	2525	300	74	19.7
	60	58.2	1	950	2400	250	80	20.8
	100	97.0	1	1400	1950	175	91	33.6
	140	135.8	1	1450	1600	125	96	53.9
FEF Black	20	19.4	1	500	3225	375	70	34.8
Super Multifex Whiting	20	28.6	2	1050	3250	220	80	12.4
Blanc Fixe	20	47.4	2	900	2350	230	75	8.0
Iceberg Clay (Calcined Clay)	20	28.4	2	775	1925	225	75	20.8
Hi-Sil 233/LM-3 (Fine Silica – Silicone Oil) (100/20)	20	25.2	2	900	2525	350	86	37.6
Silene EF	20	21.6	2	700	3125	350	80	32.8

aging, low temperature flexibility, and to some extent, compression set. Lower amounts of curing agent and less complete oven post cure, consistent with compression set requirements, permit higher loadings with best heat resistance. The silica-containing pigments show the greatest loss in elongation upon heat aging.

Other than the compounding which relates to curing, formulation changes relating to plasticity have received the greatest attention. Many new additions to the polymer line have been made for this purpose alone. In general, the trend has been toward the development of raw gums that provide lower viscosity characteristics. The primary reason for this trend has been to escape the necessity for using volatile plasticizers or hydrocarbon type additives which do not support the basic heat and chemical resistance of these elastomers. Therefore, it is advisable to select the raw polymer, or blends of raw polymer, which appear to best meet total processing requirements before considering the addition of plasticizers. Liquid plasticizers are, nevertheless, used effectively in small amounts. Table 16.18 shows the effect of hydrocarbon materials which have been used as processing aids.[12] The best general purpose softener is Harflex 330, generally used at 2 to 5 phr. Zinc stearate and low molecular weight polyethylene improve surface smoothness. Diak #3 has an excellent softening effect upon FPM elastomers, and Copper Inhibitor No. 50 is also a processing aid; however, the use of these materials as plasticizers is limited to their function as curatives. The addition of 5 to 10 parts of diisobutyl ketone has proven to be an excellent way to develop building tack in roll covering procedures, and the gradual expulsion of this particular ketone during the oven post cure leaves no effects upon the cured properties.

Several fluorocarbon type liquid plasticizers are also available. "VITON" LM, a waxy, opaque semisolid at room temperature, which becomes a clear fluid above 150°F, may be added at the 5 to 25 part level to maintain a good balance of physical properties and processing ease. Heat resistance is affected to some extent, while solvent chemical resistance is only slightly affected. Shrinkage during oven post cure at 400°F increases from 0.7 to 2.2% above the normal shrinkage. Fluorosilicone fluid FS-1265 (10,000 centistokes) has been used for developing low durometer compounds.[9]

Compounding and Processing – "KEL-F" Brand Elastomers. The successful compounding and processing of "KEL-F" elastomers is confined to the peroxide or blocked amine cures and compression molding or extrusion. "KEL-F" Elastomer 5500 is slightly lower in viscosity than "KEL-F" Elastomer 3700, with both being regarded as being extremely tough. With a typical formulation, the secret to making successful parts lies in shaping the uncured preform as near as possible to the size of the finished part, and then molding without introducing further strains. Chemical resistance is affected more by the curing system than the grade of elastomers; for example, the amine crosslinks are more susceptible to hydrolysis than those produced by peroxides.

Compounding and Processing – Fluorosilicone Rubbers. The fluorine-containing silicone rubbers have been made commercially available as raw gums of different viscosities, both with and without curing agents or reinforcing pigments. Most

TABLE 16.18. EFFECT OF PROCESSING AIDS ON COMPOUND VISCOSITY AND CURED PROPERTIES OF "FLUOREL" 2140 ELASTOMERS

Basic Compound

"FLUOREL" Elastomer 2140	100	
Thermax	15	Press cure: 20 min/320°F
Maglite K	20	
HMDA-C	1	Oven Post Cure: 24 hr/400°F

	A-18	*B-18*	*C-18*	*D-18*	*E-18*	*F-18*	*G-18*	*H-18*
	Raw Polymer Control	Basic Compound Control						
Diisobutyl Ketone			5					
Ansul Ether 181				5				
Harflex 330					5			
Polyethylene 8416						5		
Copper Inhibitor #50							0.65	
"DIAK" #3								3 Omit HMDA-C
Mooney Viscosity ML4 at 212°F								
Initial reading	208	210+	145	158	168	208	210+	200
4 Minute Reading	122	160	78	90	103	94	148	132

Stress Strain Properties								
Original								
Tensile Strength, psi		2495	2350	3200	2200	2420	3300	2550
Elongation, %		310	275	290	440	350	300	295
100% Modulus		490	610	550	260	540	560	510
Oven Aged 16 hr/600°F								
Tensile Strength, psi		1280	1170	1600	1330	1300	1600	1470
Elongation, %		145	150	80	115	130	100	100
100% Modulus		1020	985		1210	1130	1600	1470
180° Bend		Pass	Pass	Pass	Pass	Pass	Pass	Pass
Hardness, Shore A								
Press Cure Only	40	61	62	57	58	62	61	60
Original		66	66	66	61	66	65	69
Oven Aged 70 hr/500°F		71	71	73	70	73	71	76
Oven Aged 16 hr/600°F		77	78	83	80	80	83	86
% Weight loss during								
Oven post cure 24 hr/400°F		1.1	5.7	5.2	1.6	1.5	1.3	2.3
% Weight loss during oven aging								
70 hr/500°F		2.7	2.9	3.1	4.9	3.8	3.3	4.2
16 hr/600°F		7.7	8.4	9.8	10.9	8.4	9.7	10.3

fabricators work with these custom-made products which are usually compounded with silica pigments and peroxides that are commercially available also in paste form. The most common modification is to add fillers for obtaining increased hardness. Most ramifications have been discussed in previous sections and indicated in tables relating to them.

Compounding for Specific Properties – Processing. It is well to remember that most changes made for the purpose of adapting fluorocarbon rubbers to processing conditions tend to detract, even if only slightly, from some of the important cured physical properties. Since small changes in formulations sometimes produce great effects on vulcanizates operating under extreme conditions, careful attention must be paid to this matter when compounding fluorocarbon rubbers.

Chemical Resistance. As indicated in Table 16.16, the type of oxide used is the main determinant of chemical resistance. For example, "VITON" B when compounded with magnesium oxide exhibited a volume swell of 61% in red fuming nitric acid; with zinc oxide-Dyphos (dibasic lead phosphite), a swell of 45%; and with litharge, a swell of 24%.[14] Similarly, a general purpose "FLUOREL" 2141 compound containing magnesium oxide swells 110% in concentrated hydrochloric acid at 158°F; a litharge cure swells only 2%. A "FLUOREL" 2140 compound containing 60 parts of Dyphos swelled less than 3% in 250°F steam after four months.

Heat Resistance. Compounding for improved heat resistance has taken three directions: resistance to embrittlement upon long term aging, greater strength when measured at elevated temperatures, and nonflammable characteristics. Compounds containing calcium oxide and Diak #1 have the best long-term heat resistance, while "VITON" AHV filled with Cabolite P-4 produces the best hot tensiles. The development and application of nonflammable fluorocarbon elastomeric compositions began with a search for materials which could be used safely in 100% oxygen atmospheres of the Apollo Spacecraft Program. Compositions based on "FLUOREL" 2140 were the first to meet all requirements of being nonflammable, nontoxic, and having low odor. About 90 pounds of these proprietary nonflammable compositions were used on Apollo Flights 11, 12, and 13. They were molded into solid parts, chemically blown into sponge, and used as fireproof coatings for electrical components and other flammable substrates.

Low Temperature Flexibility. Except for the influence of minor differences in composition of the FPM grades – for example, the lowering of brittle point at some sacrifice in flexibility or vice versa – negligible changes have been made in the performance of these elastomers at low temperatures by compounding with or without plasticizers. An elastomer similar to "VITON" A in most other characteristics and marketed as LD-487 is reported to function, on the average, at temperatures about 20°F lower.[15]

Adhesion. Obtaining adhesion of uncured compounds to metal during the press cure and retaining good adhesion following the oven post-cure are important factors in the application of the fluorocarbon rubbers. Two coats of a 50% solution of Chemlok 607 in anhydrous methyl alcohol applied to properly prepared surfaces have been widely used and work successfully with most compounds. Under more difficult bonding conditions, for example, to high nickel alloys or porous castings with compounds containing Diak #3 or the less absorptive oxides, a 2% solution of the amino silane primer, Dow Corning Z-6020, has been used. Successful adhesion depends greatly upon handling the volatiles emitted by the curing mechanisms which take place as temperatures are increased. These volatiles, when generated at too rapid a rate for proper absorption by the compound or for adequate escape at the bond interface, will literally blow the bond apart. For this reason, press cures at 320°F have been found successful, but sometimes not at 350°F. Bonds which are sturdy following the press cure sometimes deteriorate during the oven post-cure at 400 to 500°F. The best bonding curing systems are those based upon Diak #1 and calcium oxide. The addition of several parts of calcium oxide to compounds designed with other oxides for reasons of physical properties has been shown to improve adhesion without detracting from end use performance. It is sometimes desirable to disassociate the factors of bonding from the development of desired properties in the cured compound. To achieve this, a rubber tie coat applied to the primed metal surface may be used. A 20% solution in MEK of "FLUOREL" Elastomer 2141 containing 60 parts of calcium oxide and one part of Diak #1 has been effective under some of the more difficult bonding conditions.

The bonding of the cured rubber to metal may be obtained with an epoxy adhesive, such as "SCOTCHWELD"* EC 1838, and the bonding of two cured pieces to each other: for example, the joining of the ends of two extruded and cured ropes to form a large diameter O-ring has been accomplished with another epoxy cement, "SCOTCHWELD" EC 2216, which produces a flexible joint. Uncured compounds have also been bonded to uncured Neoprene W, for example, in a laminated roll construction, by incorporating 40 parts of silica and 5 parts of epoxy in the Neoprene W compound.[16]

Unvulcanized Silastic LS-53U can be bonded to metal using Dow Corning A-4040 fluorosilicone primer. The vulcanized and oven cured product can be bonded to metal by using the combination of a Dow Corning A-4094 primer to the metal overcoated with Silastic 142 fluorosilicone adhesive.[9]

Resilience. The resilience of these compounds is low, and this deadness is sometimes employed in the capacity of vibration dampening. Except for compounds which can be employed with only a low level of oven post-cure, it is difficult to produce cured vulcanizates having permanently high resilience. For example, a typical compound will show a rebound of 7% using the Bashore Resiliometer before and after post-cure at 400°F for 24 hours. With ten parts of Harflex 330 added, the press cure cured rebound is 14% and the post cured is 10%. With 10 parts of Ansul Ether 181, the values are 32% and 7%, respectively, which indicates essentially total loss of plasticizer and resilience during the post-cure.

Tear Strength. Improvements in tear strength are sometimes desired, both at the processing level and for end-use performance. Thin edges of parts formed in

*Registered Trademark of 3M Company, Saint Paul, Minnesota.

undercut mold areas may sometimes tear even under the best conditions of mold lubrication. The addition of 0.5 to 1 part of zinc stearate to the compound has helped alleviate such tearing conditions by reducing the force required to separate the cured part from the mold. Defects due to poor hot tear strength are naturally accelerated as press cure temperatures are increased, especially in the 350 to 400°F range. The higher molecular weight gums, such as "VITON" AHV, sometimes process better in this respect. An effective way to improve the tear strength of cured vulcanizates for better end-use performance is to replace the commonly used medium thermal carbon black with 10 to 15 parts of super abrasion furnace black dispersed with 5 parts of Harflex 330.

Color. A typical compound containing oxide, amine, and no carbon black is to varying degrees tan in color, following either the press cure or the oven post-cure. Distinctive shades of green, yellow, orange, and red can be developed using inorganic oxide color pigments. Among the curing agents, Diak #1 appears to establish the best color tone. Titanium dioxide does not provide its usual enhancement of color tones in fluorocarbon rubbers, and it may also accelerate degradation during heat aging.

Coatings and Sealants. Since the latex is not commercially available, solvent coating systems have been developed generally using the lower molecular weight elastomer types. Table 16.19 lists effective solvents and shows the Brookfield viscosity of 15% solutions of "VITON" B.[17] Higher solution solids at the same viscosity may be obtained by starting with "VITON" A-35, "VITON" C-10, or "FLUOREL" Elastomer 2146. The solution stability of these solvent based coatings is dependent upon the type of oxide and amine. An estimation of their relative pot life can be made by referring to Table 16.16. The lower molecular weight types of these elastomers are also preferred basis for the development of sealants. Table 16.20 shows a sealant compound based on equal parts of "VITON" C-10 and "VITON" Elastomer LM.[18] The technology of coating objects using heat shrinkable tubing compound from blends of FPM elastomers and polyvinylidene fluoride resin has also been developed.[19]

TABLE 16.19. SOLVENTS FOR "VITON" ELASTOMER

Solvent	*Boiling Range (°F)*	*Flash Point (Tag-closed Cup)*	*Brookfield Viscosity (cps) (15% Viton B by volume)*
Methyl Ethyl Ketone	174–177	28	6,100
Acetone	132–134	0	4,200
Ethyl Acetate (99%)	169–174	24	7,500
Dimethyl Formamide	307	136	15,000
Tetrahydrofuran	151	6	6,600
4-Methoxy-4 methyl pentanone-2	147–163	141	21,900
4-Methoxy-4 methyl pentanol-2	164–167	140	23,300

TABLE 16.20. SEALANT BASED ON "VITON" C-10 ELASTOMER

Compound	*A-21*
"VITON" C-10	50
"VITON" LM	50
Magnesia (Low activity)	5
MT Carbon black	15
Fluorosilicone oil[a] (1,000 centistrokes)	10
Methyl Ethyl Ketone	40
Ketimine[b]	2
% Nonvolatile by Weight	70
% Nonvolatile by Volume	59
Mixed in one quart Baker Perkins sigma blade mixer	
Extrusion Rate,[c] g/sec	
Original	50
After 1 week storage at 75°F	30
After 2 weeks storage at 75°F	3 (gel
Shear Adhesion[d] – 2 in. per minute – pounds/inch (maximum reading)	
Steel/Steel	150
Aluminum/Aluminum	135
Glass/Glass	110
Physical Properties[d]	
100% Modulus, psi	500
Tensile strength, psi	650
Elongation, %	100

[a]FS-1265 was used.
[b]Epon Curing Agent H-2 was used.
[c]Semco Cartridge, 1/8 in. orifice, 50 psi.
[d]Cure: step + 24 hours at 400° F in air oven.

Sponge Rubber. Development and control of the critical factors of compound viscosity and cure rate required for producing both open-cell and closed-cell sponge have been accomplished with the raw polymers commercially available. A "VITON" A formula used for producing closed-cell sponge is shown in Table 16.21.[20]

Steam Cures. Compounding significantly affects the development of hardness of open steam cures compared with press cures. This is of practical importance with extruded configurations which are cured in open steam, then perhaps post-cured in an oven. Table 16.22 shows how the curing systems affect cured hardness under these conditions. Compounds A-22 and D-22 illustrate that when the blocked amines are used with magnesium oxide or calcium oxide, a 5 to 7 point increase in cured hardness of the final post-cured part results when the initial cure takes place in open steam. In certain instances, this can be used to advantage by developing higher cured hardness without the addition of more filler; however, where it is

TABLE 16.21. CLOSED-CELL SPONGE COMPOUND

	A-21		
"VITON" A	100		
Magnesia (low activity)	15		
MT Carbon black	25		
Petrolatum	3		
Diak No. 1	1.25		
Cellogen AZ		5	
Diethylene glycol		2	
Press Cure Temperature	325	325	342
Time, minutes	30	60	30
Final Thickness, approximately 0.5 inch[a]			
Density, lb/cu. ft	22	22	18
Compression/Deflection			
25% Deflection, psi	12	14	9
50% Deflection, psi	25	28	20
Compression Set, Method B, 50% Deflection			
22 hours at 158°F, %	48	47	55
Water Pick-up			
ASTM D-1056, Suffix L), %	0.3	0.5	0.7

[a]Mold: 6 x 6 x 0.25 in. with bevelled inside walls.
Preform: 300 g; 0.3 in. thick.

desired to use compounds interchangeably at either the press or autoclave while maintaining equal hardness, it is necessary to use the straight hexamethylene diamine, say with the safe processing "FLUOREL" Elastomer 2141, in conjunction with magnesium or calcium oxide (Compound B-22), or otherwise use zinc oxide-Dyphos or litharge as shown in compounds C-22 and E-22.

Shrinkage Control. The factors which affect shrinkage, and thereby dimension control, have previously been pointed out. Most compounds following oven post cure at 400°F show a shrinkage of approximately 3.5%. Some success has been achieved in reducing shrinkage, and thereby lessening variation, in the production of O-ring seals by post-curing on mandrils of specified diameter; nevertheless, a shrinkage figure of 3 1/2% obtained from molds which are designed to produce, say nitrile rubber, calculated for 2% shrinkage, usually calls for the cutting of new tooling in order to ensure consistent fitting and performance on an installed seal. A low shrinkage elastomer, "FLUOREL" 2150, is used to provide linear shrinkages from 1.7% to 2.2% when compounded similarly to "FLUOREL" Elastomer 2140 and post-cured at 400°F.[5]

Compression Set Resistance (LCS Types). Because of the importance of compression set resistance to their superior performance in seal applications, one of the most significant developments among these elastomers in recent years has been the introduction of new types that provide compression set values which are only 1/3

TABLE 16.22. EFFECTS OF PRESS CURE VS. OPEN STEAM CURE ON HARDNESS

	A-22	*B-22*	*C-22*	*D-22*	*E-22*
"FLUOREL" Elastomer 2141	100	100	100		
"FLUOREL" Elastomer 2140				100	100
Thermax	15	15	15	15	15
Maglite K	20	20			
Zinc Oxide			10		
Dyphos			10		
Calcium Oxide				20	
Litharge					20
HMDA-C or Diak #1	1		1		
HMD Amine		1			
Diak #3				3	3
Cure: 20 min/320°F. Oven Post-cure: 24 hr/400°F					
Press Cured Hardness					
Before Post-cure	63	64	60	59	59
After Post-cure	67	69	68	75	71
Steam Cured Hardness					
Before Post-cure	70	68	67	76	65
After Post-cure	75	67	69	80	73
Original Stress-Strain Properties					
Press Cured					
Tensile Strength, psi	2130	2140	2410	1640	2000
Elongation, %	270	245	270	185	165
100% Modulus, psi	525	620	585	1050	1030
Oven Aged 16 hr/600°F					
Tensile Strength, psi	1100	1400	1520	1180	1550
Elongation, %	175	155	165	60	50
180° Bend	Pass	Pass	Pass	Fail	Fail

to 1/2 as high as the best obtainable with previous technology. Until the advent of these new LCS type fluoroelastomers, improvements in the compression set resistance of typical compounds were dependent upon the post-cure conditions, as indicated in Table 16.9. Although increasing post-cure temperatures from 400 to 500°F, or higher, did produce some improvements, this method produced a diminishing return particularly when its effects upon other physical properties were taken ito consideration. In addition to greatly improved compression set characteristics at 400°F, significant improvement was also found at room temperature, where the FPM types have shown unexpectedly high values.

The LCS (low compression set) type of FPM elastomer is based upon an entirely new curing system. "FLUOREL" Elastomer 2160 is cured with combinations of calcium hydroxide and curing agent HC-5, which is a specially prepared hydroquinone. Table 16.23 shows the physical properties of a typical formula relating to its performance as a seal in aircraft engines. Table 16.24 illustrates that

TABLE 16.23. PHYSICAL PROPERTIES OF "FLUOREL" BRAND LCS ELASTOMER 2160

	A-24
"FLUOREL" 2160	100
Thermax	20
Maglite D	10
Calcium Hydroxide	2
3M Brand Curing Agent HC-5	1

Press Cure: 20 min/335°F
Oven Post-cure: Step-cure – 2 hr/250°, 300°, 350°; 4 hr/400°; 24 hr/500°F

Original Physical Properties:	
Tensile Strength, psi	1850
Elongation, %	185
Hardness, Shore A	67
100% modulus, psi	695
Oven Aged, 70 hr/482°F	
Tensile Strength, psi	1700
Elongation, %	185
Hardness, Shore A	68
Weight Loss, %	1.5
180° Bend	pass
Aged in Anderol L-774, 70 hr/392°F	
Tensile Strength, psi	1560
Elongation, %	195
Hardness, Shore A	62
Volume Change, %	+11
Aged in Ref. Fuel B, 70 hr/77°F	
Tensile Strength, psi	1830
Elongation, %	180
Hardness, Shore A	66
Volume Change, %	+3
Low Temperature Properties (ASTM D1329)	
Temperature Retraction, TR-10, °F	+1

Compression Set, ASTM D 395, Method B (1/2" disk)

Time Hours/Temp °F	70/77	168/347	70/392	168/392
%	4	7	12	19
On O-ring, 0.139 Cross-section			26	45

versatile curing conditions may be employed by variations in curing conditions that do not significantly affect the balance of heat resistance and compression set resistance. A notable aspect of this compounding is the apparent function of hydroquinone as a crosslinking agent and also as an important retarder of scorch.

The safe processibility, fast curing, and clean molding nature of this LCS type, combined with a similarity in shrinkage, chemical resistance, and low temperature

TABLE 16.24. "FLUOREL" BRAND LCS ELASTOMER 2160 VARIATIONS IN CALCIUM HYDROXIDE AND CURING AGENT HC-5

	A-25	*B-25*	*C-25*	*D-25*
"FLUOREL" 2160	100	100	100	100
Thermax	30	30	30	30
Maglite D	10	10	10	10
Calcium Hydroxide	2	1	3	6
3M Brand Curing Agent HC-5	1	1	0.75	0. 6
Mooney Scorch, MS @ 250°F				
Initial Reading	67	62	69	72
Minimum Reading	36	34	38	42
Minutes to 10 point rise	25+	25+	19	11
Points rise in 25 min	4	4		
Mooney Cure Rate, ML @ 335°F				
Initial Reading	56	60	64	74
Minimum Reading	34	40	48	62
Minutes to 5 point rise	3.1	2.4	1.4	1.0
Minutes to 10 point rise	3.4	2.8	2.0	1.3
Minutes to 15 point rise	3.6	3.1	2.3	1.5
Minutes to 30 point rise	4.2	3.6	2.6	1.7
Press Cure:	20 min/335°F			
Oven Post-cure:	Step-cure – 1 hr/250°, 300°, 350°, 400°; 20 hr/500°F			
Original Physical Properties				
Tensile Strength, psi	1915	2010	1965	1885
Elongation, %	160	160	190	200
Hardness, Shore A	73	73	73	72
100% Modulus	910	950	635	590
Oven Aged, 70 hr/528°F				
Tensile Strength, psi	1475	1315	1345	1200
Elongation, %	160	145	210	220
Hardness, Shore A	76	77	75	75
Compression Set, Method B, 70 hr/400°F (1/2" Disk)				
%	14	14	15	20

flexibility to previous "FLUOREL"-"VITON" elastomers has made it quickly adaptable, and many important specifications relating to sealing are being rewritten so as to expand the use of this significant development in the improvement of resistance to compression set. In addition to "FLUOREL" Elastomer LCS 2160, "VITON" E-60 is another commercially available low compression set fluoroelastomer. Both materials are marketed as medium viscosity raw gums having an ML-4 at 212°F reading of approximately 60.

Another important and more recent advancement in low-compression-set fluorocarbon rubbers is shown by Viton E-60C and "FLUOREL" FC-2170. The long-term compression set resistance of O-rings has been markedly improved over Viton A and Viton E-60 by the use of the newer curing systems.[21] In addition to achieving lower compression set, this curing system also exhibits a highly favorable balance of scorch time vs. cure rate. Both "FLUOREL" 2170 and Viton E-60C show no rise in Mooney viscosity, i.e., "scorch," in 25 minutes at 250°F. These elastomers are generally cured by the addition of 3 parts of magnesium oxide and 6 parts of calcium hydroxide. Their compounds are somewhat "tender" and have a tendency to tear when thin sections are removed from under-cut molds; however, they are gaining wide acceptance as preferred processing compounds for the fabrication of many fluorocarbon elastomer parts.

MATERIALS USED IN COMPOUNDING FLUOROELASTOMERS

Cab-O-Lite P4	Cabot Corp.
Copper inhibitor No. 50	E. I. du Pont de Nemours & Co.
Diak No. 1	E. I. du Pont de Nemours & Co.
Diak No. 3	E. I. du Pont de Nemours & Co.
Dyphos	National Lead Co.
Epon Curing Agent H-2	Shell Chemical Co.
FS-1265	Dow Corning
HMDA-C	Minnesota Mining & Manufacturing Co.
Maglite K	Merck & Co.
Thermax	R. T. Vanderbilt Co., Inc.

REFERENCES

(1) Chemical Week, March 18 (1970), p. 56.
(2) Joseph P. Kennedy and Erik G. M. Tornqvist, *Polymer Chemistry of Synthetic Elastomers* Part 1, Chap. 4.
(3) R. P. Bringer, R. J. Woessner, and D. A. Stivers, "A fluorocarbon Elastomer Which Has a Low Viscosity and is Crosslinkable," *Rubber Age* November, 1966.
(4) D. A. Stivers, "Fluorocarbon Rubbers," *Vanderbilt Rubber Handbook,* 1968.
(5) "FLUOREL" Brand Elastomer data from 3M Company Technical Brochure.
(6) "VITON" Bulletin Nos. 8 and 9 (DuPont).
(7) "VITON" Bulletin No. 13.
(8) "VITON" C-10 Fluoroelastomer.
(9) Dow Corning "SILASTIC" Brand Fluorosilicone Rubbers.
(10) 3M Company "KEL-F" Brand Elastomers 5500 and 3700.
(11) J. F. Smith, Proc. International Conference, Washington, D.C. (1959), pp. 575-81.
(12) D. A. Stivers and P. D. Lanin, "Processing and Fabrication of a High-Temperature Fluid-Resistant Fluorocarbon Elastomer," *I & EC Product Research and Development*, Vol. 3, p. 61, March, 1964.
(13) "VITON" A and "VITON" A-HV, Report No. 58-3.
(14) "VITON" Bulletin No. 59-4.
(15) LD-487 Low Temperature Fluoroelastomer (DuPont).

(16) "VITON" Bulletin No. 18.
(17) "VITON" Bulletin No. 16.
(18) "VITON" Bulletin No. 23.
(19) B. P. 1, 120, 131.
(20) "VITON" Bulletin No. 27.
(21) A. L. Horon and D. B. Pattison, "Compression Set Resistance," *Rubber Age,* July (1971), p. 39.

17

URETHANE ELASTOMERS

D. A. MEYER,
The General Tire & Rubber Co.
Akron, Ohio

Polyurethane polymers provide one of the most versatile elastomer systems available. Their versatility lies not only in unique combinations of properties, but also in the variety of processing methods available. Properties such as abrasion resistance, oil and solvent resistance, tensile and tear strength, and range of hardness or modulus are often not available at the same level or in the same combination in other elastomers. Processing versatility is related to the fact that similar properties can often be obtained in materials produced through any of a variety of routes. Liquid casting materials, vulcanizable millable rubbers, thermoplastic rubbers, or spraying materials are all possible if appropriate adjustments are made in the composition.

Most of the polyurethane elastomers available commercially are based upon low molecular weight polyester or polyether polymers that are terminated with hydroxyl groups. Recently some hydroxyl-terminated hydrocarbon polymers such as polybutadiene or polyisobutylene have also become available. The other starting materials or intermediates consist of di- or polyfunctional isocyanates and, with most systems, low molecular weight polyfunctional alcohols or amines.

The liquid starting polymer is generally in the range of 500 to 3,000 molecular weight. Variations in the characteristics of this starting low molecular weight polymer and the concentration, type, and arrangement of the isocyanate and other small molecules used for chain extension provide a broad range of different polyurethane elastomers.

The end uses for polyurethane elastomers cover such diverse application areas as solid tires for industrial trucks, seals and boots, calendered sheet, potting and sealing of electronic components, general engineering mechanical goods, shoe heels and soles, and elastic thread. Cumming and Wright,[47] Dianni,[49] and Mitchell[48] have reviewed and described many of these application areas. A recent estimate of 25-35 million pounds was made for the 1969 market for these materials with a forecast of 60 million pounds by 1975.[46]

This discussion of polyurethane elastomers will be arranged to describe first some of the chemistry and structure involved in formation of these elastomers followed by a review of some of the properties of the various classes of materials available commercially.

CHEMISTRY AND STRUCTURE

The term polyurethane refers to the chemical entity that results from the reaction of an isocyanate and an alcohol:

$$\underset{\text{isocyanate}}{R-N=C=O} + \underset{\text{alcohol}}{R'OH} \longrightarrow \underset{\text{urethane}}{R-\overset{\displaystyle H}{\overset{|}{N}}-\overset{\displaystyle O}{\overset{\|}{C}}-O-R'}$$

Using this reaction, difunctional or polyfunctional isocyanates and hydroxyl-terminated low molecular weight polymers will give high molecular weight polymers if the molar ratios are properly controlled:

$$n\,OCN-R-NCO + n\,HO-R'-OH \longrightarrow \left[\overset{\displaystyle O}{\overset{\|}{C}}-\overset{\displaystyle H}{\overset{|}{N}}-R-\overset{\displaystyle H}{\overset{|}{N}}-\overset{\displaystyle O}{\overset{\|}{C}}-O-R'-O \right]_n$$

The above scheme demonstrates how the urethane reaction is used to couple the segments designated R and R′. R and R′ segments of many different types are used in commercial polyurethane elastomers. Some examples will be described later.

Another coupling reaction that is widely used along with the isocyanate-hydroxyl reaction is that between an isocyanate and an amine to give a urea:

$$\underset{\text{isocyanate}}{R-N=C=O} + \underset{\text{amine}}{R'NH_2} \longrightarrow \underset{\text{urea}}{R\overset{\displaystyle H}{\overset{|}{N}}-\overset{\displaystyle O}{\overset{\|}{C}}-\overset{\displaystyle H}{\overset{|}{N}}-R'}$$

These reactions are far from simple, however, since there are always competing reactions possible. Close attention must therefore be given to reaction rates, order of addition of ingredients, and catalysis of certain reactions in preference to others. The isocyanate is capable of reacting with the active hydrogen on a urethane or a urea group to give branching or crosslinking:

$$\underset{\text{isocyanate}}{RNCO} + \underset{\text{urethane}}{R'NHCOOR''} \longrightarrow \underset{\text{allophanic ester}}{\underset{\displaystyle CONHR}{\underset{|}{R'NCOOR''}}}$$

$$\underset{\text{isocyanate}}{RNCO} + \underset{\text{urea}}{R'NHCONHR''} \longrightarrow \underset{\text{a biuret}}{\underset{\displaystyle CONHR}{\underset{|}{R'NCONHR''}}}$$

Proper control of the stoichiometry of the polymeric polyol, isocyanate, and low molecular weight polyol or amine, the temperature, the order of addition, and sometimes the use of catalysts can produce compositions with properties typical of a crosslinked high-molecular weight rubber.

Some of the structures typical of the liquid polymers used in polyurethane elastomers are shown in Table 17.1.

TABLE 17.1. HYDROXYL TERMINATED POLYMERS

$$HO-(CH_2)_2-[O-\overset{\overset{O}{\|}}{C}-(CH_2)_4\overset{\overset{O}{\|}}{C}-O-(CH_2)_2]_n OH$$

adipic acid-ethylene glycol polyester

$$HO-[CH_2-CH_2-CH_2-CH_2-O]_n(CH_2)_4OH$$

poly(butylene glycol)

$$HO-[CH_2-\underset{\underset{CH_3}{|}}{CH}-O]_n CH_2-\underset{\underset{CH_3}{|}}{CH}-OH$$

poly(propylene glycol)

$$HO-[CH_2-CH=CH-CH_2]_n OH$$

hydroxyl-terminated polybutadiene

The isocyanates used in polyurethane elastomers also vary considerably in structure. Some of the common isocyanate structures are shown in Table 17.2.

TABLE 17.2. ISOCYANATES USED IN POLYURETHANE ELASTOMERS

OCN–C₆H₄–CH₂–C₆H₄–NCO

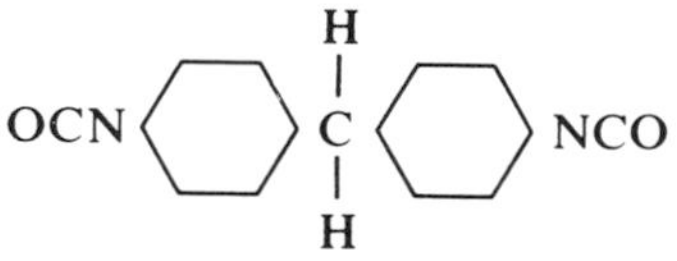

MDI
diphenylmethane-4, 4'-diisocyanate

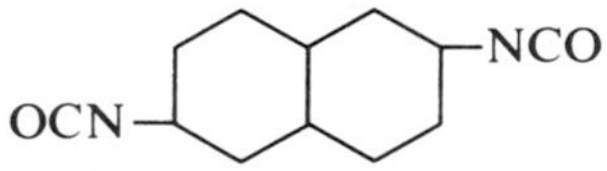

OCN–C₁₀H₆–NCO

NDI
naphtalene-1, 5-diisocyanate

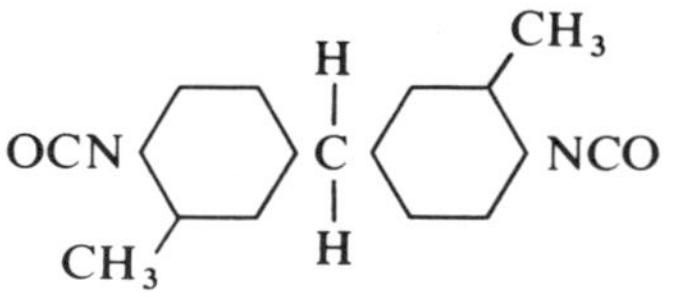

OCN–C₆H₃(CH₃)–CH₂–C₆H₃(CH₃)–NCO

TODI
3,3'-dimethyldiphenyl-4,4'-diisocyanate

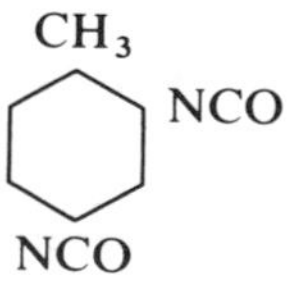

CH₃–C₆H₃(NCO)₂

TDI
2,4-tolylene diisocyanate

Some other isocyanates used in special cases such as for non-discoloring polyurethane elastomers are hexamethylene diisocyanate (HDI) and hydrogenated MDI.

The low molecular weight diol, triol, or diamine, one of the three principal starting materials, provides an additional means of regulating properties. These are usually called chain extenders. Crosslinking can be introduced with an extender that is a triol, for example, or the hardness can be increased by raising the level of isocyanate and the extender which increases the amount of rigid polar entities consisting of the urethane and urea groups in the polymer. The chain extending agents commonly used in polyurethane elastomers are shown in Table 17.3 along with the abbreviations frequently used.

TABLE 17.3. CHAIN EXTENDING AGENTS

H_2N–C₆H₃(Cl)–CH_2–C₆H₃(Cl)–NH_2

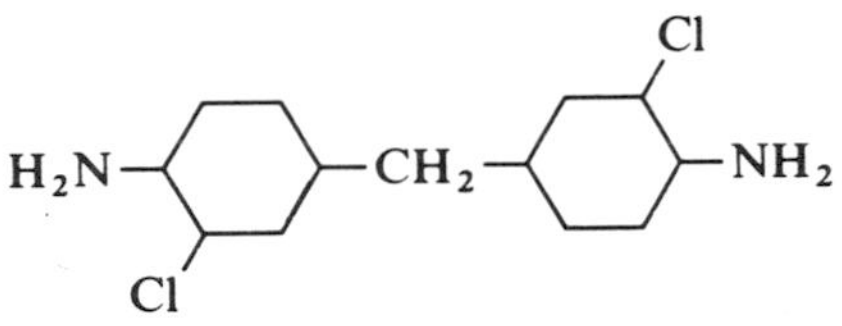

MOCA
4,4′-methylene bis
(orthochloroaniline)

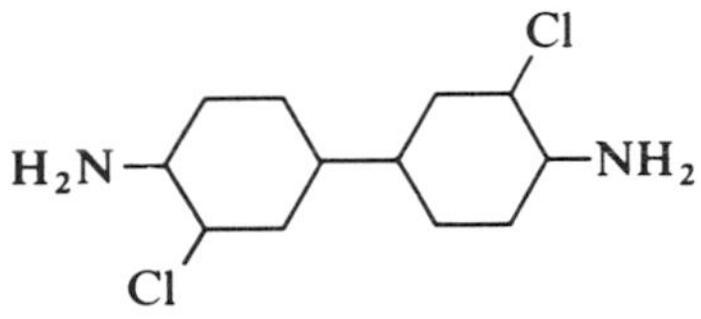

H_2N–C₆H₃(Cl)–C₆H₃(Cl)–NH_2

DCB
dichlorobenzidine

$HO\ CH_2CH_2CH_2CH_2OH$
1,4-butanediol

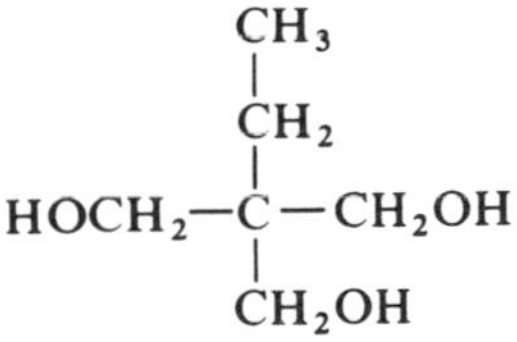

$$HOCH_2{-}C(CH_2CH_3){-}CH_2OH,\ CH_2OH$$

TMP
trimethylolpropane

In addition to the above, other materials, such as the aromatic diol N,N′- bis (2-hydroxypropyl)-aniline, and even water, can be used to couple the isocyanate terminated chains to give useful products.

When considering the preparation and properties of polyurethane elastomers, the rate of reactions must be considered along with the nature of the reactants. Some of the first reaction rate studies were carried out by Morton and Diesz.[1]

The results obtained when phenyl isocyanate was reacted with various model compounds are briefly summarized in Table 17.4.

The reaction of isocyanates with water or with an organic acid is generally undesirable for elastomer formation unless a cellular product is desired. Both of these reactions give CO_2 as a gaseous by-product but also serve to couple one

TABLE 17.4. ISOCYANATE REACTION RATES

Compound	*Structure*	*Relative Rate*
A Urethane	$R-\overset{H}{\overset{\mid}{N}}-\overset{O}{\overset{\Vert}{C}}-O-R'$	Very Slow
A Urea	$R-\overset{H}{\overset{\mid}{N}}-\overset{O}{\overset{\Vert}{C}}-\overset{H}{\overset{\mid}{N}}-R'$	Slow
An Alcohol	R—OH	Fast
Water	HOH	Fast
An Amine	RNH_2	Very Fast

polymer chain to another:

$$RNCO + HOH \longrightarrow RNH_2 + CO_2 \xrightarrow{RNCO} \underset{\text{a urea}}{RNHCONHR}$$

$$RNCO + R'COOH \longrightarrow RNHCOOCOR' \longrightarrow \underset{\text{an amide}}{RNHCOR'} + CO_2$$

Table 17.4 shows that the reaction rate of an isocyanate with water is quite rapid; therefore any available water in the system will be likely to enter into the reaction and the polymer formation.

The reaction rate of the hydroxyl group of the polyol can also be important since that of a primary hydroxyl is two to three times that of a secondary hydroxyl group. Also of importance is the variation of reactivity of the –NCO group. The –NCO group in the 2 position of 2,4-tolylene diisocyanate is, for example, only about one-tenth as reactive as the one in the 4 position. These differences in the rate of reaction are used in various ways to control the level of branching and crosslinking relative to chain extension and to regulate the processing time. Catalysts can also greatly change the reaction rates and even the rate of one reaction in preference to another. The chemistry and details of the rate phenomena have been described in detail in reviews by Saunders and Frisch,[6] by Frisch,[7] and by Wright and Cumming.[4 5]

The influence of structure on the behavior of polyurethane elastomers follows generally the same rules as for other elastomers and for the same reasons except that the special response of segmented or block structure is aded to the basic characteristics of rubbery behavior. The factors involved in the special behavior of segmented or block polymers have been the object of many research studies in the last twenty years. In recent years, the concept of domains or regions in the

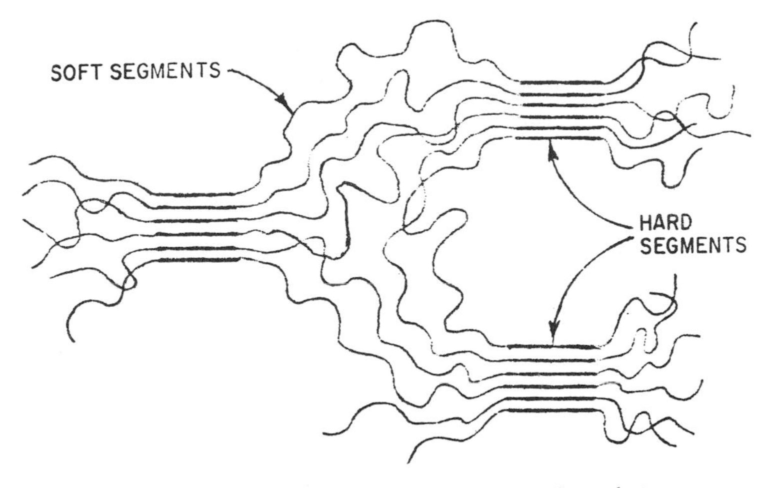

Fig. 17.1 Segregation of hard segments in polyurethane elastomers.

segmented elastomer composition consisting of localized associations of the sections of the polymer chains having strong intermolecular forces has provided new understanding of these block or segmented elastomers.

Cooper and Tobolosky[4] concluded from viscoelastic behavior that the hard segments in a segmented polyurethane associate with each other and are responsible for reinforcement of the structure. These domains of hard segments serve in many ways as filler particles.

The presence of these domains and their behavior during extension of a specimen have been observed directly in ABA polymers of the styrene-butadiene-styrene type by Beecher et al.[44] using microscopic techniques involving staining and observation of the domains in the strained state. Bonart et al.[2,3] using wide and low angle X-ray work established the presence of segregated hard (plastic) segments in polyurethane elastomers where butanediol was used as the extending agent. Figure 17.1 shows in simple form the concept of domains in a polyurethane elastomer.

CASTING SYSTEMS

Urethane polymers that are converted into end items directly from a liquid or semiliquid state are called casting systems. Commercial systems in this class are available from all of the base polymers such as polyester, polyether, and also hydrocarbons such as polybutadiene. Polyurethane casting systems generally consist of the three starting materials described earlier, namely a liquid polymer with hydroxyl end groups, a polyfunctional isocyanate, and a low molecular weight polyol or polyamine. In low hardness compositions, however, the formulation may involve only a polymeric polyol and a polyfunctional isocyanate. A general scheme of the reaction steps involved in the preparation of a polyurethane casting system is shown below:

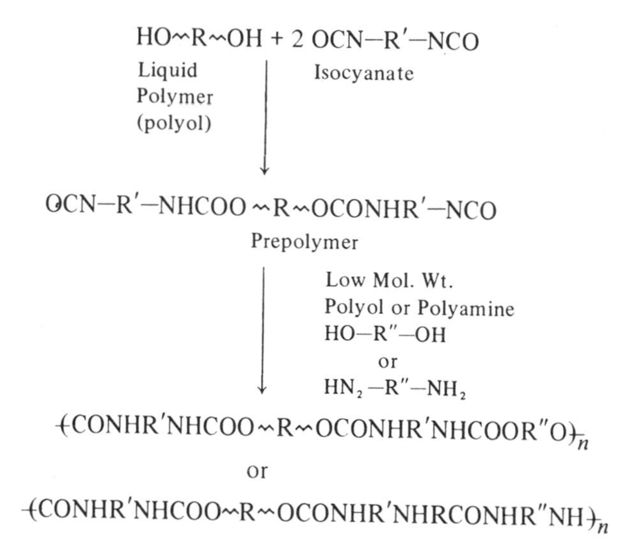

The production of rubber parts from polyurethane casting systems involves processing steps quite different from those used in conventional production of rubber items. The casting procedure generally follows one of the two techniques known as the prepolymer route or one-shot technique. The prepolymer route involves the following principal steps.

Preparation of Prepolymer. The polyol and diisocyanate are heated in an inert atmosphere to form a liquid polymer terminated in isocyanate groups as shown above. The reaction can be followed by the exotherm and/or by the —NCO number. If one of the commercial prepolymers is used as a starting material, this step will be eliminated.

Prepolymer Degassing. The prepolymer is heated to casting temperature or slightly higher and subjected to a vacuum to remove dissolved gases. Continuous degassing devices are available. This step is necessary if bubble-free castings are to be produced.

Addition of Curing Agents. A low molecular weight polyol or polyamine is added at this stage with thorough mixing. The mixing may be accomplished in a batch process or by a continuous mixing device. The temperature of the prepolymer and the curing agent is first adjusted to provide low viscosity and a satisfactory reaction rate.

Casting. The completed mix is poured or injected into a heated mold in a manner designed to prevent the entrapment of air bubbles. Some of the techniques used to obtain satisfactory parts free of defects have recently been described by Blaich.[15]

Curing. The object is cured in the mold for a period that may range from a few minutes to an hour or more. Temperatures of 200 to 300°F are usually used.

The time and temperature depends in a large degree upon the development of adequate strength to allow removal from the mold without distortion.

Postcure. Common procedures for this step consist of placing the formed part in an oven for a period of one to twenty-four hours at elevated temperatures or, in some systems, storing for one week at room temperature.

The degassing, curing agent addition, and casting can be accomplished in a batch type operation involving small containers and only a few ounces or a few pounds of materials. This technique is frequently used for experimental work and even small volume production. Large volume production and more sophisticated experimental work involves machine mixing. A diagram is shown giving the general form of the commercially available mixing machines. A detailed review of mixing machine design for polyurethane elastomers was recently presented by Copcutt,[16] who discusses some of the design variations possible in mixing heads, gear pumps, and other elements in the mixing machine.

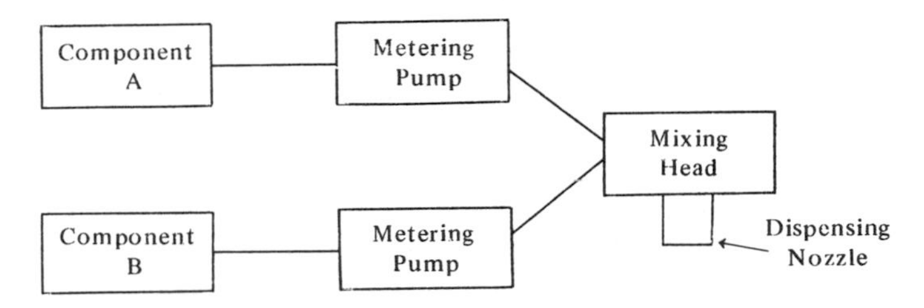

The component tanks, metering pumps, and lines generally must be heated to keep the components in a liquid state and to reduce their viscosities to a range suitable for metering and mixing. Degassing may be done as a batch operation during the preheating step on the components or by a continuous degassing device. Many variations are made on the simple scheme shown in the diagram. The number of components may, for example, be increased to three or more. Also the mixing and metering may be arranged to give an intermittent mixing that delivers discreet charges for individual mold cavities or the mixing head may deliver a continuous flow of mixed material. The selection of a specific mixer depends upon the production system in which it is to be used and the characteristics of the casting materials involved.

Polyester Casting Systems

Polyester urethane casting systems provided the first polyurethane elastomers used. An extensive description of the combination and properties of these elastomers was presented by Bayer, Muller, et al.[8,9] in the early 1950's. Later reviews were presented by Piggott et al.[10] and Saunders.[11,12] The early polyester casting systems were based on 1,5-naphthalene diisocyanate and a polyester from adipic acid and ethylene and propylene glycols. The system later developed in the United States and described by Piggott et al.[10] was based on diphenylmethane-4,4′-diisocyanate and apparently a similar polyester. Variations in hardness are readily

TABLE 17.5. PROPERTIES OF CAST POLYESTER URETHANES

Prepolymer F242	100.00	100.00	100.00	78.00
1,4 Butanediol	3.00	7.00	–	13.20
Diethylene glycol	4.40	–	–	–
Hydroquinone di-β-hydroxyethyl ether	–	–	14.40	–
MDI	–	–	–	22.00
Stannous Octoate	0.03	0.03	0.03	–
Shore Hardness	60–65A	78–83A	90–95A	50–55D
Tensile Strength, psi	4500	7500	4500	5500
Ultimate Elongation, %	650	620	550	430
Tear Strength, pli	170	260	350	650

available in these systems by changing the type and/or amount of chain extending agent. Table 17.5 gives an example of the variations in hardness and other properties than can be achieved by such changes as taken from a Technical Bulletin[13] of Mobay Chemical Company.

The processing procedure used with this system involves preheating the prepolymer to 140°F and degassing at 220°F followed by mixing with the extender and catalyst. The pot life at 220°F is on the order of 4 to 5 minutes during which time the material must be cast or poured into the mold. The parts can be demolded after the strength of the product has increased sufficiently to allow handling without damage. The exact time depends upon the composition, size, and complexity of the part, and temperature of the cure. A postcure is frequently recommended to develop full strength. This can be accomplished in an oven, for example, by heating for 4 to 6 hours at 212°F. Several options are available for preparing cast polyester urethane materials. The formulator may start with one of the polyester resins which are available from a number of different suppliers and prepare a prepolymer prior to the final mix and casting operation. A large number of suppliers offer isocyanate-terminated polyester prepolymers which can then be chain extended in different ways to give a broad range of end products. Some of the commercial

TABLE 17.6. ISOCYANATE-TERMINATED POLYESTER PREPOLYMERS

Trade Name	*% NCO*	*Hardness Range*	*Supplier*
Multrathane F66	6.4	60A to 80D	Mobay Chemical Co.
Multrathane F242	6.4	60A to 80D	Mobay Chemical Co.
Solithane 113	10.6	70A to 80D	Thiokol Chemical Co.
Solithane 291	3.0	55A to 80A	Thiokol Chemical Co.
Vibrathane 6004	4.2	80A to 95A	Uniroyal, Inc.
Vibrathane 6005	3.0	55A to 70A	Uniroyal, Inc.
Vibrathane 6006	7.0	85A to 55D	Uniroyal, Inc.
Vibrathane 6008	3.0	70A to 90A	Uniroyal, Inc.
Cyanaprene 4590	2.4	45A to 90A	American Cyanamid Co.

isocyanate-terminated prepolymers are listed in Table 17.6 along with the nominal –NCO content and the approximate hardness range for which the prepolymer was designed. A large number of additional suppliers offer prepolymers and other starting materials for use in polyurethane casting systems. Kallert[14] recently (1968) noted that twenty-three firms were active in the world market offering more than forty different systems of raw materials for the preparation of urethane elastomers.

Polyether Casting Systems

Polyurethane elastomers can be prepared from hydroxyl-terminated polyethers in the same manner described for polyester urethane elastomers. Two types of commercial systems are presently in use, one based on poly(butylene glycol) polyols and the other on poly(propylene glycol) polyols.

Poly(butylene glycol) based prepolymers are available from E. I. DuPont under the trade name "Adiprene." The early "Adiprene" liquid polyurethane casting systems were described by Athey[17,18] in 1959 and 1960. These prepolymers were described by Athey as NCO-terminated polymers based on poly(butylene glycol) and 2,4-tolylene diisocyanate. A series of these liquid polymers are offered by E. I. DuPont under the trade name "Adiprene." The properties, when formulated with various amines and low molecular weight polyols, are described in a series of technical bulletins by Athey and others.[19,20,21] This system usually uses a diamine such as 4,4′-methylene-bis-o-chloroaniline, commonly known as MOCA, for chain extension but other diamines and polyols may also be used. Some typical formulations and properties are shown in Table 17.7, as taken from the above references.

TABLE 17.7. PROPERTIES OF ADIPRENE ELASTOMERS

Adiprene L-100	100	–	–
Adiprene L-167	–	100	–
Adiprene L-315	–	–	100
Isocyanate Content	4.1	6.3	9.5
Specific Gravity	1.06	1.07	
MOCA	12.5	19.5	26.0
Mixing Temperature (°F)	212	185	170
Cure	1 hr @ 212°F	1 hr @ 212°F	1 hr @ 212°F
100% Modulus, psi	1100	1800	4500
Tensile Strength, psi	4500	5000	10000
Elongation, %	450	400	270
Hardness			
Shore A	90	95	–
Shore D	45	50	75
Tear, pli (D470)	75	150	117

The strong intermolecular attraction provided by the aromatic diamine, MOCA, gives high hardness, modulus, and tear strength. Blends of MOCA and other diamines such as methylene dianiline and cumene diamine are reported to give variations in reaction and curing times as well as some modification in properties. When short chain diols or triols are used as chain extenders with these polymers, entirely different properties are obtained as shown in Table 17.8. The data were taken from the above references on "Adiprene."

TABLE 17.8. POLYOL CURES OF "ADIPRENE"

Adiprene L-100	100	
Adiprene L-167		100
1,4 Butanediol	3.5	5.1
Trimethylol propane	0.8	1.0
Cure	16 hr @ 212°F	6 hr @ 285°F
100% Modulus, psi	275	325
Tensile Strength, psi	2750	2100
Elongation, %	470	500
Hardness Shore A	62	60

Poly(propylene glycol) based polyurethanes for casting systems were described in 1960 by Axelrood and Frisch.[30] Table 17.9 shows some properties of these polyurethanes and the variation in these properties with the molecular weight of the polyol as was described by these authors. The identifying number indicates the approximate molecular weight of the starting poly(propylene glycol). Prepolymers were prepared with two moles of TDI and one of polyol. MOCA was used as a curing agent.

TABLE 17.9. PROPERTIES OF POLY(PROPYLENE GLYCOL)-TDI ELASTOMERS

Polyol	PPG 2000	PPG 1575	PPG 1000	PPG 775
Hardness Shore A	60	65	88	–
300% Modulus, psi	400	700	2100	4300
Tensile, psi	1500	3400	5000	4800
Tear, pli	120	150	310	580

The tensile properties are generally not as high as those of the polyester or poly(butylene glycol) as can be seen by comparing the data in Tables 17.5, 17.7, and 17.9. This is especially true in the low hardness range polymers and may be a consequence of the lack of crystallizing capability in the poly(propylene glycol) polymer. The lower level of physical properties from the poly(propylene glycol) has been reported by Wright and Cumming[31] and by Ferrari.[32] The lower cost of the base polyols, however, makes these polyurethane elastomers useful for a wide variety of end items.

One-Shot Poly(propylene glycol) Casting Systems

The one-shot or one-step polyurethane casting systems require that the polyol, diisocyanate, and curing agent react at rates such that a uniform polymer is produced. This procedure eliminates the prepolymer step but requires careful use of catalysts to obtain suitable polymer structures. Axelrood and various co-workers[33,34] have described the techniques required for one-shot cast urethanes from poly(propylene glycol) polyols and have compared their properties with those obtained using the prepolymer method. These workers demonstrated the use of specific catalysts to increase the hydroxyl-isocyanate reaction rate up to 10 or 20 times faster than the amine-isocyanate reaction rate, thus favoring polymer formation from polyol and isocyanate in a manner similar to that when a prepolymer is first prepared and then reacted with a diamine. Some typical properties attainable from poly(propylene glycol) one-shot casting systems are shown in Table 17.10, as taken from the data of Axelrood.[33]

TABLE 17.10. ONE-SHOT ELASTOMERS FROM POLY(PROPYLENE GLYCOL)-MOCA-TDI

System	*Prepolymer*	*One-Shot*	*One-Shot*
Polyol	PPG 2010	PPG 2010	PPG 1010
NH_2/OH Ratio	1:1	1:1	1:1
Stannous Octoate	None	0.025	0.025
300% Modulus, psi	900	920	2350
Tensile Strength, psi	2800	2440	3900
Elongation, %	500	1020	–
Hardness			
Shore A	75	82	–
Shore D	–	–	49
Graves Tear, pli	210	305	450
Bashore Rebound	37	46	36

The polyol number indicates the approximate molecular weight of the poly(propylene glycol) polymer. According to Axelrood et al.[34] a two-component system can be used. The polyol, the diamine, and the catalyst (stannous octoate) are blended to form one component and the TDI forms the other. This one-shot system has the advantage that both components are of relatively low viscosity at room temperature so that they can be readily mixed and cast at room temperature eliminating the need for heated lines and storage tanks.

Other Casting Systems

A number of other casting systems have been developed and are used for special purposes. No attempt will be made to list all of these but mention of some and their special features seems desirable. Casting systems based on poly-ϵ-caprolactone polyesters have been described by Magnus.[35] Better hydrolytic stability and better

TABLE 17.11. PROPERTIES OF CAST POLYURETHANES BASED ON POLY-ε-CAPROLACTONE POLYOLS

Polyol D560[a]	100	100	100	100
TDI 80/20	14	14	14	14
% NCO in Prepolymer	4.20	4.20	4.20	4.20
MOCA	12	–	–	–
1,4 Butanediol	–	4.0	–	–
Triisopropanolamine	–	–	6.0	–
50/50 1,4-Butanediol and Triisopropanolamine	–	–	–	5.0
Hardness Shore A	88	91	58	52
300% Modulus, psi	2070	1060	700	420
Tensile, psi	6900	7100	3850	4800
Elongation, %	420	580	400	455
Tear, pli, ASTM D624 Die B	430	375	110	275
Compression Set, Method B 22 hr at 158°F	19	35	3	5

[a]Niax Polyol D560 2000 M.W. Poly-ε-Caprolactone Diol, Union Carbide Corp.

low temperature flexibility were shown for the poly-ε-caprolactone polyester as compared to the conventional cast polyurethanes based on adipic acid polyesters. When compared to the polyether-based urethanes, better strength and low temperature flexibility were found. Some typical properties of the poly-ε-caprolactone polyurethane and variations with type of extender as taken from the data of Magnus[35] are shown in Table 17.11.

Another casting system that has been introduced recently is based on polybutadiene polyols. Properties and formulations for this sytem have been described by Verdol et al.[36,37] and by Moore and coworkers.[38] The work reported covered many formulations, including reinforcement of the liquid polybutadiene with conventional rubber fillers such as carbon black and silica. The high reactivity of the hydroxyl group on the polybutadiene polymer chain is claimed to give a

TABLE 17.12. ONE-STEP CAST BUTADIENE BASED POLYURETHANES

Polyol R-45M[a]	100.0	100.0	100.0	100.0
Extender	4.2	11.5	16.8	33.5
Polyisocyanate	17.3	26.9	34.5	57.6
Stannous Octoate	0.2	0.06	0.23	0.27
Tensile, psi	415	1155	2205	1565
Elongation, %	285	310	315	195
100% Modulus, psi	220	455	765	1225
Tear, pli	65	105	175	145
Hardness Shore A	58	72	80	89

[a]Hydroxyl terminated polybutadiene polymer, 2500 to 2800 MW., Sinclair Oil Corporation
[b]Isonol C-100, N,N'bis(2-hydroxypropyl)-aniline, The Upjohn Co.
[c]Isonate 143L, a polyisocyanate structurally similar to MDI, The Upjohn Co.

special advantage for one-step casting systems. Some of the common highly polar chain extending agents, however, are not very compatible with the polybutadiene liquid polymer. Some properties of one-step polybutadiene cast urethane elastomers are shown in Table 17.12 as taken from the data of Moore et al.[38] The hardness is changed by varying the concentration of hard segments in the polymer, i.e., by changing the amount of N,N'bis(2-hydroxypropyl)-aniline and aromatic diisocyanate.

Similar properties were shown to be obtainable from these butadiene based polyols by using the prepolymer technique.

MICROCELLULAR CAST POLYURETHANES

Microcellular elastomeric materials at various densities can be readily formed from most of the polyurethane liquid casting systems. These microcellular materials have closed cells and densities in the range of 20 to 60 lb/cu. ft and exhibit many of the properties of solid elastomers. Several techniques have been used to create microcellular cast urethane elastomers. The CO_2 and solvent blowing systems are the two most common procedures. Kane[39] described the use of the water-isocyanate reaction to produce microcellular materials from the polyether prepolymer Adiprene L. Controlled amounts of water in the formulation react with isocyanate to give CO_2 gas and a cellular product.

Bianca[40] used methylene chloride to produce closed cell materials with densities as low as 20 lb/cu. ft. This low boiling liquid is volatilized by the heat of the reaction during the polyurethane polymer formation. The gas bubbles are trapped and a microcellular elastomeric material is formed.

Microcellular cast elastomers can be prepared from either polyesters or polyethers. One-step microcellular elastomers from various polyester polyols were described by Saaty et al.[41] Poly(propylene glycol) polyols were used for microcellular elastomers proposed for canvas shoe soles[42] while poly(butylene glycol) polyols were used to form experimental microcellular urethane elastomers for automotive fender extensions.[43]

THERMOPLASTIC POLYURETHANES

Solid polyurethane elastomers that can be milled, extruded, injection molded, and calendered on conventional plastics processing equipment and that develop high strength on cooling to room temperature are classed as thermoplastic polyurethanes. Many other processing techniques common to the plastics industry can be applied to this class of polyurethanes. Some of these are blow molding, solution coating, and heat or solvent sealing. Also included in the class of thermoplastic polyurethanes is a modification called thermoplastic-thermosetting polyurethanes which were described in 1962 by Piggott et al.[25] These materials are similar to the linear thermoplastic polyurethanes but show a change in properties when post-cured after the forming operation. Compression set decreases upon post-curing and is believed to indicate the formation of allophonate crosslinks or branch points by reaction of terminal NCO groups with urethane groups along the polymer chain.

The unique properties of thermoplastic polyurethanes were described in 1958 by Schollenberger.[24] Many reviews of their properties have been published[23,26,27] as commercial use has been established. At moderate temperatures and below, these materials exhibit the properties of crosslinked rubbers. The hard segments in the polymer chain form associations, as in Fig. 17.1, that not only serve as filler particles but also function as crosslinks. Above the softening temperature of the hard segments or in suitable solvents, the material behaves as a linear uncrosslinked polymer that can bc processed in the melt or in solution. Schollenberger[23] described the properties of some thermoplastic polyurethanes prepared from the random polymerization of a diisocyanate, a polyglycol, and a low molecular weight glycol. A variety of diisocyanates and other components were examined. Properties were influenced by changing any of the three starting materials. As in the casting polyurethane systems, segments having symmetry, capability of crystallizing, or a rigid aromatic structure gave polymers with high hardness and/or high strength. Generally, polymers based on polyesters gave higher strength than those based on polyethers. Thermoplastic elastomers from poly(butylene glycol) and poly(propylene glycol) were examined.

The thermoplastic polyurethanes have become the second most important class of urethane elastomers with important applications in wire and cable jackets, calendered film, and adhesives. Some of these processing techniques and applications have been reviewed in detail by Mitchell,[22] by Schollenberger,[23] and by Esarove.[27] Processing conditions vary somewhat from one grade to another but calendering and extrusion temperatures of 300 to 350°F are typical. The polyester-based polymers are presently the most widely used thermoplastic

TABLE 17.13. THERMOPLASTIC POLYURETHANE ELASTOMERS

Type	*Trade*	*Hardness Range*	*Supplier*
		Thermoplastic	
Polyester	Estane 5701	88 Shore A	B. F. Goodrich
	Estane 5702	68 Shore A	B. F. Goodrich
	Estane 5740X070	50 Shore D	B. F. Goodrich
	Estane 5740X100	78 Shore A	B. F. Goodrich
	Vibrathane S-4	40 Shore D	Uniroyal, Inc.
	Vibrathane S-5	50 Shore D	Uniroyal, Inc.
	Vibrathane S-6	60 Shore D	Uniroyal, Inc.
Polyether	Vibrathane E-6	65 Shore A	Uniroyal, Inc.
	Vibrathane E-9		Uniroyal, Inc.
		Thermoplastic-Thermosetting	
Polyester	Texin 480A	83 Shore A	Mobay Chemical Co.
	Texin 192A	91 Shore A	Mobay Chemical Co.
	Texin 355A	55 Shore D	Mobay Chemical Co.

TABLE 17.14 PROPERTIES OF ESTANE THERMOPLASTIC POLYURETHANES

Property	*Estane 5702*	*Estane 5740X100*	*Estane 5701*	*Estane 5740X070*
Hardness Shore A	70	76	88	95
300% Modulus, psi	500	1200	1300	3500
Tensile, psi	5000	6000	5800	5800
Elongation, %	700	550	500	450

polyurethanes. Commercial polymers are available with a wide range of properties and include some polyether-based thermoplastic polyurethanes, as shown in Table 17.13. The properties of a series of these thermoplastic polyurethanes are given in Table 17.14 as taken from technical service bulletins on "Estane."[50] The compression set, method B, of Estane 5740X100 was given as 18% after 22 hours at 75°F and 86% after 22 hours at 158°F. Esarove[27] and Schollenberger[23] have described a number of actual and potential applications for the thermoplastic polyurethanes.

VULCANIZED MILLABLE RUBBERS

The vulcanizable millable polyurethane rubbers are those that process on conventional rubber machinery at temperatures of 150 to 250°F. They process and handle in a manner similar to other millable rubbers. Generally the addition of fillers, vulcanizing agents, and other compounding ingredients is used to develop optimum properties for a given application. The general method used for preparation of millable polyurethane rubbers consists of reacting a hydroxyl-terminated polymer with a diisocyanate.[28,29] A slight excess of the polymeric diol is used so that the polymer molecules are terminated in hydroxyl groups, thereby producing material with a stable viscosity. A bulk viscosity suitable for milling can be achieved by carefully controlling the amount of excess polyol. Unsaturation can be introduced into the polymer in various ways when sulfur vulcanizable rubbers are desired. Polyester and polyether types are available.

Some of the polyurethane millable gums presently available are shown in Table 17.15. Those rubbers designed for use with a sulfur vulcanization system can generally also be cured with a peroxide system.

The millable crosslinkable urethane rubbers listed in Table 17.15 can be modified with a variety of compounding ingredients such as fillers and plasticizers to meet specific processing and end-use requirements, just as with other rubbers. They can be reinforced by fine-particle blacks and silicas to give high tensile and tear strength and excellent abrasion resistance. Typical properties of three different rubbers are shown in Table 17.16.

The black-reinforced compositions described show high tensile strength, excellent oil resistance, and good low temperature flexibility. The formulations shown here are not strictly comparable in that two are formulated with HAF black and

TABLE 17.15. POLYURETHANE MILLABLE GUMS

Type	*Trade Name*	*Vulcanization System*	*Supplier*
Polyester	Genthane S	Peroxide	General Tire & Rubber Co.
	Genthane SR	Peroxide	General Tire & Rubber Co.
	Vibrathane 5003	Peroxide	Uniroyal, Inc.
	Vibrathane 5004	Peroxide	Uniroyal, Inc.
	Elastothane 455	Sulfur	Thiokol Chemical Co.
	Elastothane 625	Sulfur	Thiokol Chemical Co.
	Elastothane ZR651	Sulfur	Thiokol Chemical Co.
Polyether	Adiprene C	Sulfur	E.I. DuPont de Nemours & Co.
	Adiprene CM	Sulfur	E.I. DuPont de Nemours & Co.

one with SAF black; however, some general relationships can be stated. For example, the peroxide-cured elastomer shows less compression set and better heat resistance than the sulfur-cured elastomers but requires greater care in handling to prevent contamination. The polyester urethane elastomers show less swelling in oils and fuels than the polyether elastomers but are more susceptible to degradation by

TABLE 17.16. MILLABLE CROSSLINKABLE RUBBERS

Type	*Polyester*	*Polyester*	*Polyether*
Cure System	*Peroxide*	*Sulfur-Acc.*	*Sulfur-Acc.*
Trade Name	*Genthane S*	*Elastothane 455*	*Adiprene C*
Formulation	*(1)*	*(2)*	*(3)*
Reference	#51	#52	#53
300% Modulus, psi	2900	–	2475
Tensile, psi	5300	5100	5150
Elongation, %	510	520	540
Hardness Shore A	70	72	64
Comp. Set, Type B, %			
22 hr @212°F	12	30	20
Volume Swell, %			
ASTM Oil No. 3	+3[a]	–2.5[a]	+20%[b]
Reference Fuel B	+15	–	+29%
Low Temperature Stiffness			
Temp. at G=10,000 psi	–40°F	–24°F	–35°F

Formulations:

(1) Genthane S 100.0, HAF Black 30.0, Stearic Acid 0.2, DiCup 40C 5.0.

(2) Elastothane 455 100.0, SAF Black 30.0, Cadmium Stearate 0.5, MBTS 4.0, MBT 2.0, ZC –456 Activator 1.0, Sulfur 2.0.

(3) Adiprene C 100.0, HAF Black 30.0, Cumar W-2 1/2 10.0, Sulfur 0.75, MBTS 4.0, MBT 1.0, RCD-2098 Activator 0.35.

[a] 3days @ 212° F
[b] 7 days @ 212° F

moisture at elevated temperatures. Carbodiimides are very effective as additives to improve the hudrolytic stability of polyester urethane elastomers. There are of course other differences in processing and physical properties that will influence the selection of a particular elastomer for a given end use situation. All of these materials, however, show the characteristic urethane properties of high tensile strength, oil resistance, and excellent abrasion resistance.

REFERENCES

(1) J. H. Saunders and K. C. Frisch, *Polyurethanes: Chemistry and Technology, Part I. Chemistry,* Interscience Publishers, New York, N. Y., 1962, p. 205.
(2) R. Bonart, *J. Macromol. Sci.,* **B2**, No. 1, 115 (1968).
(3) R. Bonart, L. Morbitzer, and G. Hentze, *J. Macromol. Sci.,* **B3**, No. 2, 337 (1969).
(4) S. L. Cooper and A. V. Tobolsky, *Text. Res. J.,* **36**, 800 (1966).
(5) S. L. Cooper and A. V. Tobolsky, *J. Appl. Polym. Sci.,* **11**, 1361 (1967).
(6) *Polyurethanes: Chemistry and Technology, Part I. Chemistry,* Interscience Publishers, New York, N. Y., 1962, Chaps. III and IV.
(7) K. C. Frisch, in *Polyurethane Technology,* P. F. Bruins, Ed., Interscience Publishers, New York, N. Y., 1969, Chap. 1.
(8) O. Bayer, E. Muller, S. Peterson, H. F. Piepenbrink, and E. Windemuth, *Rubber Chem. Technol.,* **23**, 812 (1950).
(9) E. Muller, O. Bayer, S. Peterson, H. F. Piepenbrink, W. Schmidt, and E. Weinbrenner, *Rubber Chem. Technol.,* **26**, 493 (1953).
(10) K. A. Piggott, B. F. Frye, K. R. Allen, S. Steingiser, W. C. Darr, J. H. Saunders, and E. E. Hardy, *Chem. Eng. Data,* **5**, 391 (1960).
(11) J. H. Saunders, *Rubber Chem. Technol.,* **33**, 1259 (1960).
(12) J. H. Saunders, *J. I.R.I.,* **2**, No. 1, 21 (1968).
(13) Technical Information Bulletin No. 87-E26, Mobay Chemical Co., Pittsburgh, Pa.
(14) W. Kallert, *J. I.R.I.,* **2**, No. 1, 26 (1968).
(15) C. F. Blaich, Jr., in *Polyurethane Technology*, P. F. Bruins, Ed., Interscience Publishers, New York, N. Y., 1969, Chap. 9, p. 185.
(16) D. P. T. Copcutt, *J. I.R.I.,* **2**, No. 1, 41 (1968).
(17) R. J. Athey, *Rubber Age,* **85**, 77 (1959).
(18) R. J. Athey, *Ind. Eng. Chem.,* **52**, 611 (1960).
(19) R. J. Athey, J. G. DePinto, and J. M. Keegan, "Adiprene L-100," Bulletin No. 7, Elastomer Chemicals Department, E. I. DuPont de Nemours & Co.
(20) R. J. Athey, "Adiprene L-167," Bulletin No. 12, Elastomer Chemicals Department, E. I. DuPont de Nemours & Co.
(21) R. J. Athey and J. G. DePinto, "Adiprene L-315," Bulletin No. 1, Elastomer Chemicals Department, E. I. DuPont de Nemours & Co.
(22) D. C. Mitchell, *J. I.R.I.,* **2**, No. 1, 37 (1968).
(23) C. S. Schollenberger, *Polyurethane Technology,* P. F. Bruins, Ed., Interscience Publishers, New York, N. Y., 1969, Chap. 10, p. 197.
(24) C. S. Schollenberger, H. Scott, and G. R. Moore, *Rubber World,* **137**, 549 (1958).
(25) K. A. Piggott, J. W. Brittain, W. Archer, B. F. Frye, R. J. Cote, and J. H. Saunders, *Ind. Eng. Chem.,* **1**, 28 (1962).
(26) P. Wright and A. P. C. Cumming, *Solid Polyurethane Elastomers,* McClaren and Sons, London, 1969, Chap. 8.

(27) D. Esarove, *SPE Tech. Pap., Vol XI,* 21st Annual Technical Conference, March, 1965, Session XXVI-3, p. 1.
(28) N. V. Seeger, T. G. Mastin, E. E. Fauser, F. S. Farson, A. F. Finelli, and E. A. Sinclair, *Ind. Eng. Chem.*, **45**, 2538 (1953).
(29) E. E. Gruber and O. C. Keplinger, *Ind. Eng. Chem.*, **51**, 151 (1959).
(30) S. L. Axelrood and K. C. Frisch, *Rubber Age,* **88**, No. 3, 465 (1960).
(31) P. Wright and A. P. C. Cumming, *Solid Polyurethane Elastomers,* McClaren and Sons, London, 1969, Chap. 6.
(32) R. J. Ferrari, *Rubber Age,* **99**, No. 2, 53 (1967).
(33) S. L. Axelrood, C. W. Hamilton and K. C. Frisch, *Ind. Eng. Chem.*, **53**, No. 11, 889 (1961).
(34) S. L. Axelrood, L. C. Smith and K. C. Frisch, *Rubber Age,* **96**, No. 2, 233 (1964).
(35) G. Magnus, *Rubber Age,* **97**, No. 4, 86 (1965).
(36) J. A. Verdol, P. W. Ryan, D. J. Carrow, and K. L. Kuncl, *Part I, Rubber Age,* **98**, No. 7, 57 (1966).
(37) J. A. Verdol, P. W. Ryan, D. J. Carrow, and K. L. Kuncl, *Part II, Rubber Age,* **98**, No. 8, 62 (1966).
(38) R. A. Moore, K. L. Kuncl and B. G. Gower, *Rubber World,* **159**, No. 5, 55 (1969).
(39) R. P. Kane, *Rubber World,* **147**, 35 (1963).
(40) D. Bianca and R. E. Knox, *Rubber Age,* **98**, No. 5, 76 (1966).
(41) N. N. Saaty, L. W. Abercrombie, H. E. Reymore, Jr., and A. A. R. Sayigh, *J. Elastoplast,* **1**, 170 (1969).
(42) K. Murai and K. Fukuda, *J. Elastoplast.,* **1**, 150 (1969).
(43) W. K. Fischer, *J. Elastoplast.,* **1**, 241 (1969).
(44) J. F. Beecher, L. Marker, R. D. Bradford, and S. L. Aggarwal, *J. Polym. Sci.,* **C(26)**, 117 (1969).
(45) P. Wright and A. P. C. Cumming, *Solid Polyurethane Elastomers,* McClaren and Sons, London, 1969, Chap. 2.
(46) *Rubber World,* **160**, No. 5, 50 (1969).
(47) A. P. C. Cumming and P. Wright, *J. I.R.I.,* **2**, No. 1, 29 (1968).
(48) D. C. Mitchell, *J. I.R.I.,* **2**, No. 1, 37 (1968).
(49) J. D. Dianni, *J. I.R.I.,* **2**, No. 3, 125 (1968).
(50) Technical Data Sheets, on Estane Resins 5702, 5740X100, 5701 and 5740X070, B. F. Goodrich Chemical Co.
(51) Bulletin "Genthane S" GT-S4, Chemical Division, The General Tire & Rubber Co.
(52) M. M. Swaab, *Rubber Age,* **92**, No. 4, 567 (1963).
(53) H. G. Schwartz and W. V. Freed, Bulletin "Adiprene C," Elastomer Chemicals Department, E. I. DuPont de Nemours & Co.

18

LATEX AND FOAM RUBBER

T. H. ROGERS and K. C. HECKER
Goodyear Tire & Rubber Co.
Akron, Ohio

Latex was used by the South American Indians long before being "discovered" by European explorers in South America early in the Fifteenth Century. These natives used the "liquid rubber" for reinforcing moccasins and making ball-like objects and other useful articles.

Early man obtained the latex by felling the trees, stripping the bark, and cutting into the inner layer, thus permitting the latex to ooze out and be collected. The Europeans found that by systematically tapping the tree the latex could be extracted without destroying the tree. With the development of plantations in the Far East, it was found that latex could be preserved by adding ammonia to it as soon as possible after coming from the tree. This marked the beginning of our commercial latex technology. This preserved natural rubber latex was commonly referred to as normal latex and had a total solids content of about 38%. It was developed into commercial applications for adhesives, cloth waterproofing, dipped goods, and other small volume products. It was not until technology was developed to produce high solids latices that latex really began to assume the stature of a valuable item of commerce.

Although the principles of creaming and centrifuging had been known for many years, it was not until 1923 that these methods of concentration began to take on commercial importance. In that year Traube patented the idea of adding certain gums, such as Irish moss, to natural rubber latex to accelerate the natural process of creaming. In the same year Utermark obtained a patent for the treatment of a "milk sap" in a centrifuge machine of the cream-separator type. Dunlop Rubber Company developed this process on a large scale in the Far East. This method accounts for most of the concentrated latex in use today. Natural latex concentrated by the creaming process has some commercial use and an evaporated concentrated latex stabilized with fixed alkali and soap has some limited commercial applications.

Emulsion polymerization of synthetic rubber has been known for many years—since 1912 when a patent was issued to Bayer and Company for the emulsion polymerization of isoprene. Although there was much interest in synthetic rubber through the years, as evidenced by the great number of patents on the subject after 1930, it was not until the advent of World War II, with the resulting short supply of natural rubber, that synthetic rubber came into its own. The use of synthetic rubber latex increased rapidly after 1943. However, because the first general purpose synthetic rubber latices were limited to hot recipes (polymerization temperature about 50°C), small particle low solids latices resulted which could only be used for adhesive, impregnation, and spreading applications. After 1948 the development of cold recipes (polymerization temperature about 5°C) greatly increased the types of latex available.[1]

It was not until large particle size high solids synthetic rubber latices were made available that the synthetics were used to any great extent in foam rubber, and then only in blends with natural rubber latex.[2] Improvements in high solids SBR latices during the 1950's resulted in latices which enabled foam producers for the first time to make many foam articles from 100% synthetic rubber latex.[3]

Specialty latices have been known for many years. Neoprene latex (polychloroprene) was first produced in 1934.[4] Other specialty synthetic latices include NBR, PVC, vinyl acetate, acrylic, and *cis*-polyisoprene. Carboxylated versions of several of the above latices were subsequently developed, resulting in many improved properties.

Latex technology is a highly specialized field that is not too familiar to most polymer chemists and even many rubber compounders. The major reasons for these are two: (1) products made directly from latex constitute only about 12% of total rubber usage; and (2) the art and science of handling latex problems is more involved than regular rubber compounding and requires a good background in colloidal systems.

LATEX vs DRY RUBBER

While a latex differs in physical form from dry rubber, the properties of the latex polymer differ only slightly from its dry rubber counterpart. Unlike the dry polymer which must undergo mastication before use, the latex polymer need not be broken down for application, thus retaining its original high molecular weight which results in higher modulus products. Other advantages enjoyed by applications involving latex are lower machinery costs and lower power consumption, since the latex does not have to be further processed into dry form and compounding materials may be simply stirred into the latex using conventional liquid mixing equipment. This is the major reason why latex products can be made by many small companies with low capitalization. However, latex does have some disadvantages when compared to its dry-polymer counterpart. The latex polymer cannot be reinforced with low cost pigments in the manner that carbon black reinforces dry polymers. Water must be removed to obtain a product and shrinkage occurs during this water removal. Because of the water removal problem, applications are generally limited to products having thin-wall structure or to cellular forms. However, a great variety of products fall into these categories as will be demonstrated later.

GENERAL LATEX PROPERTIES

Latex can be defined as a stable aqueous dispersion of a polymeric substance having particle diameters in the range of about 500 Å to 50,000 Å (0.05μ to 5μ).[5,6]

Because the polymer deposited from the latex during application is in the form of discrete particles, the ability of the particles to coalesce under the conditions of application may be more of a factor in determining ultimate tensile strength than the molecular weight of the polymer. The fusing of polymer particles can be aided by the inclusion of plasticizers with the polymer. These are usually liquids added to latices during compounding and may be volatile or nonvolatile, depending on whether or not it is desirable to have them remain in the polymer after drying. Internal plasticization may be obtained by copolymerizing the primary monomer with a second monomer which results in a softer polymer. Thus, while polystyrene latex will not form a continuous film on drying, a copolymer of styrene with at least 30% butadiene will form a continuous film. This is one example of permanent or nonmigrating plasticization.

The selection of a latex for a particular end use is determined not only by the type of polymer, but also by such other factors as particle size and shape, pH, type and amount of stabilizer, and electrical charge on the particle.

A latex may consist of particles of essentially the same diameter[7] or it may consist of a wide range of particle sizes from very small (500 Å) to very large (>10,000 Å) diameters.[5] During polymerization, all of the particles are usually very small. They may be agglomerated to larger size by one of several methods, such as precisely controlled conditions in the reactor,[8] post-reactor chemical addition,[9] freezing,[10] and high pressure impingement.[11]

The particles are usually not evenly distributed over the size range but rather have greater concentrations at one or more locations in the range. These are represented in Fig. 18.1 by the peaks in the particle distribution curves of several common latices.

Latices used for binders or saturants must have small particles to properly penetrate the fibers. On the other hand, latices used for foam rubber applications must have some large particles and a particle size distribution such that a high polymer content is obtained in the latex with over 60% total solids being desired.

For a given latex, as the polymer content increases, the viscosity also increases and the rate of increase becomes very rapid near the theoretical high solids limit for the system (Fig. 18.2). Ideally, for maximum solids concentration each of the interstices between particles should be filled by a smaller particle. If this were able to be accomplished, a very wide range of specific sizes would be required to

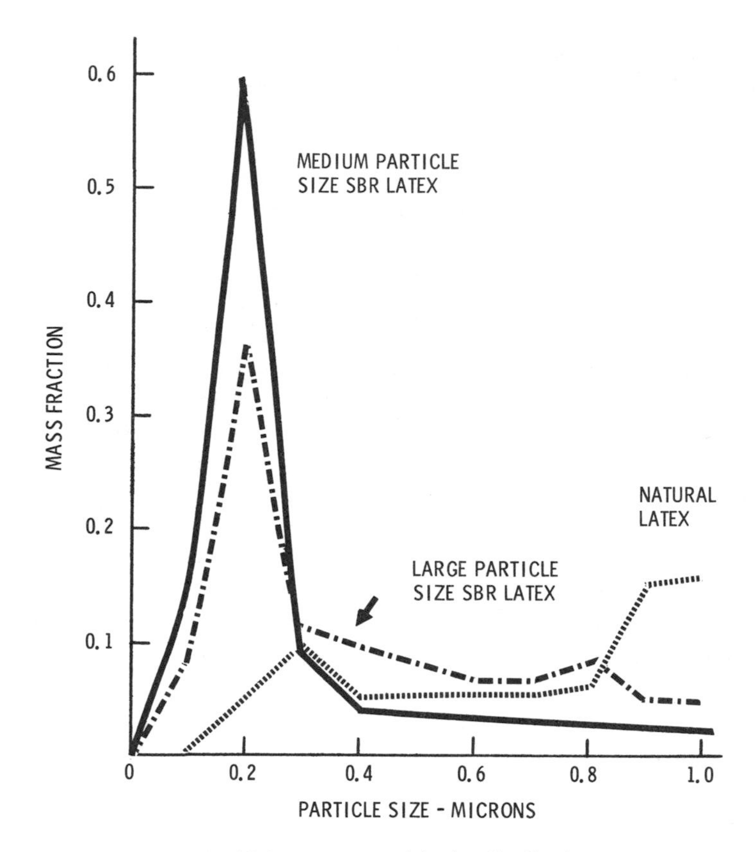

Fig. 18.1 Latex particle size distribution.

maximize the solids/serum ratio. Commercially, this is not feasible, but effort is made to come close to it.

An application such as foam rubber requires a high solids latex because excess water results in severe shrinkage, accompanied by foam structure breakdown, and with other applications, the excess water lengthens drying time. Because particle size distribution is so critical, the reliance on a single value representing the average particle size can be very misleading. Particle size distribution studies are very time consuming. A solids/viscosity curve, as shown in Fig. 18.2, is easily obtained and can provide valuable information if a relationship is established between the shape of the curve and applications characteristics and requirements.

In all latex applications stirring or pumping, and usually addition of compounding materials, are involved. Thus, mechanical and chemical stability are essential. Stirring of latex causes a rapid movement of the polymer particles with many collisions between them. Even when not being stirred, the particles because of their colloidal size undergo Brownian motion and move rapidly in a limited sphere with frequent collisions.[12] The large particles with their slower motion will either cream

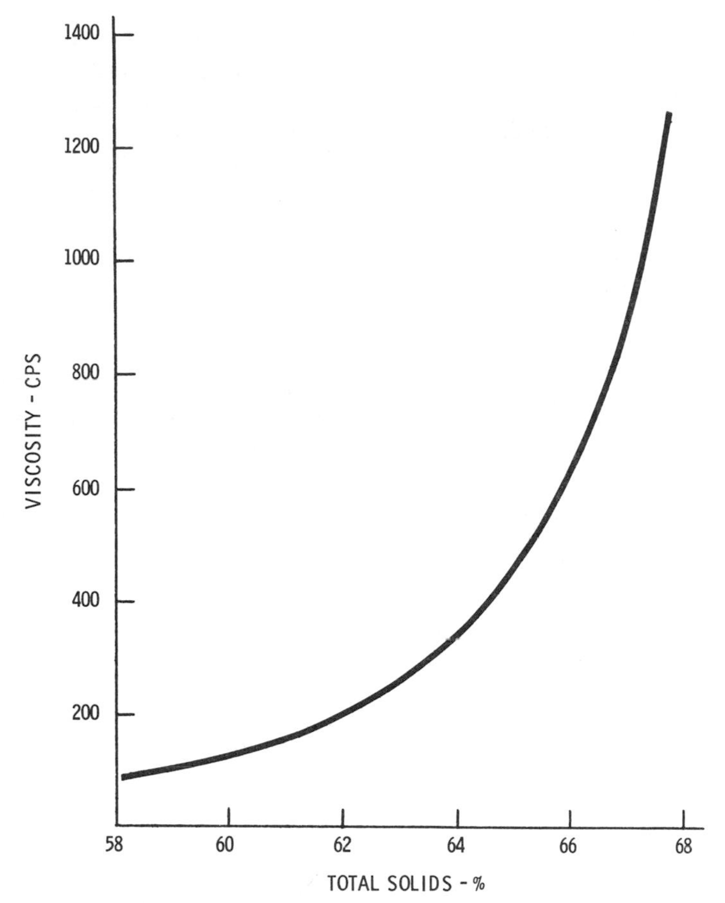

Fig. 18.2 Solids-viscosity relationship for SBR foam latex.

to the top or settle to the bottom, depending on the specific gravity of the polymer relative to that of water. Thus, they become tightly packed. Obviously the particles, if not properly stabilized or separated from each other, tend to adhere together and floc or microcoagulum forms. In extreme cases, the particles may coagulate into a solid mass.

On the other hand, at some point in any latex application, the particles must be destabilized and formed into a continuous polymer mass. Thus, a balanced stabilization system becomes a very important and sometimes frustrating operation.

Polymer particles are usually stabilized by adsorbed surface active agents (soaps and surfactants) which confer a charge to the particles resulting in mutual repulsion of the particles. Some surfactant is necessary to stabilize the particles during polymerization and additional surfactant of the same or different type may be post-added, depending on the intended application. Water molecules are hydrogen bonded to the surfactant molecules, forming an additional hydrated protective layer around each particle.[13] The surfactant molecules have limited solubility in water and those not adsorbed on the particles usually exist in the water phase as micells, i.e., fairly large, charged aggregates of long chain electrolyte.[1] The critical micell concentration (cmc) is the concentration of electrolyte in the water phase at which micelles begin to form.[14] The surfactant molecules in the water phase are in equilibrium with the adsorbed surfactant molecules on the particles. Any appreciable dilution of the latex will, therefore, rob surfactant from the particles, resulting in possible destabilization of the latex if the adsorbed surfactant, before dilution, is just barely adequate for stabilization.

A simple method of estimating at what point the particles are completely saturated with adsorbed surfactant is shown in Fig. 18.3. The surface tension of the latex is plotted against the amount of soap added. As soap is added, surface tension will drop, rapidly at first, and then more slowly when the particles are saturated. The intersection of the extension of straight lines from the vertical and horizontal portions of the curve indicate the saturation point.

Surface active agents fall into three categories, as shown below, with many subdivisions within each category:

Type	*Charge Conferred to Particles*	*Examples*
Anionic	Negative	Sodium lauryl sulfate Potassium oleate
Cationic	Positive	Stearyl dimethyl benzyl ammonium chloride
Nonionic	Neutral	Ethylene oxide-alkylphenol condensation product

The majority of commercial latices are stabilized with anionic surfactants, although mixtures of different types of anionic and anionic-nonionic surfactants are sometimes used.

The choice of surfactant is largely dependent on the application involved. Too much or the wrong kind of surfactant might interfere with the knitting together of the particles during application, resulting in poor cohesion or adhesion.

The pH of the system can have an important effect on stability. Although many anionic surfactants will impart stability over a wide pH range, certain compounds added to the latex may cause destabilization unless added within a narrow pH range. Many bactericides and fungicides are effective only in a certain pH range. Soaps such as potassium oleate lose their stabilizing ability as the pH is lowered and they revert to fatty acids. Latices having carboxyl groups in the polymer must be polymerized at low pH levels. Anionic surfactants with a wide pH stabilizing range, such as the benzene sulfonates, or cationic surfactants are used to stabilize the latices during polymerization.[15]

Destabilization, or coagulation, can be achieved by dehydration of the particles, i.e., removal of water of hydration associated with the surfactant molecules,

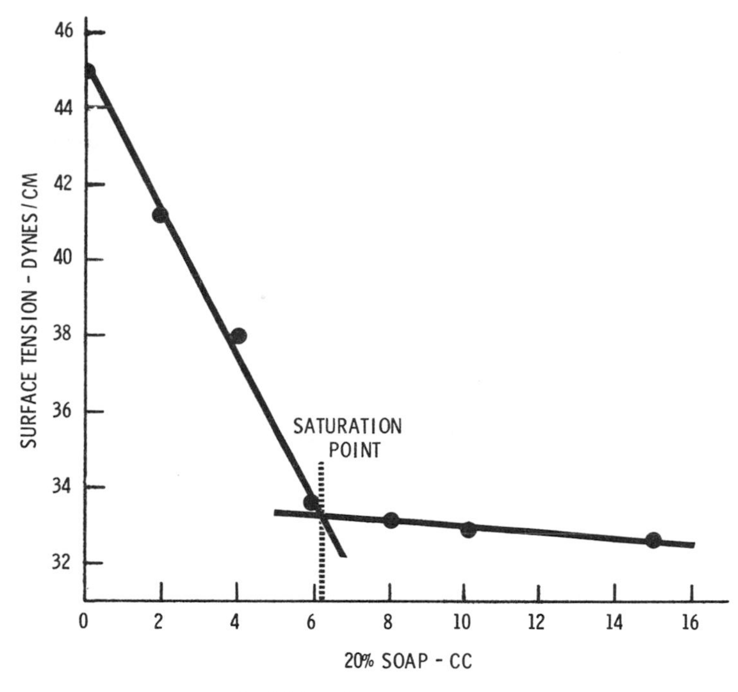

Fig. 18.3 Soap titration.

neutralization of the charge on the adsorbed soap, or depletion of the surfactant from the particle surface. The simplest form of dehydration is air drying of the latex to form a polymer film. Latex may also be added to alcohol, which adsorbs water, causing coagulation by dehydration. If formic or acetic acid is added to latex, the pH is lowered, causing the soap to revert to its acid form, thereby neutralizing the charge on the particles and nullifying the mutual repulsion of the particles. This is the common method of coagulating natural rubber latex on the plantations to produce commercial forms of dry rubber. Other acids are also effective for natural latex and soap emulsified synthetic latices.

Soap can be depleted from the surface of polymer particles by adding other small particle materials, such as colloidal silica, having high surface area. The soap is then in equilibrium with the water and both polymer and silica particles. Part of the soap is thus desorbed from the polymer particles resulting in some destabilization of the colloidal system. Sodium silicofluoride, used as a gelling agent for foam rubber, hydrolizes in the foam compound to form both hydrofluoric acid, which lowers the pH, and colloidal silica, which adsorbs soap from the polymer particles.[16]

Many heat gelling agents, such as polyvinyl methyl ether and polyglycols, owe their effect to their ability to precipitate from solution in the form of colloidal-sized particles when the temperature is raised. Ammonium salts lose volatile ammonia on heating, leaving an acid which lowers the pH and causes gelling of the latex. Still others, such as potassium silicofluoride, increase greatly in solubility as the temperature is increased, resulting in faster hydrolysis and destabilizing action.

Destabilization can also be achieved by the action of divalent metal ions, such as that of zinc. In a latex containing zinc oxide, ammonia, and amine salts, a soluble zinc ammonium complex is formed. The zinc ion is thought to be associated with a maximum of four ammonia molecules in this complex. Increasing the temperature or lowering the pH reduces the number of ammonia molecules coordinated to each zinc ion. In the presence of fatty acid soaps, the more reactive lower zinc ammines thus formed react with the soap forming insoluble zinc soaps.[17] The latex thickens and eventually coagulates.

LATEX TYPES

Latices can be divided into two general types—natural and synthetic. Natural latices can be subdivided according to methods of concentration and preservation. Synthetic latices can be subdivided according to polymer type, with different methods of polymerization and stabilization possible for each. Natural latex consumption in the U.S., after a peak in 1955, fell to a low level, but has grown steadily for the past 6 years to 73,058 long tons in 1971. S-type synthetic latex consumption also has grown steadily in volume through the years to 119,302 long tons in 1971. N-type synthetic consumption has been fairly constant at 12,000 to 15,000 long tons for a number of years.

Natural Latex

Natural latex is produced in special vessels outside the cambium layer of the *Hevea brasiliensis* tree. The reason why the rubber tree makes rubber is still not known. One theory holds that it is a waste product of the tree. However, the mechanism by which the tree makes rubber has now been fairly well determined.[18] Sugars, produced by photosynthesis, undergo many chemical transformations which result in a five-carbon atom building block. These building blocks are then linked together to form a chain containing about 40,000 carbon atoms. The catalyst for this reaction is a highly specific enzyme. A pyrophosphate group is added to the building unit to form isopentyl pyrophosphate. The pyrophosphate groups are eliminated when rubber is formed from the building units.

Latex as it comes from the tree has a solids content of about 36%, a surface tension of 40.5 dyn/cm (30°C), and a pH of 6.[19] The polymer is primarily *cis*-1,4-polyisoprene. Latex fresh from the tree is stabilized by naturally occurring proteins and phospholipids. It also contains other materials such as resins, sugars, mineral salts, and alkoloids. The protein stabilizer is very susceptible to bacterial action and would be destroyed within a few hours if ammonia were not immediately added. Soap-forming fatty acids are formed when the ammonia hydrolizes the lipids.[20] This ammonia soap then becomes the primary stabilizer, displacing the adsorbed protein from the particle surface.

If latex is preserved entirely by ammonia (high ammonia latex), a portion of the ammonia is usually removed before application. Other preservatives, such as sodium pentachlorophenate, sodium salt of ethylenediamine tetracetic acid, boric acid or zinc alkyl dithiocarbamates, may be used with smaller amounts of ammonia. This is known as low ammonia latex and has advantages of lower cost and elimination of the need to deammoniate the latex before processing into products.

Latex is concentrated to greater than 60% rubber solids before leaving the plantation, and this is accomplished either by centrifuging or creaming. Creamed latex is usually produced by treating the dilute latex with small amounts of sodium or ammonium salts of alginic acid.[19] The migration or creaming of particles to the upper portion of the latex takes place over a period of several days, increasing the rubber solids in this area. The lower portion of the latex, containing some very small rubber particles and other chemicals mentioned above, is then separated from the cream and is either discarded or is coagulated to salvage the small quantity of rubber remaining in it. The upper portion is usually referred to as the concentrate and the lower portion as the skim or serum.

The American Society for Testing and Materials (ASTM) classifies natural rubber latex into four categories according to method of concentration and preservation (Table 18.1). Some specifications for each group are given in Table 18.2.

TABLE 18.1. MAJOR TYPES OF NATURAL LATEX

Type	*Concentration Method*	*NH_3 Level*	*Other Preservatives*
I	Centrifuge	High	–
II	Cream	High	–
III	Centrifuge	Low	sodium pentachlorophenate sodium salt of ethylene-diamine tetra acetic acid, boric acid, zinc diethyldithiocarbamate
IV	Cream	Low	Same as III

TABLE 18.2. SELECTED ASTM NATURAL LATEX REQUIREMENTS

	Type I	*Type II*	*Type III*	*Type IV*
Total Solids, min, %	61.5	64.0	61.5	64.0
Dry Rubber Content, min, %	60.0	62.0	60.0	62.0
Total Alkalinity (calculated as ammonia; expressed as a percentage of water in the latex), %	2.0	2.0	2.0	2.0
KOH Number, max	0.8	0.8	0.8	0.8
Mechanical Stability, min, sec	540	540	540	540

Most of the natural rubber latex is produced in Malaysia, Indonesia, and Liberia. Latex concentrate constitutes slightly more than 8% of the world natural rubber supply, and about 90% of this is centrifuge concentrated.[20] Principal outlets for natural rubber latex are foam rubber, dipped goods, and adhesives.

Synthetic Latices

Synthetic latices can be generally classified by type of polymer structure. Major suppliers of synthetic latices are shown in Table 18.3.

Styrene-butadiene (SB) latices are the most common and best known of the synthetic types. This family of latices consists of rubbery polymers with styrene levels below 50% and resinous polymers with styrene contents above 50%. They are produced with various polymerization systems using a large number of different soaps and surfactants with the resulting latices having a wide variety of properties.

Principal applications for the rubber latices are foam and textile coatings and for the resin latices, paint and textile coatings. These latices are generally low in cost and have good vulcanized physical properties. They do, however, lack chemical and solvent resistance, have poor laundering and dry-cleaning resistance, and only fair light stability.

Vinyl acetate homopolymer and copolymer latex is another large volume item. Its principal uses are in paint and adhesives, being particularly noted for its excellent adhesive properties.

The acrylics make up a third class of large volume consumption latices. These are generally produced as copolymers of methyl or ethyl methacrylate and other monomers such as styrene and vinyl acetate. They can be made both rubbery and resinous. They exhibit excellent chemical and light resistance and find major application in paint and textiles. They are higher priced than SBR latices and are used only where superior physical properties are needed.

Nitrile latices have a much lower consumption volume than the above latices, but they have unique properties which make them useful. They are used where oil and solvent resistance are needed. Primary applications are for paper and textiles. They can be compounded and processed like SB latices. Principal disadvantages are high cost, poor cold properties, and low resilience.

Polychloroprene latex (neoprene) is produced in the US by Du Pont. The polymer has excellent chemical resistance and relatively high green strength, making it suitable for dipped goods. It may also be used in paper, foam, and adhesive applications. Because of its high chlorine content, it has advantages of reduced compounding requirements to meet flame retardance standards. Disadvantages are its high price, slow curing, bad odor, and inferior low temperature values exhibited by products containing it.

Polyvinyl chloride latex (PVC) usage is only minor as compared to that of PVC organosol and plastisol systems. It is used primarily as a chemical resistant saturant and coating latex. It must be plasticized to be film forming unless it contains a comonomer such as vinyl acetate.

Cispolyisoprene latex is a relative newcomer to the latex field. It is commercially produced only by the Shell Chemical Company. The polymer is solution

TABLE 18.3. MAJOR US PRODUCERS OF SYNTHETIC LATICES

Hot SBR Latices

Dewey & Almy (Darex)
Firestone (FR-S)
General
Goodyear (Pliolite)
Hooker
Shell
Standard Brands, Inc (Tylac)
Uniroyal (Naugatex)

Polystyrene Latex

Bordon
Dow
Koppers
Monsanto
Morton Chemical
Polyvinyl Chemicals
UBS

Nitrile Latices

Firestone (FR-N)
Goodrich (Hycar)
Goodyear (Chemigum)
Polymer Corp (Kryvac)
Standard Brands, Inc (Tylac)
Uniroyal (Nitrex)

Acrylic Latices

Bordon
Catalin
Celanese
Dow
General Tire
Goodrich
Monsanto
Morton Chemical
National Starch
Polyvinyl Chemicals
Reichhold
Rohm & Haas
UBS
Union Carbide

Solution Isoprene Latex

Shell

Cold SBR Latices

Copolymer (Copo)
Firestone (FR-S)
Goodyear (Pliolite)
Shell
Uniroyal (Naugatex)

SB Resin Latices

Bordon
Dewey & Almy
Dow
Firestone
General
Goodyear
Koppers
Uniroyal

Vinyl Pyridine Latices

Firestone (FR-S)
General (Gen-Tac)
Goodrich (Hycar)
Goodyear (VP)
Uniroyal (Pyratex)

Polyvinyl Acetate Latices

Bordon
Celanese
Colton
Dewey & Almy
Du Pont
National Starch
Reichhold
Seidlitz
Shawinigan

Polyvinyl Chloride Latex

Bordon
Dow
Firestone
Goodrich
National Starch

Polychloroprene Latex

Du Pont (Neoprene)

polymerized and must be made into a latex by first emulsifying the rubber-in-solvent cement, stripping off the solvent, and concentrating the resulting latex. It has very good vulcanized properties and finds use in foam rubber and dipped goods.

Many of the above latices can be polymerized with monomers that build carboxyl groups on the polymer chain. Up to 10% of carboxylated monomers such as acrylic, methacrylic, fumaric, or itaconic acid may be copolymerized with other monomers to form polymers which are self-curing with heat or may be cured with such materials as divalent metal oxides, polyamines, epoxides, diisocyanates, organic peroxides, in combination with or in place of sulfur.[21] These latices have many desirable properties, such as improved aging, high filler level acceptance, improved stability, and ease of application.

LATEX COMPOUNDING

Latex compounding involves not only the addition of the proper chemicals to obtain optimum physical properties in the finished product but also the proper control of colloidal properties which enable the latex to be transformed from the liquid state into final product form. Much of the latter has been discussed under the section on general latex properties.

Viscosity control in the latex is very important. As discussed previously, the particle size of the latex has a great effect on viscosity. Large particles generally result in a low viscosity. Dilution with water is the most common way to reduce viscosity. Certain chemicals such as trisodium phosphate and sodium dinaphthylmethane disulfonate are effective viscosity reducers.

Thickening may be accomplished with either colloidal or solution thickeners. Small particle size materials such as colloidal silica will thicken latex when added to it. Solutions of such materials as alpha protein, starch, casein, sodium polyacrylates, and polyvinylmethylether will also thicken latex. The type of thickener chosen depends largely on the type of latex, the application involved, and experience of the compounder.

A major problem in many applications is foam control. Not only does excessive foam reduce application speed, it has a detrimental effect on the quality of the finished product. On the other hand, some applications, such as foam rubber and the new methods of applying carpet backing compound in the foamed state, require an easily foamed mix.

Care must always be taken during latex compounding not to draw air into the stirring latex mix. Latex stabilizers vary greatly in their ability to foam. Materials such as silicone antifoam agents and octyl alcohol are effective foam repressors but may be harmful if used in excess. For example, excess silicone may result in the formation of tiny pits or "fish eyes" in latex films.

The particle size of solid materials added to latex must usually be made as small as possible to insure intimate contact with the rubber particles. Solid materials are usually added to latex as dispersions. The material to be added is mixed with dispersing agents and suspending agents in water and ground to a small particle size in a ball mill or attritor. In these devices stones or other hard pebble-sized materials are made to tumble and mix with the chemicals causing them to be reduced to very small size. Liquids that are not water soluble are emulsified in dilute solutions of stabilizers using high speed and high shear mixers. Curing or vulcanizing unsaturated polymer may be achieved with sulfur and appropriate accelerators, as is common in dry rubber applications. In the case of the various carboxylated polymers, crosslinking may be obtained with a variety of other materials mentioned previously. This discussion will be limited to the sulfur system.

While the sulfur is the same as that used in dry rubber compounding, the accelerator system may be quite different. The latex compounder is not faced with the mill scorch problem experienced in dry rubber compounding. The cure cycle may be, of necessity, short and at relatively low temperature as in low pressure steam vulcanization of foam rubber. Therefore, ultra-accelerators such as zinc diethyldithiocarbamate alone or in combination with thiazoles, polyamines, and guanidines are used. The latter two also function as gel sensitizers, or secondary gelling agents, in the preparation of foam rubber.

As with dry rubber, zinc oxide is used as a cure activator. It is also used as the vulcanizing agent for neoprene. Stearic acid, usually used with zinc oxide, is not needed. Its function may be taken over by the fatty acids of the stabilizer soaps. Zinc oxide also functions as a gelling aid, especially when used in ammonia or ammonia salt systems.

Because of the great surface area exposure of most latex products, protection against oxidation is very important. Many applications involve light colored products which must not darken with age or exposure to light. Nonstaining antioxidants such as the hindered phenols must be used. Where staining can be tolerated, amine derivatives such as phenylenediamines may be used. These have good heat stability and are also effecitve against copper contamination, which causes rapid degradation of rubber.

Loading materials are used primarily as fillers or extenders, although some stiffening may be obtained with certain small particle clay fillers. Most of the nonblack fillers may be used in latex compounds. Carbon black does not reinforce latex in the manner that it does dry rubber, and is used only in small amounts in latex for color, as are various other dyes and colored pigments.

Oils may be used as softeners and waxes as detackifiers. Starch and its derivatives are used as reinforcing materials.

LATEX FOAM

Latex foam owes its beginning to Untiedt, who in the late 1920's stabilized natural latex of about 40% total solids content with soap, beat air into it, refined the resulting foam until a fine celled froth was formed, and dried it in an oven, forming a dense spongy cushioning material.[22]

Latex foam is a flexible cellular material. The term "cellular material" is defined by ASTM as "a generic term for materials containing many cells (either open, closed, or both) dispersed throughout the mass." The ASTM definition for a "flexible cellular material" is "a cellular organic polymeric material which will not rupture when a specimen 200 x 25 x 25 mm (8 x 1 x 1″) is bent around a 25 mm mandrel at a uniform rate of one lap in 5 seconds at a temperature between 18 and 29°C."[6]

There are currently two major methods of producing latex foam: the Dunlop process and the Talalay process. Each has been in use for many years and each has unique characteristics that makes it suitable for cushioning applications.

The Dunlop process traces its origin back to the late 1920's with commercial development during the 1930's.[23] Production of foam by this method consists of the following steps:

1. Beating air into compounded latex.
2. Adding the gelling agent to the foam latex.
3. Pouring the foam into a mold.
4. Gelling and vulcanizing the foam rubber.
5. Stripping the foam rubber from the mold.
6. Washing and drying the foam rubber.

Early foam production was by batch process, using large bakery-type mixers equipped with 120-qt bowls and wire whips which turned in a planetary motion, beating air into the compounded latex. A method for foaming latex was developed using the Oakes frother, which is essentially a modified marshmallow mixer. In this machine, air is injected into a stream of latex forming a coarse foam which is forced between toothed rotor and stator plates to refine the foam.[24] Multiple units may be used to reduce the foam to sufficient fineness. The proper pressure is also essential. The gellant is injected late in the frothing stage. The foam is conducted through a hose or pipe to conveyorized molds, or to a spreader blade system.

The mold is filled with foam, the lid is placed on the mold, and after a time interval of several minutes for complete gelling, the closed mold is passed through a steam-filled tunnel to vulcanize the foam. The lid is then removed, the foam stripped from the mold and passed through a hot air drying tunnel. The mold, while still hot, is sprayed with a release agent and passed through a temperature conditioning tunnel. The mold may be conditioned as high as 55°C while the foam is usually 20-27°C, but in some cases may be as high as the temperature of the mold. The mold is constructed in the shape of the final product. The lid, which may be lowered and raised automatically, contains many metal lugs, which result in the familiar core holes found in foam products. Heat is transferred through the metal lugs into the foam, thus aiding the gelling process, which will be discussed in some detail later. The cores result in relatively thin foam sections, even in thick products such as mattresses. In the finished foam, core skin surface and dense foam adjacent to the skin act as reinforcing members to support the load. Also, as foam density increases, load support becomes more efficient. The greater the core hole volume, the higher the foam density that can be used for the least net weight. Thus, a light "apparent" density results from a high density foam. A lighter "apparent" density core stock cushion can support the same load (compression resistance) as a heavier slab stock foam.

The foam may be spread to a thickness of about two inches onto a moving belt to form slab stock. An improved surface results if steam is applied to the surface.[25] It may also be apread to a thickness of about one inch onto a fabric substrate, such as a carpet. In practice, the gauge on fabric is limited to 1/2 inch or less. This latter application is the fastest growing segment of the foam industry at the present time. Molded latex foam usage has decreased in the US in recent years, but the overall latex foam production has increased due largely to the thin gauge carpet applications.

The classical method of gelling latex foam is by use of a delayed action gelling agent such as sodium silicofluoride. This system is rather complicated and involves several of the methods of destabilization discussed earlier. The basic chemical reaction is as follows:

$$Na_2SiF_6 \rightleftharpoons 2Na^+ + SiF_6^=$$

$$SiF_6^= + 4H_2O \rightleftharpoons Si(OH)_4 + 4H^+ + 6F^-$$

The primary destabilizing agent is hydrofluoric acid which lowers the pH of the system, thereby destroying the soap. In addition, there are secondary factors which contribute to the gelling process. The fine particle size silicic acid desorbs soap from the latex particles. Also, as the pH is lowered, conditions are more favorable for the formation of soluble zinc ammonia complex, from which follows the formation of insoluble zinc soap. Because of these contributing factors, gelation takes place at a pH of about 8.5 rather than at a lower level.

Before gelation, the stabilized rubber particles are distributed through the soap bubble matrix. As the particles are destabilized, gelation starts, knitting the particles together within the soap matrix. As gelation continues, all of the soap is eventually destroyed, leaving in its place a matrix of tightly knit rubber particles which form the basic open cellular type of foam structure.

A very delicate balance of conditions is required to bring about successful foam gelation. A material such as diphenylguanidine raises the pH of gel, thereby increasing the time between gelation and soap break, preventing foam collapse. Figure 18.4 shows pH-gelation curves. The solid line shows a normal condition with gelation taking place at a pH of about 8.5 and about 5 minutes after the gelling agent is added. This is commonly referred to as gel time. About 3 minutes more elapse before the soap is completely destroyed as indicated by a lack of crackle sound when the foam is touched. This is called the soap time. If there is very little difference between the soap time and the gel time, collapse of the foam structure occurs before the particles are well knit together (dotted line). If the soap time is too long in relation to the gel time, a broken stringy foam structure occurs (dashed

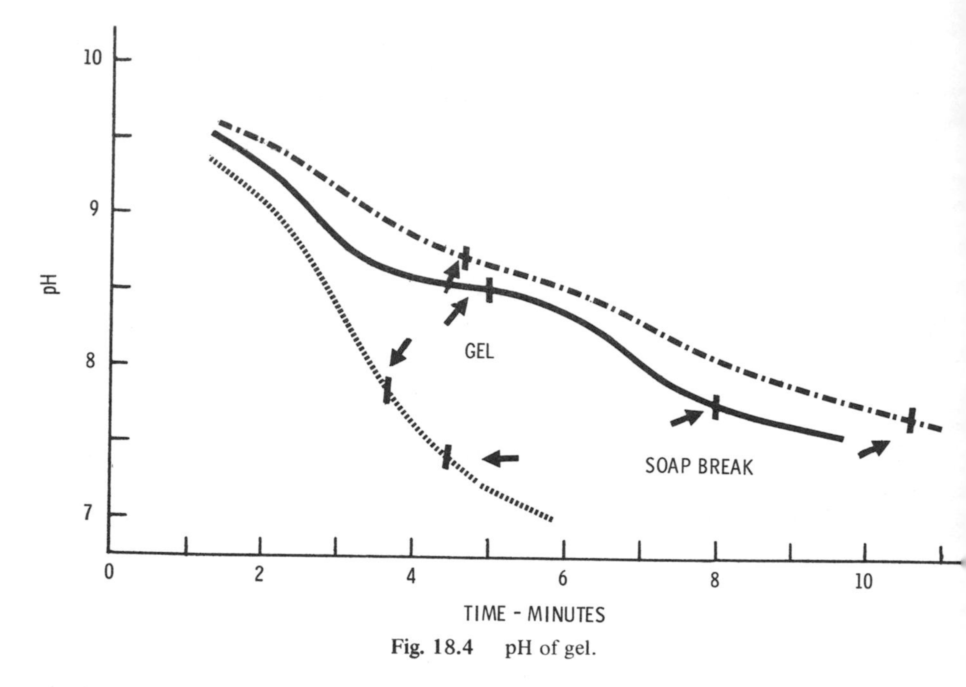

Fig. 18.4 pH of gel.

line). Both of these extremes are to be avoided in making an acceptable type of latex foam rubber.

The foam shrinks after gelation, resulting in minute tears and breaks in the structure. A slightly coarse, thick-walled, spherical cell structure with a minimum of breaks results in the best load supporting foam. Slab stock, where free shrinkage occurs, has fewer breaks but has a flattened or elliptical structure which results in poorer compression resistance.

The gelling agent primarily used in the foam-on-carpet application is ammonium acetate. Heat is used to drive off the ammonia, with the resulting acetic acid causing gelation.

Another method, mentioned above, for making foam-on-carpet is the no-gel process. No gelling agent is used. However, the stabilizer must be carefully selected to maintain the foam structure when heat is applied to dry and cure the foam. Problems associated with the use of a gelling agent are eliminated. However, because the foam must be sufficiently dry before flexing can take place, this process is currently limited to ovens with longer straight passes before any foam bending takes place.

The Talalay process, the second major method of latex foam production, had its beginnings with Joseph Talalay in the mid 1930's.[26] Foaming of the compounded latex was accomplished in the mold by the catalytic decomposition of hydrogen peroxide. In the late 1950's this method of foaming was replaced by a vacuum foaming method. Basic steps are as follows:

1. Compounded latex partially foamed and injected into a closed mold.
2. Vacuum expansion of the foam latex to fill mold.
3. Freezing of the liquid foam latex by refrigerating the mold with glycol/water mixture.
4. Permeation and gelation of the foam latex with carbon dioxide.
5. Heating the mold to vulcanize the foam rubber.
6. Stripping of the foam rubber from the mold.
7. Washing and drying the foam rubber.

A peripheral semipermeable gasket allows vacuum expansion of the liquid latex foam without sucking it into the vacuum line.[27] Foam overflow is eliminated and waste is held to a minimum. Vulcanization takes place at about 110°C. Rapid heat exchange is accomplished by means of long, narrow, closely-spaced pins in both top and bottom mold surfaces. The major problem faced in the Talalay process is one of getting the most efficient and economical heat transfer through large masses of metal.[28]

Talalay foam is characterized by relatively large spherical, fairly uniform-sized thick wall interconnecting cells and an almost skinless surface. This cell type and the presence of the pinholes results in a highly efficient load supporting structure whose resilience is improved by the rapid expelling and taking in of air during compression and release. The Talalay process is by nature limited to molded foam. It has an advantage over cored stock Dunlop foam in this application because of the small diameter pins. It can be molded into standard rectangular shapes called "block foam" and these may be cut into various size cushions without the small diameter pin holes showing through the cover fabric. The trimmed scrap may be used for stuffing arms and springs. Dunlop foam is usually molded to the exact size and shape, resulting in high mold fabrication costs when cushion designs are frequently changed.

Latices primarily used in foam rubber are SBR and natural. In the US, much 100% SBR foam is produced, especially in thinner gauge molded and slab foam. Natural/SBR latex blends are used in very thick gauge products such as mattresses where extra strength and elongation is required to facilitate stripping the vulcanized foam rubber from molds containing large lugs. In Europe the tendency is to use greater amounts of natural latex, up to 50% or more, in blends with SBR. Typical foam formulations are found in Table 18.4.

New innovations in SBR foam include coagglomeration of two or more latices. Thus, SBR and polystyrene latices are coagglomerated by the freeze/quick-thaw method or the pressure impingement method. Improved foam stress/strain properties result.[29]

Thin gauge slab foam can be prepared with carboxylated SBR latices. These latices permit the use of nonsulfur cure systems which result in improved resistance to discoloring, improved aging, and resistance to deterioration from laundering. The latices cost more than noncarboxylated SBR types and this has been a deterrent to their general acceptance.

Foam with oil and solvent resistance has been prepared from nitrile latices.[30] However, this foam has poor low temperature properties and resilience.

Neoprene latex has been used to make foam rubber where solvent resistance or flame retardancy is needed. Mattresses consisting of neoprene foam rubber are purchased on military specifications for use on naval vessels. High cost and high odor level have discouraged most commercial applications. *Cis* polyisoprene latex has been used to replace natural latex in some applications. It results in a more fluid froth and, like natural latex, it contributes to cigarette burn resistance of foam rubber.

TABLE 18.4. LATEX FOAM FORMULATIONS

	Molded Foam	*Fire Retardent Molded Foam*	*Foam-on-Carpet*
	Dry Parts	*Dry Parts*	*Dry Parts*
SBR Latex (68%)	100.0	100.0	70.0
Low Ammonia Natural Latex (62%)	–	–	30.0
Potassium Oleate (18%)	1.0	2.0	2.0
Antioxidant Dispersion (50%)	1.0	1.0	1.0
Sulfur Dispersion (60%)	2.0	2.0	2.0
Zinc Mercaptobenzothiazole Dispersion (50%)	1.5	1.5	1.0
Zinc Diethyldithiocarbamate Dispersion (50%)	0.5	0.5	0.5
Clay Filler (100%)	10.0	–	150.0
Aluminum Trihydrate (100%)	–	10.0	–
Antimony Trioxide/Silica (100%)	–	10.0	–
Antimony Trioxide (100%)	–	2.0	–
Chlorinated Paraffin Dispersion (65%)	–	20.0	–
Tris (2,3-dibromopropyl phosphate) (100%)	–	8.0	–
Tetrabromabisphenol A (100%)	–	4.0	–
Trimene Base Solution (50%)	0.2	0.2	0.3
Diphenyl Guanidine Dispersion (40%)	0.6	0.6	–
Zinc Oxide Dispersion (25%)	3.0	3.0	3.0
Sodium Silicofluoride Dispersion (25%)	2.5	2.5	–
Ammonium Acetate Solution (15%)	–	–	2.0

The fire retarding properties of latex have become a very important consideration in recent years. There are over 540,000 dwelling fires annually in the US, claiming about 6,500 lives. Government legislation is expected under the Flammable Fabrics Act, which may eventually require most dwelling interior material and clothing to have fire retarding properties. Paraffin wax has been used for many years to impart cigarette burn resistance to latex foam.[31] Formulations have been developed (Table 18.4) using halogen-containing additives, antimony oxide, hydrated alumina, and phosphate compounds which impart a higher degree of fire resistance to latex foam.[32]

URETHANE FOAM

Without doubt, the most spectacular family of polymers developed during the last two decades is polyurethane; and the recent accelerated growth of polyurethane foams may be modestly described as fantastic. All of us have, to some degree, the mind of the historian; and when a significant occurrence takes place, we want to find out how it all began, what it is all about, and what the future holds. In order to present this story as completely as possible using the least amount of space, we will in the following order tell you about the essential chemistry involved, something about the history of this development, the types of urethane foams and how they differ, the technology involved, where they are used, and what the future looks like in this industry.

Chemistry Involved

The chemistry of polyurethane foams is very complex and involves many reactions taking place simultaneously and then gradually terminating. The major ingredients consist of a hydroxyl terminated resin, usually called a polyol, and a polyisocyanate. The reaction of these materials results in a polyurethane:

$$\underset{\text{Polyol}}{HO-R-OH} + \underset{\text{Diisocyanate}}{O{=}C{=}N-R'-N{=}C{=}O}$$

$$\downarrow$$

$$\underset{\text{Polyurethane}}{-O-R-O-\overset{\overset{\displaystyle O}{\|}}{C}-NH-R'-NH-\overset{\overset{\displaystyle O}{\|}}{C}-}$$

with the polyurethane being characterized by the urethane linkage:

$$-O-\overset{\overset{\displaystyle O}{\|}}{C}-NH-$$

and this represents the chain propagation reaction.

The second major reaction in the process of making urethane foam is that involved in the reaction of isocyanate and water which releases CO_2 gas to expand the liquid into a foam and which also enters into the crosslinking and chain propagation functions. For simplicity, it is usual to describe this reaction using monofunctional components as follows:

$$R-N{=}C{=}O + H_2O \longrightarrow \underset{\text{Carbamic Acid}}{\left[R-NH-\overset{\overset{\displaystyle O}{\|}}{C}-OH\right]} \longrightarrow R-NH_2 + CO_2\uparrow$$

The amine then reacts with additional isocyanate to form a substituted urea:

$$R-NH_2 + R-N{=}C{=}O \longrightarrow R-NH-\overset{\overset{\displaystyle O}{\|}}{C}-NH-R$$

The isocyanate also reacts with the urea and the urethane to yield biuret and allophanate linkages.

Although water to a greater or lesser degree is used in practically all urethane foams, relatively inert low boiling liquids such as fluorocarbons (monofluorotrichloromethane) and chlorocarbons (methylene dichloride) are used as auxiliary agents to expand the foam. The advantages of these auxiliary gasifiers are many. The isocyanate/water reaction is very complex and in making low density foams the quantity of water would be so high in order to get sufficient CO_2 for blow that crosslinking would be increased to such a degree as to yield a low resilient boardy type of foam. Economics also dictates blowing by inert gas, because it is lower in cost than CO_2 generated by isocyanate. In making slab foam of great thicknesses–of the order of three feet or more–blowing by inert gas reduces the chance of the highly exothermic reaction taking place in such an efficient thermal insulation so as to char or burn the center of the foam. The evaporation of the inert gas tends to moderate the foam processing temperature. Other advantages of inert gas are greater simplicity and control of the reaction. The isocyanate/water reaction is affected by temperature, pressure, mixing efficiency, and other variables that are somewhat decreased with the use of an inert gas blowing agent.

The two major components used in making polyurethane foam are the polyhydroxy resin commonly called polyol and polyisocyanate, of which toluene diisocyanate is the most popular. Also involved in producing foam are catalysts, antioxidants (usually present in the polyol), surfactants such as organosilicones, auxiliary blowing agents, fillers, dyes, and plasticizers. Before going into the story of how these materials are brought together to produce the various types of urethane foam, some of the early and up-to-date history of this development may be of value.

History of Development

The actual chemistry of urethane foam began in 1937 with Otto Bayer and his co-workers of Farbenfabriken Bayer (usually referred to as Bayer) who discovered a new method for the synthesis of high-molecular weight compounds by reacting hydroxyl compounds (glycols) with diisocyanates.[33] In the early fifties several USA companies including Du Pont, Allied Chemical, and Mobay, the last originally being a joint venture of Monsanto and Bayer, produced polyols and polyisocyanates which were used mostly for flexible foams. Several of the rubber companies, including Goodyear and General, began using these chemicals to produce flexible and rigid foam in addition to several industrial products. In 1971 a total of 890M pounds of urethane foam was produced in the USA of which 73%, or 650M lb, were of the flexible type. This is expected to grow at the rate of 15% per year for the next five years.[34]

Types of Urethane Foam

Urethane foams are divided into two main general classifications: flexible and rigid. Although reference is frequently made to semirigid foam to handle materials such as shock absorbent and low resilient products used, for example, in making automobile crash pads, a recent action by ASTM D-11.33 committee on Flexible Cellular Materials has developed a definition for cellular materials that classifies them as either flexible or rigid. A flexible material is described as "a cellular organic polymeric material that will not rupture when a specimen 200 mm x 25 mm x 25 mm (8″ x 1″ x 1″) is bent around a 25 mm (1″) diameter mandrel at a uniform rate of 1 lap in 5 seconds at a temperature between 18°C and 29°C." Obviously, a rigid material is one that will rupture using this method. With this new definition, practically all of the semirigid foams will now be classed as flexible.

Flexible slab foams are divided into two major types: Slab Urethane Foam and Molded Urethane Foam; and two separate methods of testing these foams have been established by ASTM. For slab foam ASTM D1564 is used; and for molded, ASTM D2406 is recommended.

Flexible slab urethane foam is subdivided into many different types of products with varying densities, cell structure, size, and hardness. Molded flexible urethane foam which just recently was a single type is now called hot molded foam to distinguish it from two new arrivals – one called cold molded foam, and more recently referred to as highly resilient foam; and the other called integral skin foam.

Rigid urethane foam is made in a variety of shapes and designs; such as slab, sandwich types as used between metal or wood plates, molded types involving contour shapes for furniture paneling, molded forms for encapsulating products for shipping, sprayed compounds for temperature insulation, and a host of others.

Slab Urethane Foam. Inasmuch as slab foam is the oldest and still the most dominant of the urethane foams, it will be handled first. As George Gmitter and E. M. Maxey disclose,[35] flexible slab foam accounts for two-thirds of the present production of flexible foam. Although polyesters represented the polyol base of early urethane foam production, polyethers in the past fourteen years have taken over practically the whole market. The only polyester foam being made today is used for clothing, filters, and a few low volume items. Several factors were responsible for the change from polyester to polyether including economics (polyethers are lower in cost), quality (polyether results in a foam of great resiliency, and hence improved cushioning), and aging properties, especially humidity aging which is far superior in the polyether urethane foams.

At this time we should give an example of how polyether urethane slab foam is made. As stated previously, the polyol represents the major weight of the composition, and it usually is stated in a weight unit of 100:

Polyol (3000 Molecular Weight)	100.0
Silicone	1.0
N-Ethylmorpholine	0.40
Triethylene diamine (Dabco)	0.05
Stannous octoate	0.45

Water	3.30
TDI 80/20	42.50
Freon 11	Varied
Methylene chloride	Varied

These ingredients are very accurately metered into a mixing head in two or more streams. The polyol may be premixed with the amine activators, the water, and the silicone, resulting in one stream. The stannous octoate, which is a very powerful activator and quite reactive, may be premixed with polyol to form a second stream and the liquid TDI may constitute the third stream going into the mixer head. The fluorocarbon and chlorocarbon may be mixed into one of the polyol streams, or it may be injected directly into the mixing head. In any case, these ingredients are brought together, efficiently and intimately blended; and in a small fraction of a second the slightly milky mixture exits from the mixer onto a moving belt where it completes all of the complex chemical reactions described previously. Very quickly the liquid transforms from a fluid to a more viscous foamy material, and as the exothermic reaction continues the gas continues to expand very rapidly and uniformly until the foamed mass is changed into a nonliquid "bread-like" consistency. As the material expands, the foam at the surface collapses a little, forming a tough surface skin. This is an exothermic reaction and, depending upon the formulation used and the total thickness of the slab or block, the temperature may well exceed 132°C in the center.

The belt travels on an inclined angle away from the mixer-dispensing unit which travels sideways back and forth across the belt to distribute the urethane liquid. The purpose of the inclined belt is to keep the rapidly rising foam from flowing back upon the liquid and partially blown foam that has not as yet reacted to a gelled state. This inclined plane technique results in block foam of good structural properties.[36] The foam then reaches a horizontal stage in the conveyorized system, where it is cut into blocks of varying lengths. The thickness of the blocks may be from a few inches up to six feet, depending upon the capacity of the equipment and the throughput of the mixer. The latter may go up to 1200 lb/min. Many of the industrial machines operate at 500 lb/min mixing capacity.

Density of the foam is determined by the quantity of water and organic blowing agent used in the formulation. By far, the most popular density is between 1.8 and 2.0 lb/ft^3, and this is largely used for seating. Some small quantity of low density foam, around 1.0 lb/ft^3 density, is made for pillow back chair applications and other uses.

The same equipment used for making polyether foam is used for polyester foam, resulting in the so-called reticulated foam used in gasoline containers to diminish sloshing and in air conditioners as filters. The size of the foam cells may be varied by changing the pressure in the mixing head. As the pressure is increased, the color of the liquid flowing out of the head becomes more water clear and the initial foaming is delayed, which results in larger pore size of the finished foam. Air may be bled into the mixing head and this accelerates the foaming reaction, resulting in a cloudy liquid coming from the mixing head, and small-pore foam results.

The blocks may be sliced into desired thicknesses, and practically all of the block is used. The skin is usually sold at a discount and is used in furniture to block out springs. The trimmings find their way into other parts of the chair, and some are ground and used in hassock tops, undercarpet padding, stuffing for inexpensive pillows, toys, etc. The market price of slab urethane foam may vary from 5c/board foot up to 20¢/board foot, depending upon quality and market conditions. Scrap foam has a seasonal variation in value, but presently the price is 12 ¢/lb.

Fillers of various types may be added to urethane foam to raise the density for improvement in cushioning. The use of fillers for reducing cost is not too practical, because the density increase more than offsets the savings brought about by lower compound cost. Plasticizers may be used as well as dyes and coloring pigments in urethane foam.

Molded Hot Foam. A typical molded hot foam formulation is as follows:

Polyol (3000 Mol Wt Triol)	75.00
Polyol (2000 Mol Wt Diol)	25.00
Silicone	2.00
Triethylene diamine (Dabco)	0.20
N-Ethyl morpholine	0.50
Stannous Octoate	0.20
Freon 11	7.00
Water	4.00
TDI (≈ TDI Index of 105)	50.40

As in the slab foam method, these ingredients may be metered into the mixing head in two or more streams. If two streams are used, one would be the TDI and the other the polyol containing the remaining ingredients. Inasmuch as the stannous octoate and the amine react, the mixture must be used immediately. This is accomplished by mixing the chemicals in a small holding tank immediately prior to pumping into the mixing head. A three-stream system would allow the stannous octoate to be mixed with some of the polyol and added separately from large holding tanks. Metal molds, usually aluminum mounted on iron frames and carried by a conveyor pass under the mixing head dispenser, and a metered quantity of the polyol containing all the ingredients is automatically poured into the cavity part of the mold. The liquid is distributed lengthwise along the cavity, as the conveyor moves along. The lid automatically comes down at a specified place on the line to seal the mold. The liquid inside the mold expands rapidly, becoming viscous as it rises to fill out the mold; and following the chemistry earlier described, polymerization and crosslinking take place to make a flexible foam. The mold continues on through a hot air oven; and when it emerges at the stripping station, the mold is automatically opened, the cushion is stripped from it, mold lubricant usually made from polyethylene and wax is sprayed over the two sections of the mold, and it continues on to the pouring section to begin again the cycle.

Automotive cushions may be large, full volume units used for the complete seat of a car, and frequently fabric reinforcing elements and metal inserts have to be positioned in the mold before it is filled. This operation is performed manually by workers stationed immediately before the pouring area. These workers also inspect the mold and remove any small pieces of foreign material that may be present.

After the cushion is inspected and the overflow trimmed, it is ready for the trip to the automobile plant. On an efficient operation, 90% shippable cushions are obtained, with 6% that require minor repairs before being classified as shippable, and 4% that are beyond economical repair and are scrapped. Scrap foam has value, however, and is bulked with the trimmings and sold the same as is done with block foam scrap. Molds before being filled are conditioned to temperatures of 32°C and may be as high as 52°C. If the molds are too warm, the reaction goes too fast and localized blisters are formed which result in an imperfect cushion. Also, if foreign material adheres to the outside of the mold resulting in hot spots, imperfections in the cushion result.

In a normal hot molding line, many different sizes and shapes of cushions are programmed on the unit. Densities may also be varied by programming the ingredients going into the mixer. Following each pour, the mixer is automatically cleaned by solvent being sprayed within the mixing chamber and through the pour pipe. It usually takes seven or more days at ambient temperature before a cushion obtains its maximum physical properties such as tensile, elongation, permanent set retention, and hardness. It usually takes this long before the cushion arrives at the assembly plant to be installed, so there is no need to store cushions at the foam plant for this maturation to take place. For maximum efficiency, the plant manager likes to move cushions from his plant to the assembly plant as quickly as possible; and as soon as they pass inspection, they are on the shipping dock.

Cold Foam or "High Resilient" Polyurethane Foam. Very recently a new foam has been developed which is called cold foam or high resilient foam. The combination of materials used to make this type of foam is basically the same as used for block and hot molded foams with the exception that the polyol is more reactive, being capped with ethylene oxide to establish more reactive sites, and the isocyanate is a polymeric polyisocyanate having higher functionality. This results in a faster buildup of network structure and a higher overall crosslink density in the finished foam. Initially, this foam required warm molds, and high oven temperatures were not required; so it was termed cold foam. New methods have been developed that use some oven cure. Because the outstanding positive characteristic of this foam is its high resiliency, it is now being referred to as "high resilient polyurethane foam." The cushions are usually made at higher densities (3 to 4 lb/ft^3) as compared with hot polyurethane foam (2 1/4 to 3 lb/ft^3). The major raw materials, the polyol and the polymeric polyisocyanate, are also somewhat higher in cost. These two factors result in a somewhat higher manufacturing cost. However, the resilience and breathability of the foam are superior to hot foam and it closely resembles the properties of the still higher cost latex foam.

A typical formulation for cold foam is:

1. Polyol (Molecular Weight 6500)	100.00
2. Water	2.00
3. Tetramethyl ethane diamine	0.20

4. Dimethyl ethanol amine	0.50
5. Freon 11	7.00
6. Polyisocyanate (Mondur MRS) (Isocyanate Index 1.05)	38.30

Items 1, 2, 3, 4, and 5 may be blended together and pumped into the mixer at 27°C. Item 6 is pumped in at 38°C. The mold is maintained at 49°C and after mixing and pouring into the mold, both mold sections being tightly clamped, the flexible cushion may be extracted in 5 minutes or less. As with other foams, maximum physical properties are not obtained for several days. This type of formulation lends itself to the pouring of huge molded sections such as a complete furniture divan. The process may also be used for making block foam.

The development of high resilient foam makes the production of an integral foam cushion for seating more feasible. Presently, the foam core is made separately and then upholstered with fabric or plastic materials. This operation involves quite a bit of labor which is transferred into the ultimate cost of the product. Many forward-looking people in the industry have always envisioned the foaming of urethane in a mold already containing the cover or skin. High resilient foam which requires very little post heating and which has improved cushioning properties offers possibilities here.

Integral Skin Foam. This is a relatively new type of foam that uses essentially a formulation similar to impact resistant foam, except that less blow is used and the mold is relatively cold when filled with the liquid polyol/TDI blend. The cooler surface tends to collapse forming a tough skin of high density foam which becomes an integral part of the foam cushion. The density of the foam decreases toward the center of the cushion. Applications for this foam are armrests, horn buttons, visors for cars, and many small items. In Europe, some of the smaller automobiles use this type of foam for safety crash pads. The major advantage is lower cost of finishing the foam. It may be painted and does not require a cover which usually results in greater labor costs. The major disadvantages are (1) higher overall density, usually between 12 to 20 lb/ft^3, which results in higher raw material cost; and (2) surface discoloration which relegates the use of the product to black or other dark-colored products. If the product is painted, small surface bubbles which are usually present are accentuated. A limited quantity of this foam is finding small application in the automotive industry.

Semirigid Foam. Using such a complex chemical system as is done in making urethane foams, one would naturally expect a large gray area between the flexible and rigid types, and there is. This area encompasses the so-called semirigid foams or high impact resistance materials. The basic chemistry remains the same, but the major building blocks, the polyol and isocyanate, are slightly changed. Whereas the polymer network structure of flexible foam largely consists of long molecules with few crosslinks and that of rigid foam of shorter length molecules with many crosslinks, the semirigids fit somewhere in between. There are two major methods of making semirigids, one being by the prepolymer route and the other being the one-shot process. The latter is by far the more popular method today.

In all of the flexible foams, which according to the ASTM definition includes the semirigids, the prepolymer method may be used; but because of economy in manufacture and versatility in being able to change from one product to another, the one-shot process is preferred. Prepolymer technique involves the mixing together of some of the polyol and all of the isocyanate—and these materials undergo a chain extension chemical reaction—which the foam chemist calls a "prepolymer." This prepolymer, along with the rest of the polyol-containing catalyst, surfactant, etc., are then mixed in the normal manner to make foam. The prepolymer method involves lower exotherm in the foam reaction, and it is somewhat easier to control. Because of the higher viscosity of the prepolymer as compared with that of raw polyol, flow patterns are not as good and large bubbles may be trapped in the foam. When these bubbles are close to a surface, the product is defective.

Most of the semirigid foam produced is used in automobile crash pads and head restraints. The foam has very low resilience resulting from its high hysteresis, and when a fast moving object comes in contact with it, much of the shock is absorbed. The story of crash pad manufacture is an interesting one, and new developments are occurring daily. All new cars sold in the USA are required to have padded dash boards, and all of these are presently being made from semirigid urethane foam. Usually an ABS (Acrylonitrile/Butadiene/Styrene) sheet is vacuum-formed over a mold to the exact dimension of the crash pad desired. This is placed in a mold cavity and the foam is next poured in. The operator then places a metal or rigid plastic frame on top of the mold and the lid is positioned. As in the case of "cold foam," no heat is used to accelerate the cure because this would result in ABS skin shrinkage in places. In a few minutes the crash pad is taken from the mold, trimmed, inspected, and boxed for shipment. You will notice in the cars in which you have ridden that crash pads come in all sizes and intricate shapes. Foam used in this type of product must be sufficiently stable to flow long distances into very thin feather-end sections. It must also adhere well to the skin in order to give good performance. Low temperature and high temperature physical properties must be satisfactory in order to accommodate the vehicle under varying climatic conditions. Presently, the whole interior of the automobile is being considered for impact resistant padding—the door panels, the roofing area, etc. This offers new potential business for crash pad manufacture.

Rigid Polyurethane Foam. Rigid foam is not an elastomeric material, so in this chapter very little space will be devoted to it. The chemistry is essentially the same as that used for making flexible foams, and the equipment is very similar also. It may be processed in slab form, sprayed, and molded into all types of shapes. Simulated wood parts used for cabinet doors, cabinets, and decorative paneling are becoming very popular. Complete furniture units such as divans, chairs, and automobile seats may be constructed where the rigid members are high density microcellular foam, and the cushioning parts are soft flexible foam. The insulation value of rigid foam is excellent, and it is used for making low density insulation panels for refrigerators and buildings. It can be sandwiched between two or more metal slabs to make structural members. During the early history of urethane foam, the rigid type was predicted to grow to tremendous volume because of its ease of application. However, flexible foam outdistanced the rigid foam and only about 1/3 as much rigid foam is presently produced as compared with flexible foams. Presently the future of rigid foam looks extremely good because of the increasingly higher cost of making wood counterparts. Also, rigid urethane foam can be compounded to make it relatively nonflammable; and this factor is becoming much greater in importance.

Equipment Used to Make Foam

The basic equipment used for making urethane foam consists of (1) storage tanks to hold adequate quantities of components used in making the foam; (2) pumps of sufficient accuracy and volume to meter and transport the components into the mixing head; and (3) a mixing head that will very quickly and thoroughly continuously mix the components which then flow onto a belt or into a suitable mold cavity.

There are basically two major types of foam machines; a high pressure type, which is the older, and a low pressure type, which is very popular in the USA.

High pressure foam machines are so named because the foam materials are moved through high pressure pumps and into a static mixing head through special nozzles at pressures as high as 2000 psi. The material streams are thoroughly mixed by an impingement action. Additional mixing may be obtained by the addition of a power-driven agitator.

With low pressure foam machines, the raw materials are introduced into a mixing head at pressures less than 100 psi. Mixing then occurs through mechanical agitation of the two or more streams. When the foam is not being poured, the raw materials are recirculated back through the storage pots.

Textile Coating

Latex carpet and upholstery backsizing consists of a film forming latex adhesive applied to the backside of the carpet or fabric in order to increase the weight, produce a stiffer fabric, overcome raveling of cut edges, improve dimensional stability, and, in the case of carpets or rugs, impart anti-skid properties, lock the pile yarns to the backing fabric, laminate and secondary jute backing, and improve the laying characteristics.[37] SBR type latices are generally used in the applications along with some natural latex. A nonsulfur cured natural latex mix has been developed which gives acceptable backing properties.[38]

Blends of low styrene latices and high styrene, or resin latices, result in the desired softness or firmness, commonly referred to as hand. The latex is compounded with varying amounts of sulfur, accelerators, antioxidants, stabilizers, and fillers. In the past few years, carboxylic modified SBR latices have been used in greater amounts than noncarboxylated. These have good physical properties both in the cured and uncured state. They have excellent tensile strength, dry filler acceptance, aging, binding power, and chemical and mechanical stability. They are

TABLE 18.5. CARPET BACKSIZING FORMULATION

Ingredient	*Dry Parts*
Carboxylated SBR Latex (50%)	100
Antioxidant Dispersion (50%)	1
Clay (100%)	300
Water	To 70% Total Solids
Sodium Polyacrylate Solution (5%)	To desired viscosity

self-curing, and thus need no sulfur or accelerators. If good launderability is a requisite, zinc oxide may be used as a curing agent. Antioxidants, stabilizers, fillers, thickeners, and antifoam agents are also used with carboxylic modified SBR latices. A typical carpet backsizing formulation is found in Table 18.5.

The latex compound is usually applied to the carpet or fabric by means of a roll, knife, or spray. In a typical setup, the compound is applied by a dip roll and is then spread to the desired thickness with a doctor knife. The carpet or fabric then passes over drying rolls or through a drying oven where moisture is removed and the latex size is cured.

This is a large and growing field with a market of about 50,000 long tons per year of SBR latices for carpet backsizing and 6,000 long tons per year for woven fabric.[39]

Nonwoven Fabric

A nonwoven fabric has been defined by the joint Nonwoven Committee of the American Society for Testing and Materials and the American Association of Textile Chemists and Colorists as a "Structure produced by bonding or the interlocking of fibers, or both, accomplished by mechanical, chemical, thermal, or solvent means and the combinations thereof. The term does not include fabrics which are woven, knitted, tufted, or made by the wool felting process."

Nonwovens are not an industry, but rather they are specialized segments of many industries broadly grouped under the following market categories;[40]

1. Consumer market (wearing apparel, carpet backing, poromeric shoe uppers, disposable diapers, blankets).
2. Industrial-commercial market (needled felt products, filters, wiping cloths, carpet backing).
3. Hospital market (disposable sheets, pillowcases, surgical dressings, gowns and caps, uniforms and bandages).
4. Military and other markets (disposable cargo parachutes, tents, sleeping bags, clothing, towels, washcloths).

Nonwoven items may be either disposable or durable. They are made from such fabrics as cotton, nylon, viscose rayon, acetate rayon, polyester, and polypropylene.

There are many methods used to bond the fibers, the use of latices being the most popular. Most of the latices used are of the carboxylated type with acrylics leading the list due to their overall superiority in adhesion, color and light fastness, water and solvent resistance, and availability in a variety of softnesses. Other latices used are nitrile, SBR, polyvinyl acetate, and PVC copolymers.

The following steps constitute a typical process:

1. Formation of the web from the fibers.
2. Saturation of the web with a binder.
3. Drying the web and crosslinking the binder.

The web, formed by any one of a number of processes, is passed through the saturant on supporting conveyor screens and into a hot air oven for drying and setting–about 5 minutes at 150°C. A typical saturant formulation is given in Table 18.6. The carboxylated latices are generally crosslinked by means of melamine resins or zinc oxide.

TABLE 18.6. NONWOVEN FABRIC SATURANT

Ingredient	*Dry Parts*
Medium Nitrile Latex-Carboxylic Modified (45%)	100
Melamine-Formaldehyde Curing Agent Solution (50%)	5
Diammonium Phosphate Acid Catalyst Solution (5%)	0.5
Nonionic Surfactant Solution (10%)	2
Polysiloxane Heat Gelling Agent Solution (10%)	1
Water	To 10–20% Solids Content

The object of the binding process is to coat the fibers only at the point of contact between them. Binder pickup levels may vary widely but are usually in the range of 20-40% of the fabric weight. At less than 30% binder pickup the fiber properties predominate. Above this level it is the binder properties which predominate.[41]

Paper

There are three basic processes involved in the treatment of paper with latex. These are: (1) beater and wet end addition of latices and resin emulsions; (2) saturation of paper and paper plastics with latices and emulsions; and (3) coating of paper with latices and emulsions.

In the beater addition process, fibers are treated with a special pulp beater until specific fiber properties are developed. A stabilized latex of about 2% solids is then added to the pulp so that 3-20% of rubber on the weight of the paper is incorporated. Alum is then added, as the coagulating agent, to bring the system to a pH of 3.5–4.5. The stock is then allowed to stand for 2-3 hours while the rubber is deposited on the fibers. The last step consists of running the fiber onto a paper-making machine and finishing. A number of different latices may be used in this operation, such as styrene-butadiene, acrylonitrile, acrylic, polyvinyl acetate, natural rubber, polyvinyl chloride, and neoprene. If desired, vulcanization agents, plasticizers, colors, antioxidants, and fillers may be incorporated. Zinc oxide is used as an acid acceptor and curing agent and it also increases wet gel strength.

This process is especially useful in making tough, leather-like products, such as imitation leather, industrial filter paper, disposable diapers, shoe parts, tough pliable fiberboard, gasket paper, and packaging paper.[42] Beater-treated papers have properties which are quite different from papers saturated in the dry condition. Because of the greater number of fiber-to-resin bonds produced in beater addition, the papers produced by this method are tougher and better able to retain their strength after creasing. There are, however, a number of disadvantages to this application. These include excessive loss of materials during treatment, slow drainage on the machine, two-sidedness of the paper, foaming, and fouling of machine parts and calenders due to tackiness of the resin. The introduction of new types of resins has in part overcome these difficulties. However, there are still cases where beater addition is unsatisfactory.

Saturation of paper with latices is a method intermediate between the internal treatment of beater addition and the external coating of the paper. Internal treatment causes a change in physical properties while external treatment is used when a surface-protective film is desired. The choice of methods will depend on available equipment and end use of the treated paper.[43]

Because the fiber-fiber bonds are not interrupted as they are in beater addition, saturated papers are in general softer and more flexible than are those produced by beater addition. As with beater addition, a great variety of elastomeric and resin latices may be used, depending on the desired end products. The latices may be compounded with stabilizers, thickeners, antioxidants, water-soluble thermosetting resins, plasticizers, wetting agents, and softening agents.

The latex compound at 5-35% solids can be applied to either wet or dry paper. In a typical process, the latex may be applied by dipping, roll coating, or spraying. An air knife or doctor blade is used to remove the excess resin. The pickup averages about 30% of dry solids on the weight of the paper. Heat is applied to drive off water and increase the adhesion between the resin and fibers. Heat may be increased to the softening point of the resin to obtain maximum burst strength, tensile strength, and elongation.

Coating of paper represents the third type of paper treating process. Paper is coated to improve its appearance, to protect it from soiling, to overcome permeability problems, to impart scuff and abrasion resistance, and to act as a base for printing inks.

Latex is used, together with casein or starch, as a binder for pigments. As with the other two processes, a number of latices may be used, the principal type being styrene-butadiene. Fillers generally used are clay, satin white, zinc oxide, Paris white, blanc fixe, lithopone, and barytes. Clay gives a low dull finish, and satin white results in the highest finish. Other compounds commonly used are

TABLE 18.7. LATEX-STARCH PAPER COATING FORMULATION

	Dry Parts
Coating Clay Slurry (65%)	100
Starch Slurry (20%)	12
SBR Latex (44%)	8
Water	To 40% Total Solids

vulcanizing agents, stabilizers, paraffin emulsions, and plasticizers. Table 18.7 is a typical latex-starch coating formulation.

The latex compound coating may be applied to one or both sides of the paper, which is usually a high density sized sheet. It may be applied by a transfer belt, rotating long-haired brush or in double coating, by submerging the paper under a roll in a trough containing the latex compound. The film is smoothed out with a series of reciprocating brushes, dried with an air blast and festooned. After drying it is calendered to give the desired finish.

Protective Coatings

One of the most significant developments in the protective coatings field since World War II has been the rapid growth of water-base paints. They have many superior qualities such as ease of application, ease of cleanup, good leveling, quick drying, low odor, good durability, and good color retention.[44] They can be used in a great variety of applications such as interior and exterior wall finishes, emulsion masonry paints, semigloss and gloss paints, concrete floor coatings, and roof paints.

A latex paint formulation is made up generally of five groups of compounds: (1) pigment dispersion, (2) protective colloid, (3) preservative, (4) polymer latex, and (5) water. The three major types of latex used in paint compounds are polyvinyl acetate, acrylic, and styrene-butadiene. The polymer in the latex is the main film forming constituent. The use of a plasticizer is necessary with polyvinyl acetate. Vinyl acetate copolymers, as formed with ethylhexyl acetate comonomer, may be used without a plasticizer. The emulsifying agent may be either anionic or nonionic. The protective colloid stabilizes the emulsion, particularly against pigmentation. Since latex paints do not have the pigment binding power of oil-based paints, a high opacity pigment, such as rutile titanium dioxide, is necessary.

It possesses excellent hiding power, durability, and ease of wetting. Other pigments are often used as extenders to lower cost, improve consistency, and improve water resistance. Thickeners are added to increase the consistency of the paint and thereby affect the flow, ease of brushing, and leveling properties. They may also act as protective colloids. A typical formulation for an interior wall paint is shown in Table 18.8.

TABLE 18.8. INTERIOR WALL PAINT

	% by Weight	
	Dry	*Actual*
Rutile Titanium Dioxide	20	20
Whiting	10	10
Sodium Metaphosphate	0.2	2
Methyl Cellulose	0.2	10
Vinyl Acetate Copolymer Latex	20	36
Organo Sulfur Blend	0.1	0.5
Butyl Carbitol Acetate	2	2
Water	–	19.5
	50.50	100.0

Dipped Goods

A great variety of articles such as surgeon's gloves, translucent drug sundries, footwear, flexible squeeze toys, and metal coated compounds are made by dipping processes. These processes include simple dipping where one or more coats are applied with no coagulant being used; the Anode process, where a form is first dipped into coagulant and then into the latex compound; and the Teague process, where the form is first dipped into the latex compound and then into the coagulant,[4 5]. The Anode process enjoys the most popularity.

The latices used are primarily natural rubber and neoprene. These have high wet gel strength, which inhibits "mud cracking" (large random cracks) and cracking in the interior angles during drying on the forms. The latices must be well stabilized and they usually contain antioxidants, sulfur, accelerators, and zinc oxide. Various fillers, oils, and thickeners may be used, depending on the desired product. Commonly used coagulants are calcium chloride, calcium nitrate, zinc nitrate, and acetic acid.

In the Anode process, the form, which is usually aluminum, porcelain, or stainless steel, is first dipped into the coagulant and then into the latex compound which is contained in a dipping tank provided with mechanical agitation and a temperature controlled jacket. The form may be dipped by manual control or automatic operation. It is essential to have uniformity in immersion and withdrawal rates. Care must also be taken to avoid trapping air, which causes pin holes and blisters. After withdrawal of the form, flow may be controlled in many ways, but generally by rotating the form to ensure even distribution of the deposited latex. Leaching, drying, and preliminary finishing operations such as beading or trimming then follow. The latex products may be vulcanized in circulating hot air, steam, or hot water. Depending on the nature of the article, vulcanization may take place on or off the form. If cured on the form, they may be stripped wet or dry. Finishing operations include washing and drying. A typical dipping compound for household gloves is found in Table 18.9.

TABLE 18.9. HOUSEHOLD GLOVE DIP[a]

Ingredient	Dry Parts
Centrifuged Natural Latex (60%)	100
Nonionic Stabilizer Solution (20%)	0.5
KOH Solution (10%)	0.25
Sulfur Dispersion (50%)	1.75
Zinc Mercaptobenzothiazol Dispersion (50%)	1
Antioxidant Dispersion (40%)	0.5
Zinc Oxide Dispersion (40%)	2

[a]Cure: 40 min at 120-127°C in hot air.

Molding and Casting

One of the earliest known practical applications of latex is the molding of rubber articles by causing the latex to coagulate in some predetermined form. It is widely used today in the production of advertising displays, figures, and toys.

A high solids latex, usually natural rubber, is used to prevent excess shrinkage. The latex is compounded with stabilizers, fillers, sulfur, and accelerators, color pigments, and in heat-sensitized compounds, with materials such as ammonium acetate chloride, sulfate or nitrate, or sodium silicofluoride. The amount of filler used, such as clay or whiting, will depend on the rigidity required in the finished article.

Two methods are generally used in the molding process. The first consists of filling a porous mold, such as plaster of Paris or dental stone, with the latex compound. The latex compound is allowed to stand in the mold for a predetermined length of time to build up the required thickness on the walls of the mold. The surplus is then poured out and used again. The deposit is allowed to remain in the mold until sufficiently dry to handle. It is then removed, dired at 60-70°C, and vulcanized at 94°C.

A second method that is widely used is the Kaysam Process, which allows minute reproduction of the interior surface of a nonporous, closed mold by rotational heat gelation.[46] A typical compound used in this process is illustrated in Table 18.10.

TABLE 18.10. SLUSH MOLDING COMPOUND

Ingredient	Dry Parts
Centrifuged Natural Latex (60%)	100
Nonionic Stabilizer (20%)	1.5
Antioxidant Dispersion (50%)	1
Sulfur Dispersion (50%)	2
Zinc Diethyldithiocarbamate Dispersion (50%)	1
Zinc Oxide Dispersion (50%)	4
Kaolinite Clay Slurry (60%)	15
Casein Solution (10%)	0.5
Ammonium Acetate Solution (10%)	2

The latex compound is poured into an aluminum or magnesium alloy mold which is rotated about several axes at the same time. Hot water or steam may be used to heat the mold. The mold is rotated for about 4 minutes at 85°C and then cooled in water. The wet gel is then stripped from the mold and placed in a leach tank for 24 hours to remove soluble material and strengthen the gel. The wet gel is then dried in a controlled temperature and humidity room and cured in an oven for about 1 hour at 105°C. Finishing operations include buffing, painting, and packing.

Adhesives

Latex can be used as adhesives in a wide variety of applications, such as cementing soles to shoes, cementing paperboard to itself and to fabric and wood, cementing secondary backing to carpets, and bonding fabric to leather.

A great number of elastomeric latices may be used for adhesives depending on the end use. Nitrile, SBR, natural, and neoprene latices are commonly used. Natural, nitrile, and neoprene adhesives have good strength. Natural has excellent initial tack. Nitrile and neoprene have good oil and gasoline resistance.

A good adhesive material must have good adhesive strength, i.e., adherence to the material to be bonded, and good cohesive strength, i.e., resistance to internal rupture. Most elastomers have good cohesive strength but may require tackifiers such as terpenes, terpene-phenolics, rosin esters, and resorcinol-formaldehyde resins to improve adhesive strength. These resins also lengthen open dry tack time and reduce materials cost. They are especially necessary in dry combining where the adhesive is applied to the surfaces to be bonded and allowed to dry before compressing the two materials being bonded.

Nonlatex rubber adhesives often contain solvents which greatly improve adhesion. However, the rubber must be masticated in order to combine with the solvents. This mastication reduces molecular weight film strength of the rubber. Latex adhesives may also be combined with solvents, usually added as emulsions, to improve tack without reducing film strength. The solvent cements contain about 12-20% rubber, whereas a latex cement contains a minimum of 35% rubber and will cover 4-5 times as much surface.

Stabilizers, antioxidants, and thickeners may be used in addition to solvents and tackifying resins. If a high temperature application is involved, vulcanizing ingredients are also used. Wetting agents can be used to improve penetration and anchorage on porous surfaces. Sodium silicate and colloidal silica are frequently added to latex cements to promote adhesion on nonporous surfaces such as tile and glass. Alcohol may be added, if solvents are used, to partially destabilize the latex so that the solvent may act directly on the particles. A typical adhesive compound used to bond aluminum foil to paper is represented in Table 18.11.[47]

The processes used are either wet combining or dry combining. In the wet combining processing, a high viscosity latex compound is applied to one of the materials to be bonded with a spreader bar or coating roll. The second material to be combined is then laminated to the coating material and dried, usually on drying cans. The dry combining process consists of applying latex adhesive containing

TABLE 18.11. ALUMINUM FOIL–PAPER ADHESIVE[47]

Ingredient	*Dry Parts*
Neoprene Latex	100
Zinc Oxide	15
Antioxidant	2
Ammonium Caseinate	15-25
Sodium Silicate	0.25
Hardening Agent	As Required

25-50% tackifier to both materials to be combined and letting the surfaces dry. The materials are then laminated by passing between squeeze rolls.

Tire Cord Adhesive

Styrene-butadiene type latices are used to bond textiles to rubber in such applications as automobile tires, rubber belts, hose, and rubberized fabrics.

In tire cord adhesives, the fabric, usually rayon, nylon, or polyester, is bonded to the rubber carcass with a resorcinol-formaldehyde latex (RFL) adhesive. Usually a small particle styrene-butadiene latex such as type 2108 is used. Varying amounts of vinyl pyridine-styrene-butadiene terpolymer latex may be blended with the styrene-butadiene latex to improve adhesion. The resorcinol-formaldehyde resin solution (5-15%) is either preformed or prepared in-situ. The formaldehyde resorcinol mole ratio is usually 1.5-2.5/1. The catalyst for the resin solution is sodium hydroxide. The optimum pH range is 7-9. A blend of sodium and ammonium hydroxide with the pH adjusted to above 10 is reported to be better than sodium hydroxide alone.[48]

A typical formulation for bonding nylon cord to rubber is shown in Table 18.12. This RFL compound is a two-part system consisting of the resorcinol-formaldehyde resin and the latex blend. The RF compound is aged at room temperature for 4 hours and then added to the blend of the two latices, to form the RFL compound. It is mixed thoroughly, but carefully to avoid air entrainment.

The RFL compound is aged for 24 hours at room temperature before use. This

TABLE 18.12. NYLON TIRE CORD DIP FORMULATION

	Parts by Weight	
Ingredient	*Dry*	*Actual*
1. Water	–	304
2. Resorcinol–Technical Grade	11.6	11.6
3. Formaldehyde Solution	4.7	12.8
4. Sodium Hydroxide Solution	1	10
5. 2108 Latex	50	125
6. Vinylpyridine-Styrene-Butadiene Latex	50	125
Total Solids about 20%		

compound is now referred to as the "dip," and it is applied by passing the tire cord through a bath of the adhesive and then through a drying oven. It is then woven into suitable fabric to be used in tire building. The RFL adhesive is refrigerated when not in use. Storage life of the compound is one to two weeks. Adhesive systems for polyester tire cord have been developed using blocked isocyanate-RFL,[49] isocyanurate-RFO,[50] and epoxy-RFL.[51]

Miscellaneous

Rubber molds for casting plaster are made by spraying, brushing, or dip coating a suitable latex compound on a form, then drying and vulcanizing the resulting mold. As little as 3% SBR or NBR latex added to bitumen greatly improves the resulting paved road surface.[52,53] Animal or plant fiber batting can be sprayed with natural, NB, or SB latex compound to result in low cost cushioning material.

REFERENCES

(1) W. D. Harkins, *J. Polymer Sci.*, **5**, 217 (1950); O. Gellner, *Chem. Eng.*, **73**, 16, 74 (1966); H. Merken and D. Phillips, *Rubber Age*, **96**, 863 (1965).
(2) R. M. Pierson, R. J. Coleman, T. H. Rogers, D. W. Peabody, and J. D. D'Ianni, *Ind. Eng. Chem.*, **44**, 769 (1952); *Rubber Chem. Technol.*, **25**, 983 (1952).
(3) T. H. Rogers and K. C. Hecker, *Rubber World*, **139**, 387 (1958).
(4) J. C. Carl, "Neoprene Latex," E. I. du Pont de Nemours & Co, Del., 1962, Preface.
(5) D. C. Blackley, *Rubber J.*, **148**, 7, 78 (1966).
(6) ASTM D1566, Part 28, Rubber, Carbon Black, Gaskets.
(7) E. B. Bradford, J. W. Vanderhoff, and R. Alfrey, Jr., *J. Colloid Sci.*, **11**, 135 (1956).
(8) M. S. Kolaczewski, R. W. Hobson, and J. J. Hamill, US 3,080,334 (1963).
(9) L. H. Howland, E. J. Aleksa, R. W. Brown, and E. L. Borg, *Rubber Plast. Age*, **42**, 868 (1961).
(10) L. Talalay, *Proc. Fourth Rubber Tech. Conf.*, 443 (1963).
(11) B. D. Jones, *ibid.*, 485 (1963).
(12) R. D. Dean, *Modern Colloids*, D. Van Nostrand Co., N.Y., 1948, p. 35.
(13) D. C. Blackley, *High Polymer Latices*, Vol. 1, Palmerton Publishing Co., N.Y., 1966, p. 20.
(14) R. D. Dean, *Modern Colloids*, D. Van Nostrand Co., Inc., N.Y., 1948, p. 217.
(15) D. C. Blackley, *High Polymer Latices*, Vol. 1, Palmerton Publishing Co., N.Y., 1948, p. 97.
(16) E. W. Madge, *Latex Foam Rubber*, John Wiley & Sons, N.Y., 1962, pp. 23-27.
(17) H. C. Jones and C. A. Klaman, *Rubber Age*, **73**, 63 (1953).
(18) J. Bonner, *Proc. Nat. Rubber Res. Conf., Kuala Lumpur*, 11 (1960); F. Lynen, *Rev. Gen Caout.*, **40**, 83 (1963); *Aust. Plast. Rubber J.*, **21**, 29 (1966).
(19) T. H. Rogers, in *Encyclopedia of Chemical Technology*, 2nd Ed., Interscience Publishers, N.Y., **17**, 660.
(20) E. M. Glymph, *Rubber World*, **160**, 57 (1969).
(21) H. P. Brown, *Rubber Rev.*, **30**, 1347 (1957).
(22) F. H. Untiedt, US 1,777,945 (1927).

(23) W. H. Chapman, D. W. Pounder, and E. A. Murphy, Br 332,525 (1929), US 1,852,447 (1932).
(24) E. W. Madge, *Latex Foam Rubber*, John Wiley & Sons, N.Y., 1962, pp. 78-81.
(25) T. H. Rogers, US 2,469,894 (1949).
(26) J. A. Talalay, BP 455,138 (1935).
(27) J. A. Talalay, 2,731,669 (1950).
(28) C. Jennings, *Rubber J.*, **148**, 66 (1966).
(29) K. O. Calvert and J. L. M. Newnham, *Rubber Age,* **98**, 73 (1966).
(30) S. N. Angrove, E. S. Graham, G. Hilditch, R. A. Stewart, and F. L. White, *Trans. Inst. Rubber Ind.*, **42**, T1 (1966).
(31) T. H. Rogers and W. T. L. TenBroeck, US 2,594,217 (1952).
(32) K. G. Hecker, *Rubber World,* **159**, 3, 59 (1968).
(33) *Angowandto Chemie,* 1947, A59, No. 9, 257-72.
(34) Chem. Eng. News, **50**, 8, 10 (1972).
(35) G. T. Gmitter and E. M. Maxey, "One Shot Slab Polyether Urethane Production," in *Polyurethane Technology*, John Wiley & Sons, 1969, Chap. 2.
(36) T. H. Rogers, N. R. Bender, and T. R. TenBroeck, US 2,827,665.
(37) D. C. Blackley, *Rubber J.,* **148**, 8, 82 (1966).
(38) T. A. Mursalo, *Rubber Dev.,* **22**, 55 (1969).
(39) P. E. Hurley, *Rubber World,* **160**, 2, 52 (1969).
(40) G. E. Millman, *American Dyestuff Reporter,* **58**, 7, 32 (1969).
(41) D. C. Blackley, *High Polymer Latices*, Vol. 2, Palmerton Publishing Co., N.Y., 1966, p. 649.
(42) J. P. Casey, *Pulp and Paper Chemistry and Chemical Technology*, 2nd Ed., Vol. 2, Interscience Publishers, N.Y., 1960, p. 971.
(43) J. P. Casey, *ibid* Vol. 3, 1961, p. 1944.
(44) C. R. Martens, *Emulsion and Water Soluble Paints and Coatings*, Reinhold Publishing Corp., N.Y., 1964, p. 7.
(45) G. G. Winspear, *The Vanderbilt Latex Handbook*, R. T. Vanderbilt Co., N.Y., 1954, p. 179; R. J. Noble, *Latex in Industry*, 2nd Ed., Palmerton Publishing Co., Inc, 1953, p. 482.
(46) G. G. Winspear, *The Vanderbilt Latex Handbook*, R. T. Vanderbilt Co., N.Y., 1954, p. 210.
(47) J. C. Carl, *Neoprene Latex*, E. I. du Pont de Nemours & Co., Del., 1962, p. 99.
(48) M. I. Dietrick, *Rubber World,* **136**, 6, 847 (1957).
(49) G. W. Rye and W. D. Havens, US 3,268, 467 (1966).
(50) R. G. Aitken, US 3,318,750 (1967).
(51) M. S. Shepard, US 3,308,007 (1967).
(52) E. A. Sinclair and K. E. Bristol, *Rubber World,* **161**, 3, 67; E. A. Sinclair and K. E. Bristol, *Rubber World,* **161**, 4, 66.
(53) E. A. Sinclair, US 3,254,045 (1966).

19

RECLAIMED RUBBER

JOHN E. BROTHERS
Midwest Rubber Reclaiming Company
East St. Louis, Illinois

HISTORY

It has been quite obvious to every one associated with the rubber reclaiming industry that many changes have occurred in this industry over the past 20 years. Equipment has not radically changed over the past 10 years but new thinking has modified the application of work principles, and changes in the raw material (scrap rubber) continue to pose new and different production problems. These changes have not been too obvious, and to those outside the industry it is possible that these changes have not been observed. Fortunately for the reclaim industry, tire production has tended to stabilize on the use of SBR or natural, or combinations of SBR and natural rubber. As new or modified elastomers have been developed these have been evaluated as tire polymers but none has yet been established as an ideal tire elastomer. Means have been found to live with natural and SBR and coblends of these during our reclaiming operation and each new elastomer developed does raise a question as to "how can this one be reclaimed and processed if it proves to be the ideal tire elastomer?" This is accepted as one of the industry's big problems, when and if a new tire elastomer is developed, and it must be resolved quite rapidly.

Prior to World War II we had natural rubber as the basic source of supply for reclaimable scrap. Since that time the scrap supply has consisted of natural or SBR or coblends of the two with just enough Hypalon, Butyl or EPDM to cause periodic problems on devulcanization and refining. Synthetic rubber (SBR) does not reclaim in exactly the same way nor does it produce the same end product as natural rubber, but it does produce a reclaim that is as satisfactory as that from natural rubber. Should the tire industry ever move to the use of the more saturated polymers like Butyl or EPDM, a sorting program would need to be started due to contamination with or by the highly unsaturated polymers. The use of polybutadiene as a tread rubber does create some problems but these can be resolved.

With the discovery of vulcanization of natural rubber by Charles Goodyear in

1839 there did develop an immediate need for the reclaiming of cured or vulcanized rubber. This was due to the limited amount of rubber available at that time and also due to its high cost. The manufacture of boots and shoes was one of the big applications for rubber, and naturally this was the type that was first reclaimed in high volume. Since the state of cure was not carried to a maximum, this rubber was fairly easily reclaimed by limited heat and some work.

The next high volume user of rubber was bicycle tires and tires for wagons and carriages. This developed into a more difficult reclaiming problem, especially since the reinforcement fabric had to be removed. Woven fabric or cord fabric made from cotton was used in pneumatic tires to strengthen the sidewall and minimize shape distortion in service. The fabric was removed by cooking in a wooden or lead vat using diluted sulfuric acid followed by the reclaiming methods available at that time. The heater or Pan process was patented in 1858 and the Digester process was patented in 1889. In order to produce the best reclaim, the fiber was first removed and then the rubber was heat softened and milled. It was a combination of the Digester process and the availability of bicycle tires that led to the Marks-alkali process around 1890-1900. Around 1900-1910 rubber tires were placed on pleasure vehicles (buggies) and the stage was set for the final development of a rubber tire, for the horseless carriage. From then on there was a steady and growing supply of a uniform raw material for the reclaimed rubber industry. As the automobile industry developed, the demand for reclaimed rubber developed and the stable source for scrap, from tire and inner tubes, did result in major uniformity and opportunity to develop and maintain specifications. Also, the tire and tube scraps became more available until today one of our major problems is to dispose of the excessive quantities of discarded tires rather than searching for the type scrap desired for reclaiming.

It seems that there have always been problems relative to the removal of fiber from the volume scraps, and today this is accomplished either by hydrolysis or mechanical separation. As the automobile industry expanded, tire scrap became more and more plentiful and the competition between tire manufacturers did produce a more and more standardized automobile tire. As a result of this standardized source, the reclaimed rubber industry developed so that it could produce an end product capable of meeting rigid standards. Current practice and current scrap availability does permit us to ignore specialty items as a source unless our customer requests some of these specialties for a unique application. It is an accepted truism that as any raw material becomes more and more uniform, and as competition enters into the picture, there will always be a gradual upgrading of quality. The rubber reclaiming industry has also benefited by such historical experience.

In the early stages of reclaiming, cotton fabric was the only fabric used and this was separated from the rubber by air or by acid treatment. The acid treatment did not remove free sulfur in the compound, so the exposure was maintained at such a temperature that cure was not continued. After the alkali process was introduced this method of defibering was discontinued except for specialty reclaims.

In the late 1920's and early 1930's acid cooks were used for the following specialties. The first was to remove the fabric from an uncured rubber compound in order to salvage the rubber portion in the uncured state, and the second was to produce a very dry and smooth reclaim for automobile topping and insulated wire. The latter was a two-stage operation requiring first the fabric removal from the fully cured scrap followed by the Pan Heater method. By this time rubber compounding techniques had been improved to the point that there was very little free sulfur left in the scrap to be reclaimed, so that the alkali added neutralized any residual acid and produced a very dry devulcanized scrap that could be refined many times to develop an extremely smooth reclaim. At the present time the automobile topping application for reclaimed rubber has completely disappeared from the production scene and very little reclaim is used in insulated wire.

Another innovation that has developed in the industry over the past two decades has been the gradual replacement of cotton cord or woven fabric in automobile tires by rayon, nylon, polyester, and glass fiber, to mention a few of the fabrics that have shown the greater activity. All of these fibers were studied and production methods developed to remove these fibers from top grade reclaims. It will be rather obvious that not all of these fibers can be completely removed from the rubber during the devulcanization and refining process. Some of the latter (like glass) will pulverize and be present in finished reclaim as an additional load on the rubber hydrocarbon.

As more types of synthetic rubber became available, the industry tried to determine where these could be used to advantage in regard to both cost and quality. It was quickly noticed that butyl rubber had an ability to do a far better job of confining air than did natural rubber. This advantage was exploited in a shift of elastomer in the manufacture of inner tubes from natural rubber to butyl. Shortly thereafter the tire industry developed and introduced the tubeless tire for passenger cars. With overall acceptance of this tire, it was thought that the reclamation of inner tubes would drop off to a very low volume. This did not prove to be the case since reclaimed butyl tubes did find a major end use in the compounding of the inner liner of the tire. Due to the difference in unsaturation between butyl and SBR or natural the two types of elastomer cannot be coblended. This resulted in a problem relative to the inner liner compound that was eventually resolved by a shift to chlorinated butyl or reclaimed butyl. Both of these materials will coblend with natural or SBR. The use of the saturated polymer, in the tire, created some problems but these were not too serious due to the small quantity present. With the bus and truck tires still using butyl inner tubes, and with passenger car owners still buying some inner tubes, there appears now to be an adequate supply of butyl scrap to meet the demand.

It has been established that a time period of four to five years will lapse between the manufacture of a given tire, its sale, its use, and then its availability to the reclaimed rubber manufacturer as scrap. Of course, if factory rejects are used, the reclaimer will be able to process and still consider those problems that will present themselves when today's production changes show up as available scrap several years hence. This does enable the reclaimer to have some of his problems resolved in part before the volume production problems develop. A case in point could well be the belted tire using glass fiber as well as polyester fiber. As the industry looks forward into the 1970's, there are many changes coming along that will modify some of our methods, and probably some of our present end products.

Even though we assume that the cycle on the average will take between four and five years, we do know that in many instances it will require twenty or more years for the scrap to find its way back to the reclaimer. Production on both black and red natural rubber tubes, in any volume, was discontinued many years ago, yet the reclaimers have purchase orders for these materials that are active and reissued even now. They cannot be classed as high volume items, but these scraps are still available in the trade.

Prior to World War II, as mentioned earlier, our raw material was natural rubber. With the advent of World War II and the tremendous upsurge in synthetic rubber production, there was an interim period where tires were made in part from natural and in part from SBR. This did develop tremendous problems for the reclaimer and much work was done to resolve these problems in our softening and defibering processes. Different chemicals were used and on some occasions mechanical processes were changed.

The above gives a bit of history relative to changes that have occurred in the industry, but might not be too readily recognized since they have been gradual with no real drastic change at any one time.

In addition to natural rubber, SBR, and butyl, there have been a large number of special elastomers developed for uses other than tires and tubes. We have proven that most of these are salvageable and can be processed and revulcanized, but they do not have a major place in our industry, since the volumes are either small or the original end product so variable due to specification that no high volume standardized scrap is available. Most of these elastomers are processed on a job basis for special reuse or application.

DEFINITIONS

If the terminology used in the industry is considered, it might be concluded that devulcanization is the reverse of vulcanization. Such conclusions can not rightfully hold true since none of the combined sulfur from the rubber compound is removed as the scrap is devulcanized. One explanation offered is that a break in the crosslinked rubber molecule is developed as the scrap is depolymerized. This means that a shorter chain structure is probably produced with additional double bonds that are then more readily available for further sulfur crosslinkage as the reclaim is used. Vulcanization or crosslinkage results from the interaction of the elastomer with curing agents, and this will normally occur at a location in the rubber molecule that has a double bond (unsaturated linkage) in the molecular structure.

When the original compound was developed, there was a definite relationship established between the compound, the time and temperature of cure, and the size and thickness of the part produced. If there are several compounds used as an assembly to produce one part, the rate of cure is so controlled that all compounds will reach the same state of cure at the same time. A similar problem does develop as vulcanized scrap is depolymerized. Dependent upon type of scrap, and size and shape, it will need to be chopped or ground to reasonably uniform size, so that the heat penetration can be controlled throughout the mass that is being devulcanized. Naturally, the smaller the size developed the more uniform will be the heat penetration, and the more uniform will be the devulcanization. There are commercial limits to which such grinding can be carried. It is also known that rubber is a good heat insulator and the heat must penetrate the rubber uniformly from all sides. During the devulcanization process, the heat will penetrate the swollen rubber scrap and differences in softness will develop between the outside of the ground scrap and the center. During refining this center portion is the source of tailings. The larger the chunk of rubber to be devulcanized, the greater will be the difference between the surface and the center.

The original definition for reclaimed rubber as written by J. M. Ball in the first edition of this book will be quoted: "It is the product resulting from the treatment of vulcanized scrap rubber tires, tubes and miscellaneous waste rubber articles by the application of heat and chemical agents, whereby a substantial 'devulcanization' or regeneration of the rubber compound to its original plastic state is effected, thus permitting the product to be processed, compounded, and vulcanized. Reclaiming is essentially depolymerization; the combined sulfur is not removed. The product is sold for use as a raw material in the manufacture of rubber goods, with or without admixture with crude rubber or synthetic rubber."

METHODS OF RECLAIMING

In order to clarify the various methods currently used in softening or depolymerizing cured rubber, it is felt that each one of these should be given a limited amount of attention. It is probably best that these be described at this time, then followed by an overall description of the manufacturing process. The main ways of depolyermizing rubber scrap are as follows:

1. Digester
2. Heater or Pan
3. High Pressure Steam
4. Banbury
5. Reclaimator
6. Dynamic Steam Devulcanization

The first two methods of manufacturing, termed long-cycle processes, might take all the way from six to twelve hours to devulcanize, while the latter four might require times of five or six minutes to two hours. Some of these are batch methods and some are continuous methods. It is axiomatic that the larger the particle size of the rubber to be devulcanized, the longer the time required to soften or penetrate. It must also be realized that like every other chemical reaction, the higher the temperature of reaction the shorter will be the time required. Economics will enter into the picture at this time and decisions must be made relative to how much money can be spent on size control or reduction as measured against time and effort spent in subsequent manufacturing operations. It is well recognized that the finer the scrap is ground, the more uniform will be the devulcanization, and the less work will need be applied in the mill room—and the mill room is the point at which the major portion of the cost is expended. Included in the preliminary stage prior to devulcanization must also be the matter of fiber separation. At this point it must be resolved whether a reclaim will be produced having all of the fiber retained and refined into the rubber as a portion of the reclaim, or whether the scrap will be chemically treated to remove the fiber, as is normally done in the digestion method, or thirdly, whether the fiber will be mechanically removed prior to fine grinding for the other devulcanization systems. If fine ground rubber is to be devulcanized, the grinding operation can be speeded to a very major extent by mechanically removing the fiber. This will also upgrade the final reclaim. Originally the digester would hydrolyze all of the fabric and remove this fabric during the depolymerization, but this method will not completely remove all of the fibers currently being used.

The Digester Process. In the Digester Process, coarsely ground scrap is used which normally contains all the fiber in the tire. The digestion method is a wet process using scrap that has been ground to around 3/8 to 1/4 inches thick. If scrap is used that is too finely ground, too much of it can be lost in the subsequent washing and dewatering operations. As the scrap is loaded into the autoclave, chemicals are added that will swell the vulcanized rubber permitting easier and more uniform depolymerization, and some additional chemicals are added that will dissolve a major portion of the fabric. As pointed out earlier not all fabric can be completely removed. To accomplish this, a dilute solution of sodium hydroxide is used with natural rubber and a dilute solution of a metallic chloride is used with SBR. The discovery that these chlorides could be used in conjunction with SBR scrap to meet this requirement was a major breakthrough since the use of sodium hydroxide did produce an unsatisfactory reclaim with SBR by this method. The problems of grinding in the manufacturing process will be touched upon but at this point it is enough to say that the tire bead is removed to a major degree, leaving only a small residual amount of metal that is removed by magnets and fine wire screens later on in the process. After the scrap has been chopped or ground to the point that it will pass through the sorting screens that control maximum size, the scrap is conveyed to the digester. The digester is a large jacketed autoclave that will accommodate a high volume of scrap (around 5000# to 6000#), chemicals, peptizers, and the water and defibering agents. The digesters have a series of paddles spaced along the drive shaft, that are continuously driven at a slow speed throughout the cook, to maintain the charge constantly in movement for uniform heat penetration, as well as to keep all of the charge under the established water level. These autoclaves are jacketed, and this prevents dilution of chemicals by water condensate as the process goes forward. Established times and temperatures are specified and this, in conjunction with scrap size and chemicals, results in a uniformly softened rubber particle. The digester process is strictly a batch process. The steam pressure in the autoclave can be varied due to factory conditions, but generally pressures around 200 psi are used with a digestion time of eight to twelve hours. During this time the softener and peptizer have been given an opportunity to react with the rubber at this temperature and the fiber becomes hydrolyzed.

At the end of the specification cook, the charge is blown down under pressure into a blow-down tank containing a given amount of water for dilution and washing. From there the charge is pumped onto vibrating screens which permit a major flow-through of water and hydrolyzed fiber, and allow a water spray to wash further. The scrap is then conveyed to a dewatering press and from here it is conveyed into a dryer to remove the water down to a controlled percentage. A bone dry scrap is not produced, since this will cause major difficulties in the later refining operations. However, water must be controlled to an established limit. A small amount of residual water is necessary to prevent build up of excessive heat in the refining step. This acts as a control of oxidation and actually assists in pulling together the mass of devulcanized scrap. In many instances the material that passes through the vibrating screens as a discharged slurry is in turn picked up and filtered through a smaller mesh screen to save the rubber fines that would originally pass through the coarse separating wash screen. Finely ground scrap can be handled by this method but finer screens are required to minimize the loss.

The alternate depolymerizing methods will be described at this point, while the preparation and finishing operations in producing the finished product will be described later.

The Heater or Pan Process. This particular process is reasonably simple but the tire scrap must be ground to a smaller particle size than in the digester. After grinding, the scrap is treated in a blender with the required chemicals and conveyed into pans which are of such size that they will fit the diameter of the heater used, when stacked on top of each other. Pans are filled with the treated scrap to a depth of six to eight inches. An alternate to this is to devulcanize in a container having approximately the size and shape of the interior of the heater, and equipped with vertical separators having perforations approximately every six or eight inches. These separators are so established that they will permit steam to circulate from the bottom to the top of the scrap through the interior of the scrap, so that all of the treated ground scrap will be processed to approximately the same end softness. Since this is not a method whereby the scrap is continually moved, all the precautions necessary to permit the steam to permeate the entire mass uniformly must be taken, thereby preventing layers of scrap to develop which will result in nonuniform softness. If the latter situation is permitted to develop, variable reclaim quality will result.

In this method of devulcanization, live steam is introduced to the heater to make direct contact with the scrap. Normally, steam in the range 200-225 psi is used but this will vary depending upon factory steam available and vessel strength. After devulcanization, the heater is blown down to a zero pressure, unloaded, and the scrap removed from its container. Normally, this scrap will be in chunks similar to the layers of a cake. After these are removed they are subjected to equipment that will break up the cake and the scrap then is conveyed to the mill room for refining. No drying is required in this method of devulcanizing and the small amount of water remaining will assist in the refining operation.

High Pressure Steam Process. This devulcanization process is one wherein the scrap has been ground to size approximating that of the digester method. The scrap is loaded into a high pressure vessel and subjected to steam pressure in the range of 600 to 1000 psi. The scrap is maintained in the vessel for periods varying between one and ten minutes and then the pressure is quickly released with the scrap blown into a cyclone, to dissipate the pressure. It then falls into an area whereby it can readily be subjected to a squeezing, dewatering pressure and dried. This again is a case where the steam is in direct contact with the scrap, but there is no effort made to keep the scrap in motion while it is being devulcanized.

Banbury Process. The Banbury Process of reclaiming rubber is one wherein fine ground scrap is charged into the Banbury and by use of high air pressure on the ram and a high speed shearing action between rotor and shell, temperatures in the range of 450 to 550°F are developed. Again this devulcanization is dependent upon work and temperature. The cycle will normally require 5 to 12 minutes depending upon the types of scrap and other Banbury operating conditions. Modifying agents can be added if desired. The scrap is cooled and dropped into refiners for the finishing operations. Quality of the end product can vary depending upon the type of scrap used. If fiber-free whole tire scrap is used, a high quality reclaim can be produced, but if the fiber is left in the scrap a lower quality will result.

The Reclaimator Process. Aside from the digester and heater processes the Reclaimator Process is the method used to produce the major volume of reclaim. It is the newest method used in the industry, although it is approximately 17 years old. This method uses fine ground scrap and produces reclaim by working the scrap at elevated temperatures between a screw and a barrel type body and a delivery head. This equipment has quite close clearance and very limited overall size tolerance. The depolymerization relies on working the scrap at high temperatures and can be classed as a continuous or nonbatch method. It takes advantage of a rapid initial drop in plasticity and in turn controls the work quite closely to take advantage of this initial dip. Temperatures generated are in the range of 350 to 400°F, and time requirements range between one and four minutes. The quality produced is comparable to that from any of the other methods but time and temperature tolerances must be controlled far more rigidly. After passing through this depolymerization cycle the scrap is finished in a normal manner.

Dynamic Devulcanization. This title is used here for lack of a better name. This method is essentially the same as the Heater or Pan process but many of the disadvantages of the heater have been eliminated. It has long been recognized that one of the major problems in the Pan or Heater method was one of nonuniform heat penetration throughout the mass of ground scrap as it remained static in the container pans as devulcanized in the heater. This tends to develop layering or nonuniform devulcanization in the pans. At times the steam and heat would fail to penetrate the large mass of scrap, and this resulted in harsh stock that caused poor refining. Also, at times, softeners could concentrate in the center of the cake, causing an extra soft layer. These factors would result in nonuniformity.

Efforts have been made by many people to correct this situation, and several remedies are now available. The same general method of devulcanization is being used whereby softeners or peptizers are added to the fine ground scrap, and this scrap is then devulcanized in an autoclave in a steam atmosphere. The ground, treated scrap is kept moving throughout the cycle to eliminate layering, and higher steam pressures or temperatures can be used in order to reduce the time required.

Two general approaches are used in regard to the scrap movement. In one approach, the autoclave is mounted in such a manner that it will rotate throughout the cook. This has also been modified so that the direction of rotation can be changed if desired. As the autoclave is rotated, the stock is baffled so that it can change position as it is cooked.

Another approach is to keep the autoclave stationary but to have a series of screw type agitators on the inside that will keep the ground scrap constantly moving. Screws are so set up that the scrap is moved to the ends of the chamber by one series and then returned to the center by a second series. The autoclave is not completely loaded and, as a result, a waterfall effect does develop which will change position of the scrap, continuously, in a still different manner.

Extremely uniform depolymerization is effected when fine ground scrap is used. Usually this equipment is built to withstand steam pressures up to 500 psi so that almost any reasonable time cycle can be developed once temperature has been raised to operating efficiency. The reclaimer can also control the size of this autoclave so that the most efficient sized batch can be used. With proper control of scrap size, chemicals, time, and temperature, a scrap can be devulcanized that will for all practical purposes produce no tailings in the mill room. As described earlier, tailings are the center cores of the devulcanized rubber chunks that are not as soft and plastic as the outside. This results in a lump that will not join the main matrix of the refined sheet but ends as an unsmooth area in the sheet.

Comments on Scrap for Devulcanization. Each individual reclaimer has resolved which scrap is the best type for the production equipment available in any particular plant. It is recognized that as of today there is available to the reclaimer a far greater quantity of all desirable scrap than will be consumed. As a result the reclaimer can be critical of the scrap that is received and can so specify the types desired. This means that production may concentrate on passenger tire scrap up to a given size if desired, or truck and bus tire scrap up to a given size. In addition to this, there are tractor tires and other off-the-road type tires in limited quantities that can be considered for specialty type reclaims. By experience it has been learned that certain types of scrap contain either natural or SBR blended together or all natural or all SBR. Coblended scrap containing these two basic elastomers can also be readily devulcanized. If part of a tire is made of 100% natural rubber and a part of the tire is made from 100% SBR, major devulcanizing problems will occur and an effort is made to stay away from this type of scrap. It is known that factory rejected tires will result in manufacturing problems unless they are handled as a specialty item. The reason for this is that this type scrap has never been exposed to aging or oxidation and does have a far heavier tread on the tire that can cause problems relative to devulcanization and stain characteristics. At this point it should be mentioned that our customers are extremely conscious of stain and nonstain characteristics and this requirement must always be kept in mind as reclaimed rubber is produced.

OVERALL MANUFACTURING DETAIL

Cured rubber scrap, as discarded by society, is purchased and scheduled into the rubber reclaim plant. As many labor saving devices as possible have been introduced into the manufacturing operations to keep hand labor at a minimum. A decided effort has been made to schedule tire scrap into reclaim plants so that incoming cars or trucks of tire scrap can be unloaded directly into the feed lines to start the reclaiming operation. In this manner, double handling of scrap into the yard and then back into the manufacturing area can be minimized. Even so, each plant must maintain a given inventory of used tires in their yard so that operations can be continued regardless of various situations that may occur. Shipments of tires are delayed in winter time, and yard tires occasionally must be brought in to satisfy production requirement and these in turn can have a major amount of snow, water, or ice trapped on the inside of the casing. When this occurs, correction must be made for moisture in order to properly produce specification devulcanized scrap.

All tires do contain materials that must be removed sometime during the manufacturing process. It is established routine to remove all metal present either as bead wire or tramp metal and all fabric is removed on the first line tire reclaims.

Whole tires are fed into large cracking mills. These are similar to the typical mills used in the rubber industry except that they are more rigidly reinforced to withstand the initial thrust of the cracking operation, and in addition corrugated rolls are used to permit the grabbing of the tire without the slippage that might be expected in the bite of the mill. There is a major difference in the speed of these two rolls as they work against each other. Whole tires are dropped into this cracking unit and after one pass they are conveyed onto a screening shaker. This machine has a sizing screen and all ground rubber small enough to pass this screen is conveyed to a second grinding unit that is identical to the first but with rolls pulled closer together for finer grinding.

There is a sizing screen following the second grinding operation and the mesh opening is of such a size that scrap through this screen is small enough to be delivered either to the digesters for devulcanization or to another area where the fabric is mechanically separated from the rubber. (See the following section: "Mechanical Separation of Rubber and Fiber.") The ground scrap scheduled to the wet digesters is fine enough at this point to permit the chemicals to penetrate and chemically defiber. If the scrap is scheduled to be fine ground, the fabric is mechanically removed at this point to facilitate the grinding operation.

At each of the screening operations mentioned above, all scrap too large to pass through the sizing screen is automatically returned to the particular grinder and is continually recirculated until small enough to pass the sizing screen. In addition to the grinding and sizing operation described, there is an operator stationed behind the first grinder to manually remove the major volume of metal that was in the original tire as the bead. Any small pieces that go through the screen will in turn be separated by one of the various magnets in process or in the final screening operation given all reclaimed rubber prior to the finishing refining.

Mechanical Separation of Rubber and Fiber. The ground rubber and fiber mixture from the first surge bin is now conveyed to a shaker table that has three separating screens in place. All of the ground scrap that will not pass the first of the three screens is classed fiber with slight rubber contamination. This is low percentage and is usually air conveyed to a container that will eventually be routed to a truck for hauling away and scrapping. The material that goes through the third (or fine screen) is practically fiber-free rubber, and is conveyed to a holding or surge tank pending delivery of scrap from the center screen. Scrap from the center screen is now routed to a vibrating separation table. By air pressure, tilt of screen, and vibration, this portion is separated into three parts. The loose fluffy fiber is permitted to move across the screen and collected with that portion already designated as "scrap." The heavier portion (rubber) is vibrated off into another channel and routed over to the all-rubber portion already collected, while the

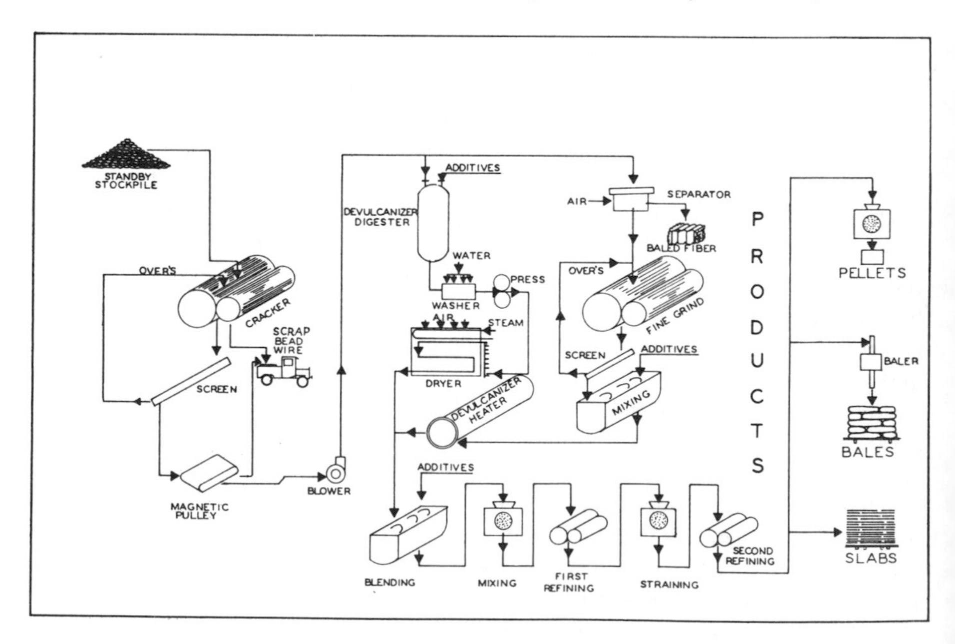

Fig. 19.1 Flow diagram.

center cut is rerouted back over the separation table. This is a continuous operation with fresh scrap being added to the rerun at all times.

The scrap rubber from the surge tank that has been fiber separated is now routed over to the fine grind equipment. Since the major portion of the fiber has now been mechanically removed, screens with much smaller openings can now be used and more efficient grinding can proceed, since the fiber will not keep plugging the openings of the separating screens. As indicated earlier, as the depolymerizing action moves away from the digesters, finer ground scrap is used since this is more conducive to uniform devulcanization and also the potential waste in rubber need not be of major concern, since the devulcanizing systems using fine ground scrap do not require the wet process with subsequent washing and drying. The fine grinder is essentially the same piece of equipment as the original cracking unit except it is lighter in weight and will normally run at a higher speed. This particular grinder is again connected with vibrating screens and the material through the screen has been ground fine enough for devulcanization. The material that does not go through the screen is again rerouted back through the cracking unit, and will be permitted to recirculate with new scrap as long as required to grind it.

This fine ground scrap is then conveyed to a Baker-Perkins type mixer, or some other similar type mixer, and the required chemicals are added for devulcanization. If no chemicals are to be used, the scrap can move directly to the devulcanizing equipment for further processing.

Mill Room Operations. After devulcanization the scrap is then conveyed to the mill room. The first step in the mill room is one of warm up of the scrap or the addition of certain pigments that will enhance the quality. This can be done in Banbury mixing equipment, open mills, screw blenders, mill strainers, or mixer strainers. All of these will plasticate the scrap and prepare it for the first, or breaker refiner.

A breaker refiner is a two roll mill similar to a mixing mill except that the rolls are limited to around a maximum of 36″ long so that when the rolls are pulled tight together there will be a minimal distortion due to the pressures developed. The rolls are smooth and accurately crowned to do the job required. The speed is high (around 40 rpm) and the two rolls are of different diameter and rotate at different speeds so that there is a high friction ratio between the two which tends to iron out the stock to a smooth clean sheet. These rolls are water cooled so temperature can be controlled. Rolls are set to produce a sheet of reclaim approximately .012 inches thick.

The sheet is dropped into a screw conveyer which carries the reclaim to a strainer. The strainer is similar to a rubber tubing machine except that it is built so that all reclaim is pushed through a wire screen to remove the last of the metal. After straining, the reclaim is conveyed to a second refiner called a finisher. This is the same type machine as the breaker, described above, except that it is normally set to deliver a sheet of clean reclaim between .004 and .006 inches thick. Again cooling water is required to maintain operating temperature.

This thin sheet of reclaim is scraped from the face of the roll and allowed to build up on a wind-up roll, layer on layer, to a thickness of approximately one inch. At that time the sheet is cut the length of the wind-up roll and removed. This then is a slab of reclaimed rubber weighing approximately 30 to 35 pounds. The slabs are piled on skids until the total weight is 1500 to 2000 pounds. Each slab is dusted with soap stone to prevent slab adhesion. The product is then ready to ship to a customer, after quality control approval.

As an alternative to this slab process the sheet can be air conveyed to a cyclone and permitted to drop into a baler where bales of controlled weight are produced, dusted, skidded, and shipped. A third alternative is to route the refined sheet into another strainer (pelletizer) where the reclaim is pelleted and dusted and then bulk shipped. A fourth method is to convey to a strainer or tuber and produce an extruded log or slab for shipment.

Due to recent improved methods of devulcanizing it is possible, with the fine ground scrap, to develop a clean reclaim with only one refiner pass rather than two.

ADVANTAGES OF RECLAIMED RUBBER

In reviewing the various advantages relative to the use of reclaimed rubber it is felt that the first step to take is to outline the area where the major quantity of reclaimed rubber is consumed. The most recent figures are those that have been accumulated by industry use in 1968. Table 19.1 indicates the end products and the percentage of the total reclaimed rubber used in each market. This table was obtained from a talk presented by Mr. Allen E. Crapo of Uniroyal Chemical Division, Uniroyal, Inc., in 1967 but updated in 1969:

TABLE 19.1. RECLAIMED RUBBER MARKET

End Product	*1968*
Tires and Tire Repair Materials	67.0
Inner Tubes	5.2
Hard Rubber Battery Boxes, Covers, Steering Wheels	1.3
Auto Mats & Auto Mech. Goods	10.0
Heels, Soles, & Rubber Footwear	2.0
Cements & Dispersions	3.2
Hose, Belting & Packing	4.3
Mech. Goods–Other than Auto	4.1
Rubber Surfacing Material	0.9
All other uses–toys, proofing & insulated wire	2.0
	100.0%

It can readily be observed that the major portion of reclaimed rubber is used in automobile tires, and the next area of high use is in automobile mats and automobile mechanical goods. At one time one of the prime reasons for the use of reclaimed rubber was to reduce the raw material costs of rubber compounds. With the advent of SBR the wild fluctuation in the price of natural rubber has been controlled, and with the potential world wide production capacity of SBR, and modified SBR, at its present level, a major price increase in this commodity cannot be visualized. At the present cost level of natural rubber and SBR it can no longer be said that there is a pound cost reduction possible in the utilization of reclaimed rubber, but the sale of reclaimed rubber is continuing and the advantages accrued are in the area of processing. This would naturally exhibit itself in the finished cost of an article and still not show as a saving in the pound or pound volume cost. With the increased use of oil-extended elastomers, which does enable the synthetic rubber industry to produce in volume at a lower cost, the advantages of the use of reclaimed rubber need be reevaluated.

In June 1968, the Rubber Reclaimers Association, Inc. issued Technical Bulletin #10, which in part covered the advantages that could be obtained by use of reclaimed rubber. The summary of these advantages as covered in the bulletin are shown in Table 19.2.

TABLE 19.2. ADVANTAGES OF RECLAIMED RUBBER

Low Material Cost

Low stable price
Reduced accelerator levels

Uniformity

Low Processing Costs

Rapid break down and mixing
Short curing cycles
Low power consumption and minimum peak power demand

Improved Processing

Low nerve with minimum shrinkage and swell
Increased calendering and extrusion rates
Blister-free calendering at heavy gauges
Low thermoplasticity
Fast solution in high solid mastics

Reduced Scrap

Nonscorch compounds
Minimum stock recycling
Fewer rejected parts

Good Product Performance

Good aging
Minimum reversion
Excellent air retention (butyl)

The low material cost advantage is used not necessarily to achieve a lower pound cost but to keep the compound in the same general cost range and still maintain the other advantages. The price of reclaimed rubber has been quite stable for the past 20 years, showing far less variation than any other general purpose elastomer. The point should be made at this time that reclaimed rubber should never be classed as a low cost or low quality rubber. It must always be considered as a compounding ingredient in a rubber compound and be allowed to stand or fall on its own merits. It does contribute certain factors to the compound that no other raw material will do. On many occasions tire compounders have decided that a higher quality and lower priced tire could be manufactured using no reclaim. These conclusions did stand up in the winter season but, just as soon as the hot weather arrived, reclaim was reintroduced in order to process the compounds within established production and processing standards. Use of reclaimed rubber does permit factory processing to proceed at established rates, but at a lower operating temperature, thereby reducing risk of scorched stock particularly in summer, when cooling conditions are not at an optimum. Lower temperatures are developed and this in turn allows safer warehousing of processed compounds with less scorch probability. The heat history of the compound is improved while still maintaining the desired rate of cure.

Reclaimed rubber is manufactured to close tolerances and established standards and is uniform from shipment to shipment. In many instances, due to a high volume usage, a buyer will insist on two or more sources of supply on the reclaim purchased. Reclaim manufacturers can so control their end product that reclaims from different suppliers can be interchanged with no ill effect observed in the end product. It is accepted by the industry that compounds containing reclaim mix faster with lower power consumption. It is also possible to develop reclaims that are soft and plastic enough to completely disperse in the second pass of a two-pass mix. When this is done a considerable saving will result since the internal mixer is a high cost piece of equipment to operate.

Use of reclaimed rubber will tend to minimize nerve and shrinkage in the mixed compound as it is being prepared for curing. The preparation can be made closer to tolerance since there will be less shrinkage for which allowances must be made. Calendered blanks will hold size and thickness more accurately and extruded goods might be somewhat rougher but, if this is permitted, in many cases forms can be eliminated. Forms are strictly an expense item. When rubber is uncured and hot, or when extruded items are in the early stages of curing, the compound can soften and flow and change shape. If the part is intricate and size tolerances extremely rigid, dimensions might be out of specification when the part is cured. To minimize this possibility a cured piece of rubber is produced in which the part can nest and maintain dimensions. This is called a form. When a compound containing reclaim is calendered, fewer plies are required, since the calender can be set at a heavier gauge and fewer blisters and fewer calender lines or "fish eyes" will develop on the sheet. Equipment can be run at a higher speed, and less shrinkage will develop. Reclaim also permits faster calendering at a lower temperature.

Compounds containing reclaim will usually cure faster than those with no reclaim, provided the same acceleration ratio is used in the two compounds. Reclaim can be produced on the rough side to permit easier air escape during the molding operation, thereby reducing the number of air vents required in an automobile tire, for example. Even though the reclaim-containing compound does cure faster, it seems to be safer as far as scorch is concerned.

Due to some residual antioxidant, the reclaim-containing stock appears to age somewhat better and it is felt that there is less reversion. The use of butyl reclaim in the innerliner stock in tubeless tires does improve air retention. It is accepted compounding technique that butyl elastomer is used to improve the air retention characteristics of a compound. Butyl reclaim does have similar air retention characteristics to butyl elastomer, and in addition, is compatible with SBR and natural rubber while butyl elastomer is not. Chlorinated butyl is also compatible with natural and SBR.

It is acknowledged that the use of reclaimed rubber will reduce the overall abrasion resistance of a rubber compound, and improved abrasion resistance can be developed by using new rubber and carbon black loading. Some boot and shoe manufacturers feel that the advantages gained by use of a limited amount of reclaim outweigh the disadvantages of lowered abrasion resistance. For this reason some is still used in boots and shoes. Reclaimed rubber is no longer used in automobile tire treads, due to its inability to give abrasion characteristics required by current high speed driving. Reclaimed rubber is used in tread stocks to some degree on off the road tires. Here, some of the advantages of the original pigment load from the reclaim can be utilized and a loss in abrasion is not a major factor due to the low speed at which these off the road vehicles move.

There does seem to be a steadily growing demand for fiber-free, finely ground cured rubber in a number of applications. As this demand grows, new and more rigid specifications will be developed. This type product could become a factor in the industry.

For years there has been a demand by industry for a completely nonstaining reclaim. Much research work has been done on this subject but to date the reclaimed rubber industry has not been able to produce a truly nonstain compound. Reclaims are sold as "nonstainers" but this is a relative term. It is recognized that the sidewall compounds in many tires are made from nonstain compounds and the reclaimed rubber industry will use white sidewall tires as a basic scrap. However, all of these do have treads, and the best tread stocks are permitted to be staining compounds. The problem for the reclaimer is to eliminate the effect of the tread stock on the nonstain reclaim. There are several ways that the staining effect can be minimized. Certain pigments can be added to the reclaim that will tend to adsorb the staining chemicals and most of the "nonstain" reclaims are made in such a manner.

As the rubber industry grows, so must the reclaimed rubber industry grow. Over the years there has been a tremendous increase in the number of rubber-like materials that in no way resemble the original rubber. For this reason, the rubber industry has tended to use the generalized name of elastomers rather than rubbers. As these elastomers are developed they normally command a special price and as a result the reclaim industry has been called upon to develop some means of salvaging the scrap of these expensive elastomers. The industry has been reasonably successful in this endeavor and it is believed that all can be reclaimed, although some methods are either too dangerous or too expensive to produce a commercial

TABLE 19.3. FORMULATIONS WITH RECLAIMED RUBBER

Carcass

Ingredient	Amount
Natural Rubber #1 Rib S.S.	40.0
SBR 1707	62.0
Whole Tire Reclaim Non Stain	30.0
N660-GPF Black	42.0
Stearic Acid	1.5
Zinc Oxide	3.0
Naphthenic Process Oil	4.0
Petroleum Resin	3.0
Sulfur	3.0
Antioxidant	1.0
D.P.G.	.15
Benzothiazylsulfenamide Type Accelerator	.9
	190.55

Butyl Inner Liner

Ingredient	Amount
SBR 1500	50.0
Natural Rubber #1 Rib S.S.	20.0
Butyl Inner Tube Reclaim	55.0
Petroleum Plasticizer	3.0
N550-FEF Black	35.0
Antioxidant	1.0
Zinc Oxide	3.0
Stearic Acid	1.0
Benzothiazylsulfenamide Type Accelerator	.8
Sulfur	2.0
	170.8

Butyl Inner Tube

Ingredient	Amount
Butyl 218	81.0
Butyl Inner Tube Reclaim	35.0
Paraffinic Plasticizer	20.0
Zinc Oxide	5.0
Tetramethylthiuram disulfide	1.0
Mercaptobenzothiazole	.5
Sulfur	1.75
N660-GPF Black	60.0
	204.25

TABLE 19.3. *(continued)*

Automobile Mat	
Non-Stain Whole Tire Reclaim	208.0
Mineral Rubber	13.0
Hard Clay	45.0
Whiting	95.0
Zinc Oxide	3.5
Stearic Acid	1.0
Sulfur	3.0
Benzothiazyl disulfide	1.2
Lime	1.0
Zinc Dimethyldithiocarbamate	.3
	371.0

Semi Pneumatic Tire	
Whole Tire Reclaim	208.0
N774-SRF Black	61.5
Aromatic Hydrocarbon Resin	24.3
Sulfur	3.5
Stearic Acid	1.0
Keystone Filler	24.3
Whiting	40.0
Yellow Crude Wax	1.7
Petroleum Plasticizer	16.3
Benzothiazyl disulfide	1.4
Zinc Oxide	3.5
Antioxidant	1.0
Tetramethylthiuram disulfide	.25
	386.75

product. Reclaims from these elastomers are all specialty items since there is not enough available scrap of one given standard quality to manufacture and have stock reclaims on hand. As a result, all of these elastomers are processed on a "job" or individual service basis. Customers will save enough of one general type compound of one elastomer to make a single run commercially feasible. This lot will then be shipped to the reclaimer as the manufacturer's scrap and will be processed on a cost-plus basis. The manufacturer retains title to the scrap and the reclaimer will use his "know how" and equipment to depolymerize this special elastomer and return the finished product with only a processing charge and a profit. In this manner, the valuable scrap is not lost and additional production can be scheduled using the advantages of the reclaimed scrap.

In Table 19.3 are a few examples of formulations using reclaimed rubber, modifications of which are in productive use. Each individual compounder might need to modify further in order to fit quality requirements, processing advantages, and curing limitations.

REFERENCES

(1) Technical Bulletins No. 1 through No. 10, The Rubber Reclaimers Association, Inc., 1963-1968.

(2) J. M. Ball, "Reclaimed Rubber," Chap. 17 of *Introduction to Rubber Technology*, edited by Maurice Morton, Van Nostrand Reinhold, 1959.

(3) J. M. Ball, "Reclaimed Rubber," The Rubber Reclaimers Association, Inc., 1947.

20

RUBBER-RELATED POLYMERS
I. THERMOPLASTIC ELASTOMERS

W. R. HENDRICKS
R. J. ENDERS
Elastomers Technical Center
Shell Development Co.
Torrance, California

INTRODUCTION

The thermoplastic elastomers are a unique new class of polymers in which the end use properties of vulcanized elastomers are combined with the processing advantages of thermoplastics. Because of their unique molecular configuration they may be processed using the same techniques utilized with other thermoplastics, but the mechanical properties of the final articles are essentially indistinguishable from those of similar articles fabricated from conventional vulcanized elastomers. Utilizing such conventional processing techniques as milling, injection molding, extrusion, blow molding, and vacuum forming, these polymers yield useful articles having true elastomeric properties without compounding or vulcanization.

The thermoplastic elastomers can best be appreciated as a class when their properties are compared with those of other polymeric materials. Such materials can be characterized as either thermosetting or thermoplastic, and also as either hard or rubbery. Six classes of polymeric material result: thermosetting and hard; thermosetting and flexible; thermosetting and rubbery; thermoplastic and hard; thermoplastic and flexible; thermoplastic and rubbery. Of these six classes the first five are well known. The thermoplastic elastomers constitute the sixth class (Fig. 1).

The technical and commercial importance of a development such as the thermoplastic elastomers should not be underestimated. Kesser[1] has said: "The revolutionary technical possibilities of an elastomer that can be injection molded, or handled by any other thermoplastic technique, to yield a product with the

	THERMOSETTING	THERMOPLASTIC
RIGID	EPOXY PHENOL- FORMALDEHYDE UREA- FORMALDEHYDE HARD RUBBER	POLYSTYRENE POLYVINYL CHLORIDE POLYPROPYLENE
FLEXIBLE	HIGHLY LOADED AND/ OR HIGHLY VULCANIZED RUBBERS	POLYETHYLENE ETHYLENE- VINYL ACETATE COPOLYMER PLASTICIZED PVC
RUBBERY	VULCANIZED RUBBERS (NR, SBR, IR, etc.)	THERMOPLASTIC ELASTOMERS

Fig. 1 Classification of polymers.

properties of vulcanized rubber, without any kind of cure or after treatment, almost stagger the imagination." Kesser was referring to stereoblock polymer of propylene, but we now see the self-vulcanizing feature as a general characteristic of a broad class of block copolymers.

THE CHEMICAL AND PHYSICAL NATURE OF THE THERMOPLASTIC ELASTOMERS

Various methods of preparing thermoplastic elastomers have been reported in the literature.[2,3] In general, these preparation techniques result in thermoplastic elastomers which are ordered, block copolymers of the general structure A-B-A, where A represents a polymer segment which is glassy or crystalline at service temperatures but fluid at higher temperatures, and B represents a polymer segment which is elastomeric at service temperatures. Choice of segments type and length and the weight fractions of A and B are crucial in achieving elastomeric performance. In principle, A can be any polymer normally regarded as thermoplastic (for example polystyrene, polymethylmethacrylate, polypropylene, etc.), and B can be any polymer normally regarded as elastomeric (for example, polyisoprene, polybutadiene, polyisobutylene). Our discussion will concern itself primarily with the styrene-butadiene-styrene (S-B-S) or styrene-isoprene-styrene (S-I-S) polymers which represent the only commercially available thermoplastic elastomers. Examples of these materials are the KRATON® Thermoplastic Rubber made by the Shell Chemical Company.

Conventional elastomers obtain their rubbery properties as a result of the long chain polymer molecules being joined by covalent (and irreversible) chemical crosslinks formed during the process of vulcanization. The unique behavior of the thermoplastic elastomers, on the other hand, is attributed to the structure of the linear plastic-elastomer-plastic structure. Thus, in S-B-S or S-I-S thermoplastic elastomers, rigid polystyrene end segments are joined by an elastomeric polybutadiene or polyisoprene center section. The polystyrene segments associate with each other to give large aggregates referred to as "domains." At normal service temperatures these domains are hard and glasslike and immobilize the ends of the rubbery polybutadiene or polyisoprene segments. This end segment immobilization in conjunction with chain entanglements creates a "physical infinite network" or "physical crosslinks." At higher temperatures these polystyrene domains soften and may be disrupted by applied stress, allowing the polymer to flow. This process, like other physical phenomenon, is reversible and on recooling, the polystyrene domains again harden and the polymers regain their high tensile strength, high elongation, and resilience. A diagrammatic representation of the suggested molecular arrangement is given in Fig. 2. It consists of a number of polystyrene aggregates, referred to as "domains," dispersed in a continuous matrix of elastomer segments.

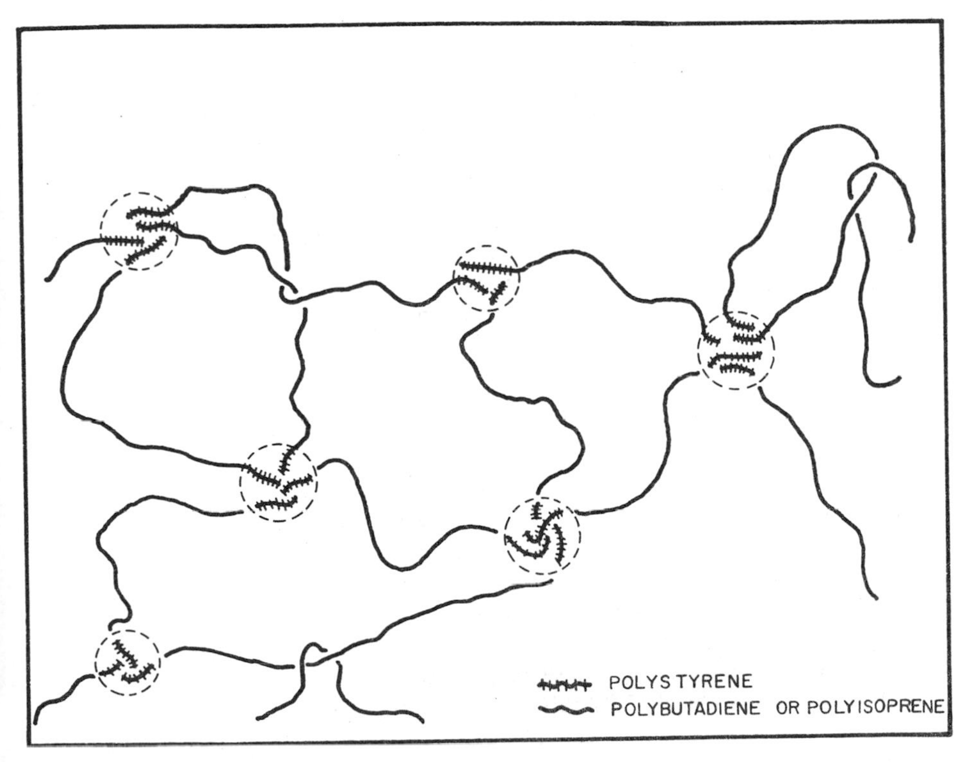

Fig. 2 Phase arrangement in S-B-S and S-I-S block copolymers (schematic).

TABLE 20-I-1. COMPARISON OF S-B-S AND B-S-B POLYMERS

A. Composition		
Segmental Molecular Weights ($\times 10^{-3}$)	10S-52B-10S	28B-20.5S-28B
Styrene Content (%)	27.5	27
Total Molecular Weight-($\times 10^{-3}$)	73	76
B. Mechanical Properties[a]		
100% Modulus, (lb/sq in.)	240	70
Extension at Break (%)	860	120
Ultimate Tensile Strength (lb/sq in.)	3950	70
Shore A Hardness[b]	65	17

[a]Tensile specimens were cut from compression molded sheets using a die having a 1″ constricted length. Extension rate was 2″/in.
[b]ASTM 1706-61.

Block copolymers having structures such as A-B or B-A-B do not show the tensile behavior characteristic of thermoplastic elastomers since for a continuous network to exist both ends of the elastomer segment must be immobilized in the nonelastomeric domains. Comparison of the stress-strain properties of a styrene-butadiene-styrene type thermoplastic elastomer with those of a corresponding butadiene-styrene-butadiene polymer show the former composition to have the high strength, modulus, and elongation of a true thermoplastic elastomer while the latter composition is a weak material which does not develop appreciable strength unless vulcanized (see Table 20-I-1). Similarly, polymers such as I-B or B-I-B have tensile properties characteristic of unvulcanized conventional synthetic rubber.

In addition to the properties and arrangement of the block copolymer segments themselves, two other parameters influence the physical behavior of thermoplastic elastomers. They are total molecular weight and the relative proportion of the two types of segments present in block copolymer molecules. A low molecular weight of polystyrene end block (less than about 8000) results in poor tensile strength. Too high a center segment molecular weight results in a low degree of thermoplasticity and poor processability. If the ratio of polystyrene end segment molecular weight to center segment molecular weight is too high (above about 35%), nonelastomeric behavior results.

PHYSICAL AND MECHANICAL PROPERTIES

The major difference between the thermoplastic elastomers and conventional thermoplastics is, of course, the true elastomeric behavior of the thermoplastic elastomers. The ASTM definition of an elastomer is "a material which at room temperature can be stretched repeatedly to at least twice its original length and upon immediate release of the stress will return with force to its approximate original length." The uncompounded, three-block thermoplastic elastomers readily pass this test, whereas the flexible thermoplastics do not.

The elastomeric nature of the three-block thermoplastic elastomers can be appreciated by reference to Fig. 3. The stress strain curves for a variety of materials

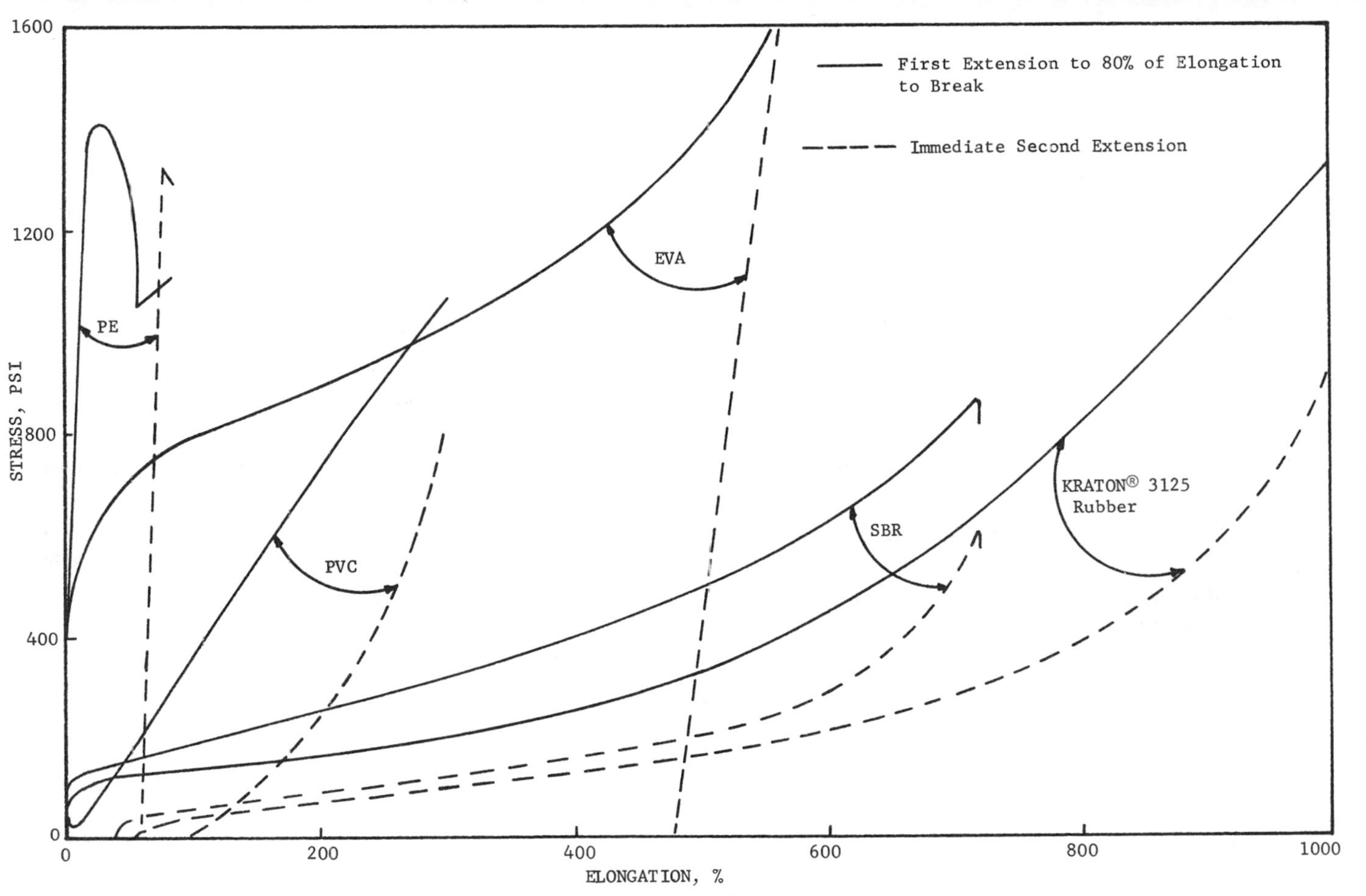

Fig. 3 Stress-strain behavior of elastomers.

TABLE 20-I-2. COMPARATIVE TENSILE PROPERTIES

	Natural Rubber Gum Stock	*Silica Reinforced SBR*	*KRATON® 3125 Rubber*	*KRATON® 3200 Rubber*	*Ethylene Vinyl Acetate*	*Plasticized Polyvinyl Chloride*		*Low Density Polyethylene*
Tensile Strength[a] at Break, psi	3000	2100	3000	1800	1900	1200	2100	1100
Modulus at[a] 300% Extension, psi	500	300	200	500	950	950	–	–
Elongation at[a] Break, %	600	900	1100	800	660	360	250	120
Set at Break, %[a]	10	30	30	40	500	150	120	65
Elongation/Set	60	30	37	20	1.3	2.4	2.1	1.9
Hardness, Shore A	55	45	55	65	91	40	67	94
Specific Gravity[b]			0.95	1.02	0.94	1.3	1.23	0.93
Melt Index g/10 min[c]								
G			15	7	32		136	
E					6	52	12	25

[a] "D" die specimen extended at 200%/min, 23°C.
[b] Measured using air pycnometer, 23°C.
[c] ASTM D-1238.

are compared to that of the commercially available thermoplastic elastomer KRATON® 3125 Rubber. The various polymers were extended to 80% of their maximum elongation, allowed to relax, and immediately re-extended. The high extensibility and elastic recovery of the truly rubbery thermoplastic elastomer contrasts vividly with the much lower elongations and higher permanent sets shown by even the most flexible of the conventional thermoplastics.

In addition to possessing excellent elastic recovery, the thermoplastic elastomers also exhibit the tensile properties typical of rubber vulcanizates. These and other properties are presented in Table 20-I-2 and again clearly show the marked resemblance of the physical properties of the thermoplastic elastomers to those of vulcanized rubbers. As with other thermoplastics the thermoplastic elastomers are anisotropic and show evidence of orientation when processed as a melt. This orientation is a function of the shear applied to the molten polymer and is manifested by greater stiffness and modulus in the direction parallel to that of the applied shear. As would be expected, this orientation effect increases with increased molecular weight and with increased proportion of polystyrene end block. The latter can also lead to the cold drawing which occurs when polystyrene content exceeds about 35% and is believed to result from the formation of a second continuous phase.

Another characteristic of the thermoplastic elastomers is resilience. This can be expressed as the Yerzley resilience (a measure of the damping in a material preloaded to 20% deflection), or as the falling ball rebound (a measure of the rebound when a steel ball is allowed to fall on a slab of the rubber). Some comparative data obtained by both methods are given in Table 20-I-3.

The compression set of thermoplastic elastomers is high by vulcanized rubber standards and increases rapidly with temperature. The high set values obtained are not surprising, however, since these materials are thermoplastic and retain their mechanical characteristics by means of the physical crosslinks provided by the polymer structure, and such crosslinks become increasingly labile and ineffectual as temperature increases.

At low temperatures, all rubbery polymers exhibit a sharp rise in elastic modulus and become rigid. However, the modulus of the thermoplastic elastomers having

TABLE 20-I-3. COMPARATIVE RESILIENCE AND REBOUND

	KRATON® 3125 Rubber	*Vulcanized Natural Rubber*	*Plasticized PVC*	*EVA*	*PE*	*ASTM Test Method*
Hardness, Shore A	55	50	67	91	94	
Yerzley Resilience at 20% Deflection, %	75	75	0	–[a]	–[a]	D945
Falling Ball Rebound, %	60	60	10	52	36	

[a] Not measurable–20% deflection not obtainable on this apparatus.

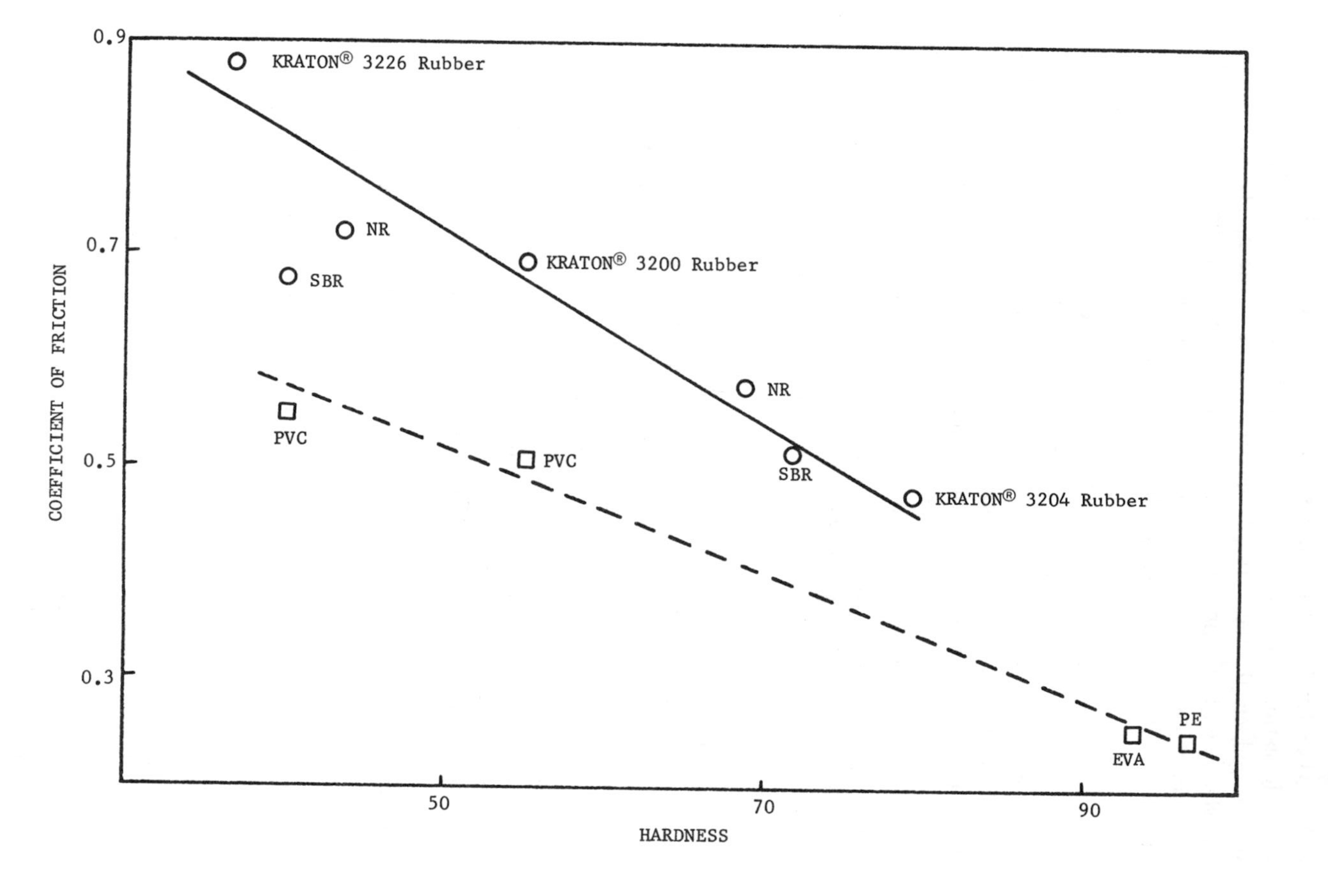

Fig. 4 Frictional properties of elastomers.

either butadiene or isoprene as the center block is relatively insensitive to temperature until conditions far below conventional service temperatures are reached. The effect of temperature on the modulus of torsional stiffness of the thermoplastic elastomers is less than that of most thermoplastics and these materials retain their rubbery character at temperatures far below those at which conventional flexible thermoplastics become stiff and brittle. On the other hand, at temperatures somewhat higher than ambient, thermoplastic elastomers containing rigid thermoplastic end blocks soften and eventually melt. This imposes a service temperature ceiling similar to that for flexible thermoplastics and considerably lower than that for vulcanized elastomers.

Generally, rubbery materials are distinguished by a surface having high frictional properties. In this respect, the thermoplastic elastomers are similar to conventional vulcanized rubbers. This behavior can be shown by measurements of the coefficients of friction, some of which are presented in Fig. 4. The coefficients of friction against polished steel were measured and plotted as a function of the hardness of the material tested. The materials shown fall into two groups: vulcanized rubber and thermoplastic elastomers in which the coefficient of friction is high, and the conventional thermoplastics in which it is low. In general, at a given hardness, vulcanized rubber and thermoplastic elastomers have coefficients of friction about 50% higher than thermoplastics of comparable hardness.

As would be expected, the chemical resistance of the thermoplastic elastomers is determined by the nature of the various block components. Block copolymers of polystyrene-polybutadiene-polystyrene, or polystyrene-polyisoprene-polystyrene are not affected by water, alcohol, or dilute acids and alkalis. Because of the absence of chemical crosslinks, however, they are soluble in ketones, esters, and many hydrocarbons. They are also swollen or dispersed by contact with fats or oils. The chemical resistance of thermoplastic elastomers of the general structure A-B-A will be most profoundly influenced by the chemical resistance characteristics of the A component.

The electrical properties of the thermoplastic elastomers are good, generally falling between those of low density polyethylene and compounded poly(vinyl chloride), materials widely used in electrical insulation.

The thermoplastic elastomers as produced are water-white and transparent. Their elastomeric properties are not dependent upon the presence of plasticizers, and "bleed-out," or plasticizer migration, is therefore not a problem. Since they contain no sulfur or sulfur residues, they do not strain or tarnish metal surfaces. The absence of sulfur or accelerators also leads to a product with little odor or taste, particularly valuable in food or drug applications.

PROCESSING

Introduction

In general, the thermoplastic elastomers can be readily processed in both conventional rubber and thermoplastics processing equipment. Selection of the best

TABLE 20-I-4.

	KRATON® RUBBER SERIES				
	1000	*2000*	*3000*	*4000*	*5000*
USE	Adhesives, Caulks & Sealants	Pharma-ceutical & Medical Goods	General Purpose Mechanical Goods	General Compounding & Polymer Blending	Footwear
FORM	Crumb	Pellet	Pellet	Crumb	Pellet

equipment depends on the ultimate application, and on the type of thermoplastic elastomer being processed. A list of the commercially available KRATON® thermoplastic rubbers, along with some of their intended end uses and physical forms, is given in Table 20-I-4. It can be seen that the thermoplastic elastomers are available commercially in pellet or crumb form. The pellet form, of course, gives the best hopper flow due to its uniformity, and as such is best for processing in thermoplastics equipment. The crumb form is intended for customer compounding. Crumb, however, can also be processed in some conventional thermoplastics equipment, especially equipment having large feed throats, and pellets can be used in rubber operations.

In the general rubber processing equipment area, it is important to remember that the materials are thermoplastic and as such require heat to soften the glassy domains. Typical rubber operations would involve some sort of a compounding operation using a Banbury internal mixer and/or a mill, with the mix then being fed into extrusion or calendering equipment. Compression molding is also feasible, although both heating and cooling operations are required.

The plastics processer must take into account the extreme flexibility, and also the high surface friction of the thermoplastic elastomers. These elastomers' characteristics often call for special methods of part stripping or ejection of the finished article.

Rubber Processing Equipment

The primary equipment within the rubber industry that is useful for processing thermoplastic elastomers are mills, calenders, extruders, and compression molding presses. These types of equipment will be considered in more detail in this section as well as a final section covering vulcanization. Vulcanization, as previously stated, is not necessary with the thermoplastic elastomers; however, it has been selectively used to modify certain polymer properties.

In general, all types of rubber mills have been used in the processing of thermoplastic elastomers. Optimum mill roll temperature is in the range of 230 to 250°F. To minimize the work time on the mill it is generally recommended that the materials be first preheated. The simplest preheating technique involves piling the thermoplastic elastomer and other compounding ingredients into the nip of the roll and allowing them to sit there for a period of two to three minutes. This will preheat enough of the material so that it will readily band. The banded sheet can be easily stripped from the mill, although it is well to remember that this sheet will have higher green strength than conventional SBR. If compounding is involved on the mill the higher melt flow of the thermoplastic elastomers will allow shorter mixing cycles than with conventional rubbers.

Processing of thermoplastic elastomers on calenders is similar to processing of other elastomers. Roll temperatures in the range of 175-210°F will provide best results.

Thermoplastic elastomers can be extruded in conventional short-barrel rubber equipment. Again, extruding will require the addition of heat. Temperatures in excess of 250°F will be required. Normal extrusion appears to be best with barrel temperatures around 300°F. Conventional plate dies have been used as well as other dies typical of the rubber industry. Just as in the processing of vulcanizable elastomers, the longer the barrel length is relative to the screw diameter (L-D ratio), the better will be the control of the extrudate. The thermoplastic elastomers are characterized by very low die swell which simplifies die design and allows for the extrusion of very small cross-sections without the marked increase in orientation characterized by high drawdown. No vulcanization is required to reach the stated properties of the thermoplastic elastomers and so this additional step is eliminated.

In the area of compression or transfer molding, generally a sheet or billet or sometimes even pellets of the material can be used for molding. Preheating will generally be required in order to provide proper melt flow in the mold. Once the material is up to temperature (200°F), pressure can be applied to the press in order to fully fill the cavity. Shortly after the pressure reaches its maximum, cooling will have to be provided in the area of the mold in order to harden the mechanical crosslinks (domains), and establish again the physical properties of the molded part. In general, warming temperature of around 150 to 200°F are optimum and cooling down to 140°F will be necessary. Very short press cycles will be typical because it is unnecessary to wait for vulcanization of the elastomer in the mold. Again, because of a lack of vulcanizing agents and accelerators in the system, there is no problem with scorch of the material.

Thermoplastics Processing Equipment

Injection Molding. Useful articles can be made from the thermoplastic elastomers using conventional injection molding techniques. As with all thermoplastics, the thermoplastic elastomers form free flowing melts when subjected to heat and pressure, and on cooling the elastomeric properties of the molded product reappear. When compared with conventional thermoplastics, the thermoplastic elastomers have slightly different molding properties. The principal differences exhibited by the thermoplastic elastomers are low mold shrinkage, high surface friction, high flexibility, high compressibility, and high extensibility. These differences call for modifications of molding conditions, and in many cases of mold design.

Mold Design. The most important area of consideration is that of mold design. Although thermoplastic elastomers have been molded in many of the molds designed for other widely used thermoplastics, it is generally better to design a mold specifically for use with thermoplastic elastomers.

In designing such a mold it should be kept in mind that the mold shrinkage of these materials is substantially less than that of other thermoplastics. For most molded parts, mold shrinkage will be less than .005 inches per inch (some thermoplastic elastomers approximate zero shrinkage). This low shrinkage factor combined with the high surface friction inherent in elastomers can make part ejection or sprue pulling more difficult than for normal thermoplastics. For this reason, it is desirable to maximize the draft in most areas. Also, whenever possible, stripper plates or sleeve ejection should be employed to remove the molded parts from the mold. If ejector pins are used they should be large and should work against the thickest section of the molded part. Whenever possible, it is recommended that air assist systems be incorporated. This provides a break in the vacuum and will assist in part ejection. In many cases, parts have been totally ejected by air. Further modifications that have improved the ejectability of the molded parts are the use of honed mold surfaces and Teflon coated molds.

One of the advantages of the thermoplastic elastomers comes from an entirely elastomeric nature. When stripping parts that have an under cut or other construction which would normally require a cam opening or collapsible core, the thermoplastic elastomer can generally be blown off or peeled off because of the elasticity of the rubber.

Mold filling has been accomplished using all of the conventional techniques. This includes hot or insulated runner systems and sprue gates. Balancing of the system in multi-cavity molds, however, is recommended. In order to get optimum flow in the cavity, it is generally recommended that the material flow into the cavity be directed at a surface. This impingement directly onto a surface will cause the material to mass and provide more uniform flow in the cavity. If this is not done, jetting of the polymer stream can occur, which results in a finished part with poor surface appearance, due to premature cooling of the injection polymer. It is also recommended that the diameter of the runner be relatively large in order to minimize the effects of a rapid viscosity change on cooling.

A variety of gating systems can be used in molding of thermoplastic elastomers. The gate should be as large as is compatible with the part design or mold type. Sprue fan, slot, and tab gates are most satisfactory. Pin and submarine gates can also be used; however, in these cases, gate positioning is important. This refers back to the requirement of impinging upon a mold surface. Diaphragm gates are, in general, unsatisfactory due to the frictional heating generated as the material passes through the gates. Diaphragm gates that have been successful have been at least .030 inches thick. This thickness, however, will require a secondary trimming or punching operation.

The final area of consideration is mold lubrication. In general, it is undesirable to use lubricants on the mold surfaces since these may adversely affect the surface of the finished part and may in some cases be detrimental to the final physical properties of the molded part. However, if part design or mold characteristic requires the use of some lubricant, it is inadvisable to use mold releases which contain silicone because of their detrimental affect on the physical properties of the polymer.

Injection Molding Conditions. Most types of injection molding machines have been used to process the thermoplastic elastomers. This includes ram machines, reciprocating screw machines, screw/ram accumulator combinations, and two-shot injection molding machines. Typical processing temperatures are in the range of 300 to 400°F. Injection pressures and injection rates should be as low as possible in order to minimize the shear orientation during molding. This orientation is a function of the shear applied to the molten polymer and manifests itself in greater stiffness or modulus in the direction of greatest shear. Shear can be minimized by the same techniques used for conventional thermoplastics, i.e., decreasing ram pressure or ram speed, increasing the temperature of the mold, or annealing of the molded part. Other factors that will decrease molecular orientation are large gate size, centrally positioned gate, and honed rather than polished mold surfaces.

In ram type machines it is recommended that the ram bottom at the end of each stroke. This will minimize both the cushion of material and the gate scar caused by overfilling.

Extrusion. The thermoplastic elastomers can be extruded on most conventional extruders currently in use. Best extrudates can be obtained with tapered and polished dies, and with screws having deep metering flights and low compression ratios.

The viscosity characteristics of the thermoplastic elastomers require some alteration of conventional extrusion conditions. The thermoplastic elastomers may be extruded at melt temperatures ranging from 275 to 400°F. A conventional profile with the lowest temperature in the feed sections and in the transition section, and with the highest temperatures in the metering zone will give the best output; however, this profile causes the most surging within the system. Raising the general profile, especially in the transition section, will reduce output slightly, but provide the most controllable extrusion. Lowering the extrudate temperature is best accomplished in the metering section of the extruder, but has also been successfully accomplished by cooling the die.

The thermoplastic elastomers can be handled on take-off equipment essentially like that used with conventional thermoplastics. It is well to remember, however, that the material is an elastomer, and that use of the tension roll to provide thickness control may result in necking down of the extrudate. This can be overcome by maintaining a sufficiently high extruder temperature to ensure minimal distortion and loss of cross-sectional integrity. Excessive drawdown, particularly at lower melt temperatures, produces a high degree of molecular orientation in the machine direction and as such, changes the physical properties of the extruded product.

In extruding sheet or film, the optimum temperature for polish or calender rolls is usually less than for other thermoplastics. This is because the extrudate tends to stick to metal rolls at higher temperatures. Generally, the temperatures recommended for polish or calender rolls are in the range of 115 to 175°F.

Due to the high coefficient of friction of the thermoplastic elastomers, extrusion rates are generally higher than are those of conventional thermoplastic materials. Output rates 20 to 30% higher than with polystyrene are not unexpected.

The thermoplastic elastomers commercially available are thermally stable for normal extrusion; however, extrusion at high temperatures, or at extremely low temperatures with a high compression ratio, high shear screw, can cause degradation which will lead to abnormal product behavior. The products of degradation, however, are neither toxic nor corrosive.

Thermoforming. Thermoforming can be used on most of the thermoplastic elastomers available on the market. The thermoplastic elastomers will give good deep draw and surface definition. The heating cycles and temperatures recommended are shorter than those used with impact polystyrene, and care must be taken not to overheat the sheet.

Blow Molding. Conventional blow molding techniques can be used for forming the thermoplastic elastomers. Melt temperature setting of between 250 and 400°F have been used to give good parison definition. At lower temperatures melt fracture is possible while at higher temperatures parison sag is exaggerated. Within the indicated temperature range blow ratios of up to 3.0 to 1 have been obtained. In general, lower than normal blow pressures are needed. The low blow pressure is recommended because of the good plasticity of the materials at normal processing temperatures. Parison die swell is again lower than expected with conventional thermoplastics.

The thermoplastic elastomers can be formed in any of the commercially available blow molding machines, but those having an intermittent fast drop parison are particularly well suited for these materials. In screw operated machines the same design considerations apply as with normal extrusion, that is, screws designed for PVC or ABS allow a broader range of processing conditions without fear of overheating or degrading the polymers.

Typical land widths and normal pinch off configurations work satisfactorily. It is recommended that the internal edges of the land be slightly broken to prevent tearing of the molded article on mold openings. The elastomeric nature of the thermoplastic elastomer permits the use of undercuts in the mold surface. Parts are generally flexible enough to be removed without the use of cams or parting sections of the mold.

COMPOUNDING

Although the thermoplastic elastomers are supplied in the market as ready-to-use materials, it is sometimes desirable to compound the polymeric system in order to cheapen, soften, harden, stiffen, or in some other way modify the properties of the thermoplastic elastomer.

This process can be accomplished by Banbury or mill mixing, and in many cases during injection molding or extrusion, etc.

The commercially available thermoplastic elastomers exhibit three unique features:

1. The products need no vulcanization to develop the final compounded rubbery properties.
2. The materials are completely thermoplastic and their high melt flow allows easy mixing.
3. The materials have very high tolerances for fillers, processing oils, resins, and other diluents.

The following description will include information on the methods that have to date been used for the compounding of thermoplastic elastomers, the effects of various compounding ingredients on the thermoplastic elastomers, and the effects of various stabilizers on the products.

The final topic discussed in this section will refer to the special use of the thermoplastic elastomers as minor volume compounding ingredients in other commercially available polymer systems.

Mixing and Processing Methods. In general, the thermoplastic elastomers can be compounded in mill or Banbury mixing equipment, Farrel Continuous Mixing equipment, injection molders, and extruders. The general area of mixing is differentiated from the mixing of conventional SBR rubber by the fact that the thermoplastic elastomers process better at higher temperatures; thus all of the equipment will have to be heated in some fashion in order to lower the viscosity of the thermoplastic elastomer to the point where the material can be readily compounded. In the case of the Banbury mixer, the chamber and rotor temperatures are generally set greater than 150°F. For mill mixing, this temperature is more nearly 230°F. For other types of equipment, it is preferable to be able to go above 300°F to get a homogeneous mix. In general, the thermoplastic elastomers can be mixed in shorter cycle times with less power consumption than conventional vulcanizable rubbers. The usual technique of warming up mill rolls by the frictional heat generated during milling of natural rubber can be used for the compounding of thermoplastic elastomers. This technique has also been used in short barrel, low compression ratio rubber extruders; however, generally a better product will be generated if a warm feed is used in rubber extruders. Satisfactory compounding results have also been obtained by first dry tumbling together various ingredients that will be in the final thermoplastic elastomer compound. The ingredients are then fed to the molder or extruder and the shear mixing occurs during the plasticizing cycle of the machine. For extruders, this shear mixing will be better in twin screw or two-stage mixing screw devices although some types of mixes can be satisfactorily prepared in single stage extruders fitted with mixing screws.

Effects of Compounding Ingredients. The types of fillers that have been used to advantage in thermoplastic elastomers are clay, silica, calcium carbonate, talc, glass beads, asbestos, and barytes. The general effect of these fillers is to increase the hardness and stiffness of the compounded products and to reduce their cost. In general, reinforcement of the thermoplastic elastomer is unnecessary, since this is provided by the domain structure of the polymer. Thus, reinforcing carbon blacks or silica fillers are not required in thermoplastic elastomers. They may be used, however, if some other property is contributed. Examples of this will be pigmentation, conductivity, etc. High levels of reinforcing fillers will cause a decrease in melt flow and tensile strength.

Calcium carbonates and clays are the most commonly used fillers. Calcium carbonates in general have faster incorporation times than clays and can usually be loaded at higher levels. Products filled with hard clays have a higher hardness than corresponding formulations filled with calcium carbonate.

Rubber extending oils can also be used to reduce cost. The general effect of an extending oil is to increase the processibility and to decrease the stiffness and hardness of compositions based on thermoplastic elastomers. Highly aromatic extending oils should not be used with thermoplastic elastomers. Since the rubbery properties of the thermoplastic elastomers are due to the physical crosslinks formed by the polystyrene end blocks, highly aromatic extending oils will tend to plasticize the end blocks, thus reducing the physical properties of the final compound. In general, paraffinic and naphthenic oils are satisfactory. Use of naphthenic oil is particularly recommended and will give compounds with good processibility and good physical properties, with no bleedout.

Many different thermoplastic resins can be blended with thermoplastic elastomers to achieve a variety of effects such as hardness modification, improved processibility, improved weatherability, or decreased cost. For example, compounding with polystyrene is a useful means of increasing the hardness and stiffness of compounded products without a substantial decrease in elongation or melt flow. Ethylene-vinyl-acetate is a thermoplastic copolymer which has been used in minor amounts to subtantially improve the ozone resistance of thermoplastic elastomers. Other thermoplastic polymers or resins which have been used to achieve a balance of properties are low molecular weight polystyrene resins, polyethylene, ethylene-ethyl-acrylate, and polyurethane.

Stabilization. In general, the thermoplastic elastomers for compounding contain a processing stabilizer, usually IONOL® to protect the polymers during manufacture, shipment, and storage. Additional antioxidants will have to be added to these elastomers in order to protect them during compounding and subsequent processing. The typical antioxidants which are satisfactory with the thermoplastic elastomers are generally hindered phenols such as Antioxidant 330 from Ethyl Corporation, Irganox 1010 from Geigy, and Plastanox 2246 from American Cyanamid. Any of these antioxidants at about the .3 to .5 phr level, together with an equal amount of dilauryl thiodipropionate (DLTDP), will generally provide satisfactory stabilization during compounding and subsequent processing.

Like most other elastomers, thermoplastic elastomers contain an unsaturated polymer chain and as such are susceptible to degradation by ozone. This is particularly true when the molded parts are under stress. The general result is one of increased surface hardening and eventual cracking. The most effective conventional antiozonants are NBC (Nickel dibutyldithiocarbamate) from duPont, and Ozone Protector 80 from C. P. Hall. These antiozonants are generally added in less than 2% quantities. Addition of up to 25% of ethylene vinyl acetate copolymer as previously mentioned is a more effective means of improving the ozone resistance and oil splash resistance of thermoplastic elastomers. Using EVA, however, will have more of an influence on the properties of the thermoplastic elastomers, giving a compound having properties falling somewhere between those of the ethylene vinyl acetate copolymer, and those of thermoplastic elastomers. This later technique requires a high shear processing after the ingredients are mixed together in order to develop fully the ozone resistance. This phenomenon is true of most compounds even if they do not contain ethylene vinyl acetate, that is, the higher the shear during the final processing operation the better will be the ozone resistance.

The ultraviolet radiation resistance of the thermoplastic elastomers is comparable to that of other styrene-butadiene rubbers. Therefore, in the majority of non-outdoor applications UV degradation is not a problem. If direct exposure to sunlight is expected, the thermoplastic elastomers will need to be protected. In order to minimize UV degradation, one or more of the UV stabilizers given in Table 20-I-5 at about .5 phr should be added to the compound.

A combination of .3 phr each of Uvinul 400 and Tinuvin 326 is a particularly effective UV stabilizer system for most products. In white stocks, this combination can cause slight discoloration; however, Tinuvin P at .5 phr is a satisfactory substitute. All of the UV stabilizers given in Table 20-I-5 are more effective in the presence of reflective or light absorbing pigments. Therefore, addition of some (up to 15 phr) of either TiO_2 or carbon black is recommended whenever possible.

Polymer Modification. The previously referred to topic involves the use of the thermoplastic elastomers as a minor modifier of commercially available polymer systems. Areas of interest fall into the following categories:

1. *Compounding with Vulcanizable Elastomers*
   a) Improvement in green strength.
   b) Improvement in melt flow.

TABLE 20-I-5. UV STABILIZERS

Material	*Chemical Composition*	*Supplier*
Uvinul 400	2,4-dihydroxy-Benzophenone	General Aniline & Film Corp.
Tinuvin 326	Substituted Hydroxyphenylbenzotriazole	Ciba-Geigy Corporation
Tinuvin P	Substituted Hydroxyphenylbenzotriazole	Ciba-Geigy Corporation

2. *Compounding with Rigid Thermoplastics*
   a) Improvement in impact properties.
   b) Improvement in melt flow and processing.
   c) Improvement in gloss.
3. *Compounding with Flexible Thermoplastics*
   a) Extended flexibility.
   b) Better feel.
   c) Improved frictional properties.
   d) Improved processing.

Adhesive Compounding. The unique structure of the thermoplastic elastomers provides a useful range of properties in adhesives, caulks, and sealants. The block polymers possess the excellent strength of conventional rubber vulcanizates, can be formulated with a number of other polymers and tackifying resins, and they are readily soluble. The thermoplastic nature of the block polymers also makes them suitable for hot melt or extrusion applied adhesives, sealants, and coatings.

The principal advantages of the thermoplastic elastomers in adhesive systems are as follows:

No premastication required
Soluble in a wide range of solvents
Rapid dissolution
Low solution viscosity
Processible as a melt
High cohesive strength
Highly elastic
Flexible at low temperatures
Easily formulated

VULCANIZATION

Although the thermoplastic elastomers do not require vulcanization to achieve their elastomeric properties, vulcanization has been attempted and crosslinks will occur in the polydiene middle block.

In general, the by-product of the vulcanization cycle would be a harder, stiffer, less elastomeric material which cannot be recycled except as a rather hard filler. Also, the products of vulcanization would be a higher temperature stability and increased solvent resistance. Conventional sulfur and peroxide systems have been used as well as radiation crosslinking. Little to date has been done in this area due to the general lack of need for vulcanization, but it would be normal to expect that systems that are used for the polydienes would also be used for the thermoplastic elastomers (i.e., polyisoprene systems for the S-I-S thermoplastic elastomers, and polybutadiene systems for the S-B-S thermoplastic elastomers).

REFERENCES

(1) Theodore O. J. Kesser, *Plastics Applications Series*, Reinhold, 1960, p. 73.
(2) L. J. Fetters, *J. Poly. Sci.*, Part C, Vol. 26, p. 1.
(3) R. Zelinski and C. W. Childers, *Rubber Chem. & Tech.*, **41**, 161 (1968).

20

RUBBER-RELATED POLYMERS II. POLY(VINYL CHLORIDE)

WAVELAND D. DAVIS
The Goodyear Tire & Rubber Company
Chemical Materials Development Department
Akron, Ohio

INTRODUCTION

Poly(vinyl chloride) is a member of the large family of polymers and copolymers which have in common the vinyl group, $CH_2 = CH-$. Through common usage, the word "vinyl" frequently refers to poly(vinyl chloride) homopolymers and copolymers. The designation "PVC" is an accepted abbreviation for poly(vinyl chloride) homopolymer.

Polyethylene, PVC, and polystyrene are the three leading plastic materials produced in the United States. Only PVC, however, closely resembles the elastomers in the ability to be plasticized and formulated into a wide range of materials. In 1970, 3.2 billion pounds of PVC were produced in the United States. How has PVC achieved this successful growth? There are several reasons:

(1) By compounding PVC with plasticizers and fillers, a wide range of softness and hardness can be obtained. Thus, from the same molecular weight resin, either a soft food wrap film or a hard, rigid pipe can be produced.

(2) Articles made from PVC have excellent and unusual physical properties. These products have replaced many natural materials such as leather in upholstery, glass in bottles, rubber in wire and cable coverings, wood in building products, and metal in pipe.

(3) PVC is easy to process. A blend of the resin with compounding ingredients can be fed directly to processing equipment, such as an extruder or injection molder. From a milled strip, PVC can be calendered easily.

(4) The low material cost of PVC resin has enabled it to replace more expensive materials and enter new markets. The average market price of general purpose PVC resin has declined from 32 cents per pound in 1956 to about 11 cents per pound in 1971.

This chapter will examine the monomer production processes and the polymerization methods that have led to these unique properties and low cost. It will then review the properties of PVC and the compounding methods that are used. Finally, it will examine the fabrication processes and resulting market applications.

VINYL CHLORIDE MONOMER

Vinyl chloride monomer, $CH_2 = CHCl$, is a colorless gas at normal room conditions. Its boiling point is $-13.4°C$. Although vinyl chloride is generally considered nontoxic, it will cause dizziness and anesthesia at concentrations above 5%. The major hazard to the use of vinyl chloride is fire and explosion. The open cup flash point is $-78°C$ and the explosive limits in air are 4–22% by volume.

Manufacturing Processes. The major process used for many years to produce vinyl chloride reacts acetylene and anhydrous hydrogen chloride in the presence of mercuric chloride:

$$CH \equiv CH + HCl \rightarrow CH_2 = CHCl$$

In recent years, this process has diminished in importance because of more favorable economics in using ethylene as a feedstock. The bulk cost of acetylene is about 8 cents per pound and hydrogen chloride is 3 cents per pound. Ethylene is available at petrochemical sites for about 2.5 cents per pound and chlorine for 3 cents per pound. The basic process employing ethylene is as follows:

$$(1)\ CH_2 = CH_2 + Cl_2 \rightarrow CH_2Cl{-}CH_2Cl$$
$$(2)\ CH_2Cl{-}CH_2Cl \rightarrow CH_2 = CHCl + HCl$$

In the first step above, ethylene is chlorinated to produce 1,2-dichloroethane. The second step is a pyrolysis reaction which yields vinyl chloride and hydrogen chloride. In order to utilize the hydrogen chloride, balanced processes are used.[1] The simplest procedure is to react the hydrogen chloride with acetylene to form more vinyl chloride. A balanced process using acetylene and ethylene is as follows:

$$CH \equiv CH + CH_2 = CH_2 + Cl_2 \rightarrow 2\ CH_2 = CHCl$$

Another method to utilize the hydrogen chloride is to oxychlorinate ethylene to produce 1,2-dichloroethane:

$$CH_2 = CH_2 + 2\ HCl + 1/2\ O_2 \rightarrow CH_2Cl{-}CH_2Cl +.\ H_2O$$

The oxychlorination is a key step for developing a balanced process, which can be represented as follows:

$$2\ CH_2 = CH_2 + Cl_2 + 1/2\ O_2 \rightarrow 2\ CH_2 = CHCl + H_2O$$

A third method to utilize the hydrogen chloride is to oxidize the hydrogen chloride to chlorine by the Deacon process:

$$2\ HCl + 1/2\ O_2 \rightarrow Cl_2 + H_2O$$

A balanced process with this reaction can be devised also.

Industry Capacity. The development of the new balanced processes using ethylene has made the original acetylene process uneconomical in this country. Furthermore, to an increased extent, only large vinyl chloride plants can be competitive. In the United States, the capacities of the newer plants are at least 300 million pounds per year.[2] In 1971, there were 10 producers in this country with a total capacity of about 4.7 billion pounds per year.

TABLE 20-II-1. 1971 UNITED STATES POLY(VINYL CHLORIDE) CAPACITY

Producer	*Location*	*Capacity, millions of pounds*
Air Products and Chemicals	Calvert City, Ky.	120
Allied Chemical Corp.	Painesville, Ohio	200
American Chemical Corp.	Long Beach, Calif.	125
Borden Chemical	Illiopolis, Ill.	240 (combined total
	Leominster, Mass.	
Conoco Plastics	Aberdeen, Miss.	150
Diamond Shamrock Corp.	Delaware City, Del.	250 (combined total)
	Deer Park, Tex.	
Escambia Chemical Corp.	Pensacola, Fla.	50
Ethyl Corp.	Baton Rouge, La.	150
Firestone Tire & Rubber Co.	Perryville, Md.	240 (combined total)
	Pottstown, Pa.	
General Tire & Rubber Co.	Ashtabula, Ohio	100
B. F. Goodrich Chemical Co.	Long Beach, Calif.	650 (combined total)
	Henry, Ill.	
	Louisville, Ky.	
	Pedricktown, N.J.	
	Avon Lake, Ohio	
Goodyear Tire & Rubber Co.	Niagara Falls, N.Y.	190 (combined total)
	Plaquemine, La.	
Great American Chemical Corp.	Fitchburg, Mass.	40
Hooker Chemical Corp.	Burlington, N.J.	60
Keysor Chemical	Saugus, Calif.	60
Monsanto Co.	Springfield, Mass.	150
Pantasote Co.	Passaic, N.J.	120 (combined total)
	Point Pleasant, W. Va.	
Stauffer Chemical Co.	Delaware City, Del.	80
Tenneco Chemicals Inc.	Burlington, N.J.	260 (combined total)
	Flemington, N.J.	
Thompson Plastics	Assonet, Mass.	150
Union Carbide Corp.	Texas City, Tex.	320 (combined total)
	South Charleston, W. Va.	
Uniroyal Inc.	Painesville, Ohio	130
Total		3,835

As a result of the new monomer technology, the price of vinyl chloride has decreased from about 10 cents per pound in 1960 to 5 cents per pound in 1971. This, of course, has led to a lower selling price for the resin and resulting plastic products.

POLY(VINYL CHLORIDE) MANUFACTURING METHODS

PVC is formed from vinyl chloride monomer by addition polymerization (discussed in Chap. 20–I). The polymerization methods used to manufacture the resin commercially are (1) suspension, (2) bulk, (3) emulsion, and (4) solution.

Many of the producers of PVC resin also manufacture the monomer. As in other industries, there is a trend towards integration in the PVC industry, both backward to monomer production and forward into the fabrication of plastic goods. A number of PVC producers have gained captive outlets for resin through mergers and acquisitions. The 22 domestic producers of PVC are listed in Table 20-II-1.[3]

Polymerization Initiation. Initiators which produce free radicals at a relatively low temperature are used to polymerize PVC. In emulsion polymerization of the resin, water soluble initiators such as potassium persulfate are employed. For suspension, bulk, and solution polymerizations, vinyl chloride soluble initiators are used.

The common monomer initiators are lauroyl peroxide, isopropyl peroxydicarbonate (IPP), and azobisisobutyronitrile. IPP has been used widely because of its low material cost and the fast polymerization times that can be obtained. The IPP is added to the polymerization reactor as a solution in an inert solvent such as hexane.

To overcome the potential instability of IPP, peroxydicarbonate initiators can be prepared in situ.[4] The preferred initiator, diethyl peroxydicarbonate, can be produced from ethyl chloroformate and sodium peroxide as follows:

$$2\,C_2H_5{-}O{-}\overset{\overset{\displaystyle O}{\|}}{C}{-}Cl + Na_2O_2$$

$$\downarrow$$

$$C_2H_5{-}O{-}\overset{\overset{\displaystyle O}{\|}}{C}{-}O{-}O{-}\overset{\overset{\displaystyle O}{\|}}{C}{-}O{-}C_2H_5 + 2\,NaCl$$

The ethyl chloroformate is added to the vinyl chloride whereas hydrogen peroxide and sodium bicarbonate (to form sodium peroxide) are mixed in the water phase. Since the diethyl peroxydicarbonate is formed during the polymerization, the shipping and handling problems of an unstable initiator are avoided and a more uniform, controlled reaction rate occurs. In addition, the raw material cost of the in situ initiator is less than that of IPP.

Suspension Polymerization. Suspension polymerization, also referred to as pearl, bead, or granular polymerization, is used in the majority of general purpose PVC manufacturing processes in the United States. Starting in 1950, the process was introduced commercially and quickly replaced the emulsion process in this country for producing general purpose resin.

In suspension polymerization, monomer is added to a reactor containing water and is dispersed into tiny droplets by the reactor agitation. The droplets are stabilized by what is called a "suspending agent." This is usually a protective colloid, such as gelatin, polyvinyl alcohol, or methylcellulose. The suspending agent is mixed in the water phase before the monomer is charged. In some cases, inorganic salts, buffers, or surface active agents are used to impart special properties.

As discussed previously, a monomer soluble initiator is used, such as IPP or an in situ peroxydicarbonate type. With the initiator dissolved or formed in the monomer phase, each dispersed droplet can be considered as a tiny bulk polymerization site surrounded by the suspending agent and water. The continuous water phase is a heat transfer agent between the droplets and the cooled walls of the reactor jacket. Efficient transfer of the heat is necessary since the polymerization is highly exothermic at about 500 BTU per pound of vinyl chloride.

The water also carries the suspending agents and other chemical additives in the recipe which control the resin particle size and structure. These ingredients are varied to produce either a fine particle size resin that can be used to prepare plastisols, an intermediate particle size "dry blending" resin, or a relatively coarse resin with high porosity. This is why suspension polymerization is so versatile.

The reaction temperature used to produce a medium molecular weight resin is about 55°C with a resulting pressure of 115 psi. At a conversion of around 90% and a reaction time of 6–8 hours, the polymer and water are in a slurry form. The slurry is transferred to a flash tank where the unreacted monomer is removed to the monomer recovery system. The polymer usually is isolated by centrifuging and then dried in a rotary dryer.

Bulk Polymerization. Bulk polymerization converts the monomer to polymer with an initiator but without water or suspending agents. Thus, water removal is not required in the finishing operation. The primary advantages of this process are (1) the process is potentially more economical than other methods and (2) the final polymer does not contain suspending agent or surfactant residues.

The commercial bulk polymerization process was developed as a one-stage process by the French firm, Saint Gobain. Later, the merged company of Pechiney-Saint Gobain developed a two-stage technique.[5] The process has been licensed to other firms, including three United States companies.

The first stage in the bulk process is carried out in a vertical reactor fitted with a flat blade turbine. About one-half of the total vinyl chloride feed is charged to the first stage vessel and is polymerized to about 10% conversion. The resulting "seed" prepolymer is gravity transferred along with additional monomer to the second stage reactor.

The second stage vessel is positioned horizontally and is equipped with incurvated blades which rotate at slow speed. After the first stage products are charged to the second stage, additional initiator and monomer are added and the reaction proceeds to 70–85% conversion. Heat of reaction is removed by the reactor jacket, agitator shaft cooling, and a reflux condenser. Since the second stage reaction is about four times longer than the first stage, several second stage vessels are provided for each first stage reactor.

The reaction is terminated by evacuation of the remaining monomer. The resin is discharged under agitation from the second stage reactor and is pneumatically conveyed to the finishing operation for screening and storage in silos.

Resin produced by the bulk process is free from polymerization residues and is very porous. It therefore has high purity and displays excellent clarity and fusion characteristics. This enables the polymer to be especially suitable for producing blown bottles, clear films, rigid pipe, and fluidized bed coatings.

Emulsion Polymerization. As stated previously, emulsion polymerization is not employed now in this country to produce general purpose resin; however, it is used to produce fine particle size dispersion resin. Emulsion polymerization follows the basic mechanisms of addition polymerization that are described in Chap. 20-I. The monomer is dispersed in water by a surfactant and is polymerized with a water soluble initiator.

The emulsion process is employed to produce dispersion resin because the desired small particle size of around 1 micron cannot be achieved with conventional suspension polymerization. In normal emulsion polymerization, however, the particle size range of about 0.1 micron is too small. To increase the particle size, the reactor may be "seeded" with small amounts of pre-made latex so that polymerization of new monomer proceeds on the seed particles.[6]

The amount of surfactant used in polymerizing dispersion resins varies from 0.2 to 0.5 percent. During the course of reaction, additional vinyl chloride and surfactant usually are added. The additional monomer continues the particle growth while the extra surfactant stabilizes the emulsion system.

Once the reaction is completed in about 18 to 24 hours, the latex is degassed in a vessel and dried in a spray dryer. The drying is conducted under close control to maintain the desired particle size and degree of particle agglomeration.

Solution Polymerization. Solution polymerization is actually bulk polymerization in a solvent. The solvent chosen is usually one which dissolves the monomer but precipitates the polymer when it reaches a certain molecular weight. However, the process is not true solution polymerization since the polymer does not remain in solution.

Solution polymerization is used to produce high quality copolymers, such as vinyl chloride-vinyl acetate. As with bulk polymerization, resins of good clarity and purity are obtained, free from polymerization residues.

A monomer-soluble initiator is employed in solution polymerization. Solvents such as benzene, cyclohexane, n-butane, and chlorinated aliphatics are used. After polymerization is completed, the slurry is filtered to remove the polymer and the filtrate is returned to the reactor. A major disadvantage to the process is, of course, the cost of the solvent and the required recovery equipment.

Solvent polymerized PVC is used to manufacture fibers.[7] Such fibers are popular in Europe because they are less expensive than nylon, polyester, and acrylic fibers. Low temperature polymerization is employed to obtain a highly syndiotactic polymer that is resistant to boiling water and dry cleaning solvents.

Copolymerization. Copolymers of vinyl chloride can be made by any of the processes that have been discussed. However, suspension and solution polymerization are the methods most commonly used to produce copolymers. Important comonomers that are used with vinyl chloride are (1) vinyl acetate, (2) vinylidene chloride, (3) maleic esters, (4) vinyl ethers, and (5) propylene.

Some comonomers, such as the maleic esters, have a reactivity similar to vinyl chloride. However, others differ in reactivity and require incremental or continuous monomer addition during the course of polymerization in order to produce a random copolymer composition.

Copolymers are useful because of the improved melt flow of the resulting compound and their ability to bind large quantities of filler. They are used widely to produce coatings, phonograph records, and flooring products. Films made from copolymers have good flexibility, toughness, and ability to "cling." Although the copolymers are more expensive than PVC homopolymers, their place in the market continues to grow.

POLYMER PROPERTIES

Particle Size and Distribution. The particle *size* of a particular PVC resin is designed to suit the intended use of the polymer. In the laboratory, particle size can be determined by microscopy or by screening methods. The three major types of resins can be classified according to particle size as follows:

	Average Particle Size	*Measurement Method*
Dispersion resin	1μ	Electron and light microscopy
Dispersion-modifying resin	20μ	Light microscopy and wet screening
General purpose resin	130μ	Wet and dry screening

In addition to the average particle size, the *distribution* of particle size is important. For example, the particle size distribution of dispersion and dispersion-modifying resins influences the resulting paste viscosity of plastisols. Flow of a plastisol cannot occur until the free void space between the resin spheres is filled with plasticizer. If the spheres are of the same size and are packed closely, a 26% theoretical void space exists for plasticizer filling. If, however, the spheres are of a range of particle sizes, the available void space decreases because the smaller spheres fill the void space between the larger spheres. This is desirable for dispersion resins since more plasticizer is available as the dispersing medium and lower paste viscosities result.

For general purpose dry blend resins, however, it is usually desirable to have a narrow distribution with a minimum of fines or oversize particles. Excessive fines can generate dust and cause uneven absorption of plasticizer during dry blending. Oversize particles also can cause uneven plasticizer absorption and subsequently appear as "gels" or "fisheyes" in the finished product.

The general particle *shape* of all resin types should be spherical so that free rolling motion occurs in the handling and processing of the final compound. With dispersion resins, it is desirable also that the particles are free from agglomeration so that plasticizer is not trapped in clustered particles.

Porosity and Plasticizer Absorption. The interior of dispersion and dispersion-modifying resins usually is dense and nonporous. Thus, when compounded, all of the plasticizer should be in the surrounding liquid phase so that minimum paste viscosity is obtained. On the contrary, a dry blend resin is designed with high interior porosity to permit good absorption of plasticizer so that the resulting blend is free-flowing. Resins which are produced with extremely high porosity are called "blotter" resins because they absorb high amounts of plasticizer even without the application of heat.

The total porosity of a PVC resin can be measured by mercury intrusion.[8] The inerior volume (porosity) of a dry blend resin is about 0.3 cc per gram of resin. The actual plasticizer absorption characteristics of a resin can be determined by a "spatula rub-out" test[9] or by using a "powder-mix" in a torque rheometer test.[10]

Particle size, distribution, and porosity also affect the apparent or bulk density of a resin. Plastic fabricators usually prefer high density resins to achieve maximum production rates in their processing equipment. The density of a resin can be high if the particle size distribution is wide to minimize the void space between the particles. If the particles are comparatively large and spherical, then good dry flow results. Bulk density and dry flow must be controlled, however, to maintain uniform plasticizer absorption and freedom from fisheyes.

Molecular Weight. The molecular weight of PVC is controlled during polymerization by the reaction temperature: the higher the temperature, the lower the resulting molecular weight. Molecular weight modifiers may be employed but they are not used in PVC polymerization to the extent that they are in synthetic rubber production.

The common industry method used to characterize the molecular weight of PVC is dilute solution viscosity.[11] ASTM has standardized a test that is used by the industry.[12] The dilute solution viscosity term obtained by this method is called "inherent viscosity." The IUPAC term for this same expression is "reduced logarithmic viscosity." Commercial homopolymer resins range in inherent viscosity from 0.5 to 1.4 as follows:

Molecular Weight	*Inherent Viscosity*	*Molecular Weight*	*Inherent Viscosity*
Low	0.5–0.8	High	1.0–1.2
Medium	0.8–1.0	Ultra high	1.2–1.4

The low viscosity resins are used in molding compounds, particularly those of high hardness where low to moderate levels of plasticizer are employed. Resins lower than 0.5 inherent viscosity tend to have inferior strength and have not found wide commercial use. The medium viscosity resins are used primarily for calendering and in rigid extrusions, such as pipe.

The high viscosity resins find greatest use in extrusion processing of wire coatings. Resins in the ultra high range are difficult to process and are always used with plasticizer. As a general rule, the user will select the highest viscosity resin which is compatible with his process so that the optimum physical properties result.

Melt Flow Rheology. The processing of compounded PVC usually involves either calendering, extrusion, or molding. These operations depend on the melt rheological characteristics of the base resin and the compounded stock. As discussed previously, the molecular weight determination of a homopolymer is useful to predict its processing characteristics compared with other homopolymers made by the same polymerization system. However, the melt flow characteristics of copolymers and the effect of different compounding ingredients must be studied by measuring actual melt viscosity in a laboratory instrument.

For a laboratory study of plastic melts, a viscometer is selected that covers the range from 1 to 10^4 reciprocal seconds shear rate.[13] The range of shear rates encountered in processing is as follows:

Compression molding	$1-10\ sec^{-1}$
Calendering	$10-10^2\ sec^{-1}$
Extrusion	$10^2-10^3\ sec^{-1}$
Injection molding	$10^3-10^4\ sec^{-1}$

"Rotational viscometers" are useful for measuring the paste viscosities of plastisols but are limited in use to below 100 reciprocal seconds. The "extrusion capillary rheometer" is a more versatile instrument and can be designed for use up to 10^6 reciprocal seconds shear rate.[14]

Table 20-II-2 shows the apparent viscosity of several PVC homopolymers and copolymers measured with a Sieglaff-McKelvey capillary rheometer at 205°C. The following formulation was used:

	Parts
Resin	100.0
Advastab T360*	2.0
Advastab TM180**	1.0
Stearic acid	0.6
Butyl stearate	0.5

The data show the degree of reduction in apparent viscosity when the molecular weight is lowered and when a comonomer is introduced.

*Solid organotin mercaptide stabilizer, Cincinnati Milacron Chemicals, Inc.
**Liquid organotin mercaptide stabilizer, Cincinnati Milacron Chemicals, Inc.

TABLE 20-II-2. MELT RHEOLOGY OF TYPICAL PVC HOMOPOLYMERS AND COPOLYMERS

Resin	Inherent Viscosity	Apparent Viscosity, poises x 10^{-4}		
		Shear Rate of 10 sec^{-1}	*Shear Rate of 100* sec^{-1}	*Shear Rate of 200* sec^{-1}
Poly(vinyl chloride) homopolymer	1.08	25.87	too viscous	too viscous
Poly(vinyl chloride) homopolymer	0.83	11.03	2.97	2.00
Poly(vinyl chloride) homopolymer	0.70	4.62	1.72	1.31
Vinyl chloride – propylene copolymer	0.69	2.55	1.24	1.03
Vinyl chloride – lauryl vinyl ether copolymer	0.64	2.28	1.10	0.90

While the molecular weight and melt flow rheology can predict the processibility of a compounded PVC resin, it is often desirable to evaluate a compound under simulated processing conditions. The "torque rheometer" is an instrument that is used to study resin fusion characteristics at relatively low shear rate in the laboratory.[15] This instrument consists of a small mixing chamber which is equipped with rotors that are driven by a dynamometer. The resistance torque that is generated from the test material in the mixing head is recorded on a strip chart as torque versus time.

The torque rheomoeter is used sometimes by the PVC resin producer to classify molecular weight of resins in a standard compound.[16] In this manner, dilute solution viscosity measurements are not required. The greatest use of the torque rheometer, however, is to the plastics fabricator who can compound in the laboratory and predict Banbury and calendering performance.

COMPOUNDING OF POLY(VINYL CHLORIDE)

Since PVC itself is tough, brittle, and has poor heat stability, it must be "compounded" with other chemicals to produce the properties required for the end application. Plasticizers, stabilizers, fillers, pigments, lubricants, and modifiers are the general types of chemicals added to the resin to form the final compound.[17]

Plasticizers. Plasticizers are high boiling point, chemically stable liquids which, when mixed with PVC, reduce the intermolecular forces and lower the glass transition temperature of the polymer. The resulting compound has increased flexibility, softness, and elongation, but decreased tensile strength. The melt viscosity is lowered and the temperature required for processing is reduced.

Based on their compatibility with PVC, plasticizers may be classified as either *primary* or *secondary*. This classification is arbitrary and is used only as a guideline. Primary plasticizers are very compatible with PVC while secondary plasticizers have limited compatibility and are not used as the sole plasticizer in a compound. The secondary plasticizers are employed primarily to decrease the cost of a compound.

The simple lower molecular weight plasticizers, such as the esters produced from phthalic, adipic, phosphoric, and sebacic acids, are referred to as *monomeric* plasticizers. The higher molecular weight plasticizers which have repeating molecular units are called *polymeric* plasticizers. Usually, they are produced from the difunctional adipic, azelaic, and sebacic acids, and the propylene and butylene glycols. They range in molecular weight from 1000 to 4000.

The most widely used plasticizer is di-2-ethylhexyl phthalate (also referred to as dioctyl phthalate or DOP). DOP plus the other phthalates, such as butyl, benzyl, diisoctyl, and diisodecyl, account for over half of the total amount of plasticizers used in the PVC industry. DOP possesses optimum solvating power with fairly good low temperature properties and volatility resistance. The longer chain esters, such as diisodecyl phthalate, have lower volatility but reduced solvating power.

The adipate, azelate, and sebacate esters are used to impart low temperature flexibility to a compound. The adipates, such as di-2-ethylhexyl adipate, are usually selected because they are the least expensive of this group. However, because of the increased cost and relatively high volatility, the adipates are incorporated mostly in combination with the phthalates.

The phosphoric acid esters, such as tricresyl phosphate and tri-2-ethylhexyl phosphate, are used when flame retardancy is required in a compound. Unfortunately, these plasticizers have poor heat stability and are relatively expensive.

The polymeric plasticizers made from dibasic acids and glycols are incorporated in a compound to obtain low volatility, good resistance to migration and lacquer marring, and low extractability. However, the polymerics generally impart poor low temperature properties and increase the compound cost.

Epoxidized soybean oil, linseed oil, and epoxy stearates not only plasticize PVC, but also help stabilize against the effects of heat. Unfortunately, they are not highly compatible with PVC and must be compounded carefully to avoid "spewing" (bleeding). However, because of their stabilization effect, they are included in almost all formulations.

Stabilizers. Like rubber, polystyrene, polyethylene, and other high polymers, PVC is sensitive to heat and light. The effects of degradation lead to discoloration and eventual embrittlement. A compound therefore must be heat stabilized to withstand the effects of processing.

PVC is essentially a linear polymer with a configuration as follows:

$$-CH_2-CHCl-CH_2-CHCl-CH_2-CHCl-$$

If the structure is heat activated, the chlorine atoms will split from the chain, combining with hydrogen atoms to form hydrochloric acid. The dehydrochlorination alters the polymer to the following structure:

$$-CH=CH-CH=CH-CH=CH-$$

The dehydrochlorination proceeds down the chain in a zipper-like effect.[18] The resulting conjugated double bond or diene structure is believed to be responsible for the color development as the polymer degrades. The liberated hydrogen chloride has a catalytic effect on the dehydrochlorination reaction.

In order to stabilize PVC against the effect of heat, it is necessary to incorporate a metal salt or soap to react with the liberated hydrogen chloride to form a metal chloride. Lead salts of the weak acids were the first stabilizers used with PVC. Later, the cadmium and barium salts of various organic acids were developed to provide good long term heat stability. Calcium and zinc salts are also used, especially when nontoxic compounds are required. Commercially, combinations of cadmium, barium, and zinc compounds are preferred for general processing because a synergistic effect is achieved. Also, the zinc salt replaces part of the more expensive cadmium salt and provides sulfide stain resistance by reacting to the white zinc sulfide before the yellow cadmium sulfide can be formed.

Organo-tin compounds are used where severe processing is employed or when high clarity is desired, such as in clear rigid products. Certain octyl tin stabilizers are regulated by the FDA for use in food contact applications.

Organo-phosphites were developed for use in PVC when it was discovered that they serve as chelating agents to sequester the metal chlorides formed from the evolved hydrogen chloride and stabilizing metal. In addition, the phosphites prevent oxidation of the conjugated double bond to form carbonyl groups which would contribute additional color to the system.

As mentioned in the discussion of plasticizers, epoxidized resins and oils are almost always incorporated in a formulation. They add to the overall compound stability by providing acid acceptance.

To summarize, the compounding of PVC for heat stability usually employs a balance of three additives:

1. A metal soap acid acceptor.
2. An organo-phosphite chelator.
3. An epoxidized resin or oil.

In addition to heat stabilization, ultraviolet light resistance must be provided if the finished article is exposed to prolonged sunlight. In pigmented compounds, opacity from the pigment can block the light and thereby prevent decomposition of the compound. In translucent compounds, light stabilizers must be used. Effective ultraviolet "screens" are the benzophonones, the benzotriazoles, and the aryl esters of resorcinol. Because these materials range in cost from \$3.50 to \$5.50 per pound, they are used sparingly. However, applications such as automobile seat covers, rainwear, and building products require some protection against the effects of ultraviolet light.

Fillers. Fillers are incorporated in PVC formulations to produce opacity and hardness in the finished article and to reduce costs. Additional advantages include the improvement of electrical properties, ultraviolet light resistance, and dent resistance. By reducing the elasticity of the melt, fillers also improve processing characteristics and the dimensional stability of the finished material.

Unfortunately, the reinforcing effect obtained from using carbon black in rubber does not occur in PVC except at very low loadings. On the contrary, carbon black (and other fillers) normally lower the tensile strength, elongation, and tear strength of a PVC compound. Low temperature performance and abrasion resistance also are impaired.

The common fillers used in PVC compounding can be classified as follows:

Inorganic		*Organic*
clay	talc	carbon black
mica	diatomaceous earth	wood flour
antimony oxide	titanium dioxide	
asbestos	calcium carbonate	

The most widely used fillers are the clays and carbonates which have an average particle size of less than three microns. Asbestos is important for its use in vinyl asbestos tile.

Since the specific gravity of most fillers is 2–3 times that of an unfilled compound, the density of the compound increases with loading. Cost is therefore judged on a volume basis and low density fillers are advantageous. All types of specially coated and grades of fillers are available. Fillers with light color, fine particle size, and high purity command premium prices.

Colorants. As with fillers, a wide range of colorants is available to the PVC compounder. In addition to the effect of color, the pigment chosen is considered for cost, processing temperature, and service requirements. Usually, a pigment is selected that is insoluble in the plastic compound but is easily dispersible. The soluble dyes are more likely to cause cracking and migration and may be deficient in heat and light stability.

The colorants may be classified as follows:

Inorganic	*Organic*
titanium oxide	phthalocyanines
chromium oxide	benzidines
molydate orange	quinacridones
ultramarine blue	oxynaphthoic reds

The inorganic pigments are dense chemicals characterized by excellent heat stability, good hiding power, and low cost. Although the inorganics impart good tinting strength, their masstones are not brilliant. Titanium oxide is the most commonly used colorant since it possesses optimum physical properties at relatively low cost.

The organic pigments are characterized by their excellent brightness and transparency. They also have low specific gravity and high tinting strength. Disadvantages are higher cost and poorer heat stability than the inorganic pigments.

Normally, only small quantities of colorants are used; therefore, special dispersions or masterbatches frequently are made to insure uniformity. Masterbatches may be prepared in a separate operation by grinding the pigment in a plasticizer or by mixing it with several of the compound formulation ingredients. Masterbatches or color concentrates may be purchased from the pigment suppliers or custom mixing companies.

Lubricants. To reduce the tendency of a plastic stock to adhere to surfaces of the processing equipment, a lubricant may be incorporated in the formulation. A lubricant also reduces the internal friction of the plastic mass as it is being fluxed in the process. Both of these effects increase the calendering or extruding rate.

Stearic acid and the metallic stearates are the most common lubricants used. Waxes, oils, low molecular weight polyethylenes, and silicones also are employed. Normally, less than one percent is incorporated because excessive lubrication in a compound can cause surface bloom and exudation.

Modifiers. Compatible polymers which are added to a PVC compound are referred to as modifiers. If the effect is to reduce the melt viscosity of the compound, the modifier is called a "processing aid." If the effect is to increase the impact strength, it is called an "impact modifier."

The most commonly used processing aids are acrylic polymers, styrene-acrylonitrile resins, and chlorinated polyethylene.[19] These materials lower the compound melt viscosity which gives faster and more complete fusion, faster processing rate, and improved product forming. Levels of about 5 to 10% are used.

Impact modifiers include the acrylic polymers, acrylonitrile-butadiene-styrene resins, chlorinated polyethylene, and certain graft copolymers. To improve the impact strength, levels of from 10 to 20% are used. The use of impact modifiers has contributed to the growth of PVC in rigid applications such as pipe, building products, bottles, and injection molded parts.

Unfortunately, the use of modifying polymers to improve either processing or impact strength increases the cost of the compound. Also certain properties may suffer, such as chemical resistance, tensile strength, ultraviolet light resistance, and heat distortion temperature. As with other compounding ingredients, the physical properties and compound cost must be optimized to meet the end-use requirements.

GENERAL PURPOSE RESIN FABRICATION PROCESSES

Compound Preparation. Before a resin can be fabricated into a finished article, all of the compounding ingredients must be premixed with the resin. Either the formulation can be "dry blended" in a blender or it can be mixed and fluxed in a high-shear internal mixer such as a Banbury. If the compound is dry blended, then it is fed as a power to the fabricating equipment. If it is fluxed in a Banbury, it is dropped onto a two-roll mill and strip fed to the final process or diced into cubes for later use.

The simplest type of premixing equipment is a ribbon blender. The resin and liquid ingredients are added to the blender and heat is applied through the jacket. At approximately 95°C, the liquid ingredients are absorbed by the resin particles. The other dry ingredients of the formulation are added periodically during the blending cycle. After cooling to about 60°C, the final blend is a free-flowing material which can be fed directly to an extruder or other processing equipment. The dry blend mixing time is around 40 minutes.

Dry blends may also be prepared in a high intensity mixer which generates heat from the shearing action of the blades and the mix. External heating is not required but is sometimes employed at the start of the cycle. Mixing time is short for this operation, around 15 minutes.

When a compound is to be calendered, usually it is first fluxed in a Banbury type mixer. The fluxed compound then is fed to a warm-up mill and then to the calender. A Banbury mixing operation also may be employed if the compound is to be processed into pellets or cubes for later use. With this operation, the fluxed sheet is fed to a pelletizing extruder or a dicer. Many companies who injection mold and extrude purchase pelletized or diced compound from polymer manufacturers who provide this service.

Calendering. The calendering of PVC compound is the mechanical process of transforming a thick mill sheet into a thin, uniform-gauge sheet of continuous length. The calenders used for plastic processing are similar to those used for rubber. Plastic film calendering, however, requires higher operating temperature, greater roll pressures, and more exacting gauge tolerance.

The conditions of calendering influence the finished properties and appearance of the finished film or sheet. Higher processing temperature usually favors increased tensile strength and tear strength; however, film shrinkage can be adversely affected. Shrinkage can be controlled by reducing the tension on the sheet as it is stripped from the calender roll.

If desired, an embossing unit may be included in the calender train immediately after stripping. Following this, a series of internally water-chilled drums cool the sheet. The last operation consists of edge trimmers and wind-up rolls. Modern calenders produce up to 5000 pounds per hour or 300-400 feet per minute, depending on the film thickness and compound type.

Since a calender may cost roughly $500,000, the initial investment is relatively high. However, for mass production it is the most economical process for producing medium to thick gauge film and sheeting. For thin gauge film, extruders are more economical.

Extrusion. Extrusion is the largest single process used for fabricating PVC. Approximately 35% of the resin used domestically is extruded into wire coating, pipe, hose, or profiles. Although the principle of extruding plastics is identical to that of extruding rubber, the plastics extruder has become highly specialized to accommodate a wide range of formulations.[20]

Plastic extruders are available with either single or multiple screws. *Single screw* machines are widely used in the United States because of their low cost, simplicity, and high available power. The *twin screw* extruder has two screws positioned side-by-side which rotate either in the same or opposite direction. It mixes more intensively than the single screw machine and more efficiently removes volatiles.

However, the twin screw extruder has less load capability than the single screw type and requires the application of more external heat. In Europe, the twin screw extruder has been highly popular and, because of its intensive mixing ability, is suited especially for processing rigid compounds.

To remove volatiles from a compound as it is being extruded, a vacuum hopper and a vented barrel may be employed. Forced feed hoppers often are used in conjunction with vented extruders to increase the output.

In addition to temperature control of the extruder, the design of the screw is important so that it is matched to the fusion rate of the compound. For modern single screw extruders, the ratio of the length of the barrel to the diameter of the screw is about 24:1. The ratio of the volume in the first flight to the volume in the last flight, called compression ratio, is about 4:1 for extruding a dry blend compound and about 2:1 for a pelletized or diced compound. The dry blend requires a higher ratio in order to compact and fuse the material.

PVC film can be "blown" using an extruder with a special tubular die. The film is blown to two or three times the diameter of the die by introducing air pressure through the die. The die usually is designed so that the film is extruded vertically. The film may be stretched to orient the film for use in shrink wrap applications.

Molding. Compression molding, injection molding, and extrusion blow molding are the common processes used to mold articles from PVC. The latter two methods employ a plunger or a screw as part of the operation.

Simple compression molding of plastics, as with rubber, uses male and female dies with 200 to 2000 psi pressure. This is the earliest molding technique used with PVC and is still employed when large sections must be formed or when fine detail is required, such as in phonograph records. The usual procedure is to place a preheated slab or "biscuit" of fluxed compound into the mold, close the mold, apply pressure, cool, and then remove the article.

For mass production of plastic articles, the most economical method used is injection molding. This operation converts either powder or pellet to a finished article in a single operation. The injection molding machine consists of two parts: the injection unit and the molding unit. The injection unit heats and melts the plastic and transfers it under pressure to the molding unit. The molding unit consists of two mold halves and a clamping unit to hold the molds during injection and open them for removal of the finished article.

The simplest type of injection molding machine has a plunger unit which fluxes and delivers the plastic "shot." Plastic pellets are fed from the hopper into the plasticizing chamber, which is located between the plunger and the nozzle. Each stroke of the plunger forces material into the chamber, out of the nozzle, and into the mold.

Most plastics injection molding machines produced today are *reciprocating screw* types. The fluxing step is similar to the single screw extruder in that the shear developed from the rotation of the screw is largely responsible for heating and melting the compound. With the reciprocating screw rotating in the forward position, material is fed from the hopper and advanced to the front of the barrel.

The pressure of the melt at the front of the barrel forces the screw to move back in the barrel. When the screw reaches a position corresponding to the volume of melt required for the shot, the screw rotation stops. The screw then is advanced to inject the melt ahead of it into the mold. Rotation then resumes to prepare the melt for the next shot.

Reciprocating screw injection molders develop pressures as high as 20,000 psi. This permits the use of dry blends and the processing of rigid PVC. However, the operation is relatively severe and requires that the compound have good stability and proper rheological characteristics.

The extrusion blow molding process is used with PVC to produce bottles and other containers. The equipment extrudes a tube called a "parison" which is fed while hot into a water-cooled mold. By introducing air pressure, the parison is inflated to fill the walls of the mold and then cooled to retain the intended shape. The limited heat stability of PVC bottle compounds requires fast movement of the material through the extrusion and blowing operations.

Coating. General purpose PVC can be applied to a substrate from either a solution or from a powder blend. Because copolymers are readily soluble, they normally are used with the solution method. Solvents such as methyl ethyl ketone, tetrahydrofuran, or cyclohexanone may be employed to dissolve the resin and liquid compounding ingredients. Fabrics and metal articles are coated in this manner. If a solution is cast on a belt or a roll, it can be stripped to form a continuous, unsupported clear film. High quality food wrapping film is made in this manner.

The powder blend technique employs high porosity resins that fuse evenly in a "fluidized bed." The equipment consists of a vibrating open box which has an air-permeable barrier in the bottom. The compounded dry blend is placed in the box and air is blown from underneath through the barrier. This fluidizes the dry blend into a "bed" which is kept uniform by the vibrating box. The metal article to be coated is preheated to the proper temperature, dipped into the fluidized powder for 5 seconds or longer, and then removed.

Coatings applied to iron or steel by the fluid bed technique have attractive finishes which are much less expensive than stainless steel or brass. PVC is very adaptable because of the wide range of flexibility and pigmentation that is available. Kitchen appliances, marine fittings, and building coatings are examples of articles that can be attractively coated by fluidized bed techniques.

DISPERSION RESIN FABRICATION PROCESSES

A small particle size dispersion resin can be blended with plasticizers and other compounding ingredients to form a liquid "plastisol." This fluid can be poured into molds or coated onto substrates and fused into a solid, homogeneous plastic at 175-190°C. Toys, gloves, coated metals, coated fabrics, flooring, and foam are produced by these processes.

Compound Preparation. A plastisol may be prepared in a high shear, low speed, paddle-type mixer by mixing the resin with the compounding ingredients. Usually, some of the plasticizer is withheld during the initial mixing so that a thick paste is produced for maximum shearing action. The temperature of the mix rises because of the mechanical action; however, the rise should not exceed 5°C to minimize solvation of the polymer.

The stirring action can trap air bubbles in the compound which would be objectionable in the finished product. The formation of bubbles can be prevented by mixing under a vacuum. If this is not possible, the plastisol can be deaerated in a special vacuum chamber after mixing. After the compound is mixed and deaerated, it is stored in drums or special containers until ready for use.

To reduce the viscosity of a plastisol, volatile solvents and diluents may be incorporated in the formulation which are evaporated later in the fusion operation. Such a compound is called an "organosol." This also permits the production of low plasticizer content articles with high hardness.

Modifying resins produced by suspension polymerization also may be used to lower the paste viscosity. Modifying resins have a particle size of around 20 microns and increase the "packing" of the resin particles in the fluid system. In addition, modifying resins generally are less expensive than dispersion resins and thereby lower the compound cost.

Most plastisols are "pseudoplastic." This means that they decrease in apparent viscosity as shear rate is increased. If the viscosity increases as shear rate is applied, the paste is "dilatant."

Molding. Rotational and slush molding are commonly used to mold articles from plastisols and organosols. Rotational molding employs a battery of molds that are fitted to an arm which rotates inside a heated oven. The closed molds are filled with a measured amount of plastisol. As the molds are rotated inside the oven, heat is applied and the plastisol is fused to conform with the mold detail. Completely closed, hollow articles are produced in this manner which have very fine reproduction and no flash. Automatic equipment can be used with rotational molding to speed the operation and reduce labor costs.

Slush molding uses open metal molds which are filled with plastisol and then heated. In some operations, the molds are pre-heated before filling. When the plastisol next to the walls of the molds is fused, the excess material is poured off and the molds are heated further to complete the fusion. After cooling, the completed articles are removed from the molds. Slush molding is employed widely to produce open-end articles such as doll heads, traffic cones, and boots.

If a gelling agent such as aluminum stearate or organic bentonite is incorporated into a plastisol along with a high oil absorption filler, a "plastigel" is produced. A plastigel has a putty-like consistency and may be shaped into almost any shape at room temperature without the use of a mold. This molding technique, while unique, is used only for special applications because of the hand labor involved.

Coating. The coating of fabric, paper, and felt is the largest single application of dispersion resins. A conventional knife or roller coater may be used to apply the plastisol to the substrate. The coated material then is continuously fed into a hot air or infrared oven to fuse the plastisol. The coating may be embossed with water-cooled roll embossers or special release paper. A foam coating may be produced by incorporating a chemical blowing agent in the compound. PVC flooring can be made by coating felt backing with a foam coating, followed by the pattern print and clear top coats.

Dip coating is used to coat wire dish racks, tool handles, gloves, metal parts, and strippable coatings. The article to be coated is dipped into the plastisol, withdrawn, inverted, and heat-fused. To minimize dripping, the article may be preheated before it is dipped into the plastisol to quickly gel material on the surface. Dip coating may also be used to coat continuous lengths of wire, tubing, or yarn by drawing the material through a coating tank and then fusing the coating.

Spray coating is employed to coat large, intricate, or fixed-in-place objects. Coatings from a few mils in thickness up to 60 mils can be laid down in one pass. If required, several coatings may be applied and special spray heads can be used to produce decorative or spatter patterns.

MARKETS FOR POLY(VINYL CHLORIDE) PRODUCTS

In the preceding sections of this chapter, the reader has been introduced to the general product applications for PVC. Table 20-II-3 lists the market consumption of PVC resins in specific application areas.[21] The three largest areas are calendered

TABLE 20-II-3. POLY(VINYL CHLORIDE) AND COPOLYMER 1971 PATTERN OF CONSUMPTION

Market	*Millions of Pounds*
Calendering	
Film and sheet	540
Flooring	265
Coating	
Flooring	120
Paper and textile	117
Protective and adhesive	110
Extrusion	
Film and sheet	190
Wire and cable	370
Other	680
Molding	
Injection and blow	160
Plastisol[a]	140
Records	125
Other uses	290
Export	179
Total	3,286

[a]Includes coatings, other than flooring.

film and sheeting, calendered flooring, and extruded wire and cable. It is interesting that these were the first major applications introduced for PVC following World War II and they continue to dominate the market over two decades later.

Newer applications, however, are gaining in relative importance because of technological developments and the favorable raw material cost of PVC. The greatest market expansion is being experienced in the rigid market. It is now possible to blow mold PVC into containers that have glass-like clarity, high impact strength, and excellent chemical resistance. Food-grade PVC bottles are competing directly with glass in the food market. Applications are expanding in the nonfood packaging areas, such as containers for toiletries, detergents, and other household products.

While rigid PVC is replacing glass in the container industry, it continues to displace metal and ceramics in the pipe industry. Local building codes have been modernized to permit the use of plastic pipe and conduit in residential and commercial construction. Extruded rigid PVC house siding and profiles are gaining in popularity over wood and aluminum. Blown rigid PVC has been developed to compete directly with wood in molding and other profile applications.

Progress continues in the thin film casting and extruding areas. PVC films now have excellent clarity and freedom from imperfections. Because of these improvements and the low raw material cost, soft PVC films are widely used in food and meat wrap applications. Rigid PVC sheeting is employed in packaging applications and goes into the production of credit cards, tapes, and decorative items.

New technology has also expanded the use of plastisols. Chemical blowing agents make it possible to produce foamed articles and foam backing for carpeting and flooring. Conventional plastisols can be modified so that the finished product has a dry-like surface or a high gloss, "wet" appearance. Plastisols also can be made self-adhering to metals and other substrates. Finished articles of almost any shape and appearance can be produced from the wide choice of molding techniques.

These new markets for PVC have developed because of a combination of factors. New compounding developments and fabrication techniques are playing key roles. The low material cost of PVC resin has opened many areas for competition. Equally important, however, is the availability of high performance resins from the PVC industry. Thus, all of the technology discussed in this chapter is contributing to the important growth of this unique polymer.

REFERENCES

(1) L. F. Albright, *Chem. Eng.*, **74**, No. 7, 123 (1967).
(2) P. H. Spitz, *Chem.Eng.Progr.*, **64**, No. 3, 19 (1968).
(3) *Modern Plastics*, **48**, No. 3, 60 (1971).
(4) E. S. Smith, "Polymerization Process with Peroxydicarbonate Initiator Formed In Situ," U.S. Pat 3,022,281 (Feb 20, 1962).
(5) J. C. Thomas, *Hydrocarbon Processing*, **47**, No. 11, 192 (1968).
(6) J. R. Powers, "Polymerization of Vinyl Compounds," U.S. Pat 2,520,959 (Sept 5, 1950).

(7) *Chem. Eng. News.*, **45**, No. 11, 36 (1967).
(8) "New Improved Porosimeter," Bulletin 2405-A, American Instrument Company, Silver Spring, Maryland, 1965.
(9) Standard Specification D 1755-66, *1971 Annual Book of ASTM Standards*, Part 26, American Society for Testing and Materials, Philadelphia, Pennsylvania, 1971.
(10) Recommended Practice D 2396-69, *1971 Annual Book of ASTM Standards*, Part 26, American Society for Testing and Materials, Philadelphia, Pennsylvania, 1971.
(11) F. W. Billmeyer, Jr., *Textbook of Polymer Science*, Interscience Publishers, New York, 1962.
(12) Standard Method of Test D 1243-66, *1971 Annual Book of ASTM Standards*, Part 27, American Society for Testing and Materials, Philadelphia, Pennsylvania, 1971.
(13) G. G. Zahler and Murfitt, *Br. Plastics*, **35**, No. 12, 698 (1963).
(14) C. L. Sieglaff, *SPE Trans*, **4**, No. 2, 129 (1964).
(15) "Case Studies in Plastics Processibility," Bulletin 260, C. W. Brabender Instruments, Inc., South Hackensack, New Jersey, 1966.
(16) Recommended Practice D 2538-69, *1971 Annual Book of ASTM Standards*, Part 26, American Society for Testing and Materials, Philadelphia, Pennsylvania, 1971.
(17) *Modern Plastics*, 1970-1971 Encyclopedia, **47**, No. 10A, 309 (1970).
(18) J. V. Koleske and L. H. Wartman, *Poly(Vinyl Chloride)*, Gordon and Breach Science Publishers, Inc., New York, 1969.
(19) H. A. Sarvetnick, *Polyvinyl Chloride*, Van Nostrand Reinhold Company, New York, 1969.
(20) *Modern Plastics*, 1970-1971 Encyclopedia, **47**, No. 10A, 419 (1970).
(21) *Modern Plastics*, **48**, No. 1, 69 (1971).

20

RUBBER-RELATED POLYMERS
III. POLYETHYLENE*

JAMES E. PRITCHARD
Phillips Petroleum Company
Bartlesville, Oklahoma

HISTORY

Polyethylene became an important commercial product as a result of high pressure reaction studies which were undertaken in the laboratories of Imperial Chemical Industries (I.C.I.) in England during the 1930's. However, polyethylene was prepared prior to 1900 by the decomposition of diazomethane:

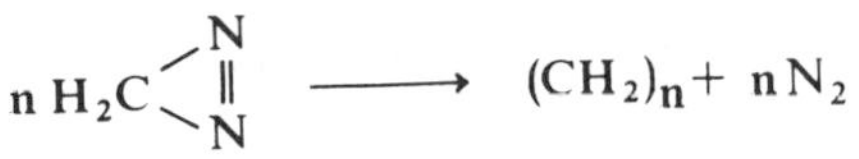

$$n\,H_2C\left\langle\begin{matrix}N\\ \|\\ N\end{matrix}\right. \longrightarrow (CH_2)_n + n\,N_2$$

This reaction was described by Bamberger and Tschirner, who stored diazomethane in ether solution over pieces of unglazed china. The white powder which separated out was found to be polymer containing repeating methylene units melting at 128°C.[1] Apparently this linear polyethylene or polymethylene was observed earlier by Pechman[2] who did not determine its composition. Subsequently, other workers have confirmed this reaction and have studied the properties of polyethylene derived from diazomethane.[3]

The I.C.I. polyethylene, which has become known as "high pressure polyethylene" or "low density polyethylene," was produced from ethylene monomer in a high pressure reaction with trace amounts of oxygen as catalyst. According to Swallow,[4] the success of the initial work at I.C.I. depended on the fact that a leaky reactor was used for the high pressure studies. Fresh ethylene, which was required to maintain pressure, contained traces of oxygen in proper catalytic amounts to ensure continuous polymerization. The subsequent development of reliable compressors capable of maintaining pressures to 3,000 atmospheres made large scale commercial production of high pressure polyethylene feasible.

Twenty years after the I.C.I. work, ethylene polymers of very different properties were developed by three different research groups. These materials, which have come to be known as high density or linear polyethylenes, also were prepared from ethylene monomer but are much closer in physical properties to the original polyethylenes of the 1900 era than to the more recent high pressure process polymers. The leading process for the production of high density polyethylene is that of Phillips Petroleum Company, which uses a chromium oxide catalyst on a silica or silica-alumina support for polymerization of ethylene in hydrocarbon solution or in hydrocarbon slurry. Polymer produced in this process may be completely linear or unbranched materials exhibiting a nominal density of at least 0.96. In addition, a wide range of polymers including low density materials (0.925–0.939) have been produced.

A second low pressure polyethylene process is based upon the work of Professor Karl Ziegler of Germany.[5] The Ziegler process uses an organometal' catalyst comprised of Group IV to Group VI metal halides such as titanium tetrachloride ($TiCl_4$) and an aluminum alkyl such as triethyl aluminum [$(C_2H_5)_3Al$] for polymerization of ethylene monomer. The original polymers produced by the various Ziegler licensees were somewhat lower in density (0.94 to 0.95) than the Phillips process material, but subsequently polymers of 0.96 density have been produced.

A third process for the production of high density polyethylene was developed by the Standard Oil Company of Indiana. The Standard process uses a catalyst comprised of molybdena on an alumina support with sodium, calcium, or their hydrides as promoters.[6]

In contrast to synthetic rubber, polyethylene has no counterpart in nature and technology was not available by direct transfer from known materials. However, some processing and fabrication techniques have been borrowed from rubber and plastics technology and modified to meet the requirements of the new highly crystalline materials. This has contributed to the rapid growth in annual consumption of polyethylene which now exceeds any other plastic material. Domestic use of polyethylene of all types in 1969 was five billion pounds. This is well above the volume of older established polymers such as polyvinyl chloride (2.8 billion pounds) and polystyrene (3.2 billion pounds).[7]

MANUFACTURE

High Pressure Process. The high pressure process ofr the polymerization of ethylene has undergone many improvements and modifications since the original work of I.C.I. However, all major producers depend upon the bulk polymerization of ethylene in molten polyethylene at pressures of 1,000 to 3,000 atmospheres and temperatures of 80 to 300°C.[8] Reactors may be stirred autoclaves or continuous tubes. At one time the products from the two types of reactors differed in physical properties and applications, but in recent years the versatility of the processes has been improved and a broad range of products may be obtained from either system. The catalyst may be trace amounts of oxygen (0.01%), organic peroxides, or other free radical sources. Certain reaction modifiers also may be used to permit production of somewhat higher density resins at the same reaction pressure.

The effluent from the reactor is comprised of molten polymer containing unreacted monomer which is removed from the polymer stream and recycled to the reactor after purification. The degassed polymer is fed to extruders for processing into pellets. In some instances strands of polymer are extruded into water baths and subsequently chopped into small pellets. In other cases molten polymer exudes from a die face and is cut by rotating knives under water to give a spherical rather than a pillow shaped pellet. The pelleted material may be used in the natural state or it may be homogenized in Banbury-type mixers and re-extruded to yield polymers exhibiting improved properties for film or other special applications. Antioxidants, slip agents, pigments, or other additives may be added in the various processing and extrusion steps.

The original high pressure process polyethylene exhibited a density of about 0.92, but more recently the scope of the high pressure process has been broadened to include polymers having densities as high as 0.93 to 0.94. One of the most direct techniques for increasing density of free radical polymerized ethylene polymers is to increase reaction pressure. In laboratory work, polymers of 0.95 density have been obtained at 7,000 atmospheres pressure.[9] However, production of medium density or high density polyethylene by the high pressure process involves sacrifice in production capacities which severely limit the feasibility of this approach. In recent years physical blends of high density and low density polyethylene have been developed to fulfill many of the demands for medium density polyethylene.

A wide variety of copolymers may be derived from the high pressure process by copolymerization of ethylene with other 1-olefins[10] or polar monomers such as vinyl acetate[11] or ethyl acrylate.[12] The 1-olefins are said to function as chain transfer agents and to yield polymers exhibiting lower melt elasticity than the ethylene homopolymer. In general, the effectiveness of the 1-olefin increases with increasing chain length. Copolymerization of ethylene with vinyl acetate leads to polymers which are rubbery in nature at 20 to 30 mole % comonomer, very tacky, soft materials at 50 to 60% comonomer, and hard glassy solids at 80 to 100% vinyl acetate. The copolymers containing minor amounts of vinyl acetate are used as wax modifiers to improve ductility. Ethylene-ethyl acrylate copolymers also are more rubber-like than the ethylene homopolymer. They may be used as tough, ductile injection molding resins and, in some instances, are excellent replacements for plasticized polyvinyl chloride.

Phillips Process. The Phillips low pressure process uses a chromium oxide catalyst on a silica or silica-alumina support for polymerization of ethylene at low pressure in paraffinic or cycloparaffinic solvent.[13,14] In one variation of the process the polymer is prepared in solution. In another modification the polymer is formed as discrete particles in hydrocarbon slurry. The catalyst is prepared by impregnation of the support with a solution of chromium trioxide or other chromium compound and then is activated by calcining in air at 400 to 850°C.

In the solution process, solvent, monomer, and catalyst slurry are charged continuously to the reactor which operates at 125 to 175°C and 20 to 30 atmospheres pressure. In typical commercial practice a thousand pounds or more of polymer are obtained from one pound of catalyst. The reactor effluent is a solution of polymer containing small amounts of catalyst which may be removed by filters or centrifuges. The solvent is removed by flashing and steam stripping or by cooling to precipitate polymer followed by filtration. The polymer is pelletized and compounded with antioxidants, pigments, or other additives as desired.

In the slurry process the reaction media, polymerization temperature, and other environmental conditions are designed to avoid dissolving of the polymer. The product is obtained in the form of a slurry of polymer particles in hydrocarbon diluent and is recovered from the reactor effluent by a simple flashing step. The productivity in the slurry process in terms of pounds of polymer per pound of catalyst is so high that the catalyst removal step may be eliminated.

A wide range of polymers and copolymers of various densities and molecular weights is available from the Phillips process. Commercial resins include density ranges of about 0.925 to 0.962 and molecular weights from perhaps 25,000 to more than 1,000,000.

Ziegler Process. Although a number of producers throughout the world have obtained licenses[15,16] for the use of the Ziegler polymerization catalyst, they have not followed identical routes to the finished polymer. In a typical Ziegler process the catalyst is prepared by combining triethyl aluminum and titanium tetrachloride in a diluent. The tetrachloride is reduced, at least in part, to titanium trichloride, which may be isolated and combined with more triethyl aluminum to yield a catalytic mixture. Alternatively, the mixture of aluminum alkyl and tetrachloride may be charged to the reactor without isolation of the trichloride. Polymerizations are carried out in batch or continuous systems and the polymer is obtained as a slurry in hydrocarbon diluent.

The hydrocarbon diluent is recovered from the reactor effluent and recycled. The catalyst is removed from the finely divided polymer by washing with alcohol, such as isopropyl alcohol, or other reagents. After the polymer cake is recovered and dried, it is pelletized and compounded with various additives by standard techniques.

The very early Ziegler type polyethylenes were of high molecular weight and were relatively difficult to process. However, various techniques have been developed for controlling molecular weight, including increasing the proportion of titanium tetrachloride in the reaction media[17] or introduction of hydrogen.[18]

Standard Oil of Indiana Process. In the Standard process[10] the molybdena on alumina catalyst with suitable promoters, such as sodium, calcium, or their hydrides, is charged to a reactor containing solvent and ethylene monomer at 230 to 270°C and 40 to 80 atmospheres. The reactor product is subjected to a flashing step to remove unreacted ethylene monomer, followed by filtration to remove catalyst. After the solvent is flashed, the polymer is steam stripped and fed as a molten stream to an extruder, where it is pelletized.

PHYSICAL PROPERTIES

Information on physical properties is used to characterize polyethylenes and to predict their performance in end use applications. In addition, some physical measurements are sensitive to structural features of the polyethylene molecule and are of value in fundamental polymer research. Because polyethylene, for the most part, is used in the natural state, free of additives and reinforcing fillers, structural features follow through to end product performance to a greater degree than is the case with an elastomer. Polymer deficiencies cannot be masked or compensated for by subsequent compounding and curing steps. Consequently, it is of greatest importance to recognize and control the composition of the polymer.

Density. The properties of polyethylene are largely dependent upon crystallinity and molecular weight. The level of crystallinity may be determined by X-ray studies or by nuclear magnetic resonance techniques.[20] Because there is a direct correlation between the degree of crystallinity and polymer density, it is convenient to measure density using a gradient column[21] or another technique.[22] As indicated in Table 20-III-1, a high density polyethylene of 95% crystallinity and 0.96 density exhibits a higher melting point, higher tensile strength, higher flexural modulus, and greater hardness than does a low density polyethylene of 65% crystallinity and 0.92 density. Chemical properties of polyethylenes also are dependent upon crystallinity and density, with high density polymers exhibiting superior resistance to oils, solvents, and permeation by water vapor and gases.[23]

TABLE 20-III-1. PROPERTIES OF POLYETHYLENES

Polyethylene	*Crystallinity, %*	*Melting Point °C*	*Density*	*Ultimate Tensile, psi*	*Elongation at Break, %*	*Flexural Modulus, psi*	*Hardness Shore D*
High Density	95	134-137	0.96	4400	25	230,000	68
Low Density	65	114-116	0.92	2000	500	45,000	52

Melt Index. The molecular weight of polyethylene, like other high polymers, may be determined by solution viscosity, melt viscosity, light scattering or ebulliometric techniques. A common method for estimating molecular weight involves measurement of melt viscosity in a simple extrusion plastometer or melt indexer. In this device, the molten polymer is extruded through a 0.0825-inch orifice at a temperature of 190°C with a dead load piston providing a force of 2160 grams. The weight of polymer in grams extruded in ten minutes is termed the "melt index."[24] Typical melt index values for commercial polyethylenes are in the range of 0.1 to 20. The polymers of low melt index or high molecular weight are preferred for heavy duty applications such as plastic pipe, while those of high melt index or low molecular weight may be preferred for high speed molding or extrusion operations. Recently it has been recognized that a simple instrument which provides a single determination of melt viscosity under an arbitrary set of conditions is inadequate to characterize fully the melt flow properties of polyethylene. In many respects the melt indexer occupies the same position in polyethylene technology as does the Mooney viscometer in rubber technology. Both instruments are of value in characterizing polymers providing the limitations of a single point determination of viscosity are recognized. An additional problem with the melt indexer resides in the fact that there are several sources of error inherent in the device which further limit its utility.[25] Nevertheless, melt index determinations are universally used in the polyethylene industry and, providing the limitations of the equipment are recognized, considerable use can be made of them.

Stress-Strain. Stress-strain properties of polyethylenes are understandable to rubber technologists although it is recognized that polyethylene is relatively inelastic beyond a few percent elongation. As indicated in Fig. 1, stress increases sharply with small increments of strain up to the point of yield after which it drops precipitously and remains more or less constant during the cold drawing step which precedes final rupture. Polyethylenes of 0.96 density are free of branching and exhibit relatively little chain entanglement. This may account for the fact that stress levels decrease regularly with increasing strain and linear polymers do not exhibit the reinforcing and orientation effect immediately preceding the break

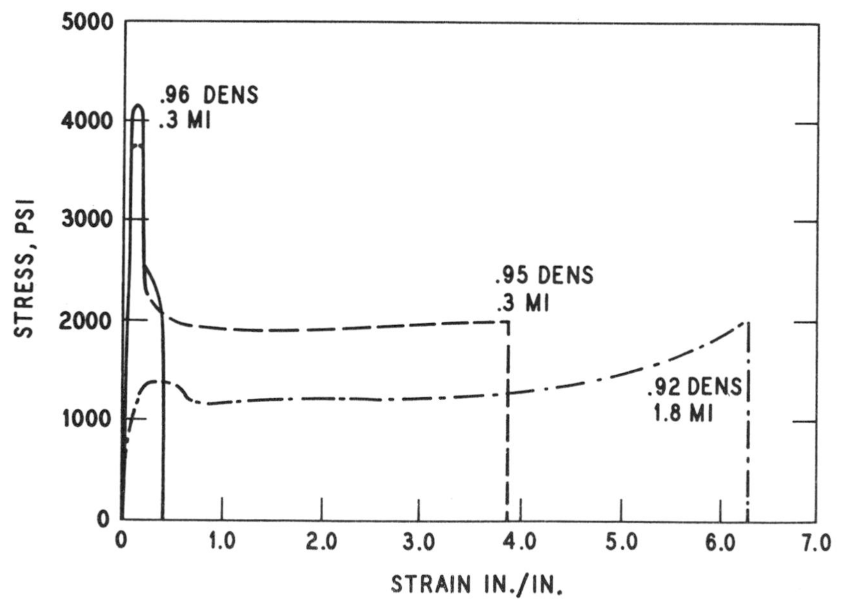

Fig. 1 Stress-strain curves for ethylene polymers. Strain rate = 5 in./in./min. Temperature = 23°C.

point as do lower density polyethylenes and other polymeric materials. The decrease in stress or stress decay exhibited by unbranched polyethylene is particularly noticeable at low rates of strain or at temperatures approaching the melting point of the polymer.

Stress-strain properties of polyethylenes, like other thermoplastic materials, are time dependent. For example, the tensile strength of 0.96 density polyethylene at a strain rate of 20 in./in./min is above 4,000 psi. However, if this polymer is subjected to a static load of 500 psi, it will break within a relatively few hours and at very low elongation. This type of failure is sometimes called "glassy" or "brittle" and is unexpected unless one is familiar with the load bearing properties of the polymer. If molecular weight is increased substantially, some improvement in load bearing properties is obtained and the failures become more ductile in nature. A more effective way of improving load bearing properties is to incorporate branches in the polymer molecule. It can be shown that three or four short chain branches per molecule will have a beneficial effect upon load bearing properties and will reduce the time dependence of the stress strain properties quite substantially. The effect of branching on polymer performance and application will be considered in more detail in the section on structural features.

Flexural Modulus. Flexural modulus values reflect the relative stiffness or rigidity of the polymer. In the determination of flexural modulus, a rectangular beam of polymer supported at both ends on a tensile testing machine is subjected to a deflecting force at the mid-point.[26] Values for low density polyethylenes are relatively low and range from 15,000 to 45,000 psi, while those for high density polyethylenes fall in the range of 150,000 to 250,000 psi. Other tests may be used to determine the rigidity of polyethylenes and one of the most common is the stiffness in flexure test.[27] In this test the force required to deflect a cantilever beam or strip of plastic is measured. Stiffness values for polyethylene are 10,000 to 30,000 psi for low density materials and 100,000 to 175,000 for high density polymers.

Impact Strength. Many laboratory tests have been devised for the determination of impact strength, but none of these is universally useful in predicting end product performance. The most common impact value is derived from the Izod test,[28] which uses a pendulum machine to break a cantilever beam of the polyethylene. The beam may be notched or unnotched, and notches of different radii may be used to provide an indication of notch sensitivity. High density polyethylenes yield Izod values of 1 to 5 ft lb/in. of notch, while low density polyethylenes, particularly high molecular weight and ductile materials, may give values above 10. Flexible, tough polymers yield meaningless values because the plastic beam simply deflects and does not break. More recently techniques for determining impact in tension have been studied.[29] In these tests the energy to break a tension impact specimen is determined by means of a pendulum machine which, in some instances, is a simple modification of the standard Izod device.

Because of the inadequacies of laboratory impact tests, it is common practice to determine impact strength directly on fabricated items such as injection molded containers, polymer films, or plastic bags. This is necessary because mold designs, extrusion rates, molding temperature, or other fabrication variables may affect product properties. A common technique for determining impact strength of injection molded containers involves dropping a steel ball on the area of the item which is most susceptible to failure. Impact strength of films may be determined by a similar technique which requires dropping a missile or dart from various heights on a stretched sample of film.[30] Blow molded items such as bottles and plastic bags often are filled with water, sand, polymer, or other material and subjected to drop tests.

Environmental Stress Cracking Resistance. Environmental stress cracking (ESC) is the term applied to the premature failure of polyethylene in the presence of soap, detergent, water, or other active environment, usually under conditions of relatively high strain.[31] This phenomenon first was recognized in polyethylene coated wire which often was lubricated with surface active materials to facilitate installation in conduits. Under these conditions, polyethylenes which appeared to perform satisfactorily in the laboratory very rapidly developed severe cracks which propagated completely through to the conductor. A laboratory test, commonly known as the Bell ESC test or bent strip test, was devised by the Bell Telephone Laboratories and has been extensively studied and widely used throughout the world. In this test, a polymer specimen 1.5 in. x 0.5 in. x 0.125 in. is provided with a razor slit 0.75 inch long, 0.020 inch deep in the center, and parallel to the longest side. The specimen is bent in a "U" shape, held in a metal channel, and exposed to detergent or other environment at elevated temperature, commonly 50°C. The time for five out of ten specimens to fail is reported as the F_{50} value. Stress cracking resistance of polyethylenes is improved by increasing the molecular weight and by including short chain branching in the molecule. Very low molecular weight materials may fail in this test in a matter of minutes, while high molecular weight materials or polymers of moderate molecular weight and moderate degrees of branching will survive for several thousand hours. This test is used widely for evaluating polyethylenes and predicting performance in wire coating applications and other uses.

Electrical Properties. Polyethylene has been of interest as an insulation material almost from the beginning of the development of the high pressure process. It offers a very low dielectric constant as might be expected from the relatively low density of the polymer. High pressure polyethylene of 0.92 density exhibits a dielectric constant of 2.28, while the value for 0.96 density materials is about 2.35. Standard techniques for determination of dielectric constant, power factor, and other electrical properties are applicable to polyethylene.[32] In general, electrical properties deteriorate very rapidly if the polymer is oxidized,[33] and every effort must be made to provide efficient antioxidant systems if the original electrical and mechanical properties are to be maintained for an extended period of time.

Effect of Additives. Antioxidants must be incorporated into polyethylene prior to fabrication of test specimens, determination of melt viscosity, or investigation of structural features if useful data are to be obtained. This is important because virgin polyethylene, particularly in the molten condition, is rapidly oxidized to yield a variety of chain scission products, network structures, and carbonyl or other oxygen-containing groups. In addition, antioxidants, and in some instances ultraviolet light stabilizers, are required if reasonable service life of end products is expected.

Antioxidants for polyethylene differ from those used in many rubber applications in that virtually all polyethylene antioxidants are food grade materials which are generally colorless and odorless. Antioxidant levels in polyethylene are relatively low, usually in the range of 0.001 to 0.1%. One of the most widely used materials is 2,6-ditertiarybutyl-4-methyl phenol:

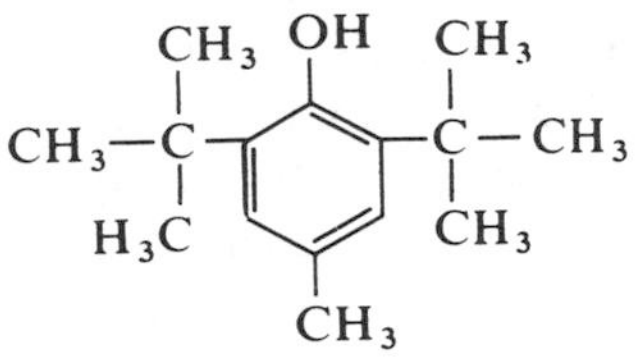

CH_3 OH CH_3
CH_3—C— —C—CH_3
H_3C CH_3
CH_3

For heavy duty applications 4,4-thiobis(6-tert-butyl)-m-cresol often is used:

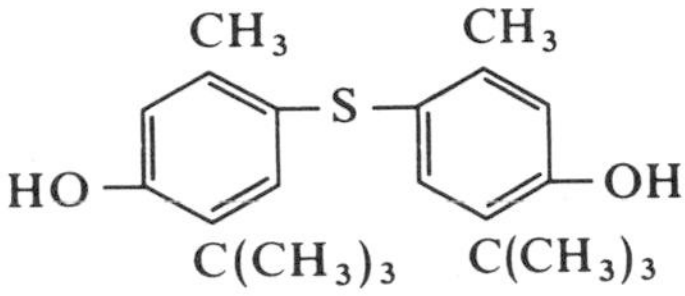

CH_3 CH_3
S
HO— —OH
$C(CH_3)_3$ $C(CH_3)_3$

One of the most common methods for determining oxidation resistance of polyethylene compounds is the oxygen uptake test.[34] In this determination a sample molten polymer is subjected to an atmosphere of oxygen at elevated temperatures such as 140°C. The absorption of oxygen by the polymer sample is measured over a period of time. When oxygen absorption is plotted against time, it is seen that very little reaction occurs initially, but within a few hours this induction period is terminated and a rapid increase in oxygen absorption occurs. Antioxidants extend the induction period and reduce the subsequent rate of oxygen absorption. In the absence of oxygen, polyethylene is relatively stable to heat, and, in fact, is said to be more stable than polyisobutylene or SBR.[35] At temperatures above about 300°C, chain scission predominates, and low molecular weight polymers may be formed. However, polyethylene does not "unzip," and very little ethylene monomer is obtained by thermal degradation of the polymer.[36]

Polyethylene does not absorb ultraviolet light and one would expect it to be resistant to ultraviolet degradation. However, traces of carbonyl, resulting from oxidation, are always present and the carbonyl groups absorb light in the ultraviolet region below 3300 Å. It has been shown that carbonyl groups exercise a very important effect on the resistance of polyethylene to ultraviolet light; and, in fact, wavelengths outside of the carbonyl absorption band (2200 to 3200 Å) do not cause polymer degradation by photo-oxidation.[37]

Unprotected polyethylene degrades on outdoor exposure within a few months to a year depending upon the hours of exposure and the intensity of the sunlight. In applications such as pipe, wire coating, and cable sheathing where a black color is acceptable, it is common practice to add two to three percent carbon black for ultraviolet light protection. Preferred carbon blacks are fine to medium particle blacks, although soft blacks at relatively high loadings will give some protection. In commercial practice the carbon black is dispersed in polyethylene in the form of a concentrate which may contain fifty percent carbon black in low density polyethylene or twenty-five percent black in high density polymer. The concentrate may be "let down" with virgin polymer to give the required black loading. Experience has shown that a good dispersion of carbon black in the concentrate virtually ensures a good dispersion of black in the let down product. Because ultraviolet light stability depends upon the quality of the dispersion as well as the particle size of the black, particular attention is paid to the composition and preparation of the concentrate when maximum resistance to outdoor exposure is required.

For nonblack polyethylene applications such as rope, fabric, or molded items designed for outdoor exposure, various types of UV absorbing organic compounds are incorporated into polyethylene. Among the most common of these are benzophenone derivatives and salicylates which are used in conjunction with antioxidants to provide outdoor life of perhaps two to five years. Although this represents a substantial improvement over additive-free polyethylene, it is substantially below the life expectancy of twenty-five years or more which may be obtained from the best carbon black stocks. Typical Weather-Ometer data for UV stabilizers in high density polyethylene at a level of 0.25% are presented in Table 20-III-2.

TABLE 20-III-2.

Exposure of High Density Polyethylene in a Weather-Ometer	*Hours to 50% Loss in*	
	Elongation	*Tensile*
2,2′-Dihydroxy-4-octoxybenzophenone	1000	>1600
tert-Butylphenyl salicylate	400	800
Control – no UV stabilizer	125	300

The condition of the polymer after various periods of exposure is indicated by measuring tensile and elongation from which the number of hours required to provide a fifty percent loss in properties may be determined. As indicated in the table, the most effective stabilizer is the benzophenone derivative. However, in some instances the lower cost salicylate material may be preferred.

STRUCTURAL FEATURES

Branching. One of the most important structural features of the polyethylene molecule is the number and type of branches. Highly branched polymers such as low density polyethylene containing 15 to 20 branches per thousand carbon atoms are low in crystallinity and density as well as stiffness and hardness because the branches reduce chain regularity. Completely linear polyethylenes, on the other hand, exhibit maximum crystallinity, density, hardness, and rigidity. The branches on low density polyethylene chains are mainly ethyl and butyl.[38] Similarly, some high density polyethylenes contain short chain branches which, in the case of Phillips process polymers are determined by the type of 1-olefin comonomer which is used. Propylene and 1-butene have been used extensively in this process to give methyl and ethyl branches, respectively. The effect of branches on the density of Phillips type ethylene polymers is illustrated in Fig. 2. It can be seen that ethyl branches are more effective in reducing density than are methyl branches. Additional studies of this type have shown that longer branches are even more

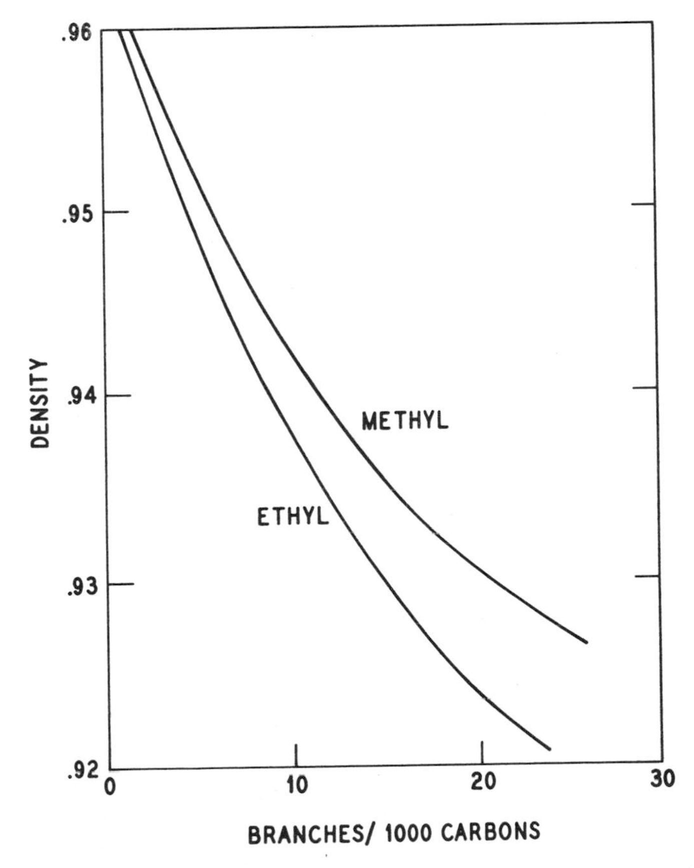

Fig. 2 Effect of methyl and ethyl branches on the density of ethylene polymers.

effective than ethyl, but the differences become less pronounced as the length of the branch increases.

With the advent of the Phillips process, completely unbranched polyolefins became available in quantity. This provided a base point for the study of branching effects which was previously unavailable. In recent years it has become quite apparent that the first two or three branches on a polymer molecule exert a very marked effect on physical properties, particularly long term load bearing properties, stress crack resistance, and cold drawing performance. For example, an ethylene homopolymer of 0.96 density may exhbit an F_{50} value in the Bell ESC test of 60 hours, while an ethylene-1-butene copolymer of 0.95 density containing four or five ethyl branches per thousand carbons will have an F_{50} value of 400 hours.[39] In the form of monofilament fiber, the copolymer will withstand a given static load, 20,000 psi for example, for ten times as long as the homopolymer. When these materials are cold drawn at temperatures below the crystalline melting point, the unbranched homopolymer necks down to smaller and smaller cross section and reaches no limiting dimensions until it eventually breaks as a fine thread. On the other hand, the copolymer, when cold drawn, necks down to a limiting dimension which is a predictable characteristic of the polymer under the conditions of test. At one time it was believed that most polymers would cold draw to a limiting dimension and would give a draw ratio which was of the order of 4 : 1. It is now known that this draw ratio is characteristic only of highly branched materials of the type which were available prior to the development of truly linear polyethylenes.

Crystallinity. Single crystals of polyethylene when grown in dilute solution are found to be flat plates in which the chain direction of the molecule is perpendicular to the plate.[40] The crystal has distinct growth layers about 100 Å thick indicating that any one molecule is confined largely to a single layer and probably is folded. However, it is suggested that some molecules may participate in more than one layer in which case the resistance to shear failure of the crystal under stress may be increased.

Although polyethylenes, particularly high density, highly crystalline materials, are relatively opaque as a result of an effective scattering of light, this does not involve polymer crystals because they are smaller than the wavelength of visible light. Light scattering and opacity result from larger structural units such as spherulites which are formed by outward growth of crystals from nucleation points. It has been noted that spherulitic growth continues until the spherulite meets an obstruction such as another spherulite or a solid surface. Spherulites may be 0.05 mm in diameter and can be seen very clearly with polarizing microscope under low power. In addition to internal diffraction of light, spherulitic structures often protrude from the surface of blown film in such a way that surface roughness from this source accounts for much of the observed opacity. The maximum growth of a spherulite requires a finite time; consequently, it is possible to limit their size and number by quenching, orienting, or applying other external or internal forces which will partially immobilize the polymer until it is completely chilled. In the quenched condition, spherulitic growth is nonexistent or so slow that it may be regarded as permanently inhibited.

As has been noted previously, there is a good correlation between crystallinity, density, and other physical properties of polyethylenes. For this reason, there is definite interest in not only providing polymers of inherently different levels of crystallinity but in modifying and controlling the crystallinity of molded and extruded products from a given type of polyethylene. Very little information has been published in this area, but it is known, for example, that crystallinity is reduced by quenching of polymer melts, and advantage of this fact is taken in the production of film by chill roll and water quench techniques. Nucleation or seeding techniques will produce many growth sites and yield a finer crystal structure. In other applications such as injection molding or blow molding it is desirable to increase the rate at which polymer freezes in the mold, thereby reducing cycle time and increasing production rates.

Molecular Weight. Determinations of molecular weight and interpretations of data for polyethylene are similar to those for rubber and other polymers. One of the most common techniques for molecular weight determination is the measurement of dilute solution viscosity, usually in tetralin at 130°C. Dilute solution viscosities may be correlated with weight average molecular weight (M_w) as determined by light scattering[41] as follows:[42]

$$n = 3.78 \times 10^{-4} M_w{}^{0.72}$$

Commercial products for the most part exhibit dilute solution viscosities of one to three corresponding to weight average molecular weights of 50,000 to 250,000.

Another commonly used molecular weight average is the number average (M_n), which may be determined by osmometry or in the case of polyethylene more satisfactorily by boiling point elevation or ebulliometry.[43] Number average molecular weights for commercial polyethylenes are in the range of 5,000 to 20,000, which gives weight average to number average ratios of 10:1 to 15:1. This ratio has been used as an indication of the breadth of the molecular weight distribution and indications are that most general purpose polyethylenes exhibit relatively broad distributions. In recent years, however, numerous polymers of relatively narrow molecular weight distribution having weight to number average ratios of perhaps 3:1 to 5:1 have become available for special uses, particularly for film paper coating and injection molding.

In general, the higher the molecular weight, the better the physical properties of a polyethylene and the more reliable the performance of the end product. However, as molecular weight increases, processibility, as measured by extrusion rates, surface quality, or other fabrication limitations, decreases. It is, therefore, important to optimize molecular weight and fabrication rates in each instance. High molecular weight polymers are particularly useful for structural and load bearing applications requiring long term reliability. Pipe, wire coating, and monofilament are applications in which these factors are important. In Table 20-III-3, for example, the effect of molecular weight on the performance of high density polyethylene pipe is indicated. It may be seen that the predicted twenty-year hoop stress, as indicated by long term stress life curves, is substantially better for the high molecular weight material.

TABLE 20-III-3. DEPENDENCE OF PIPE PERFORMANCE ON MOLECULAR WEIGHT

Polyethylene	*Predicted 20-Year Maximum Hoop Stress at 88° C, psi*
High MW (500,000)	700
Med. MW (150,000)	10

$$\text{Hoop stress, psi} = \frac{\text{Pressure (psi)} \times \text{outside diameter (in.)}}{2 \times \text{wall thickness (in.)}}$$

Molecular Weight Distribution. Although polymers of very narrow molecular weight distribution may be prepared by special fractionation or synthesis techniques, commercial polymers without exception contain molecules of widely varying size. In recent years it has been shown that a single average molecular weight value such as M_w or M_n or even a combination of these is inadequate to describe fully the molecular weight distribution of the polymer and to predict its performance in fabrication equipment and in end use applications. One technique for studying molecular weight distribution is fractionation which may be done by methods developed for other polymers except that crystalline polyethylenes require the use of hot solvents. One type of apparatus which is used is the packed column containing glass beads, clay, or similar material on which polymer is deposited.[44] Elution of the polymer from the column is achieved by adding solvent/nonsolvent mixtures with a gradual increase in the ratio of solvent to remove higher molecular weight materials as the fractionation progresses. One important deficiency of this technique is that five or ten per cent of highest molecular weight material is not recovered or at best is poorly fractionated. In many instances this is the most important component and lack of information on the high molecular weight fraction severely limits the value of the data.

The most widely used technique for fractionation of polymers is gel permeation chromatography.[45] Polymer solution is passed through a packed column containing a crosslinked microporous polystyrene. Large molecules are least readily absorbed, and therefore appear in the first fraction of effluent; the smallest molecules appear last. A correspondence between fraction number and molecular weight can be obtained by comparison with an established standard. Figure 3 indicates molecular weight distribution of two polyethylenes by gel permeation chromatography.

An important aspect of molecular weight distributions which applies directly to fabrication problems is the dependence of the apparent melt viscosity upon shear rate. As indicated in Fig. 4, polymers of broad molecular weight distribution exhibit the greatest reduction in apparent viscosity with increasing rate of shear. These materials consequently are of interest for fabrication steps such as filament extrusion where high shear rates are involved. As a corollary to this observation, it may be seen that broad distribution polymers of relatively high molecular weight

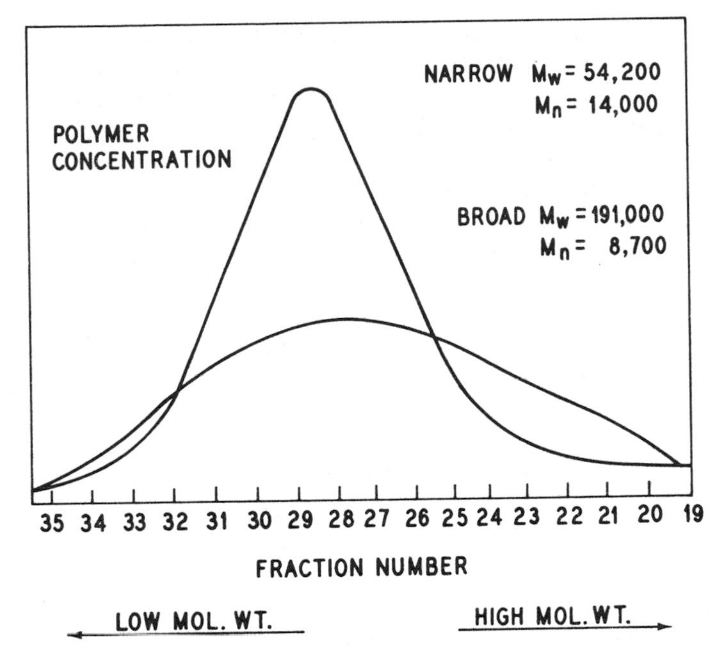

Fig. 3 Molecular weight distribution of two polyethylenes as shown by gel permeation chromatography.

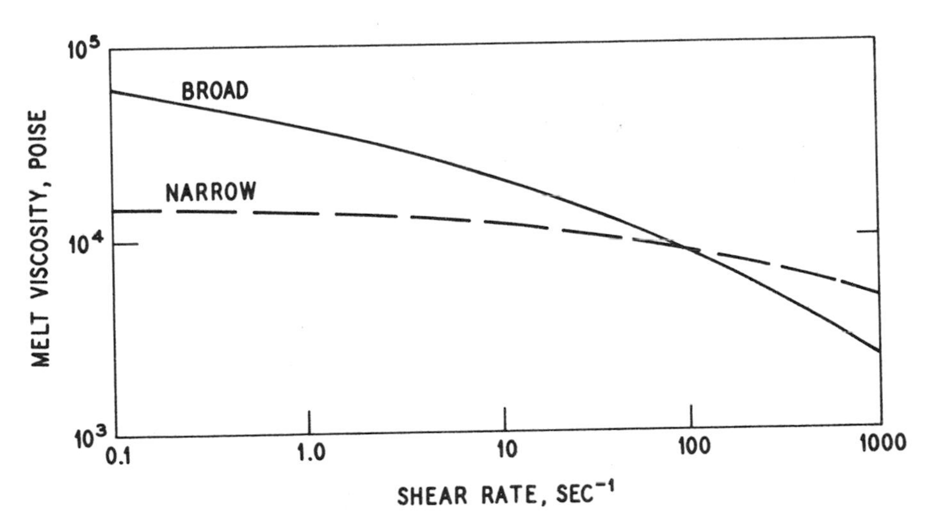

Fig. 4 Dependence of melt viscosity on shear rate for two polyethylenes of different molecular weight distribution.

may be fabricated satisfactorily if high shear rate fabrication steps are involved. On the other hand, polymers of relatively narrow molecular weight distribution exhibit low melt viscosity at low shear and are of primary interest for an operation such as paper coating in which low shear flow and spontaneous "knitting" or "healing" of the melt on the substrate is required.

Crosslinking. In general, any uncontrolled crosslinking of polyethylene resulting from inadvertent oxidation during the production of fabrication steps is undesirable. It may be shown that low order crosslinking leads to very substantial increases in melt viscosity without the concomitant improvement in physical properties normally expected from high melt viscosity (high molecular weight) material. In some instances, crosslinking may be done intentionally by addition of peroxide or other curing agents to provide resistance to creep at high temperatures or to improve environmental stress cracking resistance for special applications. Deliberate crosslinking to provide a vulcanized polyethylene compound will be considered in the section on applications. Because crosslinking exerts such a marked effect on melt viscosity, particularly when the viscosity is measured under conditions of low shear rate, it is possible to use rheological tools to detect the presence of crosslinks. One of the most effective devices for this purpose is a low shear capillary melt viscometer which operates at shear rates in the range of 0.1 reciprocal seconds.[46] The detection of crosslinking requires development of data on low shear melt viscosity

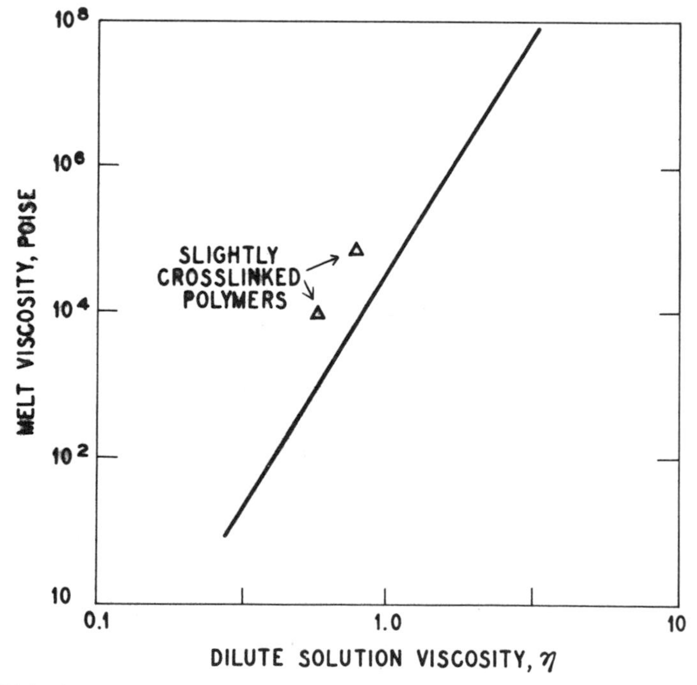

Fig. 5 Melt viscosity-solution viscosity relation for linear polyethylenes of variable molecular weight showing the effect of crosslinking.

and dilute solution viscosity for uncrosslinked polyethylenes of various molecular weights (Fig. 5). When similar data are obtained on slightly crosslinked polyethylenes, it is seen that the low shear melt viscosity is unusually high for a given solution viscosity. This technique also may be sensitive to the presence of long chain branching and, in fact, in the limiting case low order crosslinking and long chain branching may be indistinguishable.

APPLICATIONS

Fabrication Equipment. In the polyethylene industry, extruders, Banbury mixers, and injection molding machines are basic tools although roll mills and calenders are used in some instances. Extruders are used in polyethylene manufacture for pelletizing operations and in fabrication plants for wire and cable coating, production of pipe, sheet, and film, and for feeding blow molding units in the production of bottles and other containers. Because polyethylene must be extruded at an elevated temperature and because of the inherent high heat requirements for a crystalline material, polyethylene extruders must be capable of high temperature operation and high heat input rates. This is apparent from a study of a specific heat versus temperature plot.[47]

The majority of polyethylene extruders are electrically heated although some oil heated machines are in use.[48] In general, they offer length to diameter ratios of 20:1 or 30:1 compared to 8:1 or 10:1 for machines designed for rubber and noncrystalline plastics. The long barrel design ensures adequate melting, mastication and metering of the polymer, and is particularly important for high speed extrusion of high density polyethylenes. Screen packs are used to develop back pressure and for removing charred polymer and foreign particles. Stock temeratures range from 150 to 250°C or higher, dependng upon the application. Screw diameters of 1 inch to 12 inches are common and outputs range from a few pounds per hour to several tons per hour. There are a wide variety of machine designs including twin screw machines and single screw types containing high shear sections or other modifications for special blending, compounding, or homogenizing operations.

Injection molding machines for polyethylene are similar to those used throughout the plastics industry. In the injection molding process (Fig. 6) polymer is melted in the barrel of the machine and forced into a mold under high pressure. Most machines have hydraulically operated rams and mold clamps although some very small laboratory machines may have partially or totally mechanical systems. Ram pressures commonly are in the range of 10,000 to 30,000 psi with clamping forces of 30 to 2500 tons. Injection molding machines are the most common devices for making housewares such as dishes, wastebaskets, and garbage cans as well as automobile and machine parts. In recent years reciprocating screw type injection molding machines have become quite common in the plastics industry.[49] These machines offer advantages in reduced cycle time, high plasticizing capacity, improved color dispersion, low stock temperatures, and less warpage or internal stress in the molded part.

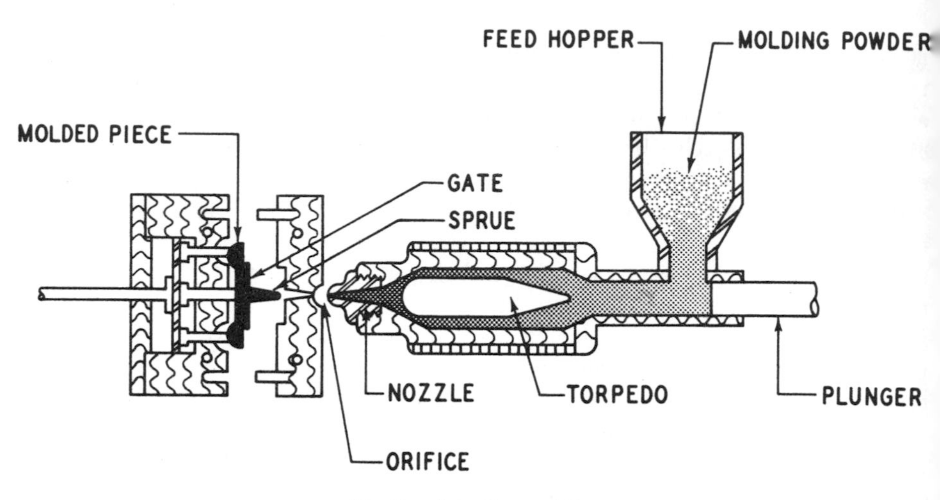

Fig. 6 Injection molding.

Banbury mixers, roll mills, and calenders are similar to their counterparts in the rubber industry. Banbury mixers often are used for homogenizing polyethylene and for incorporation of pigments or additives. Very frequently intensive mixers of the Banbury type are mounted above a pelletizing extruder which receives molten polymer from the mixer and converts it into finished resin. Roll mills are similar to rubber mills with provision for high temperature operation. Small mills are very common in laboratories, but massive units of the type found in the rubber industry are rare. Calenders in the polyethylene industry are roughly similar to rubber calenders and are used for lamination and sheeting of polyethylene. In some instances special types of film or tape are derived from calender operations. However, most laminating film or sheet production in the polyethylene industry involves extruder fed units rather than calenders.

Polyethylene Consumption. Domestic polyethylene consumption in 1971 was 6,365,000,000 pounds,[50] of which 4,480,000,000 was low density polyethylene and 1,885,000,000 high density.

Film and Sheeting. The term "film" is applied to material of less than 0.010 inch thickness, while sheet is material above this dimension. There are three basic processes for the production of polyethylene film: namely, tubular, chill roll, and water quench. In the tubular film process an extruder feeds molten polymer to a heated die designed to give a tubular extrudate. Air pressure is maintained inside the tube to provide a bubble of the desired size. The melt is cooled in air or by cooling rings, which provide an even distribution of forced air. The tubing is pulled from the die, collapsed by guide systems, and wound up on rolls. It may be used directly for making bags, or it may be slit to provide flat films up to twenty feet or more in width.

In the chill roll process the extruder feeds molten polymer to a slit type die from which the melt extrudes as a thin sheet or web. The die is placed within a few inches of the polished chill roll and the polymers pulled under and over a series of rolls onto a wind-up roll. Speeds of up to 500 to 1,000 feet per minute are common and film clarity is outstanding. In recent years the advantage for this process in film clarity has been reduced somewhat by the development of new resins which provide excellent optical properties from the tubular film process.

In general, fresh polyethylene film surfaces are nonreceptive to printing ink and must be treated by flame or electrical discharge to promote a mild oxidation of the surface. This operation may be done in line while the film is being produced or it may be part of the subsequent printing step. Very often antiblock and slip additives are incorporated in polyethylene film resins to reduce adhesion of film surfaces and permit high speed operation in packaging equipment. Fatty acid amides and finely divided silica are among the most common additives for this purpose.[51]

Low density tubular films offer good elongation and tear strength as well as relatively low haze. Materials of this type have been the utility films of the construction and agricultural industries. Medium density polyethylene films derived from chill roll techniques are used widely in the packaging of dry goods, bakery products, and other consumer items because of their excellent balance of properties and low level of haze. High density water quench film also offers good optical properties and is used in overwrap packaging where the stiffness, low moisture transmission, and tear-tape feature are desired. High density tubular film exhibits high haze and relatively low tear strength, but recent rubber-modified films of this type offer outstanding puncture and tear resistance and are used for heavy duty industrial packaging.

In the production of polyethylene sheet the most common practice involves extrusion of a molten web on a polished chill roll. This system differs from chill roll film operations in that the chill roll is heated to a temperature slightly below the freezing point of the polymer and linear speeds are very much lower. Polyethylene sheet is used for production of trays, boats, chairs, containers, and many other items by thermoforming techniques.[52,53,54] In a thermoforming operation the polymer sheet is heated above the melting point, drawn into a mold, or draped over a mold which contains numerous small holes. A vacuum is applied to propel the molten sheet against the sides of the mold. Many variations of the basic vacuum forming idea have been developed to provide low cost production of large or complex items. Thermoforming also is of interest when the demand for an item is too low to justify investment in the more expensive injection molding or blow molding equipment.

Injection Molding. Polyethylene injection molding technology originally was based on that developed for polystyrene, which continues to be an important injection molding resin. Injection molding machines commonly are rated for capacity on the basis of the number of ounces of polystyrene,which can be injected per shot. Machines of one ounce to 400 ounces capacity or greater are available. In general, equipment of this type is relatively expensive with machine prices falling in the range of $5,000 to $100,000 or more, and molds costing $500 to $10,000 each. For this reason injection molding is economical only for long continuous runs in which thousands of identical parts are produced. The choice of polyethylene for injection molding operations depends upon performance and price considerations. High density resins are gradually gaining acceptance in this area because they provide greater stiffness or permit the use of thinner walls. However, low density polyethylenes continue to be used in great volume, particularly in housewares items and general purpose containers. Colored injection molded items may be obtained by adding pigment as a dry blend on virgin pellets or by including concentrates of pigment and polymer along with the natural pellets fed to the machine. If the ultimate in color dispersion is demanded, a color compound prepared by "letting down" a concentrate with natural polymer in a Banbury mixer or extruder is used.

Injection molded items are identified by the presence of a tiny button of polymer in the "gate" area of the part. This is the point at which the polymer entered the mold cavity and very often is located at the mid-point of the bottom area in items such as dishes, pails, or other containers. Vacuum formed and blow molded items do not have this characteristic. Gates also may be located at the edge of a part and occasionally more than one gate is used. Mold design is a sophisticated art. Some designs which require close temperature control at different levels for specific areas in the mold or which contain several cavities equipped for independent, sequential, or simultaneous operation can be quite complex.

Blow Molding. Blow molding is the only major application in which high density polyethylene consumption exceeds that of low density polymer. The early technology was based on that of glass,[55,56] and some terms, such as "parison," which denotes the molten tube from which the item is blown, are used both in glass and plastics industries. In recent years most major bottle, can, and container manufacturers have entered the blow molding field. In addition, some equipment manufacturers, resin suppliers, and others are interested in special fields of blow molding such as the production of large tanks, bag and box containers, large toy cars and animals, beverage cases, and the like.

In principle, all blow molding systems require that a molten tube or parison of polymer be delivered to an open mold. The mold closes on the parison and air pressure is applied to force it against the inside surfaces. The solidified part is ejected from the mold and trimmed if necessary (Fig. 7). Although injection molding machines may be used to supply the molten parison, most producers utilize extruders for this purpose. In some designs the extruder feeds an accumulator which in turn forms the parison. This permits continuous operation of the extruder even though the parison extrusion is intermittent. In other designs, several molds are placed on a rotary table and are automatically presented under the extrusion head to receive the parison at the proper time. Continuous extrusion of tubing to be received by molds on a belt or wheel is another technique which is used widely in the United States. Blow molding is one of the fastest growing segments of the polyethylene industry. Important products derived from blow molding operations include bottles for detergent and bleach packaging, five to 50 gallon containers for home and industry, automobile ducts, and machine parts.

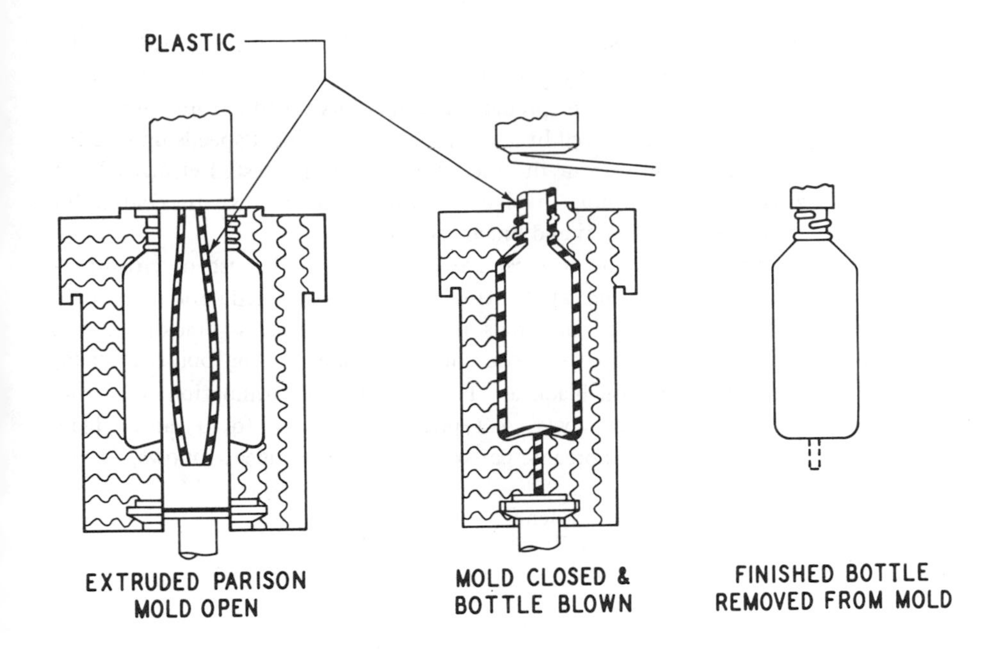

Fig. 7 Blow molding.

Wire and Cable. One of the first applications for polyethylene involved the use of low density pilot plant material for a one-mile length of submarine cable in England in July 1939.[54] Polyethylene continues to be a material of choice for this and similar applications because of its low water absorption and low power factor. Power cable is the most rapidly growing insulation application.[57] Cables handling up to 15 KV have been in service for 15 years and polyethylene is the only proved material for buried cathodic protection cable.

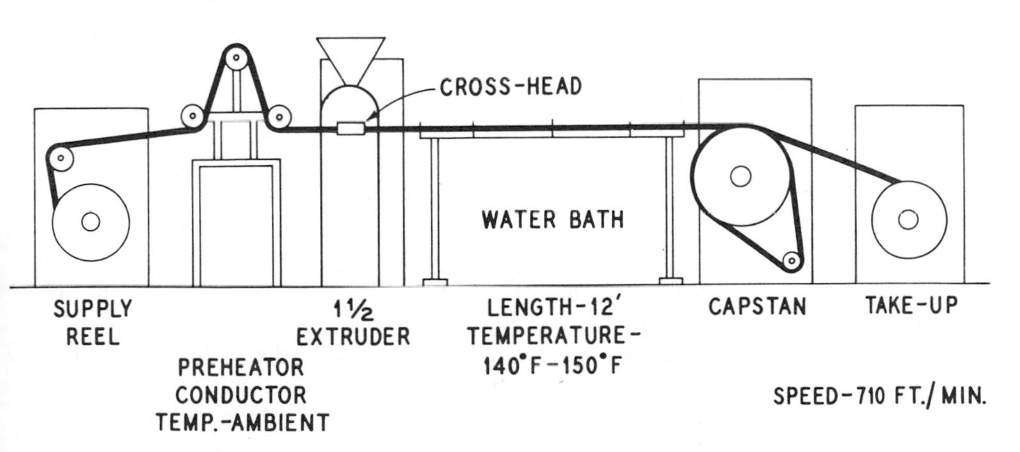

Fig. 8 Wire coating.

Polyethylene wire and cable coatings are applied by means of crosshead dies which are fed by extruders usually placed at right angles to the path of the wire[58] (Fig. 8). The wire may be cleaned, preheated, and straightened before it enters the die. It is continuously encapsulated by molten polymer at linear speeds up to 2,000 feet per minute. As it emits from the die, it is cooled and tested electrically for coating integrity. Cable sheathing involves the same principle as wire coating but diameter and thickness are greater and linear rates are lower.

In recent years considerable interest has developed in the use of crosslinked polyethylene for wire coating applications.[59,60] Crosslinked polymers offer advantages over noncrosslinked materials in resistance to stress cracking and to creep particularly at temperatures above the crystalline melting point. In many instances high levels of carbon black are included in the formulation to provide weather resistance and reduce costs. A typical formulation for a carbon black loaded polyethylene stock with organic peroxide crosslinking agent follows:

Polyethylene	100
Carbon Black	25-300
Peroxide	2
Cure	
Temperature, °C	175-200
Time, min	10-0.5

According to Dannenberg et al.[61] crosslinking occurs between the polymer and carbon black as well as between polymer molecules.

Pipe. Low density polyethylene continues to be an important raw material for the production of plastic pipe, but in recent years the trend has been toward the use of medium and high density polyethylenes for this application. The basis of choice is largely an economic one in which the lower cost of the low density polymer is weighed against the higher performance of the high density resin which permits use of thinner walls in equivalent grades of pipe. In any case, the top grades of pipe exhibiting highest pressure ratings and greatest reliability are produced from high density polymers which for this purpose must be of very high molecular weight. The major outlet for polyethylene pipe is in farm and home outdoor applications such as sprinkler systems or other cold water service. Polyethylene pipe of one to two inch diameter is most popular and is sold in coils of 100 feet or more in length. Pipe of diameters above about two inches is too rigid to coil easily and is sold in 20 to 40 feet lengths for use in industrial applications, drain lines, and low pressure service.

There are five important techniques for producing polyethylene pipe: external sizing tube, sizing rings, internal sizing, positive pressure dies, and biaxial orientation. In each case the molten polymer is forced through a suitable die by means of an extruder. In the external sizing technique[62] the molten polymer is forced against the cold walls of a sizing tube by internal air pressure or by applying vacuum. The pipe is pulled from the sizing tube into a water bath and onto a reel. This is one of the preferred methods for extrusion of high density polyethylene. In the sizing ring technique, pipe is extruded into a water bath and pulled through one or more sizing rings to obtain the desired dimensions. The internal sizing technique is used extensively for medium and low density polyethylene. The extruded pipe is pulled over a mandrel fitted with a sizing knob on the end to provide proper internal dimensions. External dimensions and wall thickness are controlled by extrusion rate and take-up speed. High quality pipe is prepared from high molecular weight high density polyethylene by a positive pressure technique in which the polymer is pushed through the die by the extruder, rather than pulled through dies or sizing equipment by the take-up unit.[63] In this process both the external and internal walls of the pipe are in contact with cold surfaces and the finished pipe exhibits a very glossy appearance. In one of the preferred methods of operation a small amount of water is used as a lubricant. High performance pipe may be prepared, preferably from high density polyethylene, by orienting the material under carefully controlled conditions.[64] In this process a thick walled, small diameter pipe is extruded in the conventional way and brought to the required orientation temperature which is usually slightly below the melting point. It is drawn down, then blown to several times its original circumference. The oriented pipe prepared in this way offers working pressures up to twice that of nonoriented pipe from the same material.

Because plastic pipe is expected to offer predictable performance over many years of service, it is necessary to develop adequate accelerated testing techniques. In general, this is accomplished by subjecting short lengths of pipe to a wide range of pressures, usually in water baths at temperatures of 20 to 100°C. The time to fail at a given pressure and temperature is recorded in the form of stress-life curves.[65,66,67] By using various established techniques, one may estimate the pressure rating for a life expectancy of 20 years or more with reasonable certainty.

Coatings. The extrusion coating of polyethylene on paper or other substrate is one of the fastest growing segments of the polyethylene industry. In commercial practice (Fig. 9) an extruder is used to feed polymer through a slit die somewhat similar to that required for chill roll film.[68] The polymer melt is laid down continuously on the substrate and then chilled by contact with a polished roll. Linear speeds of 300 to 1,000 feet per minute are common. Coatings may be 0.25 mil to 2 mils in thickness. Important substrates for polyethylene coating include Kraft paper for multi-walled bags and paper board for milk cartons and other food containers.[69] When applied to cellophane, polyethylene provides heat sealability, tear resistance, moisture barrier properties, and improved resistance to shelf aging. Polyethylenes of very low molecular weight may be applied with roll coating machines of the type used for making wax paper. Other techniques involve the use of polyethylene emulsions or micro-pulverized powders. Automobile carpet backing which can be molded to fit the contours of the floor is an important new use for micro-pulverized, low density polyethylene.

Filaments. One of the first markets for high density polyethylene was the production of monofilaments for rope, automobile seat covers, and similar applications. At one time this represented a major market for high density

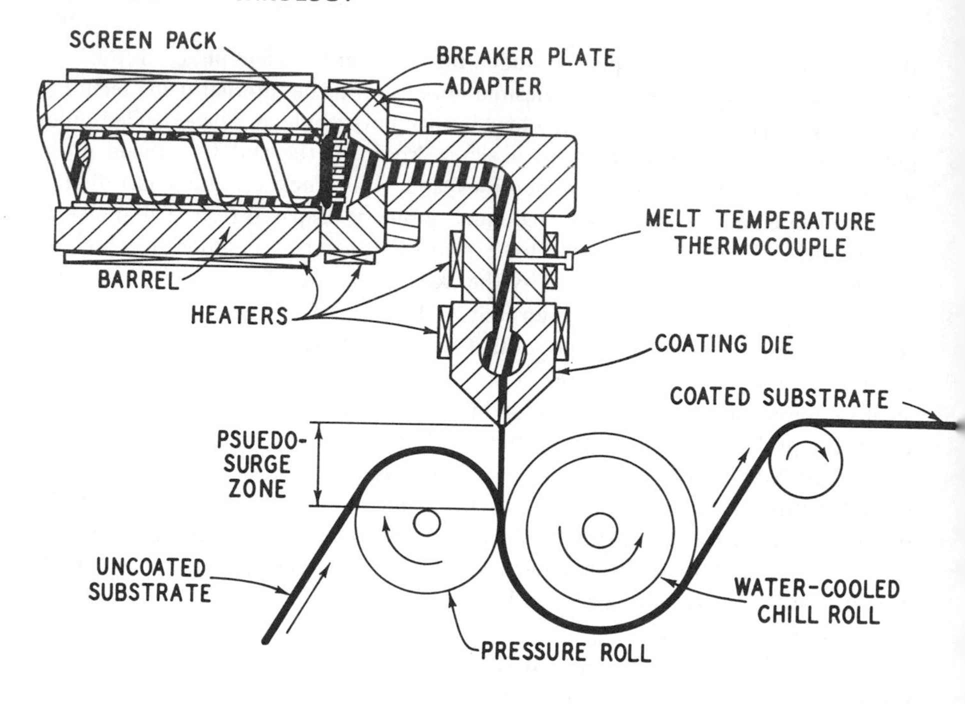

Fig. 9 Extrusion coating.

polyethylene, but the growth has been slow and the volume for monofilaments has been surpassed many times over by demands for blow molding and injection molding resins. Polypropylene has displaced polyethylene for polyolefin fibers and filaments. In the production of monofilament, several strands of molten polymer are extruded through an orifice plate and quenched in a water bath. At this point the filament has a tensile strength of only 2,000 to 4,000 psi. In the next step the strands are wrapped on godet rolls, heated in a steam bath, and subjected to draw ratios of 10:1 to 12:1. The drawing operation increases tensile strength to 50,000 to 100,000 psi.

Powder Molding. Powder molding of polyethylene consumed 95 million pounds in 1971.[50] The early techniques for powder molding of polyethylene included the Engel process,[70] the centrifugal processes of Heisler,[71] and rotational molding. In the Engel process, powder is placed in a cold mold which then is heated to 300 to 400°C. A portion of the polyethylene powder fuses on the walls of the mold. The unfused polymer is dumped out and the mold is reheated to fuse all of the remaining resin. When the mold cools, the polyethylene shrinks from the walls and is dumped out and the mold is reheated to fuse all of the remainig resin. When the mold cools, the polyethylene shrinks from the walls and is removed with ease. Large items such as boats and tanks have been prepared by this process, particularly in Europe.

In the centrifugal process, the molds are preheated and partially filled with polyethylene. The mold is rotated on its long axis to give an even distribution of polymer on the walls. This process is not widely used, but it is convenient for making objects which are open at both ends such as pipe.

Rotational molding techniques have been used for years for molding of vinyl plastisols. In this system the mold is loaded with the exact amount of polymer to be used. It is then rotated simultaneously around two axes at right angles to each other while it is being heated in an oven or by infrared lamps. After all of the polymer is fused to the walls of the mold, the mold is cooled and then opened to release the polyethylene product. In general, powder molding is of interest for (1) short runs or prototypes because of the very low investment involved and (2) large items such as 30 gallon tanks, large tubs or vats which are impractical to mold by any other technique. Commercial polyethylene molding powders are either low or high density materials which are mechanically ground to give particles in the 100 to 300 micron range.

Modified Polyethylenes and Polyethylene Compounds. Very early in the development of low density polyethylene it was found that substantial improvements in low temperature flexibility and environmental stress cracking resistance could be obtained by the addition of 10 to 15% polyisobutylene.[73] More recently in the United States these requirements have been met by the production of high molecular weight polyethylenes which do not require modification with rubber. However, high density polyethylenes containing relatively high loadings of special elastomers are being used for industrial bags and other heavy duty film applications.

Polyethylene also has been suggested as an additive for rubber where it functions as a reinforcing agent to reduce cold flow and increase hardness.[74] Low molecular weight polyethylene has been suggested as a lubricant and processing aid for butyl, natural rubber, SBR, neoprene, and chlorosulfonated polyethylene. In natural rubber tread stocks it is said to improve tear resistance and processibility and increase resistance to crack growth and abrasion. When incorporated in neoprene and chlorosulfonated polyethylene, it reduces the tendency to stock to mill rolls.[75]

Polyethylene may be chlorinated in the dry state or in solutions of carbon tetrachloride or other chlorinated solvents. At levels of 20 to 35% chlorine, the product from high density polyethylene resembles plasticized polyvinyl chloride. At 50 to 65% chlorine it offers many of the properties of rigid polyvinyl chloride. High density polyethylenes containing 40 to 50% chlorine have been suggested as additives for improving the impact strength of rigid polyvinyl chloride.[76]

Low density polyethylenes are chlorinated in the presence of sulfur dioxide to yield chlorosulfonated products which typically contain 27.5% chlorine and 1.5% sulfur.[77,78,79] The chlorosulfonated polymer may be cured in systems containing magnesium oxide or other metal oxides, litharge, or cribasic lead maleate. The vulcanizates, which are resistant to ozone and oxygen, provide good abrasion resistance, good flex life, and resistance to crack growth. Compounds of this type are also resistant to most chemicals with the exception of aromatic and chlorinated solvents.

REFERENCES

(1) Bamberger and Tschirner, *Berichte,* **33**, 955-959 (1900).
(2) Pechman, *Berichte,* **31**, 2643 (1898).
(3) Meerewein and Burneleit, *Berichte,* **61**, 1840 (1928).
(4) Swallow, *Polythene,* eds. Renfrew and Morgan, Interscience Publishers, New York, 1960 p.3.
(5) Ziegler et al., *Angew. Chem.*, **67**, 541 (1955).
(6) D'orville, *Polythene,* eds. Renfrew and Morgan, Interscience Publishers, New York, 1960, p.37.
(7) *Modern Plastics,* **47**, 70 (Jan. 1970).
(8) Dobson, *Polythene,* eds. Renfrew and Morgan, Interscience Publishers, New York, 1960, p.13.
(9) Hines, Bryant, Larchar, and Pease, *Ind. Eng. Chem.*, **49**, 1071 (1957).
(10) Davison and Erdmon, U. S. Patent 2,839,515.
(11) Perrin, Fawcett, Paton, and Williams, U. S. Patent 2,200,429.
(12) Brubaker, Coffman, and Hoehn, *J. Am. Chem. Soc.*, **74**, 1509 (1952).
(13) Hogan and Banks, U.S. Patent 2,825,721.
(14) Clark, Hogan, Banks, and Lanning, *Ind. Eng. Chem.*, **48**, 1152-5 (1956).
(15) Ziegler, Belgian Patent 533,632.
(16) Goppel and Howard, *Polythene,* eds. Renfrew and Morgan, Interscience Publishers, New York, 1960, pp. 17-27.
(17) British Patent 779,540 (1957).
(18) Belgian Patent 549,910 (1958).
(19) Field and Feller, *Ind. Eng. Chem.*, **49**, 1161 (1956).
(20) Smith, *Ind. Eng. Chem.*, **48**, 1161 (1956).
(21) Bayer, Wiley, and Spencer, *J. Polymer Sci.*, **1**, 249 (1946).
(22) ASTM D-792-60T.
(23) Jones and Boeke, *Ind. Eng. Chem.*, **48**, 1155-60 (1956).
(24) ASTM D-1238-57T.
(25) Harban and McGlamery, *Materials Research & Standards,* Vol. 3, No. 11 (1963).
(26) ASTM D-790-59T.
(27) ASTM D-747-58T.
(28) ASTM D-256-56.
(29) ASTM D-1822-61T.
(30) ASTM 1709-59T.
(31) DeCoste, Malm, and Wallder, *Ind. Eng. Chem.*, **43**, 117 (1951).
(32) ASTM D-150-59T.
(33) Biggs and Hawkins, *Modern Plastics,* **31**, 121 (1953).
(34) Hawkins, Lanza, Loeffler, Matreyek, and Winslow, *J. Appl. Polym. Sci.*, I, 43-9 (1959).
(35) Madorsky, Straus, Thompson, and Williamson, *J. Polym. Sci.*, **4**, 639 (1949).
(36) Oakes and Richards, *J. Chem. Soc.*, 2929 (1949)
(37) Haywood, *Polythene,* eds. Renfrew and Morgan, Interscience Publishers, New York, 1960, pp. 132-133.
(38) Dole, Keeling, and Rose, *J. Am. Chem. Soc.*, **76**, 4304 (1954); Miller and Willis reported by Willbourn, *J. Polymer Sci.*, **34**, 569 (1959); Boyd, Voter, and Bryant, *Abstracts,* 132nd National ACS meeting, Sept. 1957, p. 8T.
(39) Pritchard, McGlamery, and Boeke, *Modern Plastics,* **37**, 132 (Oct. 1959).
(40) Till, *J. Polym. Sci.*, **24**, 301 (1957); Keller, *Phil. Mag.*, **2**, 1171 (1957); Fischer, *Z. Naturforsch,* **12a**, 753 (1957).
(41) Debye, *J. Appl. Phys.*, **15**, 388 (1944); *J. Phys. Chem.*, **51**, 18 (1947).

(42) Stacy and Arnett, *J. Poly. Sci.*, Part A **2**, pp. 167-179 (1964).
(43) Arnett, Smith, and Buell, *J. Poly. Sci.*, Part A **1**, p. 2753 (1963).
(44) Guillet, Combs, Colner, and Slonaker, *J. Polym. Sci.*, **47**, 307-320 (1960).
(45) Moore, *J. Poly. Sci.*, A-2, 835 (1964).
(46) McGlamery and Harban, SPE Eighteenth Annual Technical Conference, Vol. 8, 1962.
(47) Wash, *Modern Plastics Encyclopedia*, **40**, 228 (1963).
(48) Kennaway and Weeks, *Polythene*, eds. Renfrew and Morgan, Interscience Publishers, New York, 1960, pp. 437-462.
(49) Elliot, *Modern Plastics Encyclopedia*, 419-430 (1969-70).
(50) *Modern Plastics*, **49**, 41-48 (Jan. 1972).
(51) Barker, Lewis, and Happoldt, U.S. Patent 2,770,608.
(52) Santer, *Polythene*, eds. Renfrew and Morgan, Interscience Publishers, New York, 1960, pp. 615-622.
(53) *Modern Plastics*, **47**, 67-69 (June 1970).
(54) Doyle and Allison, *SPE Journal*, **16**, No. 3 (March 1960).
(55) Wood, *Polythene*, eds. Renfrew and Morgan, Interscience Publishers, New York, 1960, pp. 571-579.
(56) Ferngren, U.S. Patents 2,128,239 (1948); 2,175,053 (1939); 2,175,054 (1939).
(57) Wood, *Modern Plastics Encyclopedia*, **40**, 223 (1963).
(58) *Modern Plastics Encyclopedia*, **40**, 765 (1963).
(59) Precopio and Gilbert, U.S. Patent 2,888,424.
(60) Ivett, U. S. Patent 2,826,570.
(61) Dannenberg, Jordan, and Cole, *J. Polym. Sci.*, **31**, 127 (1958).
(62) Croley and Doyle, *Plastics Technology*, **4**, 717 (Aug. 1958).
(63) U.S. Patent 3,066,356.
(64) Gloor, *Modern Plastics*, **38**, 111 (Nov. 1960).
(65) Richard, Diedrich, and Gaube, Paper presented to ACS Meeting, New York, 1957.
(66) Gloor, *Modern Plastics*, **36**, 144 (1958).
(67) Richard and Ewald, *Plastics*, **36**, 153 (1959).
(68) Wilbert and Grant, *Polythene*, eds. Renfrew and Morgan, Interscience Publishers, New York, 1960, pp. 585-597.
(69) *Modern Plastics Encyclopedia*, 181 (1969-70).
(70) Engel, U.S. Patent 2,915,788.
(71) Heisler, U.S. Patent 2,736,925.
(72) Zimmerman and Johnson, *Modern Plastics Encyclopedia*, **40**, 717-720 (1963); Zimmerman, *British Plastics*, **36**, 84 (1963).
(73) Williams, British Patent 514,687 (1939).
(74) Railsback and Wheat, *Rubber Age*, **82**, 664-671 (Jan. 1958).
(75) Bulifant, *Rubber Age*, **82**, 89 (1957).
(76) Frey, *Kunststoffe*, **49**, 50 (1959).
(77) McQueen, U.S. Patent 2,212,786.
(78) AcAlvey, U.S. Patent 2,586,363.
(79) Warner, *Rubber Age*, **71**, 205-221 (1952).

APPENDIX
THE LITERATURE OF RUBBER

RUTH MURRAY
Division of Rubber Chemistry Library
University of Akron, Akron, Ohio

KATHLEEN S. ROSTLER
The Rubber Formulary
Materials Research and Development
Oakland, California

The literature on rubbers is widely scattered through many books, periodicals, conference proceedings, technical and trade literature, standards and specifications, patents and government documents. This body of knowledge continues to increase as new materials enter the picture and as the old and new ones find new applications. It is difficult for the expert in the field to keep track of the many developments while to the one who is not so skilled in the art or to the engineer or designer, the rubber literature can be obscure. Working knowledge of what rubbers are and what they can do is necessary in order to be able to adequately understand the problem.

The references listed below are intended only to highlight some of the major sources of information, sources which are currently in print, available and understandable to the technically trained but not necessarily expert worker in the rubber field. Books provide the review of known and basic information; handbooks point out the specific facts and data; periodicals keep one up-to-date on the latest information. The newest information is usually found in conference proceedings, and the trade and technical literature provide the most practical information on a subject or product. The particular *forte* of each reference is briefly indicated, though there is much duplication of information in these sources. One generally needs to consult more than one source to find an answer to a problem.

GENERAL BOOKS ON RUBBER

P. W. Allen, P. B. Lindley, and A. R. Payne, *Use of Rubber in Engineering*, Proc. of a Conference at Imperial College of Science & Technology, London, 1966, by NRPRA, Maclaren & Sons Ltd., London, 1967.

> Papers discussing basic design considerations and the performance of rubber components in major applications in civil and mechanical engineering.

G. Alliger and I. J. Sjothun, *Vulcanization of Elastomers*, Reinhold Publishing Corp., New York, 1964.

> A series of lectures presented by the Akron Rubber Group, edited and published in book form.

G. M. Bartenev and I. S. Zuev, *Strength and Failure of Visco-Elastic Materials,* Pergamon, New York, 1968.

L. Bateman, ed., *The Chemistry and Physics of Rubber-like Substances,* Maclaren & Sons, London, 1963.

A compilation of the main studies undertaken by the Natural Rubber Producers' Research Association in the past 25 years.

C. M. Blow, ed., *Rubber Technology & Manufacture,* Butterworths, London, 1971.

An up-to-date guide for students, engineers using rubber products, and buyers of rubber goods, covering compounding principles, processing technology, and manufacturing procedures for major products, methods, and standards.

S. Boström, ed., *Kautschuk-Handbuch,* Berliner Union, Stuttgart, 5 Volumes, 1958-1962.

In German. Covers in detail types of elastomers, their processing, compounding, and properties; compounding ingredients; compounding, design, and manufacture of all types of rubber goods; and testing procedures.

S. Buchan, *Rubber in Chemical Engineering,* Natural Rubber Producers' Research Association, London, 1965.

Refers to techniques and applications involved in lining and covering chemical plant with rubber. Includes list of uses and number of examples.

C. C. Davis and J. T. Blake, eds., *The Chemistry and Technology of Rubber,* Reinhold, New York, 1937.

Originally this was the basic, comprehensive one-volume work on rubber. Now only of historical interest.

T. R. Dawson, *Rubber Industry in Germany During the Period 1939-1945,* British Intelligence Objectives Sub-Committee, Overall Report No. 7, London, HMSO, 1948.

Summarizes the technical information collected at government level after World War II concerning wartime rubber activities in Germany and Japan.

Elastomer Stereospecific Polymerization, Advances in Chemistry Series #52, American Chemical Society, Washington, D.C., 1966.

Papers presented at the Symposium on New Aspects of Elastomer Stereospecific Polymerization conducted jointly by the Division of Polymer Chemistry and the Division of Rubber Chemistry of the American Chemical Society.

A. J. Gait and E. G. Hancock, *Plastics and Synthetic Rubbers,* Pergamon, New York, 1970.

One of a series of monographs on the chemical industry, mainly in the United Kingdom, prepared as a teaching manual for senior students.

D. A. Hills, *Heat Transfer & Vulcanization of Rubber,* a Monograph of the Institution of the Rubber Industry, Elsevier Publ. Co. Ltd., New York, 1971.

Gives historical development of rubber applications and physical background of vulcanization plus uses, theories, methods of determining vulcanization isotherm, and technical aspects of vulcanization.

Werner Hofmann, *Vulcanization and Vulcanizing Agents,* Palmerton, New York, 1967.

Comprehensive survey of the methods of crosslinking and systems necessary for that purpose, geared to the rubber technician.

J. P. Kennedy and Eric G. M. Tornqvist, *Polymer Chemistry of Synthetic Elastomers,* Interscience, New York, 1968.

A comprehensive treatise on synthetic elastomers. Chemical aspects of polymer formation are emphasized rather than polymer physics.

R. E. Kirk and D. F. Othmer, eds., *Encyclopedia of Chemical Technology,* Interscience, New York, 1st Edition, 1963; 2nd Edition by H. F. Mark *et al.*

The rubber volume summarizes many aspects of rubber technology giving bibliographies to journal and patent literature.

Paul Kluckow, *Rubber and Plastics Testing,* Chapman and Hall, London, 1963.

Translation of Dr. Kluckow's survey of the most important methods for testing rubber and plastics.

Gerard Kraus, *Reinforcement of Elastomers,* Interscience, New York, 1965.

Theoretical and practical aspects of elastomer reinforcement geared to scientists, engineers, technologists, and to marketing and sales personnel.

Jean Le Bras, *Introduction to Rubber,* Maclaren & Sons, Ltd., London, 1965.

English version of popular French title covering essentials of the rubber industry. Directed to industrial employees and laymen whose education has been terminated with the high school diploma.

E. W. Madge, *Latex Foam Rubber,* Interscience, New York, 1962.

Production and marketing of foam rubber based on natural rubber latex (Part I), and on synthetic rubber latex (Part II).

Peter Mason and N. Wookey, *The Rheology of Elastomers,* Pergamon, New York, 1958.

Proceedings of the Conference on Rheology of Elastomers held by the British Society of Rheology at Welwyn Garden City in 1957.

Herman F. Mark, Norman G. Gaylord, and Norbert M. Bikales, eds., *Encyclopedia of Polymer Science and Technology: Plastics, Resins, Rubbers, Fibers,* Interscience, New York, 1964.

This covers in more detail than Kirk and Othmer the chemical substances, polymer properties, methods and processes, uses, and general background.

J. McClellan, *Rubber in Transport Engineering,* Proc. of a Conference at Imperial College of Science & Technology, London, 1970, by NRPRA, Natural Rubber Bureau, Washington, D.C.

Eighteen papers covering rubber products for automobiles, trailers, buses and coaches, railroads, boats, and hovercraft.

L. R. Mernaugh, ed., *Rubber,* Robert Maxwell and Co., Ltd., Oxford, England, 1971.

Covers all aspects of rubber which are pertinent to engineers. Part 1 gives basic characteristics and properties; Part 2 shows particular applications; Part 3 covers specific rubbers.

Maurice Morton, ed. *Introduction to Rubber Technology,* Van Nostrand Reinhold, New York, 1959.

Covers the compounding of natural and synthetic rubber, as well as testing and properties.

E. Müller, *et al.,* eds., *Methoden der Organischen Chemie* (*Houben-Weyl*), Volume 4, Parts 1 and 2, George Thieme, Stuttgart, 4th Edition, 1961 and 1963.

In German. This volume, entitled *Makromolekular Stoffe,* of this basic reference work in organic chemistry gives preparative methods for high polymers, including elastomers, on laboratory and plant scale, and analytical methods for their study.

Natural Rubber Producers' Research Association, Proceedings of the Jubilee Conference, Cambridge, 1964.

A series of chapters dealing with various topics covering the research findings at the NRPRA laboratories. Of interest to scientists and technologists engaged in rubber research.

W. G. S. Naunton, *The Applied Science of Rubber,* E. Arnold, London, 1961.

Ronald Hugh Norman, *Conductive Rubber and Plastics: Their Production Applications and Test Methods,* Elsevier, New York, 1970.

A broadened update of an earlier book covering the properties, processing of, and uses for conductive rubbers and plastics.

A. Nourry, ed., *Reclaimed Rubber: Its Development, Application and Future,* Maclaren & Sons, London, 1961.

Rubber World, *Materials and Compounding Ingredients for Rubber,* Rubber World, New York, 1968.

Physical and chemical properties of materials and rubbers available to the industry for manufacturing products. Frequently updated.

K. J. Saunders, *Identification of Plastics and Rubbers,* Chapman and Hall, London, 1966.

J. R. Scott, *Physical Testing of Rubbers,* Maclaren and Sons, London, 1965.

Research and Marketing of rubber depend upon the ability to measure physical properties of rubber. Standardized tests are described.

S. L. Rosen, *Fundamental Principles of Polymeric Materials for Practicing Engineers,* Barnes & Noble, New York, N.Y., 1971.

H. J. Stern, *Rubber: Natural and Synthetic,* Palmerton, New York, 2nd Edition, 1967.

An account of all aspects of the subject from polymer synthesis to reclaiming, suitable for anyone with basic scientific training but not necessarily any prior knowledge of rubber technology.

O. H. Varga, *Stress-Strain Behavior of Elastic Materials,* Interscience, New York, 1966.

W. C. Wake, *The Analysis of Rubber and Rubber-like Polymers,* Wiley Interscience, New York, 1969.

An assessment of the tools and analytic practice in the field of the material in question.

G. S. Whitby, C. C. Davis and R. F. Dunbrook, eds., *Synthetic Rubber,* Wiley, New York, 1954.

A comprehensive post-war publication covering all aspects of synthetic rubber technology.

George G. Winspear, ed., *The Vanderbilt Rubber Handbook,* R. T. Vanderbilt Co., New York, 1968.

Subject arrangement of fundamentals of the rubber industry. Commercial elastomers are described, including compounding information, processes, and uses.

PERIODICALS

Gummi, Asbest, Kunststoffe, A. W. Gentner KG, Postfach 688, 7 Stuttgart 1, West Germany. Monthly.

Contains articles of practical information, review of patent literature, and short abstracts of journal articles in English.

Journal of the IRI, The Institution of the Rubber Industry, 4 Kensington Gardens, London W8. Bimonthly.

Reports on the events, awards, and papers read before the IRI. Articles are of practical and current nature.

Natural Rubber News, Natural Rubber Bureau, 1108 Sixteenth St., N.W., Washington, D.C. 20036. Monthly.

Information bulletin for the natural rubber consumer and planter. Gives new developments, statistics, meetings and events, news items.

Revue Generale du Caoutchouc et des Plastiques, 42 Rue Scheffer, Paris, France. Monthly.

Theoretical and practical information is given. Includes summaries in French from the Journal of the Society of the Rubber Industry, Japan, and summarizes its own papers in English, German, Spanish, and Italian.

Rubber Age, Palmerton Publishing Co., Inc., 101 W. 31st St., New York, 1001. Monthly.

A news type journal covering new developments, business conditions, production sales, personalities, and forthcoming meetings and events.

Plastics, Rubber, Textiles (incorporating *Rubber & Plastics Age*), Rubber & Technical Press Ltd., Tenterden, Kent, England. 6 times per year in alternate months and joint subscription with *Polymer Age.*

This publication emphasizes new products, the latest processing techniques and equipment, particularly with plastics and composites.

Rubber Chemistry and Technology, Division of Rubber Chemistry, ACS, Inc., Box 123, University of Akron, Akron, Ohio 44304. Five times a year.

Articles are mostly reprints and translations of other journal articles and papers presented at the meetings of the Division of Rubber Chemistry. The original articles are of theoretical and technical nature.

Rubber Journal, Maclaren & Sons, Ltd., Davis House, P.O. Box 109, 67-77 High St. Croydon, Surrey. Monthly.

Reviews European, U.S., and Russian technical developments, standards and testing, patents, economic trends, and forecasts.

Rubber World, Bill Brothers Publishing Corp., 630 Third Avenue, New York, 10017. Monthly.

A trade journal covering new developments, business conditions, production, sales, personalities, books, and forthcoming meetings and events.

Soviet Rubber Technology, Maclaren & Sons, Ltd. Monthly.

Cover-to-cover translations of the Russian journal *Kautchuk i Rezina,* containing articles of practical information.

ABSTRACTS AND INDEXES

Bibliographies of Rubber Literature, Division of Rubber Chemistry, ACS, Inc., University of Akron, Akron, Ohio.

1. "Bibliography of Rubber Literature"
   An annual review since 1935 of the rubber periodical and patent literature grouped by subject classes and including short abstracts
2. Subject bibliographies on selected subjects of general interest to the rubber industry.

"Chemical Abstracts," Chemical Abstracts Service, American Chemical Society, The Ohio State University, Columbus, Ohio, weekly.

Papers on rubber have been abstracted in the following sections: 1912-1914, Section 26; 1915-1961, Section 30; 1962-1964, Section 46; 1965-1966, Section 49; and 1967 to date, Section 38.

"POST-J–Polymer Science and Technology Journals," American Chemical Society, 1155 Sixteenth St., N.W. Washington D.C.

Computer-based polymer information service which started in 1963 and covers journals and government report literature.

"RAPRA Abstracts," Rubber & Plastics Research Association of Gt. Britain, Shawbury. Shrewsbury, Shropshire, England.

Monthly coverage of the rubber journal and patent literature, including processing information and trade literature.

"The Rubber Formulary," Materials Research & Development, Inc., 2811 Adeline St. Oakland, Calif., monthly.

Elastomer formulations with complete test data from English language journals and from releases of suppliers, recorded individually on marginally punched 5" x 8" file cards. Published since 1948; approximately 1000 new cards per year. Selected cards for 1948-57, 1958-63, and 1964-68 available as Condensed Sets, presorted for filing.

HANDBOOKS–REFERENCE BOOKS–DIRECTORIES

J. Brandrup and E. H. Immergut, *Polymer Handbook*, Interscience, New York, 1966.

Collates polymer data including polymers, oligomers, monomers, and solvents.

J. V. DelGatto and S. R. Hague, *Machinery and Equipment for Rubber and Plastics,* Rubber World, Bill Brothers, New York, 1970.

Gives descriptions of U.S. and Canadian machines based on brochure literature.

International Rubber Directory, International Publications Service, New York, 4th Edition, 1967.

Contains alphabetical list of brand-names and trade names, manufacturers of rubber goods, a buyers guide, a company name index, and a rubber vocabulary in three languages.

Manufacturing and Suppliers of Rubber Products Used in Building, Civil Engineering and Roads, Natural Rubber Producers Research Association, Welwyn Garden City, Hertfordshire, England, 5th Edition, 1966.

Materials and Compounding Ingredients for Rubber, Bill Communications, Inc., New York, 1970 Edition.

An annual directory giving physical and chemical properties of materials, how and why they are used in the rubber product, and the specific rubbers and/or products that these materials are recommended for.

Materials Engineering, section on plastics and rubber in the mid-October issue, Reinhold Publishing Corp., New York, 1970.

Gives properties and uses of many rubbers.

Rubber Directory of Great Britain, Maclaren & Sons, London, 7th Edition, 1970.

Lists suppliers and manufacturers by name and products.

Rubber Red Book, Palmerton Publishing Co., New York.

An annual directory which includes rubber manufacturers in the U.S., Canada, and Puerto Rico, suppliers, consultants, technical and trade organizations, and compounding information.

New Trade Names in the Rubber and Plastics Industries, RAPRA, Shawbury, Shropshire, England, 1969.

An annual directory giving composition, if known, and manufacturer of new materials.

PATENTS

"Official Gazette of the U.S. Patent Office," Superintendent of Documents, U.S. Government Printing Office, Washington D.C.

Weekly abstract bulletin listing all U.S. patents issued. The large class No. 260 contains many subclasses pertaining to rubber.

"PLASDOC–Plastics and Polymers Patents Documentation," Derwent Publications, Ltd., Rochdale House, Theobalds Road, London.

Weekly coverage of U.S. and foreign patents since 1966. Service includes weekly abstract journal, IBM company code punch cards, individual country file, manual classification.

"POST-P – Polymer Science and Technology Patents," American Chemical Society, 1155 Sixteenth St., N.W. Washington D.C.

Begun in 1967 to provide a computer-prepared polymer patent information service, it covers patents in 26 countries with digests grouped according to subject classes.

GLOSSARIES–DICTIONARIES

Alexander S. Craig, *Dictionary of Rubber Technology,* Philosophical Library Inc., New York, 1969.

Some of the more important entries have been expanded into short articles and sources of further reading are given for many entries.

Glossary of Terms Relating to Rubber and Rubber-Like Materials, American Society for Testing and Materials, Philadelphia, Pa., 1956.

ASTM Special Technical Publication No. 184. This includes in an appendix a list of trade-named apparatus and chemicals for which the compositions are known.

Rubber Stitching, *Elsevier's Rubber Dictionary*, Elsevier Publishing Co., New York, 1959.

Good technical 10 language dictionary.

STANDARDS

"American National Standards Institute," 1430 Broadway, New York 10018.

Serves as a clearinghouse for nationally coordinated voluntary safety, engineering, and industrial standards. Gives USA standards to projects developed by agreement from all groups concerned in areas such as definition, terminology, materials, performance, procedures, methods of rating, and methods of testing and analysis.

Book of Standards: Part 28: Rubber; Carbon Black; Gaskets, American Society for Testing and Materials, Philadelphia, Pa., 1971.

Annual volumes concerned mostly with the development of standards, but it is a member of the American Standards Institute.

STATISTICS

"Census of Manufacturers: Major Group 30: Rubber and Miscellaneous Plastics Products," Department of Commerce, Bureau of the Census, Superintendent of Documents, U.S. Government Printing Office, Washington D.C.

Classified product statistics taken in years ending in 2 and 7.

"Current Industrial Reports," U.S. Department of Commerce, Bureau of the Census, Superintendent of Documents, U.S. Government Printing Office, Washington D.C.

Information on U.S. consumption by rubber type. This information is reprinted in Rubber World, Rubber Age, and by the R.M.A.

"International Rubber Digest," The Secretariat of the International Rubber Study Group, Brettenham House, 5-6 Lancaster Place, London.

Monthly digest giving statistical and other information on rubber.

"Rubber Industry Facts," The Statistical Department of the Rubber Manufacturers Association, Inc., 444 Madison Ave., New York, 10022.

An annual report giving current and historical rubber product data consisting of 44 reports.

"Rubber Statistical Bulletin," The Secretariat of the International Rubber Study Group, Brettenham House, 5-6 Lancaster Place, London.

Gives 56 monthly statistical reports of its 32 country members showing production, consumption, exports, stocks, major end products.

"Rubber Statistical News Sheet," The Secretariat of the International Rubber Study Group, Brettenham House, 5-6 Lancaster Place, London.

Quarterly sheet which contains information not published in the Bulletin.

"The Rubber Industry–Marketing, financial, economic investigation 1925-1970," Morton Research Corp., 1666 Newbridge Rd., Billmore, New York, 11710.

"Elastomers & Rubber Chemistry in the 1970's," Margolio Industrial Service, 634 Wood St., Mamaroneck, New York 10543.

Statistical and other marketing trends for 13 synthetic elastomer groups and for rubber compounding ingredients.

PUBLICATIONS OF SUPPLIERS

American Cyanamid Co., Rubber Chemicals Dept., Bound Brook, N.J.

"Rubber Chem Lines," bimonthly. Four-page bulletins each ordinarily including one or two brief articles on compounding, and each mentioning more extensive literature available.

"Cyanamid Rubber Chemicals." Looseleaf. Data sheets describing the company's accelerators, antioxidants, and other compounding ingredients. Updated by additions and replacements.

"Bulletins." Each bulletin covers one product, giving description, properties, and a number of formulations with physical properties illustrating its use.

Individual booklets are also issued frequently containing compounding recommendations for acrylic elastomers, urethane elastomers, and factice.

Ashland Chemical Co., P.O. Box 1503, Houston, Tex.

"Carbon Blackboard" and "Technews." Brief releases published initially as advertisements, then reprinted, giving results of compounding research. The former series deal with carbon blacks, the latter also with polymers, resins, and antioxidants.

Cabot Corp., 125 High St., Boston, Mass.

"Cabot Technical Reports." Reports of extensive laboratory studies on effects of compounding variations, principally carbon black type and loading.

Cities Service Co., Columbian Div., P.O. Box 5373, Akron, Ohio.

"Columbian Colloidal Carbons." Series of booklets, most being reprints of journal articles, giving fundamental information on carbon blacks and their use in rubber.

Copolymer Rubber & Chemical Corp., P.O. Box 2591, Baton Rouge, La.

"EPsyn Technical Data." Looseleaf. General information on compounding EPDM elastomers is supplemented by an extensive section, frequently updated, of recommendations for practical compounds with test data.

"NYsyn Technical Data." Looseleaf. Descriptions are given of the company's NBR polymers with numerous recommendations with test data for compounds for specific applications and to meet ASTM D2000 specifications.

"SBR Technical Bulletins." Recommendations for compounding SBR polymers and masterbatches to meet ASTM D2000 specifications.

"Epcar Compounders Binder." Looseleaf. Sections include descriptions of the company's EPDM polymers, general compounding, and processing recommendations, and recommendations for formulating specific rubber goods.

Goodyear Tire & Rubber Co., Akron, Ohio.

"Tech-Book Facts." Series of releases issued frequently comprising both extensive compounding studies and factory compound suggestions with the company's elastomers, resins, and compounding ingredients.

Hercules Inc., Pine & Paper Chemicals Dept., Wilmington, Del.

"Bulletins PRH Series." Properties and compounding studies of epichlorohydrin elastomers are reported.

International Synthetic Rubber Co., Ltd., Brunswick House, Brunswick Place, Southampton, Hampshire, England.

"Technical Information Sheets." Formulations and test data for recommended factory compounds with the company's SBR polymers.

"Intolan Technical Information." Looseleaf. Descriptions of EPDM rubbers and general compounding recommendations.

Midland Silicones Limited, Reading Bridge House, Reading, Berkshire, U.K. (available through Midsil Corp., 27 Bland St., Emerson, N.J.)

"Technical Data Sheets." Brief releases giving properties of the company's silicone rubber compounds and current compounding recommendations for silicone rubber bases.

Monsanto Co., 260 Springside Drive, Akron, Ohio.

"Technical Bulletins O/RC Series." Reports of compounding studies and recommendations for use of the company's rubber chemicals. Several issued annually.

Natural Rubber Bureau, 1108 16th St., N.W., Washington, D.C. (publications of The Natural Rubber Producers' Research Association and The Rubber Research Institute of Malaya).

"Rubber Developments" and "NR Technology. Rubber Developments Supplement," quarterly. Articles of general interest to users of natural rubber, with papers in the supplement dealing with specific compounding and processing problems.

"B.R.P.R.A. Technical Bulletins." Each deals with compounding or engineering design for use of natural rubber in specific applications. Issued from time to time.

"Technical Information Sheets." Recommendations for compounding natural rubber and NR latex, usually with extensive test data on a few compounds. Issued frequently.

Enjay Chemical Co., Inc., Elastomers Dept., P.O. Box 3272, Houston, Tex.

"Enjay Polymer Laboratories Technical Information Sheets." Report sheets giving practical compounding information on butyl rubber and ethylene-propylene polymers and terpolymers.

Reports in the "SYN-" and "ENJ-" series. Compounding manuals for butyl, EPR and EPDM elastomers and for poly(isobutylene), and design reports describing methods of fabricating various items. One report, "Vistalon Formulary of Rubber Compounds," in looseleaf form, continually updated, supplies recommended formulations for a wide variety of goods with extensive physical test data.

"Formal Reports." Papers on subjects connected with compounding and use of the company's elastomers, frequently with emphasis on theoretical as well as practical aspects.

General Electric Co., Silicone Products Dept., Waterford, N.Y.

"Silicone Product Data" sheets. Brief releases giving properties of silicone rubber compounds and compounding studies with silicone gums and bases.

"Technical Data Books." Booklets on silicone rubber compounds and compounding with silicone rubber gums and bases.

"Silicone Rubber Handbook." Looseleaf. Sections treat in detail properties of silicone rubber compounds, recommendations for compounding silicone gums and for fabricating silicone rubber articles, test methods, and current ASTM and military specifications.

B. F. Goodrich Chemical Co., 3135 Euclid Ave., Cleveland, Ohio (including releases issued by Goodrich-Gulf Chemicals, Inc., and Ameripol, Inc.)

"Hycar Technical Manual." Looseleaf. Sections, published from time to time, each dealing with one or a group of the company's NBR and acrylic polymers or with compounding ingredients. Included are basic compounding information illustrated by data on test series, and recommendations for factory compounds.

"Hycar Technical Supplements." Booklets giving information on new products and new compounding techniques in the same field as the manual described above.

"Hycar Latex Manual." Looseleaf. Sections, published from time to time, deal in detail with compounding and handling NBR latices.

"Hydrin Elastomers." Published as sections describing processing and compounding of epichlorohydrin elastomers.

"Ameripol CB Technical Data Reports" and "Ameripol CB Recipe Data." The former report compounding studies and recommendations for poly(butadiene) elastomers, the latter report briefly recommendations for compounding specific articles.

"Ameripol SN Technical Data Reports" and "Ameripol SN Recipe Data." The former report compounding studies and recommendations for synthetic poly(isoprene) elastomers, the latter, recommendations for compounding specific articles.

Dow Corning Corp., Midland, Mich.

"Bulletins." Some are brief leaflets, others substantial booklets, giving properties of silicone rubber compounds and compounding information on silicone gums and bases.

E. I. du Pont de Nemours & Co., Elastomer Chemicals Dept., Wilmington, Del.

"Mechanical Molded Goods: Neoprene and Hypalon," D. C. Thompson, 1955. Processing and compounding of mechanical molded goods with the two types of elastomers. The compounding section includes basic information on compounding to obtain desired properties and an extensive table, arranged by hardness and tensile strength, of recommended formulations for mechanical goods.

"Neoprene Latex: Principles of Compounding and Processing." John C. Carl, 1962. Detailed information on compounding, including description of all neoprene latices, handling recommendations, effects of compounding variables, all illustrated by extensive data in tabular and graphic form.

"The Neoprenes," R. M. Murray and D. C. Thompson, 1963. Comprehensive description of all types of neoprene with discussions, illustrated by extensive tables and graphs of vulcanizate properties, of the effects of compounding variables; numerous bibliographical references to journal articles and company reports.

"Formal Reports." Each deals in detail with the basic principles of compounding with one of the company's elastomers or compounding ingredients. None issued in past ten years.

"Informal Reports" (Blue Sheets). Brief reports, with formulations and test data, each dealing with a specific compounding problem or introducing a new material.

"Hypalon Reports," "Viton Bulletins," "Adiprene Bulletins," "Nordel Bulletins." Extensive compounding information, both general studies and practical formulations, for the company's chloro-sulfonyl-polyethylene, fluorocarbon, urethane and EPDM elastomers, respectively.

"Nordel Hydrocarbon Rubber Formulary." Looseleaf. Recommended recipes with physical properties for formulating specific types of rubber goods with EPDM.

Phillips Petroleum Co., Chemical Dept., Rubber Chemicals Div., Stow, Ohio.

"Bulletins." Series of company publications which contain both basic compounding studies and practical suggestions for factory compounds. Issued in two series; those numbered in the 100's dealing with carbon blacks, those numbered in the 200's dealing with SBR and BR polymers.

"Reports." Preliminary reports originating from the Akron Technical Center. Issued frequently, each dealing with a specific compounding problem and containing extensive test data.

"Carbon Blacks. Properties and Characteristics in Rubber Recipes." Looseleaf. An extensive study of a wide range of blacks at several loadings in seven different rubbers with extensive data on uncured, cured, and aged properties.

Polymer Corp. Ltd., Sarnia, Ont., Canada.

"Polysar Handbook," 1956. Extensive compounding information on the company's SBR, NBR and butyl rubbers, including both basic compounding studies and suggested factory formulations. Appendix contains many useful tables, including conversion tables and lists of compounding ingredients by tradenames.

"Polysar Handbook, Vol. 2," 1960. A continuation of the earlier handbook, covering new polymers developed since publication of the first volume.

"Polysar Butyl Handbook," 1966. A book similar to the first two handbooks, but devoted entirely to butyl rubbers.

"Data Sheets" and "Technical Reports." Brief company releases, several issued each year, presenting new polymers and new development in compounding.

PPG Industries, Inc., Industrial Chemical Div., Pittsburgh, Pa.

"Calcene Bulletins," "Silene Bulletins," and "Hi-Sil Bulletins." Results of compounding studies with the company's white fillers, usually illustrated by practical factory formulations.

"Technical Service Bulletins." Brief releases originating from the technical service laboratories, each dealing with a new compounding technique or recommending formulations for a specific product.

"A Silene D Formulary." Recommended compounds, with test data, for a wide variety of goods.

Standard Brands Chemical Industries, Inc., Dover, Del.

"Product Information Sheets." Descriptions of the company's NBR polymers and latices and recommendations for compounding. Those coded 1, 2, 3, 4, and 5 concern individual polymers; those coded 10 present recommendations for compounding for specific purposes.

Texas-U.S. Chemical Co., 1 Greenwich Plaza, Greenwich, Conn.

"Application Bulletins." Booklets, issued from time to time, each covering processing and compounding recommendations for use of the company's SBR and BR rubbers in a particular type of goods.

Thiokol Chemical Corp., P.O. Box 1296, Trenton, N.J.

"Bulletins." Recommendations for compounding and processing the company's poly(sulfide), acrylic, and urethane polymers, illustrated by extensive test data.

Uniroyal Chemical Div. of Uniroyal, Inc., Naugatuck, Conn.

"Compounding Research Reports." Each deals with one or a group of the company's compounding ingredients. Information includes both basic compounding information illustrated by test series, and suggestions for factory formulations.

"Paracril Technical Bulletins." Basic information on compounding the company's NBR polymers. Each bulletin deals with either one phase of compounding or one type of practical formulations.

"Findings." Irregularly issued releases reporting briefly new developments in compounding with the company's polymers and compounding ingredients, and mentioning new literature available.

Other frequently issued releases offer compounding recommendations and formulations for specific types of goods with test data. Those dealing with rubber chemicals are identified as "Forms Nos. 200–"; with latices as Nos. 420–; with NBR polymers as Nos. 510–; with EPDM as Nos. 570–; with vinyl resins as Nos. 620–; with urethanes as Nos. 740–; and with ABS resins as Nos. 820–.

R. T. Vanderbilt Co., Inc., 230 Park Ave., New York, N.Y.

"The Vanderbilt Rubber Handbook," G. G. Winspear, ed., 11th edition, 1968. A practical manual describing available elastomers, explaining basic principles of compounding, and giving recommendations for compounding specific types of goods, illustrated with numerous compound formulations and test data. Also contains a section describing frequently used physical test methods and a section of useful tables.

"Vanderbilt News." Formerly issued bimonthly, now irregularly. Each issue consists the handling and compounding of latex. Includes descriptions of latices and of compounding ingredients, compounding recommendations both general and specific, descriptions of test methods, and a section of useful tables.

"Vanderbilt News." formerly issued bimonthly, now irregularly. Each issue consists of several articles on various phases of compounding, illustrating the use of the company's various compounding ingredients, giving formulations and test data.

INDEX